Chemical Thermodynamics

Thermodynamics can never be made easy, but with the right approach and a consistent use of scientific terms, it can be made less opaque, and it can give a person, who is prepared to try, an insight into how science explains why things happen the way they do. The approach adopted in this book will give readers a better understanding of how science works together with its limitations. Unfortunately, thermodynamics, or at least some parts of it, is a subject that (apart from quantum mechanics) probably causes most confusion and bewilderment among scientists. The majority of students do not understand or "get" thermodynamics, and it is considered a "hard" or difficult subject. There are multiple reasons for this. There is of course mathematics, and many thermodynamic texts appear to be lists upon lists of differential equations. Another reason is that thermodynamics is, as often as not, poorly taught by teachers/lecturers who themselves do not understand, or appreciate, or have any interest in the subject (often all three). This results not only in a lack of scientific rigorousness in the teaching of the subject with the resulting confusion, and sometimes teachers, lecturers, and authors just get it plain wrong (this occurs surprisingly often). However, it need not be like this, and although mathematics (including calculus) is required, it can be kept to a relatively elementary level in order to obtain an understanding of this most important of subjects. No one can pretend that the subject is easy, but it can be made more accessible by a rigorous definition of terms and concepts, ensuring that a consistent use of these definitions is maintained.

Victor C. M. Freestone was a multi-award-winning first-class chemistry student. During his 30-year professional career at the Atomic Weapons Establishment (AWE), he played a key role on the research, development, and maintenance of the UK's nuclear deterrent. He also applied his unique expertise to provide specialist support to a wide range of organisations/government bodies on the nuclear and radiological threats faced by the UK and its allies.

Chemical Thermodynamics
An Introduction

Victor C. M. Freestone

CRC Press
Taylor & Francis Group
Boca Raton London New York

CRC Press is an imprint of the
Taylor & Francis Group, an **informa** business

Designed cover image: Shutterstock

First edition published 2025
by CRC Press
2385 NW Executive Center Drive, Suite 320, Boca Raton FL 33431

and by CRC Press
4 Park Square, Milton Park, Abingdon, Oxon, OX14 4RN

CRC Press is an imprint of Taylor & Francis Group, LLC

ISBN: 9781041001546 (hbk)
ISBN: 9781041001539 (pbk)
ISBN: 9781003608400 (ebk)

DOI: 10.1201/9781003608400

Typeset in Palatino
by codeMantra

This book is dedicated to the memory of Victor Charles Marshall Freestone, a loving husband, father, and grandfather. His wife, Joan, his sons, Graeme and Iain, and his grandchildren, Matthew and Jessica, were his greatest sources of joy and pride.

A special acknowledgment goes to his brother, Nigel, for his tireless efforts and unwavering support in ensuring this work was brought to publication.

Contents

Preface

It seems to me that there are two approaches in science addressing how the universe "works". The one that gets all the publicity (and money) is particle physics (more of which later in this Prologue). Unfortunately, very few are able to understand/appreciate this approach because of the mathematics. This is why publicity tends to focus on individuals (Stephen Hawking comes to mind) and large and expensive equipment (the Hadron Collider comes to mind). The other approach is and has been taught for around 150 years at schools, colleges, and universities to science-biased students. This subject is thermodynamics, and, unlike particle physics, thermodynamics is an area of value and benefit to all practical scientists and to all engineers, while particle physics is only of direct interest to theoretical physicists and mathematicians.

Unfortunately, thermodynamics, or at least some parts of it, is a subject which (apart from quantum mechanics) probably causes most confusion and bewilderment among scientists. The majority of students do not understand or "get" thermodynamics, and it is considered a "hard" or difficult subject. There are multiple reasons for this. There is, of course, mathematics, and many thermodynamic texts appear to be listed upon lists of differential equations. Another reason is that thermodynamics is, as often as not, poorly taught by teachers/lecturers who themselves do not understand, or appreciate, or have any interest in the subject (often all three). This results in a lack of scientific rigour in the teaching of the subject with the resulting confusion, and sometimes, teachers, lecturers, and authors just get it plain wrong (this occurs surprisingly often). However, it need not be like this, and although mathematics (including calculus) is required, it can be kept to a relatively elementary level in order to obtain an understanding of this most important of subjects. I do not pretend that the subject is easy, but it can be made more accessible by a rigorous definition of terms and concepts, ensuring that a consistent use of these definitions is maintained.

What is needed is a new approach highlighting the benefits of thermodynamics in practical science and a revolution in the teaching of thermodynamics. For a (very) small percentage of the money spent on particle physics, we could start to address this issue and make a difference. This book is my attempt to kick this off and to begin to educate generations of students to have an intuitive grasp of the major concepts of thermodynamics such as energy and entropy. This approach should be in conjunction with the teaching of future scientists about what science is and what science is not. This will filter down to the general population and should prove enormously beneficial, e.g., instead of the scientific meaningless phrase to "save energy", we would be told (and understand) to "minimise entropy" (or even maximise "free energy"). Having seen the mess that today's scientists have made in dealing with the COVID-19 outbreak and the lack of understanding of science by our politicians and media, this initiative is long overdue. I know that Lisa Simpson's desire for world peace and Tibet's independence is quite similar to the above, but someone has to start, and my contribution is this book, which starts with the beginnings of modern science (classical mechanics) and ends with statistical thermodynamics.

To avoid unnecessary complications, this book uses the simplest model of matter (the ideal gas) to introduce and develop the concepts of thermodynamics. To appreciate the science behind the ideal gas model, this book starts with a chapter on classical or Newtonian mechanics followed by a chapter on energy. This latter chapter concentrates on kinetic and potential energy and introduces processes that transfer energy between these two energy types. In the third chapter, the ideal gas model of matter is introduced; this allows the concept of internal energy to be introduced, and this (the existence of internal energy) is the difference between classical mechanics and thermodynamics. In addition, it is when we enter into the world of thermodynamics that the directionality of change manifests itself, we can begin to address the following question: why do macroscopic processes that occur in the real world have a directionality determined by circumstances? For example:

- Why does a hot object in a cold room always cools down but, on its own, a cold object in a cold room never spontaneously warms up?
- Why does a nail rust when left outside, but a rusty nail does not spontaneously become shiny?
- Why does ice melt when brought inside?
- Why does a bouncing ball eventually stop bouncing?

- Why does a small volume of gas always expand to fill the larger volume available to it, but this larger volume of gas does not spontaneously compress to the original small volume?
- Why do all living things eventually die, but no dead thing becomes alive?

The list is endless, and these are questions that most working scientists (and all lay people) are concerned with. Directionality suggests the passage of time, and I find it amazing that we can go from a simple model of a matter (the ideal gas) based on small "microparticles" whose movements can be modelled using classical mechanics to directionality/time.

Many scientists, most intelligent lay people (who have thought about the problem), and all the media think the answer to fundamental scientific questions like directionality/time and others, such as "what is matter composed of?", will be supplied by particle physics and its huge expensive equipment. Maybe it will, and maybe it will not, but what I do know is that for the vast majority of working scientists, the arcane mathematics of particle physics associated with quarks, leptons, and bosons is interesting, but at the end of the day, it will prove to be of little benefit to the majority of scientists and engineers, while an intuitive knowledge of thermodynamics will prove of great benefit.

The science of thermodynamics was born in the 19th century as scientists and engineers were discovering how to build and operate steam engines. Thermodynamics can be considered the polar opposite to the theories of particle physics as they are based on our experience of the macroscopic world that we can see, feel, smell, taste, and touch. One of the first principles a student learns (usually after the Principle of Conservation of Mass) is the related Principle of Conservation of Energy which tells us there is a fixed amount of energy in the universe. Although no one knows what energy is, most people appreciate that energy is a useful concept for understanding the world, for example, when petrol is used in an internal combustion engine, it provides energy to propel a car, and once this energy is used up, the car stops moving.

According to another website, there are nine forms of energy you need to know for GCSE Physics. These are (i) electrical, (ii) sound, (iii) nuclear, (iv) kinetic, (v) light, (vi) heat, (vii) gravitational, (viii) chemical, and (ix) elastic. Wikipedia adds mechanical and magnetic, plus two rather exotic forms, namely, chromodynamic wave and mechanical wave. Finally, we are told that energy cannot be created, and it can transfer from one form to another. This is the "energy" that most people think they know something about.

We know we use energy to heat our homes and move our cars, and we know we cannot create or destroy energy. This raises the question of what happens to energy once the heating of the space or the movement of the car (along a horizontal surface) is completed. As we shall see, entropy is the measure of the energy that is no longer available to heat a house or move a car. In order to understand the concept of energy, you need to understand the concept of entropy. By 'understand', I do not mean to learn the myriad of differential equations you see in thermodynamic textbooks, but to understand the concepts of energy and entropy and how these concepts allow an understanding and appreciation of how the universe works. If you ask people what entropy is, I am pretty confident that the vast majority of people (including scientists) will say they have no idea. Those with a scientific background may say something like that it is something to do with a tendency to disorder, and some may write down an equation similar to this $\Delta S = \frac{dq}{T}$ where S is entropy. When asked to explain what this equation means, they will, in all probability, say they do not know. Unlike energy, only a few scientists will state that they appreciate the usefulness of entropy in understanding how the universe works. One of the objectives of this book is to start to change this balance from "only a few scientists" to "many scientists".

Dr Victor Freestone

1 Classical Mechanics

1.1 INTRODUCTION

The study of the motion of everyday or "normal" physical objects is called classical mechanics (including the special case in which physical objects remain at rest), it is often called Newtonian mechanics because most of the subject builds on the work of Isaac Newton.[1] In this context, the dimensions of "normal" objects are typically much larger than an atom and much smaller than a star. In addition, normal physical objects are moving at speeds significantly slower than the speed of light (which is about 3×10^{8} m.s^{-1}). Although it is the oldest branch of physics, the term "classical mechanics" is relatively new. Soon after 1900, a series of revolutions in mathematical thinking gave birth to new fields of inquiry: relativistic mechanics for phenomena relating to the very fast, and quantum mechanics for phenomena relating to the very small. Broadly speaking, classical mechanics is concerned with objects whose size is in the range of 10^{-6} to 10^{8} m and whose mass is in the range of 10^{-20} to 10^{20} kg and travel at less than 10^{5} m.s^{-1}. But do not get hung up on these numbers. Normal physical objects are neither very small nor very big and do not travel very quickly (relative to the speed of light).

Despite its ubiquitous use in modern science, you may be surprised to know that Newton did not use and was not aware of the concept of energy. It is for this reason this introduction to classical mechanics ends when the concept of energy needs to be invoked: energy is covered in Chapter 2.

1.2 PHYSICAL OBJECTS AND PHYSICAL QUANTITIES

1.2.1 Physical Objects

What do I mean by a "physical object"? As far as we are concerned, the following definition will suffice: *a physical object consists of matter, occupies space, endures through time, and can move about in space*. In this definition, "matter" is used loosely as a general term for the substance that makes up all observable physical objects. This is not a rigorous definition because it is circular ("a physical object consists of matter" and matter "makes up all observable physical objects"). Matter is sometimes equated with physical objects that exhibit a rest mass.[2] In modern science, matter is considered to exhibit both wave-like and particle-like properties, the so-called wave-particle duality;[3] light does not have a rest mass. This means that as you continuously divide a physical object, you reach electrons, quarks, etc. that do not meet the definition of a physical object, i.e., we have a paradox – physical objects are composed of non-physical objects.[4]

All scientific definitions of matter that are not concerned with wave-particle duality are essentially the same – *matter is anything that has mass and takes up space (or volume)*. In other words, as stated before, we have circular definitions, and, as a consequence, matter is not a fundamental concept in modern science. Despite all this, it is still a useful concept, and the term "matter" is analogous to words such as "stuff" or "substance". If we want to move forward, we need the concept of a physical object (and a definition) and must ignore (but not forget) the logical paradoxes set out above.[5]

1.2.2 Vector and Scalar Physical Quantities

Most scientific texts consider the meaning of the term "physical quantity" to be intuitively understood and do not even consider it necessary to attempt to define it. The simplest definition of a physical quantity is:

> A physical quantity is a quantity in science that can be measured.

Physical quantities can be divided into two broad groups: vectors and scalars. Scalars are physical quantities which have only a number value or a size (magnitude). A scalar tells you how much of something there is, e.g., mass is a scalar quantity. Vectors are different because they are physical quantities which have a size and a direction. A vector tells you how much of something there is and in which direction it is acting. Force and velocity are vector quantities.

Let me illustrate some important points about vectors and scalars by considering distance (scalar) and displacement (vector): speed (scalar) and velocity (vector). The definitions for distance and displacement are[6]:

> Distance is a scalar quantity that refers to 'how much ground an object has covered' during its motion. Displacement is a vector quantity that refers to 'how far out of place an object is'; it is the object's overall change in position.

DOI: 10.1201/9781003608400-1

Speed is defined as the distance travelled divided by the total time taken, i.e., speed = $\frac{\text{distance}}{\text{time}}$. Velocity can be defined as the displacement divided by the total time taken, i.e., velocity = $\frac{\text{displacement}}{\text{time}}$. Now, consider an athlete running around a circular track. If the athlete runs faster, it would take less time to run around the track and the athlete's average speed would be higher. Conversely, if the athlete runs slower, it would take less time to run around the track and the athlete's average speed would be lower. All this seems obvious. Now let us move on to velocity because the athlete runs back to his starting position and his displacement is zero. So, since displacement is zero in this case, we can say that the athlete has an average velocity of zero which does not change with the athlete's speed. Therefore, although distance and displacement are two quantities that may seem to mean the same, they have distinctly different definitions and meanings in science.

Before moving on, let me ask the following question – is time a scalar or vector quality? The answer is not straightforward. In this book, I will consider time to be a scalar quantity. The reasons behind this decision are set down in Section 3.7.

In terms of manipulating physical quantities, all we need to know in this book is how to add scalar physical quantities and one- or two-dimensional vector physical quantities. Adding scalars is straightforward – you just add the numbers. Adding vectors is more complicated, and you must consider their magnitude and direction. This requires some knowledge of trigonometry. All I want to consider here is when vectors are in the same dimension (horizontal or vertical) or in two dimensions at right angles to each other. When the vectors are in one dimension, a simple sign convention works – a vector can have a positive or negative value depending on its direction. Conventionally, a vector whose direction is to the right or up is positive and one whose direction is to the left or down is negative. Let us consider the addition of two forces: one is 3 units and the other is 4 units acting in one dimension (horizontal or vertical), and as you can see, the directions of the vector are important because the resultant vector can be +7, –7, +1 and –1 (Figure 1.1). The resultant vector is a single vector that has the same effect as the two vectors act together.

When the vectors are in two dimensions, all we need to know to calculate the magnitude of the resultant vector is Pythagoras' theorem[7] (Figure 1.1). Unfortunately, the simple +/– sign convention does not work when the vectors are in two dimensions. Fortunately, in this book, the only time I will use two-dimensional vectors is in Section 1.9.3; otherwise only one-dimensional vectors are considered and the simple +/– sign convention will suffice.

Physical quantities can also be divided into extensive and intensive quantities:

- An extensive quantity is equal to the sum of that quantity for all its constituent subsystems. Mass is an extensive quantity; an object has mass m_1 and another has mass m_2. Then, a system simply comprising those two objects will have a mass of $m_1 + m_2$. Volume is another extensive quantity.
- An intensive quantity is independent of the extent of the system. Pressure is an intensive quantity. To illustrate, if two systems with identical pressures are combined, they still have the same pressure (not twice the pressure). Another intensive property is temperature.[8]

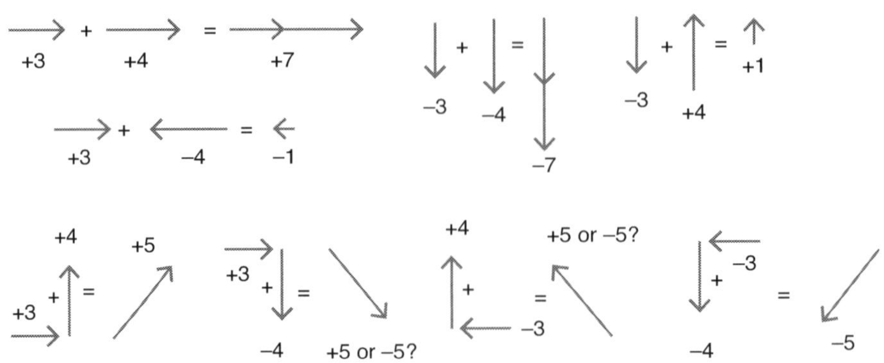

Figure 1.1 Examples of simple vector addition.

The following is a good explanation of intensive and extensive properties[9]:

> Intensive properties have an 'intensity' in a point in space. Extensive properties are undefined (or zero) in a point in space.
>
> I may point to a certain point, say 100 meters below the surface of the ocean, and the question, 'what is the pressure there?', makes sense. Hence, pressure is an intensive property. On the other hand, if I ask: 'what is the volume there?', there would be no clear answer: I would have to specify an actual volume, and cannot make do with just a point.

Dividing an extensive property by another extensive property, say, dividing the mass by the volume (i.e., density), turns it into an intensive one. Note that although an extensive property can be a vector (e.g., a force) or a scalar (e.g., a mass), an intensive quantity can never be a vector and is always a scalar quantity (e.g., pressure or temperature).

There are seven fundamental quantities in science, and the units of these fundamental quantities are known as base units. The International System of Units[10] (SI) defines seven units of measure as a basic set from which all other SI units can be derived. In this book, we will be concerned with six of them; these together with their SI base units are shown Table 1.1.[11] Please note that the contents of the "definition" content use advanced concepts not covered in this book and are for completeness only.

The SI base units have a degree of interdependency, e.g., to extract the definition of the metre from the speed of light, the definition of the base unit the second must be known.

1.3 MASS, GRAVITY, AND FORCE

The mass of a physical object is defined as the amount of matter it contains. This is not a rigorous scientific definition because we have no rigorous definition of matter, but it will have to do. The standard SI unit of mass is the kilogram (kg).[12] The Law of Conservation[13] of Mass[14] states that mass in any system,[15] closed to all transfers of matter and energy,[16] is neither created nor destroyed by any process. Chemistry has its foundations in alchemy – a subject that put much stock into magic and mysticism. The Law of Conservation of Mass took the mystery and illusion out of alchemy and brought predictability and reliability to the science of chemistry.

An inherent property of mass is that a mass attracts (i.e., exerts a pull) on any other mass. In science, any push or pull by an external influence is known as a force (a "pull" is known as an attractive force and a "push" is known as a repulsive force). All physical objects attract each other,

Table 1.1: Fundamental Quantities and Their SI Units

Quantity	Base Unit	Symbol	Description
Amount of substance	mole	Mol	The amount of substance of a system that contains as many elementary entities as there are atoms in 0.012 kilograms of carbon 12; its symbol is "mol". When the mole is used, the elementary entities must be specified and may be atoms, molecules, ions, electrons, other particles, or specified groups of such particles. In this definition, it is understood that unbound atoms of carbon 12, at rest and in their ground state, are referred to.
Length	metre	M	The length of the path travelled by light in vacuum during a time interval of 1/299792458 of a second.
Mass	kilogram	Kg	The unit of mass; it is equal to the mass of the international prototype of the kilogram.
Time	second	S	The duration of 9192631770 periods of the radiation corresponding to the transition between the two hyperfine levels of the ground state of the caesium 133 atom.
Thermodynamic temperature	kelvin	K	The fraction 1/273.16 of the thermodynamic temperature of the triple point of water. This definition refers to water having the isotopic composition defined exactly by the following amount of substance ratios: 0.00015576 moles of ^{2}H per mole of 1H, 0.0003799 moles of ^{17}O per mole of ^{16}O, and 0.0020052 moles of ^{18}O per mole of ^{16}O.
Electric current	ampere	A	The ampere is that constant current which, if maintained in two straight parallel conductors of infinite length of negligible circular cross section, and placed 1 metre apart in vacuum, would produce between these conductors a force equal to 2×10^{-7} newtons per metre of length.

i.e., every physical object in the Universe exerts an attractive force on every other physical object in the Universe.[17] This phenomenon is known as gravitation.

The velocity of a physical object is the distance moved (in a straight line) per unit of time. The SI unit of distance is the metre (m) and that of time is the second (s); therefore the unit of velocity is metres per second or in scientific notation $m.s^{-1}$.[18] When a physical object experiences a force, its velocity (which could be zero) changes, and this change in velocity is known as acceleration. Acceleration is the change in velocity per unit time, and its units are $m.s^{-1}$ per second, i.e., $m.s^{-2}$. Velocity and acceleration are considered in further detail in Section 1.7. The SI unit of force is known as the newton (N) in recognition of Isaac Newton's work (see below). A newton is defined as that force necessary to provide a mass of one kilogram with an acceleration of one metre per second, i.e., the units of force are (N or $kg.m.s^{-2}$). The concept of "force" will become more understandable when we consider velocity and acceleration in more detail.

I have already mentioned that a fundamental property of a mass is that it exerts an attractive force on all other masses in the Universe. Isaac Newton derived a general physical law of this phenomenon based on observations and inductive reasoning.[19] He realised that two physical objects attract each other with a force that is proportional to the product of their masses and inversely proportional to the square of the distance between them (see Figure 1.2). What this means is that for any two objects in the Universe, the gravitational attractive force between these two objects depends only on their mass and distance apart:

$$\propto \frac{m_1 \times m_2}{r^2} \tag{1.1}$$

The constant of this proportionality is usually designated as G, and the equation is known as Newton's Law of Gravitation[20]:

$$F = G\frac{m_1 \times m_2}{r^2} \tag{1.2}$$

Please note that it can be shown that large spherically symmetrical masses attract and are attracted as if all their mass were concentrated at their centres (see Section 1.4). This equation was first published on 5 July 1687 in Newton's "Philosophiæ Naturalis Principia Mathematica" (usually known as "the Principia"[21]).

This gravitational attractive force can be measured in the laboratory and was first carried out in 1797–1798 (a century after Newton) by British scientist Henry Cavendish using lead spheres. Cavendish's apparatus for experimentally determining the value of G involved a light, rigid rod about 2-feet long. Two small lead spheres were attached to the ends of the rod, and the rod was suspended by a thin wire. When the rod becomes twisted, the torsion of the wire begins to exert a torsional force that is proportional to the angle of rotation of the rod. Cavendish first calibrated his instrument to determine the relationship between the angle of rotation and the amount of torsional force.[22] A diagram of the apparatus is shown in Figure 1.3. Cavendish then brought two large lead spheres near the smaller spheres attached to the rod. Since all masses attract, the large spheres exerted a gravitational force upon the smaller spheres, and this twists the rod and produces a torsional force on the wire. When the rod and spheres are in equilibrium, the torque on the wire is balanced by the gravitational force between the larger and smaller spheres: this allowed Cavendish to determine the gravitational force of attraction between the masses. The force involved in twisting the torsion balance was extremely small, about 2×10^{-8} of the weight of the small balls (see next section for a discussion on "weight").

Cavendish determined G to be equal to $6.74\times10^{-11}\ m^3.kg^{-1}.s^{-2}$. This unit for G follows directly from the units for force $\left(kg.m.s^{-2}\right)$, mass (kg) and distance (m). Cavendish expressed his result in terms of the density of the Earth, and he referred to his experiment in correspondence as "weighing the world". The value I have given for G is a reformulation of his results in modern terms.

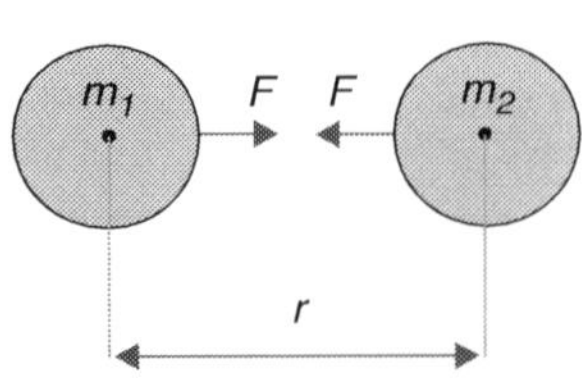

Figure 1.2 Diagrammatic representation of the attractive force between two physical objects.

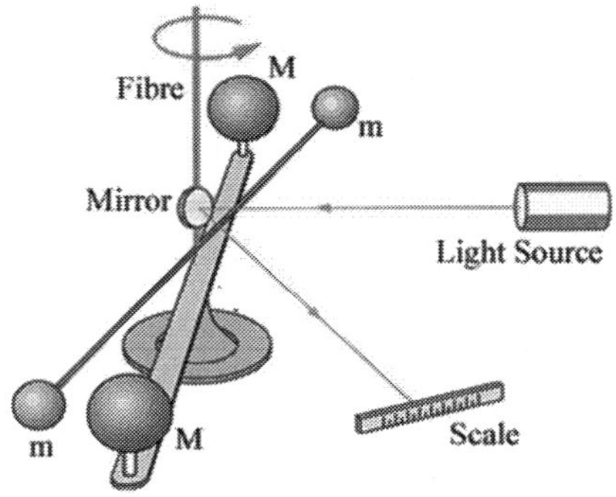

Figure 1.3 Schematic diagram of Cavendish's equipment for measurement of the gravitational constant.

Cavendish's equipment was remarkably sensitive for its time – the currently accepted value for G is 6.67384×10^{-11} $m^3.kg^{-1}.s^{-2}$, and Cavendish's result only differs by 1% from this value. The value of G is an extremely small number and demonstrates that the force of gravitational attraction is only appreciable for objects with an extremely large mass. While two lead balls exert gravitational forces upon each other, these forces are too small to be noticeable without the use of extremely sensitive instrumentation. Yet if one of the lead balls is replaced with a planet, e.g., the Earth, then the gravitational force between the other lead ball and the planet becomes perceptible without the use of scientific instrumentation. This is the subject of the next section.

1.4 ACCELERATION DUE TO GRAVITY AND WEIGHT

Newton's Law of Gravitation is rigorously written as:

> Every point mass in the Universe attracts every other point mass with a force that is directly proportional to the product of the point masses and inversely proportional to the square of the distance between them.

A point mass is a point particle with a non-zero mass and no other properties or structure. Point masses do not exist since a point has zero volume; therefore, the density of a point mass having a finite mass is infinite.[23] For a spherical mass, it can be shown mathematically that all its mass can be considered to be "concentrated" at its centre for mass times distance calculations.[24] This "point" is known as the centre of mass. The use of a centre of mass is mathematically rigorous (for a sphere) and has the advantage that the centre of mass acts the same as a point mass.[25]

Let us now consider the force (F) on a mass m on the surface of the Earth using Equation 1.2. The mass of the Earth is 5.9722×10^{24} kg,[26] and I will assume the Earth is a perfect sphere with a radius of 6.371×10^6 m.[27] We have

$$F = \frac{(6.67 \times 10^{-11}) \times (5.97219 \times 10^{24}) \times m}{(6.371 \times 10^6)^2} = m \times 9.81 \tag{1.3}$$

As the units of force are $kg.m.s^{-2}$ and the units of mass are kg then the units of '9.81' must be $m.s^{-2}$, and, as we have already seen, these are the units of acceleration. When you think about it this is not surprising as we know that a force (a push or pull) results in a change in the velocity (acceleration) of a physical object. At the Earth's surface, gravitational attraction results in an acceleration of 9.81 $m.s^{-2}$, consequently 9.81 $m.s^{-2}$ is known as the acceleration due to gravity. The acceleration due to gravity is usually denoted by the symbol "g" and Equation (1.3) is often written as:

$$F = mg \tag{1.4}$$

The force with which a physical object is attracted to Earth is known as the weight of that object, i.e., the weight (w) of a mass m is the attractive force F between the mass m and the Earth, i.e., w = F = mg. The more mass the object has, the greater its weight will be. Weight is a force, so it is measured in newton (N). On the surface of the Earth an object with a mass of 1 kg has a weight of 9.81 N (or $kg.m.s^{-2}$). The fact that the unit of weight is the same as the unit of force can be puzzling as in everyday life weight is recorded in kg: this is one of many examples where scientific rigour is at odds with normal speech. If my mass is 70 kg my "scientific weight" is $70 \times 9.81 = 687$ N, but if we introduce a new unit kilogram-force (defined as 1 kg force=9.81 N) then my weight is 70 kilogram-force. The ambiguity resulting from dropping the word "force" does not cause a problem in everyday life it just causes problems in the teaching of science.

The mass of an object stays the same wherever it is in the Universe, but its weight will change. The Moon (7.342×10^{22} kg) has less mass than the Earth and a smaller radius (1.7374×10^{6} m)[28]: the acceleration due to gravity on the Moon is 1.62 $m.s^{-2}$. Consequently, the weight of a mass of 1 kg on the Moon's surface is 1.62 N, i.e., the weight of a physical object on the moon's surface is one-sixth of its weight on the Earth's surface. It is also noteworthy that the weight of a body will vary, slightly over the Earth's surface, since the Earth is not a perfect sphere of uniform density.[29]

I have not mentioned how we can measure the mass and diameter of the Earth. The diameter of the Earth can be calculated by measuring the length of the shadow of the same object at two locations located at the same longitude at local solar noon (which is the same for the two locations) on the same day. Knowing the distance between the two locations you can use trigonometry to calculate the radius of the Earth.[30] A similar principle was used by the ancient Greeks.[31] The mass of the Earth is calculated by reversing the above calculation, i.e., knowing G and measuring g at the location in question.

1.5 INTRODUCTION TO CLASSICAL MECHANICS

The general principles of what we now call classical mechanics was first set down by Isaac Newton in 1687 in the 'Principia'. This was a complete overthrow of the Aristotelian picture of the Universe, which had prevailed for nearly two millennia.[32] What characterised classical mechanics was not only its combination of inductive and deductive reasoning based on experiment (as discussed in the Prologue), but its use of mathematics. The Aristotelian picture of the Universe was based on observation and speculation combined with deductive reasoning. The following is a good example of the Aristotelian way of thinking. Aristotle argued that the Universe is spherical, and its centre is the centre of the Earth. It is spherical, because that is the most perfect shape; and because it has a centre the Universe must be finite because a body with a centre cannot be infinite. He taught that the Earth is a sphere, because, he argued, all earthly substances move towards the centre, and would therefore have to form a sphere eventually. He also used evidence based on observation, e.g., if the Earth were not spherical then lunar eclipses would not show segments with a curved outline.

As Aristotle dominated, what we would now call, scientific thinking for nearly two thousand years it is worth setting down his views on motion. His theory of motion flows from his understanding that matter is constituted of four elements: air, Earth, fire, and water. Physical objects, being solid like Earth, would tend to clump together with other solids (Earth), so objects tend to fall to Earth, which is their natural place, and, on Earth, an object's natural state is then at rest. So, no explanation is required for physical objects that remain at rest or fall vertically, because this is natural. In contrast Aristotle considered horizontal motion an example of unnatural motion, and this did require further explanation. Making an object move usually has an obvious cause (e.g., pushing or throwing it), and therefore is not a problem. What did cause Aristotle a problem is explaining why something continues in motion. Think of a spear being thrown. At first, it is not in motion, but then the thrower's arm provides a force which accelerates it (our vocabulary, not Aristotle's). At this point, there is no big problem (in Aristotle's world view). But what keeps it going after it leaves the thrower's hand? It should fall to Earth immediately since there is nothing obvious pushing it. Aristotle's answer was that as the spear travels, it leaves a vacuum behind it. Air rushing in (the source of the cliché "nature abhors a vacuum") pushes the spear forward until its natural motion (falling) eventually brings it to Earth.

Aristotle also thought about the causes which start things moving. In the spear scenario, it is easy to say that the thrower's arm moves the spear, but what moves the thrower's arm? Aristotle said that another motion moved the arm (muscle contraction) but he also realised that some earlier motion must cause the muscle to contract and that this earlier motion must also have its own initiator. To avoid the idea that there is an infinite chain of causes, Aristotle argued that there must be an "unmoved mover" or "prime mover",[33] i.e., something which can initiate motion without itself being set in motion. Aristotle's views on motion were widely accepted because they seemed to support what people observed in nature. He also thought that weight affected falling objects. A heavier object, he argued, would reach the ground faster than a lighter object dropped at the same time from the same height.

Aristotle was revered as the greatest thinker of his time. Since neither Aristotle nor anyone else thought it a good idea to test theories by experiment, his ideas persisted for nearly 2000 years (Aristotle died in 322BC). The first big blow to Aristotle's view on motion came in the 16th century when Galileo's experiments (see Section 1.7) led him to disagree with the Aristotelian view. Aristotle's domination was such that Galileo was tried as a heretic.[34]

1.6 THE BASIC MATHEMATICS NEEDED FOR THIS BOOK

1.6.1 Introduction

This book is concerned with processes and a process is a set of activities that produces a change. A changing quantity has two properties: (i) its value at a particular time, and (ii) its rate of change at that particular time. The mathematics of change is calculus and, although this book minimises the use of abstruse mathematics, there is no escaping the fact that a basic knowledge of calculus is essential in understanding the physical sciences. Differentiation is the determination of the rate of change of a quantity when the value of the changing quantity, at all times, is known. Integration is when you determine the total change of a quantity over a given time interval, when the rate of change of the quantity is known at all times.

The basic concepts of calculus are easily understood (even if the subject quickly becomes complex and arcane). I will start with simple graphical techniques and consider the practical meanings of (i) slopes of straight lines and curves, and (ii) areas under curves. I will then move to estimating and then calculating the slopes of curves at particular points and will consider the estimation and calculation of areas under graphs. If you are interested in calculus and want to take this further, there are many textbooks and web sites dealing with introductory calculus and I suggest you consult those.[35]

Finally, there are short sections on logarithms, trigonometry, and absolute value equations.

1.6.2 Graphs

I will only be considering graphs that show mathematical relationships between two sets of data. These graphs have their axes labelled with a title, a scale, and units. Consider the following situation – you have just opened a bank account and you will pay into it £10 per week. At the end of week 1 your account contains £10, at the end of week 2 it contains £20, etc. You could show this situation graphically (Figure 1.4):

The area under the graph at a particular time shows the amount of money in your bank account, e.g., at week 8 there is £80 in your account. Note the units of the two axes also indicate this $£.w^{-1} \times w = £$. You could say this (area under graph=amount of money in bank account) is not true because at week 6.5 your account does not contain £65. This is true, but I hope you can see the concept here (and I will deal with this issue below).

Now consider that you pay into your fictitious account £1 at the end of the first week, £2 at then of the second week, £3 at the end of the second week, etc. Figure 1.5 graphically sets out the relationship between rate of payment and time. The amount in your bank account at, say, the end

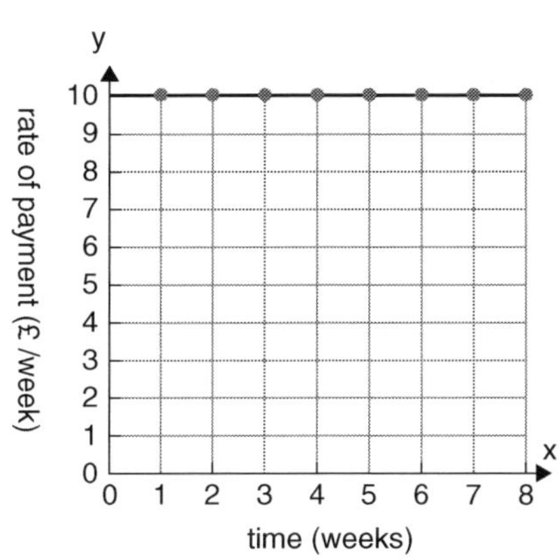

Figure 1.4 Graph of the rate of payment v. time as the rate of payment remains constant with time.

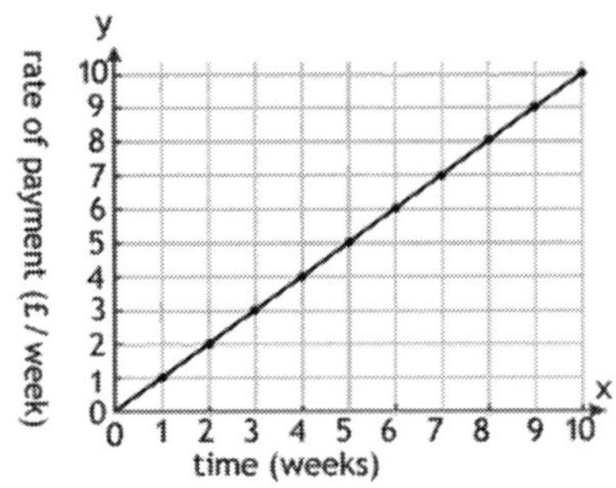

Figure 1.5 Rate of payment vs time as the rate of payment varies linearly with time.

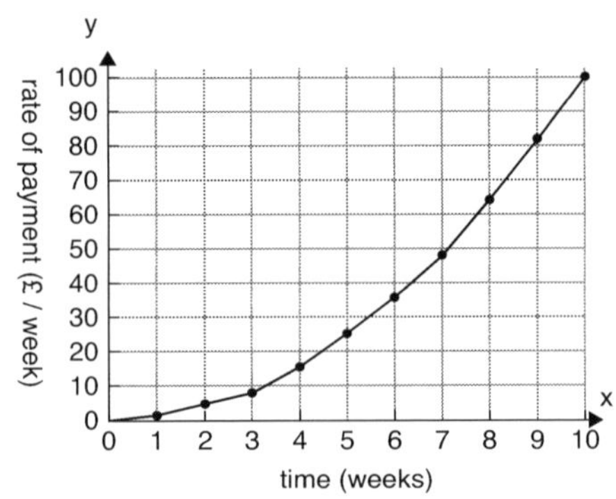

Figure 1.6 Rate of payment vs. time as the rate of payment varies nonlinearly with time.

of week 10 is the area under the graph when x=10. Knowing the area of a triangle is half the base times the height, then there is £50 in your account at the end of week 10.

Let us now consider an even more farfetched paying in system, at the end of week 1 you pay in £1, at the end of week 2 you pay in £4, at then of week 3 you pay in £9 etc. In other words, the amount paid in (in pounds) is the square of the week number that the money is paid in, i.e., at the end of week 10 you pay in £100 (£10^2). This situation is shown graphically in Figure 1.6. The determination of the area under the curve is now not so straightforward.

I admit these are rather artificial examples and would be better shown as bar charts. But we can make the examples shown in Figures 1.4 and 1.5 mathematically rigorous by using equations such as $y=x$ and $y=x^2$, where x and y represent real numbers.[36] Equations such as $y = x$ and $y = x^2$ are known as continuous functions, i.e., for any real value of x there is a real value of y (unlike our bank account graphs). Generically we say that y is a function of x (written $y = f(x)$). In the first example y (weekly saving rate) is constant and is not a function of x because $y = 10$ whatever the value of x. In the second example $y = f(x) = x$, and in the third example $y = f(x) = x^2$.

Before moving on I want to define the slope of a line and of a curve. In addition to its familiar meaning, the word "slope" has a precise mathematical meaning: the slope of a line is the change in y divided by the change in x. To find the slope of a line, pick any two points on the line. Then subtract their y-coordinates and subtract their x-coordinates *in the same order* (see Figure 1.7), i.e., if the coordinates of the first point are (y_i, x_i) and the second coordinates are (y_f, x_f) then the slope is $\frac{(y_i - y_f)}{(x_i - x_f)}$ or $\frac{(y_f - y_i)}{(x_f - x_i)}$. Also note that because the graph is a straight line you can pick any pair of coordinates on the straight line. If a line has a positive slope, then y always increases when x increases, and y always decreases when x decreases. Therefore, the graph of a line with a positive slope starts at the bottom left and goes towards the top right. If, the slope of a line is negative, then y always decreases when x increases, and y always increases when x decreases.

All of this is relatively straightforward, but how is the slope of a curve determined? You should be able to see that, unlike the slope of a straight line, the slope of a curve changes with the

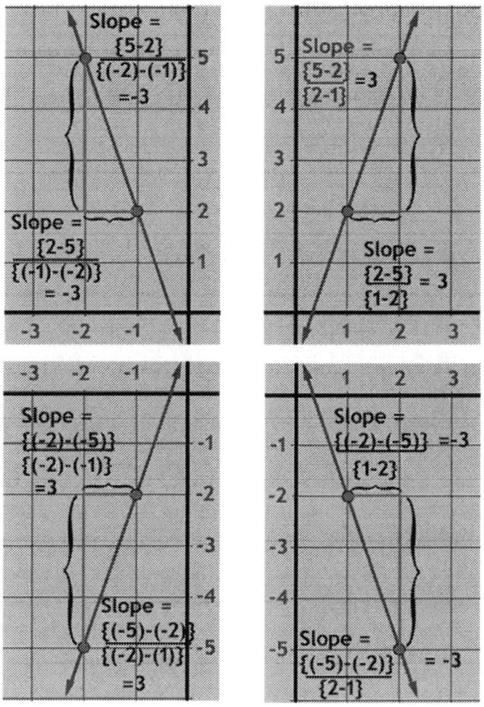

Figure 1.7 The slope of a line.

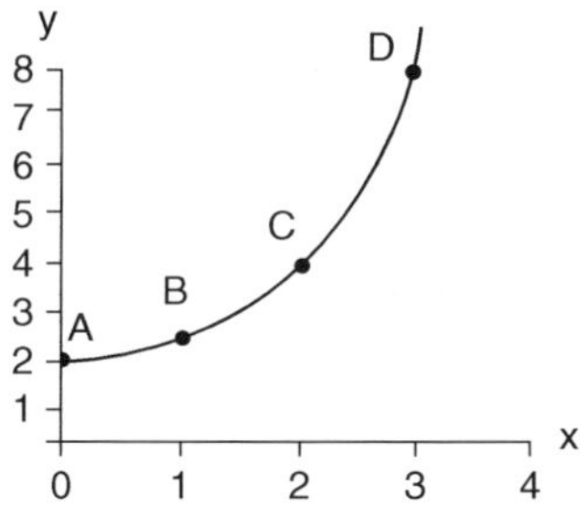

Figure 1.8 Graph of a curve.

coordinates of the curve. In Figure 1.8, even if you are uncertain on how to calculate the gradient of a curve at a particular set of coordinates, you can see that the slope of the curve increases as we move from point A through points B and C to D.

The slope of a curve at a particular point on a curve is the slope of the tangent to the curve at that point. A tangent is a straight line that touches a curve at a single point and does not cross through it. The point where the curve and the tangent meet is called the point of tangency (see Figure 1.9).

How do we draw a tangent at a particular point on a curve? It turns out that we do not need to do so because calculus does it for us (see below).

The first sentence of this section reads "I will only be considering graphs that show mathematical relationships between two sets of data". A more rigorous version of this sentence is – "I will only be considering the graphing of continuous mathematical functions between two sets of data". Now I want to set down some definitions:

1. A *real number* is a value of a continuous quantity that can represent a distance along a line. Real numbers include all the rational numbers, such as the integers and fractions, and all the irrational numbers such as the square root of 2. Real numbers can be positive, negative or zero. In fact,

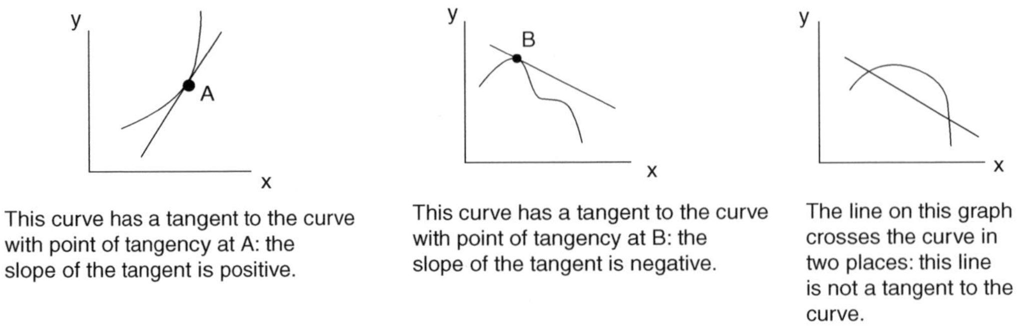

Figure 1.9 Tangents to a Curve.

nearly any number you can think of is a real number. So, what is not a real number? Well, infinity and imaginary numbers like the square root of –1 are not real numbers. Every real number can be represented by a unique point on a number line. A real number line simply called a number line, represents real numbers with unique points associated with each number on the line. The point associated with real numbers is called a coordinate. In the number line, the number 0 is called the origin. All the positive numbers or integers are represented on the right side of the origin, and the negative numbers or integers are represented on the left side of the origin.

2. *Absolute value* of a real number of a real number describes the distance of a number on the number line from 0 without considering which direction from zero the number lies. The absolute value of a number is never negative. The absolute value of a real number x is written $|x|$, e.g., $|5.3| = 5.3$ and $|-5.3| = 5.3$. I will sometimes use absolute values to clarify the mathematics (usually the arithmetic) used in thermodynamics. In Section 1.6.10 I will consider absolute numbers in a bit more detail.
3. A function is a relation between a set of input data of real numbers and a set of output data of real numbers such that each input is related to exactly one output.
4. The graph of a function is continuous, i.e., it does not have any breaks or gaps.
5. In mathematics a function is often written as $y = f(x)$.

I do not pretend these definitions are 100% mathematically rigorous but they are sufficient for the purposes of this book. Note that the graphs in Figures 1.4–1.6 are not graphs of functions because they are not continuous; I used them as they provided easily understood examples.

1.6.3 Slopes of Graphs – Introduction to Differentiation

Imagine travelling in a car with a broken speedometer and you want to know how fast you are going. One method is to use the odometer of the car and time how long it takes to travel a certain distance, e.g., the car travelled 0.8 kilometres in one minute, so the speed is $60\times0.8=48$ kilometres per hour. However, this is an average speed, and a speedometer gives you a near but not quite instantaneous speed. How can you find your instantaneous speed? The following is from "Maths is Fun" website[37]:

Alex: 'No, Sam! Not our average for the last minute, or even the last second, I want to know our speed RIGHT NOW.'

Sam: 'OK, let us measure it up here ... at this road sign... NOW! OK, we were at the sign for zero seconds, and the distance was ... zero metres!'

'The speed is 0m / 0s=0/0=I Don't Know!'[38]

'I can't calculate it Sam! I need to know some distance over some time, and you are saying the time should be zero? Can't be done.'

You would think it would be easy to work out the speed of a car at any instant in time, but it is not. Even the speedometer of a car only shows an average speed for a very short length of time (say 0.1 – 0.5 seconds). Let us now consider how we can calculate the instantaneous speed of an object. I will use as an example something like that used in the Maths is Fun website. A stone is dropped on Earth, and we want to know its speed one second after it has started falling. We will use the following equation to find the distance fallen:

$$d = 5t^2 \tag{1.5}$$

where d is the distance fallen in metres and t is the time after the stone has started falling in seconds.[39]

After 1 second the stone has fallen $5 \times 1^2 = 5$ m. Therefore, after one second speed is 5 m.s^{-1}, but this is the average speed since the stone started falling. We want to know the speed at exactly one second after the stone started falling. After exactly one second the speed is:

$$\text{Speed} = \frac{(5-5)\text{ metres}}{(1-1)\text{seconds}} = \frac{0\text{ metres}}{0\text{ seconds}} = ??? \tag{1.6}$$

Let us now consider a short time called Δt (pronounced "delta t"). At $(1+\Delta t)$ seconds the stone has fallen:

$$d = 5(1+\Delta t)^2 = 5\left(1+2\Delta t+(\Delta t)^2\right) = 5+10\Delta t+5(\Delta t)^2 \text{ metres} \tag{1.7}$$

We know after one second the stone has fallen $d = 5t^2 = 5\times 1^2 = 5$ m . Therefore, between 1 second and (1 + Δt) seconds the stone has fallen Δd metres where:

$$\Delta d = 5+10\Delta t+5(\Delta t)^2 - 5 = 10\Delta t+5(\Delta t)^2 \text{ metres} \tag{1.8}$$

To get the average speed between 1 second and (1+Δt) seconds we divide Δd by Δt:

$$\text{Speed} = \frac{\left(10\Delta t+5(\Delta t)^2\right)}{(\Delta t)} = 10+5\Delta t \text{ m.s}^{-1} \tag{1.9}$$

As Δt shrinks towards zero then speed becomes closer to 10 m.s^{-1}, and when Δt has approached zero:

$$\text{Speed} = 10 \text{ m.s}^{-1} \tag{1.10}$$

This approach is known as calculus.[40] The word "calculus" comes from Latin meaning "small stone", because it is like understanding something by looking at small pieces. Differential calculus cuts something into small pieces to find how it changes. In the above example, we used differential calculus to cut time and distance into such small pieces that a meaningful answer came out.

"Shrink towards zero" is written as a limit like this:

$$f'(x) = \lim_{\Delta x \to 0} \frac{f(x+\Delta x) - f(x)}{\Delta x} \tag{1.11}$$

Where $f'(x)$ is known as the derivative of f(x). The derivative of $f(x)$ is often written as $f'(x)$ which is usually read as "f prime of x". I will now consider differentiation in a more general sense – this will lead us towards a definition of the concept of the "derivative".

1.6.4 Differentiation and the Derivative

There is no problem finding the slope of a straight line, however, it is not so straightforward when the graph is a curve. You can find an average slope between two points, but how do find the slope at a point (see Figure 1.10)?

Consider the equation $y=x^2$, then $(y+\Delta y)=(x+\Delta x)^2=x^2+2x\Delta x+(\Delta x)^2$. The slope is:

$$\frac{\Delta y}{\Delta x} = \frac{x^2+2x\Delta x+(\Delta x)^2 - x^2}{\Delta x} = \frac{2x\Delta x+(\Delta x)^2}{\Delta x} \tag{1.12}$$

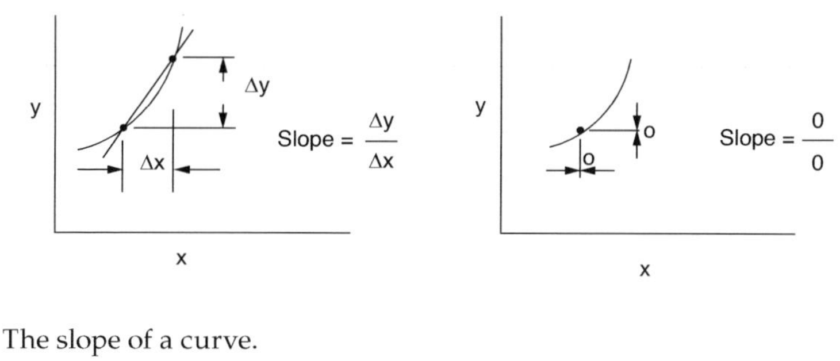

Figure 1.10 The slope of a curve.

As Δx tends towards zero we obtain the derivative of x^2 which is $2x$. The derivative of a function of a variable at a chosen value x is the slope of the tangent to the graph of the function at that point. Equation (1.12) is often written in terms of the function $f(x)=x^2$, yielding Equation 1.13.

$$\frac{f(x+\Delta x)-f(x)}{\Delta x}=\frac{x^2+2x\Delta x+(\Delta x)^2-x^2}{\Delta x}=\frac{2x\Delta x+(\Delta x)^2}{\Delta x} \tag{1.13}$$

We can now state a more general definition of derivative – *the derivative of the function* $f(x)$ *equals the limit of* $\frac{[f(x+\Delta)-f(x)]}{\Delta x}$*, as* Δx *goes to zero.*

The definition of the derivative is given by:

$$f'(x)=\lim_{\Delta x\to 0}\frac{f(x+\Delta x)-f(x)}{\Delta x} \tag{1.14}$$

The process of finding a derivative is called differentiation. There are numerous notations of differentiation in maths, the most common is $\frac{dx}{dy}$.

$$f'(x)=\frac{dy}{dx} \tag{1.15}$$

Using this notation, I can restate the above discussion in the following way. We cannot let Δx become 0 (because that would be dividing by 0), but we can make it head towards zero and call it "dx". You can also think of "dx" as being infinitesimal, or infinitely small. Similarly, Δy becomes infinitely small and we call it "dy".

To differentiate a function, we do not need to go through the palaver set out above as there are rules of differentiation which help us get the answer. The only three rules relevant to this book are listed below (remember that $x^0=1$ in mathematics[41]):

1. The derivative of a constant is zero.
2. If $y=ax^n$ then $\frac{dx}{dy}=nax^{n-1}$.
3. If $y=\ln(x)$ then $\frac{dx}{dy}=\frac{1}{x}$.[42]

1.6.5 Calculation of the Area under a Curve

So far, we have been concerned with finding the slope of the graph of $y=f(x)$ at a particular value of x, i.e., the slope of a curve. We have seen that this slope is given by the derivative $\left(f'(x)=\frac{dy}{dx}\right)$ of the given function $f(x)$. I now want to consider how to calculate the area under a curve. Areas are relatively easy to calculate if they are bounded by straight lines. However, areas bounded by curved lines are a much more difficult problem. Consider the following approach to find the area enclosed between the curve $y=f(x)$, the x-axis and the lines $x=a$ and $x=b$ (the shaded area in Figure 1.11).

We can approximate the area by dividing the interval from a to b into equal sub-intervals and drawing vertical lines at these points. We can sandwich the area of the curve by choosing the minimum value of f(x) in each sub-interval ("lower sum") and by choosing the maximum value of $f(x)$ in each sub-interval ("upper sum"). If we now increase the number of strips the area is divided into, we will get new upper and lower sums, which will be closer to one another in size and so closer to the area which we are trying to find. In fact, the larger the number of strips we take, the smaller will be the difference between the upper and lower sums, and so the better approximation either sum will be to the area under the curve, but the result will always an approximate of the true area. This is similar to problem of determining the instantaneous speed of a car or fallen stone in Sections 1.6.2 and 1.6.3 where we needed a work around approach to division by zero. In differentiation we used the concept of the derivative as given in Equation (1.14), i.e., Δx approaches but does not reach zero. If $f(x)=x^3$ then

$$f'(x)=\lim_{\Delta x\to 0}\frac{(x+\Delta x)^3-(x)^3}{\Delta x}=\frac{x^3+3x^2.\Delta x+3x.\Delta x^2+\Delta x^3-x^3}{\Delta x}=3x^2+3x.\Delta x+\Delta x^2 \tag{1.16}$$

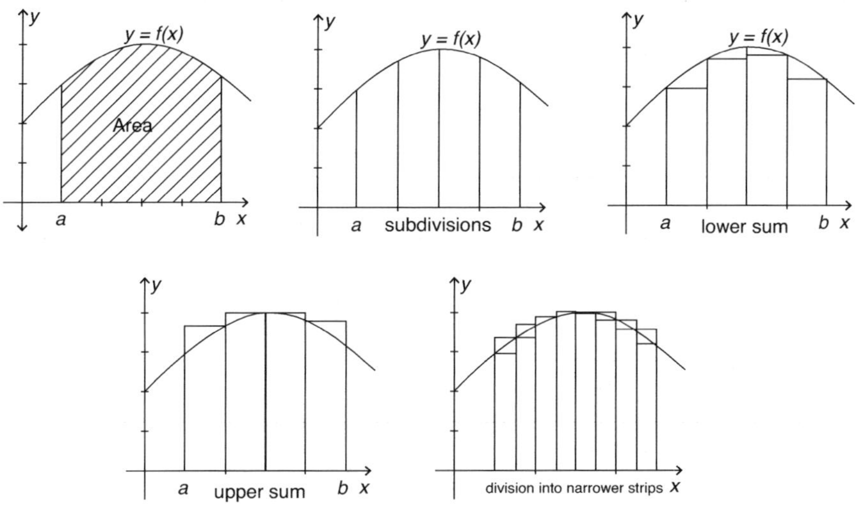

Figure 1.11 Estimation of area under a curve.

In the limit as $\Delta x \to 0$ then

$$f'(x) = \frac{dy}{dx} = 3x^2 \tag{1.17}$$

In calculating the area under a curve, we use a similar approach to avoid the "forbidden" process of division by zero. Consider a curve whose equation is $y = f(x)$, the problem is to find the area enclosed by the curve itself, the x-axis below and two ordinates (lines parallel to the y-axis). In Figure 1.12 the ordinates are $y_0 = f(x_0)$ and $y_5 = f(x_5)$. We then divide the intervening segment enclosed by the curve and the x-axis into rectangular strips of equal width, i.e., $x_1 - x_0 = \Delta x = x_2 - x_1 = x_3 - x_2$ etc.

We assume that there is a function of x which we shall call $A(x)$ ("area function") which gives us the area under the curve $y = f(x)$, from 0 to x. Remember that $y = f(x)$, is a continuous function. The function $A(x)$ may not be known, but it is a given that it represents the area under the curve. Let me reword this because it is important in appreciating the important result that follows – for a continuous function $A(x)$, each value of x has a corresponding function $A(x)$ whose value is the area beneath the curve $y = f(x)$ between 0 and x.

Examination of Figure 1.12 shows the area under the curve between 0 and x lies between two totals $\left(A_1(x) < A(x) < A_0(x)\right)$ where:

$$A_0(x) = y_0\Delta x + y_1\Delta x + y_2\Delta x + y_3\Delta x + y_4\Delta x \tag{1.18}$$

$$A_1(x) = y_1\Delta x + y_2\Delta x + y_3\Delta x + y_4\Delta x + y_5\Delta x \tag{1.19}$$

We know that $x_3 = x_0 + 3\Delta x$, or, more generally $x_r = x_0 + r\Delta x$. In maths the symbol Σ ("sigma" – the capital letter S in the Greek alphabet) stands for sum and we can rewrite Equations (1.19) and (1.20) as:

$$A_0(x) = \sum_{r=0}^{r=4} y_r \Delta x = \sum_{r=0}^{r=4} f(x_0 + r\Delta x)\Delta x \tag{1.20}$$

$$A_1(x) = \sum_{r=1}^{r=5} y_r \Delta x = \sum_{r=1}^{r=5} f(x_0 + r\Delta x)\Delta x \tag{1.21}$$

By diminishing the interval Δx (i.e., by making the strips thinner) we can reduce the difference between $A_0(x)$ and $A_1(x)$ and the "true" area $A(x)$. Table 1.2 shows the results of spreadsheet calculations using the above technique for the area under the graph $y = x^2$ between $x = 0$ and $x = 3$.[43] As $\Delta x \to 0$ then, from Table 1.2, it can be seen the area under the curve → 9.

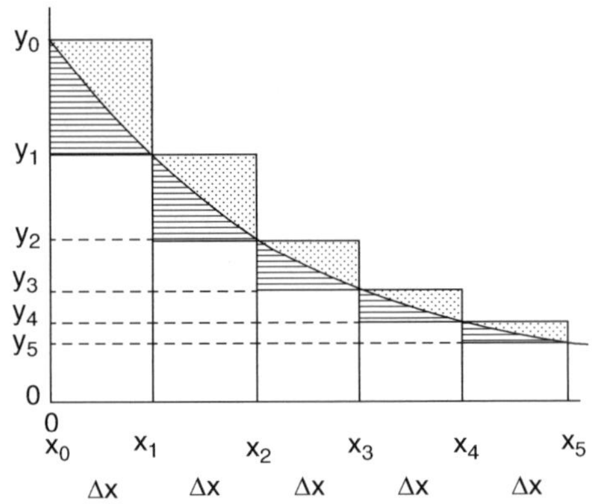

Figure 1.12 Graphical representation of the use of rectangular strips to obtain approximate value for the area between the curve and x-axis bounded by ordinates y_0 and y_5.

Table 1.2: Area under the Curve $y=x^2$ between $x=0$ and $x=3$

Δx	n	$A_0(x)$	$A_1(x)$
0.1	30	9.455	8.555
0.05	60	9.22625	8.77625
0.01	300	9.04505	8.95505
0.001	3000	9.0045005	8.9955005
0.0001	30000	9.000450005	8.999550005

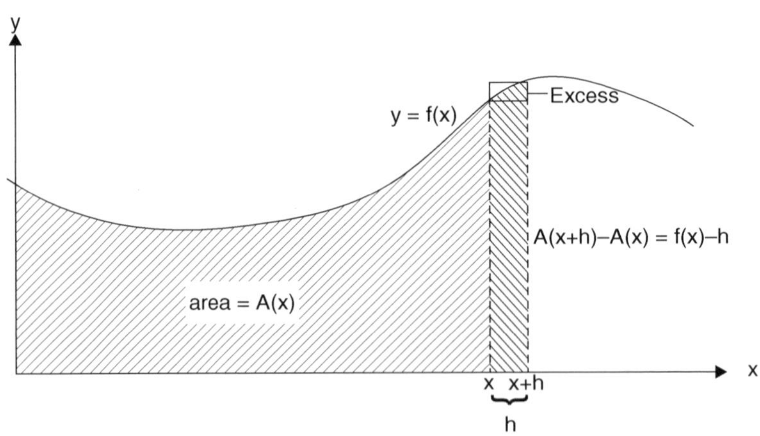

Figure 1.13 A graph of the function y=f(x).

As with finding the slope of a curve, there is a quicker way of finding the area under a curve. Consider a continuous function $y = f(x)$ whose graph is plotted as a curve (Figure 1.13). Each value of x has a corresponding area function $A(x)$ that represents the area beneath the curve between 0 and x. The area under the curve between x and $x + h$ could be computed by finding the area between 0 and $x + h$, then subtracting the area between 0 and x. In other words, the area of this "strip" would be $A(x+h) - A(x)$.

As shown in Figure 1.13 the area of this strip can be approximated by multiplying h by $f(x)$. Therefore, we can write:

$$A(x+h) - A(x) \approx f(x) \times h \tag{1.22}$$

This estimate becomes an equality if we add the triangular area of the box labelled "Excess" in Figure 1.13:

$$A(x+h)-A(x)=f(x)\times h+(\text{Excess}) \tag{1.23}$$

Rearrangement gives:

$$f(x)=\frac{A(x+h)-A(x)}{h}-\frac{\text{Excess}}{h} \tag{1.24}$$

As h approaches zero, $\frac{\text{Excess}}{h}$ must also approach zero. This is true because the area of the excess region is less than or equal to the area of the black-bordered rectangle in Figure 1.13. This indicates

$$f(x)=\lim_{h\to 0}\frac{A(x+\Delta x)-A(x)}{h} \tag{1.25}$$

Now remember, the definition of the derivative is given by Equation (1.14):

$$f'(x)=\lim_{\Delta x\to 0}\frac{f(x+\Delta x)-f(x)}{\Delta x} \tag{1.14}$$

Therefore, Equation (1.25) implies $f(x)=A'(x)$, i.e., the derivative of the area function A(x) exists and is the original function f(x). In other words, computing the derivative of a function and "finding the area" under its curve are "opposite" operations.[44] The opposite of finding the derivative of a function is, not surprisingly, called finding the antiderivative of the function. It is also known as finding the indefinite integral of the function and this is the term I will use in this book.

1.6.6 Indefinite Integration

Section 1.6.5 has shown indefinite integration is the reverse process to differentiation. The derivative of x^2 is 2x; therefore, reversing the process, an integral of 2x is x^2: this answer is easy enough to check by simple differentiation. However, we know that the derivative of a constant is zero and so any of the following will also be the indefinite integrals of 2x: $x^2+1, x^2+\sqrt{2}, x^2+2{,}801$, etc.
In fact, any real number constant will give 2x on differentiating. This is usually written as a "c", i.e., if $\frac{dy}{dx}=2x$ then the indefinite integral is $y=x^2+c$, the constant c is usually referred to as an arbitrary constant; there are an infinite number of functions that we could use as the indefinite integral. In the family of parabolas shown in Figure 1.14, $y=x^2+c$, each member has identical shape and differs only by the amount of its vertical translation. But each curve has the same derivative $\frac{dy}{dx}=2x$.

There are two rules for finding the indefinite integral:

1. The indefinite integral of ax^n is $a\times\frac{x^{n+1}}{n+1}+c$ where c is a real number constant.
2. The indefinite integral of $\frac{1}{x}$. is $\ln(x)+c$.[45]

These rules can be verified by differentiating y with respect to x. A couple of examples will prove useful.

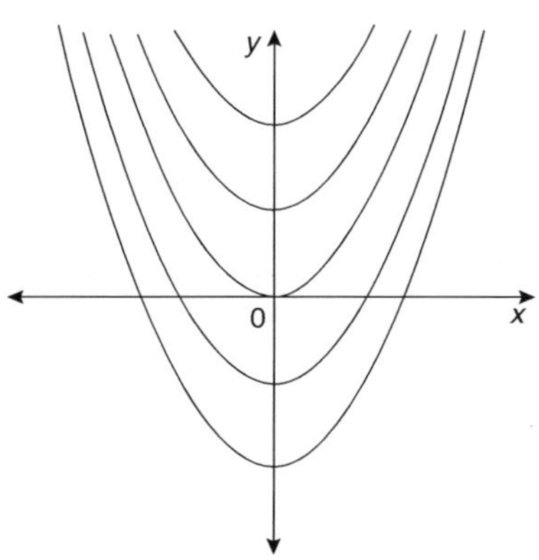

Figure 1.14 The family of curves $y=x^2+c$.

1. What is y if $\frac{dy}{dx} = 5x^7$? By definition, the derivative of y is $5x^7$. Therefore, y is the indefinite integral of $5x^7$ and is $5\times\left(\frac{1}{7+1}x^{7+1}\right)+c = 5\times\frac{1}{8}x^8+c = \frac{5}{8}x^8+c$.
2. What is the indefinite integral of $4x^2-6x-4$? $4x^2-6x-4$ can be written as $4x^2-6x^1-4x^0$. The indefinite integral is $4\times\left(\frac{1}{2+1}x^{2+1}\right)-6\times\left(\frac{1}{1+1}x^{1+1}\right)-4\times\left(\frac{1}{0+1}x^{0+1}\right)+c = \frac{4}{3}x^3-\frac{6}{2}x^2-\frac{4}{1}x^1+c ==$ $\frac{4}{3}x^3-3x^2-4x+c$.

Table 1.3 sets down the results of differentiation and integration of various functions. The differentiation of $y = f(x)$ gives $f'(x)$, the integration of $f(x)$ gives F(x)[46] and the differentiation of F(x) gives $F'(x)$. By definition $F'(x) = f(x)$.

Integration can be used to calculate the area under a curve as discussed in the next section.

1.6.7 Using Integration to Calculate the Area under a Curve – Definite Integrals

Integration is not only about undoing the effect of differentiation, but it can also be used to find areas, volumes, central points, and many useful things. All I will consider here is using integration to find the area under the two-dimensional graph of a function. In the most common form of notation, the indefinite integral of $f(x)$ is written as $\int f(x)dx$. Using this notation, we would write "the indefinite integral of 2x with respect to x equals x^2+c" as $\int 2xdx = x^2+c$. In general, for any function:

$$\int f(x)dx = F(x)+c \tag{1.26}$$

$F(x)+c$ is the indefinite integral of $f(x)$. In Equation (1.26) the $\int$ symbol is called the integral symbol, $f(x)$) is called the integrand, x is called the integration variable and c is called the constant of integration. This notation which can be confusing (and is described as "ugly" by one website) has been used for centuries.

So far there has always been an unknown constant of integration (this is the reason, such integrals are known as indefinite integrals). With definite integrals, we integrate a function between 2 points, and so we can find the precise value of the integral because the constant c cancels out. The simplest way to explain definite integrals is to consider the trivial example of finding the area under a straight-line graph. Consider the graph of $y = 2x$ shown in Figure 1.15 (alternatively $f(x) = 2x$). We know the area function $[A(x)]$ is the indefinite integral of $y = 2x$, $A(x) = F(x) = x^2+c$. Now when $x = 0: A(0) = F(0) = 0^2+c = 0$, and when

Table 1.3: Differentiation and Integration

F′(x)	F(x)	f(x)	f′(x)
x^3	$\frac{x^4}{4}+c$	x^3	$3x^2$
x^2	$\frac{x^3}{3}+c$	x^2	$2x$
x	$\frac{x^2}{2}+c$	x	$1[x^0=1]$
a	ax+c	a (a real number, a constant)	0
x^{-1}	ln(x)	x^{-1}	$-x^{-2}$
x^{-2}	$\frac{x^{-1}}{-1}+c$	x^{-2}	$-2x^{-3}$
x^{-3}	$\frac{-x^{-2}}{-2}+c$	x^{-3}	$-3x^{-4}$

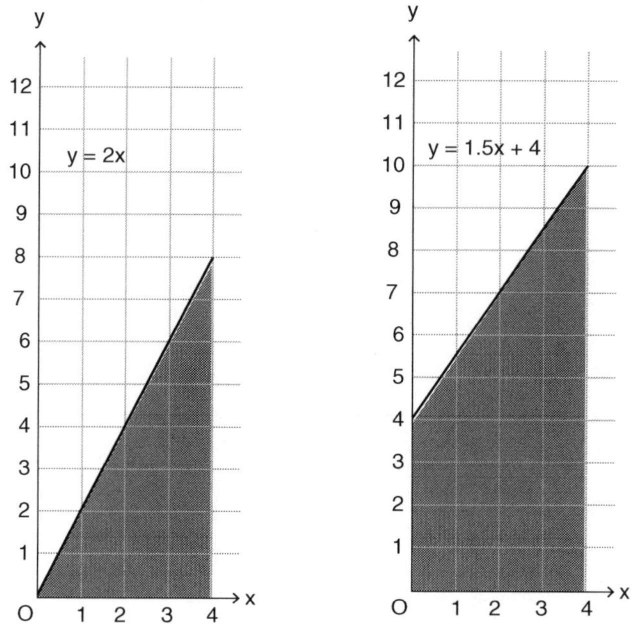

Figure 1.15 The graphs of y=2x and y=1.5x+4.

$x = 4 : A(4) = F(4) = 4^2 + c = 16 + c$. Therefore, the area of the shaded area in Figure 1.15 is $F(4) - F(0) = (16 + c) - c = 16$ units. This result agrees with the area calculated by geometry where the area of a triangle is half the base times the height (½×4×8=16).

The notation for a definite integral is $\int_a^b f(x)dx$ where a and b are the 2 points on the x-axis (usually known as limits). In mathematical texts, the above calculation is written as:

$$\int_0^4 2x = \left[\frac{2x^2}{2} + c\right]_0^4 = \left(\frac{2 \times 4^2}{2} + c\right) - \left(\frac{2 \times 0^2}{2} + c\right) = (16 + c) - (0 + c) = 16 \tag{1.27}$$

Now consider the other graph shown in Figure 1.15, i.e., $y = 1.5x + 4$ and we want to calculate the area under the graph between $x = 0$ and $x = 4$. Following the same procedure, we find the area is 28 unit², which can easily be checked to be correct by simple geometry:

$$\int_0^4 1.5x + 4 = \left[\frac{1.5x^2}{2} + 4x + c\right]_0^4 = \left(\frac{1.5 \times 4^2}{2} + 4 \times 4 + c\right) - \left(\frac{1.5 \times 0^2}{2} + 4 \times 0 + c\right)$$
$$= (12 + 16 + c) - (0 + 0 + c) = 28 \tag{1.28}$$

More generally, if we want to find the area A under a curve $y = f(x)$ between $x = a$ and $x = b$, then

$$A = \int_a^b f(x) = [F(x) + c]_{x=a}^{x=b} \tag{1.29}$$

where $F(x) + c$ is the indefinite integral of $f(x)$ and $\int_a^b f(x)$ is known as a definite integral.

Finally, I want to reinforce the above by considering two more examples (see Figure 1.16). In both cases I want to calculate the area between the curve and the x-axis, i.e., the shaded areas. The shaded area between the x-axis and the curve between the limits $x = 0$ and $x = 5$ for the function $y = x^2 - 4x + 5y$ is given by:

$$\int_0^5 \left(x^2 - 4x + 5\right) = \left[\frac{x^3}{3} - \frac{4x^2}{2} + 5x + c\right]_0^5 = \left(\frac{125}{3} - \frac{100}{2} + 25 + c\right) - (c) = \frac{50}{3} \tag{1.30}$$

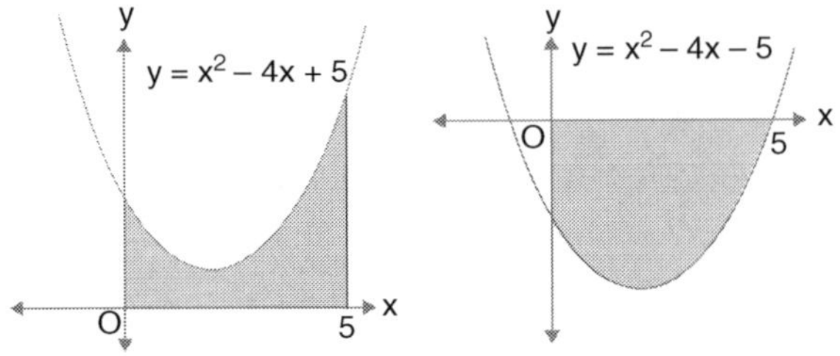

Figure 1.16 Graphs of y=x²−4x+5 and y=x²−4+5.

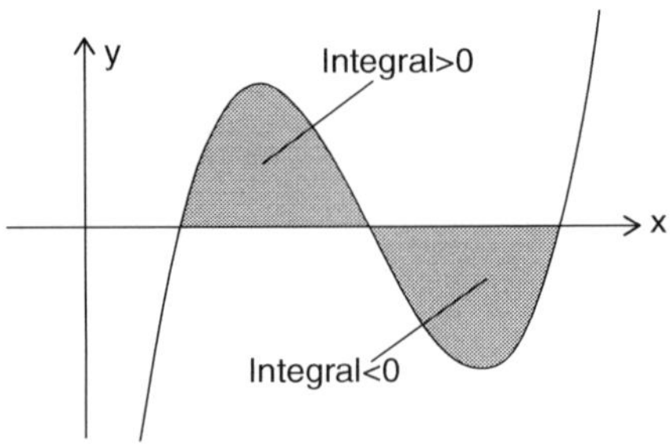

Figure 1.17 Positive and negative definite integrals.

Similarly, the shaded area between the x-axis and the curve between the limits x=0 and x=5 for the function $y = x^2 - 4x + 5$ is given by:

$$\int_0^5 \left(x^2 - 4x - 5\right) = \left[\frac{x^3}{3} - \frac{4x^2}{2} - 5x + c\right]_0^5 = \left(\frac{125}{3} - \frac{100}{2} - 25 + c\right) - (c) = -\frac{100}{3} \quad (1.31)$$

There can be no such thing as negative area, but the negative sign follows directly from the definition of a definite integral since you use the negative values of the function to calculate it. Because of this, you need to be careful when you use integration to find the area between the curve of the function and the x-axis (Figure 1.17 and use the absolute value of the calculated number. The absolute value of a number x is written $|x|$ and is always positive, i.e. $\left|-\frac{100}{3}\right| = +\frac{100}{3}$. The absolute value of a real number is the magnitude of the number without regard to its sign, e.g., the absolute value of +1 and −1 is (+)1 (see Section 1.6.10).

1.6.8 Absolute Value Equations[47]

The notation for the absolute value of x is $|x|$. Note as well that the absolute value bars are not brackets and, in many cases, do not behave as brackets. One way to think of absolute value is that it takes a number and makes it positive, and we can guarantee that $|x| \geq 0$ regardless of the value of x. There are two ways to define absolute value. There is a geometric definition and a mathematical definition. In the geometric definition, we think $|x|$ as the distance of x from the origin on a number line (distance is always a positive value) – see Figure 1.18.

From this, we can get the following values of absolute value, e.g., $|2| = 2; |-3| = 3; \left|\frac{9}{2}\right| = \frac{9}{2}$. All that we need to do is identify the point on the number line and determine its distance from the origin. Note as well that we also have $|0| = 0$. The rigorous mathematical definition for absolute value is:

$$|x| = x \text{ if } x \geq 0 \qquad \text{or} \qquad |x| = -x \text{ if } x < 0 \quad (1.32)$$

For example, from this definition, we get $|4| = 4$ because $4 \geq 0$, $|-8| = -(-8) = 8$ because $-8 < 0$, and $|0| = 0$ because $0 \geq 0$. Do not make the mistake of assuming that absolute value just makes all minus signs into plus signs. In other words, $|4x - 3| \neq 4x + 3$.

Now I want to consider solving equations that contain absolute values. Consider the following simple equation: $|x| = 4$. If we think of this from a geometric point of view, this means that whatever x is, it must have a distance of 4 from the origin, and there are only two numbers that have a distance of 4 from the origin, namely 4 and –4. So, there are two solutions to this equation: $x = 4$ or $x = -4$. We can do this for any positive number, not just 4. So, this leads to the following general formula for equations involving absolute value: if $|x| = b$ and $b > 0$ then $x = b$ or $x = -b$. Note that $|x| = -b$ is forbidden as, by definition, $|x|$ is a positive number. Consider $|2x - 5| = 9$ we know $2x - 5 = 9$ or $2x - 5 = -9$, therefore $x = -2$ or $x = 7$. Now consider $|10x - 3| = 0$ from a geometric standpoint, the equation is saying that the quantity in the absolute value bars has a distance of zero from the origin. There is only one number that has the property and that is zero itself. So, we must have $10x - 3 = 0$ which gives $x = \frac{3}{10}$. In this case, we get a single solution. Summarising we can see that if b is zero then we can just drop the absolute value bars and solve the equation. Likewise, if b is negative this is an invalid equation and there is no solution.

To this point we have only looked at equations that involve an absolute value being equal to a number, but there is no reason to think that there has to only be a number on the other side of the equal sign. Likewise, there is no reason to think that we can only have one absolute value in the problem. Consider $|x - 2| = 3x + 1$. At first glance the formula we used above will do us no good here. It requires the right side of the equation to be a positive number. It turns out that we can still use it here, but we must be careful with the answers as using this formula will, on occasion introduce an incorrect answer. So, it is essential to check our solutions to see if they are valid. If we use the above formula we can write: $x - 2 = -3x - 1$ and $x = \frac{1}{4}$; or $x - 2 = 3x + 1$ and $x = -\frac{3}{2}$. However, there is a problem with the second answer. If we plug this one into the equation, we get $\left|-\frac{3}{2} - 2\right| = 3\left(-\frac{3}{2}\right) + 1$, i.e., $\left|-\frac{7}{2}\right| = -\frac{7}{2}$ which is forbidden because by definition $\left|-\frac{7}{2}\right|$ must be a positive real number. This will happen on occasion when we solve this kind of equation with absolute values. Note that we really did not need to plug the solution into the whole equation here. All we needed to do was check the portion without the absolute value and if it was negative then the potential solution would not be a solution, and if it was positive or zero, it would be a solution. So, there is a single solution to this equation $x = \frac{1}{4}$. However, do not assume that because the first potential solution is negative it will not be a solution. Consider $|4x + 2| = 3 - x$ using the same procedure as above we obtain two answers – $x = -2$ or $x = 0$. We will only exclude a potential solution if it makes the portion without absolute value bars negative. In this case, both potential solutions will make the portion without absolute value bars positive and so both are in fact solutions.

Now consider $|2x - 1| = |4x + 9|$. This equation is different because both sides contain absolute values. The key to solving these types of equations is to recognise that the only way the two sides are equal will be if the two quantities inside the absolute value bars are equal or equal but with opposite signs, i.e., we must have $2x - 1 = 4x + 9$ or $2x - 1 = -4x - 9$, i.e., the solutions are $x = -5$ or $x = -\frac{4}{3}$.

The following formula $|a \times b| = |a| \times |b|$ can be useful when solving equations.

In this book, I will use absolute value equations to assist the reader in understanding thermodynamic formula and associated sign conventions. However, I will use absolute values in the following section on exact and inexact differentials.

1.6.9 Exact, Inexact, and Partial Differentials

The principal objective of the preceding sections is to give you the basic concepts behind calculus, and this should allow you to have some understanding of the basic calculus presented in this book. Differentiation allows the calculation of rates of change and, at least as far as this book is concerned, integration is used for finding area under a graph. Unfortunately, there is no escaping the fact that in order to appreciate certain concepts of thermodynamics (e.g., the different ways the equation representing the First Law of Thermodynamics is written down and how Clausius "invented" entropy) you do need, at the very least, to appreciate exact and inexact, and partial derivatives. To deliver what I promised in the Prologue I will generally ignore the differences between these different types of differentials (to the rage of mathematicians and physicists). However, there are occasions when I cannot escape mentioning them. The below two sub-sections are to meet this requirement. I do not pretend that the content of the following two sub-sections is mathematically rigorous, but my objective here is to give a basic understanding of the concepts.

The following two sub-sections could usefully be skipped on a first read of this book.

1.6.9.1 *Exact and Inexact Differentials*

The easiest example of the difference between an exact and an inexact differential is the difference between net distance and total distance. For example, in moving along a number line from x_A to x_B the displacement is displacement $x_B - x_A$ that is equal to the total distance walked. If one then returns to x_A, however, the displacement is now 0. Therefore, if we know the initial and final positions on the straight line, we can calculate displacement, and this is independent of the path taken to get there. If we let the differential of the displacement equal to +1, i.e., $\left(\frac{d(\text{displacement})}{dx} = +1\right)$ we can calculate the displacement by calculus:

$$\text{Displacement} = \int_{x_A}^{x_B} (+1)dx = (x_B + c) - (x_A + c) = x_B - x_A \tag{1.33}$$

Note that displacement can be a positive or negative real number depending on whether we move left to right along the number line (positive), or whether we move right to left (i.e., it is a vector having magnitude and direction). If we move from x_A to x_B and then back to x_A. We can calculate the displacement from Equation (1.33) knowing initial and final positions on the number line are both x_A.

$$\text{Displacement} = \int_{x_A}^{x_A} (+1)dx = (x_A + c) - (x_A + c) = x_A - x_A = 0 \tag{1.34}$$

In mathematics, the differential of displacement $\left(\frac{d(\text{displacement})}{dx}\right)$ is an exact differential because we can calculate displacement from a knowledge of the initial and final state.

Now let us consider total distance. The first point to note is that distance is a scalar quantity and like other scalar quantities (such as mass) its magnitude is a positive real number. It is clear that if at any time the change in the x position is negative (moving from right to left along the number line), then distance (which must be a positive real number) cannot be $x_B - x_A$. In fact, we know it must be $-(x_B - x_A) = x_A - x_B$. We can see that on the first leg of the journey along the number line (where x is increasing) the differential of distance is +1, i.e., $\left(\frac{d(\text{distance})}{dx} = +1\right)$, and on the second leg of our journey (where x is decreasing) the differential of distance is –1, i.e., $\left(\frac{d(\text{distance})}{dx} = -1\right)$. We can write:

$$\text{Distance} = \int_{x_A}^{x_B} (+1) + \int_{(+1)+}^{x_A} (-1) = (x_B - x_A) + (-x_A + x_B) = 2(x_B - x_A) \tag{1.35}$$

However, there are an infinite number of ways to move along a number line and return to the original position. This means there are an infinite number of distance magnitudes even though the displacement in each case is zero. This means that to calculate distance we need to know the path and this is the property of an inexact differential. In mathematics $\frac{d(\text{distance})}{dx}$ is an inexact differential because to calculate distance we need to know the path taken along the number line to evaluate the total distance travelled.

This is from Wikipedia[48] (I will consider heat, work, state, and path functions in Chapter 4):

> An inexact differential or imperfect differential is a type of differential used in thermodynamics to express changes in path dependent quantities. In contrast, an integral of an exact differential (a differential of a function) is always path independent since the integral acts to invert the differential operator. Consequently, a quantity with an inexact differential cannot be expressed as a function of only the variables within the differential; i.e. its value cannot be inferred just by looking at the initial and final states of a given system. Inexact differentials are primarily used in calculations involving heat and work because they are path functions, not state functions.[49]

It should be pointed out that some authors use distinctive notations to refer to an inexact differential; the most common is one of them is đ. A procedure, according to one author, is one "that becomes quite confusing and useless".[50] A differential dq that is not exact is said to be integrable when there is a function $\frac{1}{\tau}$ such that the new differential $\frac{đq}{\tau}$ is exact. The function $\frac{1}{\tau}$ is called the integrating factor, τ being the integrating denominator. The following extract may clarify matters[51]:

As opposed to an exact differential, an inexact differential cannot be expressed as the differential of a function, i.e. while there exist [sic] a function U such that $U = \int dU$, there is [sic] no such functions for đq and đw. And the same is, of course, true for any state function a and any path function b respectively: an infinitesimal change in a state function is represented by an exact differential da and there is a function a such that $a = \int da$, while an infinitesimal change in a path function b is represented by an inexact differential đb and there is no function b such that $b = \int đb$.

Consequently, for a process in which a system goes from state 1 to state 2 a change in a state function can be evaluated simply as

$$\int_1^2 da = a_2 - a_1,$$

while a change in a path function b can not [sic] be evaluated in such a simple way,

$$\int_1^2 đb \neq b_2 - b_1,$$

And for a state function a in a thermodynamic cycle

$$\oint da = 0,$$

while for a path function b

$$\oint đb \neq 0.$$

1.6.9.2 Partial Differentials

This short introduction to partial derivatives is based on that given in a Khan Academy website.[52] Remember the symbol $\frac{df(x)}{dx}$ can be interpreted as the resultant very small change in the output of the function $(df(x))$ that results from the very small change of the input dx. Now consider some function with a two-dimensional input and a one-dimensional output such as $f(x, y) = x^3 + 3xy$. We can interpret $\frac{df(x,y)}{dx}$ the same way, i.e., $df(x, y)$ is the resultant very small change in the output of the function that results from the very small change of the input dx. However, this ignores the fact that there is another input variable y. Now if we re-interpret $df(x, y)$ to represent the tiny change to the function that a small change in dy brings about, we would have a different derivative $\frac{df(x,y)}{dy}$. Neither one of these derivatives tells the full story of how our function $f(x, y)$ changes when its input changes slightly, so we call them partial derivatives. To emphasise the difference, mathematicians no longer use the letter d to indicate tiny changes, but instead use a new symbol ∂, writing each partial derivative as $\frac{\partial f(x, y)}{\partial x}$ and $\frac{\partial f(x, y)}{\partial y}$. It is also worth noting that mathematicians do not tend to use $f(x, y)$ or similar specific symbols, they use the more general f (as in $\frac{\partial f}{\partial x}$ and $\frac{\partial f}{\partial y}$).

The partial derivative of a function of several variables is its derivative with respect to one of those variables, with the others held constant. Consider the following function of three variables x, y, and z:

$$f(x,y,z) = x^3 - 2xyz \tag{1.36}$$

The partial derivatives are:

$$\frac{\partial f(x,y,z)}{\partial x} = 3x^2 - 2y \quad \frac{\partial f(x,y,z)}{\partial y} = -2xz \quad \frac{\partial f(x,y,z)}{\partial z} = -2xy \tag{1.37}$$

The derivative of $f(x,y,z)$ is then written:

$$df(x,y,z) = \left(\frac{\partial f(x,y,z)}{\partial x}\right)dx + \left(\frac{\partial f(x,y,z)}{\partial y}\right)dy + \left(\frac{\partial f(x,y,z)}{\partial z}\right)dz = 3x^2 - 2y - 2xz - 2xy \tag{1.38}$$

I presume the reason that Equation (1.38) is written in the way it is because writing $\frac{df(x,y,z)}{dx,dy,dz}$ is probably a no-no in mathematics, remember $\frac{dy}{dx} = f(x)$ can be written $dy = f(x)dx$. Equation (1.38) is often written (at least in thermodynamics) to specify the variables that are held constant:

$$df(x,y,z) = \left(\frac{\partial f(x,y,z)}{\partial x}\right)_{y,z} dx + \left(\frac{\partial f(x,y,z)}{\partial y}\right)_{x,z} dy + \left(\frac{\partial f(x,y,z)}{\partial z}\right)_{x,y} dz \tag{1.39}$$

I have done the best I can here if there are major errors I apologise. However, all I am trying to do is set down the concept of a partial derivative and not be mathematically rigorous.

1.6.10 Logarithms

The best way to think about logarithms is in terms of the relationship:

$$y = b^x \text{ is equivalent to } \log_b(y) = x \tag{1.40}$$

On the left-hand side of Equation (1.40) is the exponential statement "$y = b^x$" where "b" is the base of the power function "b^x". On the right-hand side, "$\log_b(y) = x$" is the equivalent logarithmic statement, which is said as "log-base-b of y equals x". The value of the subscripted "b" is "the base of the logarithm", just as b is the base in the exponential expression "b^x", and "y" is considered the argument of the logarithm. For example:

$$2 = 10^{0.3010} \text{ is equivalent to } \log_{10}(2) = 0.3010 \tag{1.41}$$

There are some restrictions on the base and argument values of logarithms. Only positive numbers other than 1 are allowed as the base of the logarithm. The reasons for this are:

1. When $b = 0$ then 0^x is always zero whatever the value of x.
2. When $b = 1$ then 1^x is always one whatever the value of x.
3. When b is negative then you may end up with an imaginary number (i.e., not a real number) when x is a fraction or decimal, e.g., $(-1)^{\frac{1}{2}}$ is equivalent to the square root of minus one $\left(\sqrt{-1}\right)$ which is an imaginary number.

Therefore 0, 1, and every negative number represent a potential problem as the base of an exponential function, and if those numbers cannot reliably be the base of an exponential function, then they also cannot reliably be the base of a logarithm. It is for these reasons that we only allow positive numbers other than 1 as the base of the logarithm. We now know that it does not matter what exponent we put on that base (it could be a positive number, a negative number, or 0) the exponential function is always going to be a positive number. It is for this reason that the argument of a log function is restricted to positive real numbers. In other words, the only numbers you can plug into a log function are positive numbers, i.e., you cannot have a logarithm of a negative number or zero. In summary, the base of a logarithm (b) can only be a positive number not equal to one, which means the argument of the logarithm (y) can only be a positive number, which means, we have to only allow positive arguments. The value you get for the logarithm after plugging in the base and argument (x) can be positive or negative numbers (see Table 1.4).

If you know how to manipulate exponentials, then you can manipulate logarithms (Table 1.5).

Table 1.4: Positive and Negative Values of Logarithms When Base and Arguments Are Positive Numbers

If $x=10^y$ then $\log_{10}(x)=y$

x	10^y	$y=\log_{10}(x)$
1, 000, 000	10^6	6
100, 000	10^5	5
10, 000	10^4	4
1, 000	10^3	3
100	10^2	2
10	10^1	1
1	10^0	0
0.1	10^{-1}	−1
0.01	10^{-2}	−2
0.001	10^{-3}	−3
0.0001	10^{-4}	−4
0.00001	10^{-5}	−5
0.000001	10^{-6}	−6

Table 1.5: The Four Basic Mathematical Operations of Logarithms

Exponentials	Logarithms	Comment
$b^n \times b^m = b^{n+m}$	$\log_b(xy) = \log_b x + \log_b y$	Multiplication is replaced by addition
$\frac{b^n}{b^m} = b^{n+m}$	$\log_b\left(\frac{x}{y}\right) = \log_b x - \log_b y$	Division is replaced by subtraction
$\left(b^n\right)^m = b^{(nm)}$	$\log_b x^y = y \times \log_b x$	Power is replaced by multiplication
$\sqrt[m]{b^n} = \left(b^n\right)^{\frac{1}{m}} = b^{\left(\frac{n}{m}\right)}$	$\log_b\left(\sqrt[m]{x^n}\right) = \frac{n\log_b x}{m}$	Roots are replaced by division

When the base is 10, the logarithm can be written without a subscript, e.g., log 1,000 means $\log_{10}$ 1,000. The other exception is when the base is the number e (which equals 2.718282...).[53] Such logarithms are written $\ln x$ or $\log_e x$ and are called "natural" or "Napierian" logarithms.

1.6.11 Trigonometry

Figure 1.18 describes the basic trigonometric functions of sine, cosine, and tangent. In this book the only trigonometric equation you need to be aware of is $(\sin\theta)^2 + (\cos\theta)^2 = 1$ which in mathematics is usually written as $\sin^2\theta + \cos^2\theta = 1$. This equation can easily be proved. If we let h = length of hypotenuse, o = length of opposite side and a = length of adjacent side, then from the definitions of sine and cosine given in Figure 1.19 we can write:

$$(\sin\theta)^2 + (\cos\theta)^2 = \frac{o^2}{h^2} + \frac{a^2}{h^2} = \frac{o^2 + a^2}{h^2} \tag{1.42}$$

Using the well-known Pythagorean theorem ("in a right-angled triangle: the square of the hypotenuse is equal to the sum of the squares of the other two sides") we can rewrite Equation (1.35) as

$$(\sin\theta)^2 + (\cos\theta)^2 = \sin^2\theta + \cos^2\theta = \frac{o^2 + a^2}{h^2} = \frac{h^2}{h^2} = 1 \tag{1.43}$$

This result $\left(\sin^2\theta + \cos^2\theta = 1\right)$ is used in Section 1.9.3.

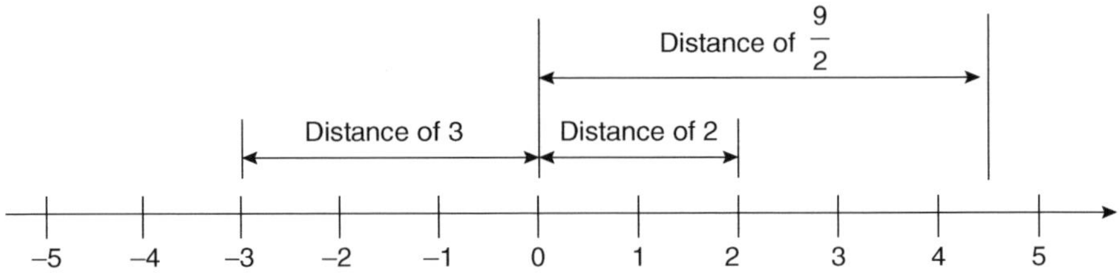

Figure 1.18 A number line.

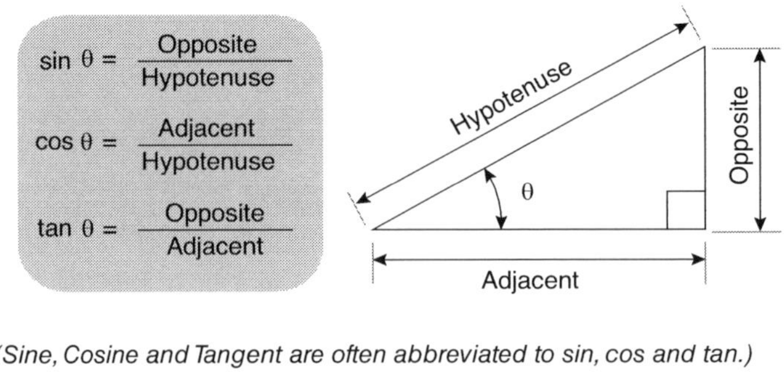

Figure 1.19 Basic trigonometric functions.

1.7 NEWTON'S EQUATIONS OF MOTION

1.7.1 Introduction

I now want to consider velocity and acceleration in more detail. In the interests of simplicity, I will only consider the mathematics of one-dimensional systems, and, in addition, I will also assume acceleration is constant (such as the acceleration due to gravity at a location on or near the Earth's surface). I will then develop a series of equations of motion and consider their graphical representations.

1.7.2 Velocity and Acceleration

When a physical object experiences a force its velocity (which could be zero) changes, this change in velocity is known as acceleration. A change in velocity could be a change in speed (i.e., magnitude of velocity) or a change in direction (remember velocity is the speed in a straight line). You know this from everyday experience. Consider travel in a car, when the car is travelling a constant velocity you feel no push or pull (force). When the velocity of the car is increasing, you are pushed back into your seat and when its velocity is decreasing you are pulled forward out of your seat. Also, when a car is turning a corner at constant speed, you are pushed sideways because the direction of velocity is changing (although speed may remain constant). When considering motion in one-dimension (or direction), acceleration can only occur with a change of speed and not with a change of direction, apart from the special case when the direction of motion changes by 180°, e.g., in the case of a bouncing ball. If in one second an object's velocity changes from 7 $m.s^{-1}$ to say 10 $m.s^{-1}$ (in the same direction), acceleration is $3\left(m.s^{-1}\right).s^{-1}$, i.e., the units of acceleration are $m.s^{-2}$.

Remember that velocity is a vector and our limitation to one-dimensional systems, allows the use of a simple sign convention as the following example demonstrates. Let us use the sign convention that travel to the right is positive and travel to the left is negative. Let the velocity of a snooker ball travelling to the right be +2 $m.s^{-1}$; when it hits the cushion at 90° and, if we assume the speed of the ball before and after collision is unchanged (this is known as an elastic collision and is rigorously defined in Chapter 2), then the velocity when it bounces off the cushion is –2 $m.s^{-1}$ (negative because it is in the opposite direction). Although the change in speed is zero the change in the magnitude of velocity is 4 $m.s^{-1}$; mathematically the velocity in the forward direction is +2 and the velocity in the reverse direction is –2, therefore the change in velocity (acceleration) is $\left[(+2)-(-2)\right]=+4$ $m.s^{-1}$ (remember acceleration is a vector quantity). This is our second encounter[54] with "sign conventions" and it will not be our last: an understanding of sign conventions is essential in understanding the subject matter of this book. If we chose the convention that the velocity of the ball is negative in the right direction, then before the collision the change in speed is still zero and the change in the magnitude of velocity is still $4\,m.s^{-1}$, but using our new sign convention and considering the vector nature of acceleration its sign is reversed $\left(\left[(-2)-(+2)\right]=-4\ m.s^{-1}\right)$. The important point is consistency – once a sign convention has been decided you must stick with it otherwise confusion reigns supreme.

To calculate the acceleration of a physical object when subjected to an external force, e.g., when two snooker balls collide, we need to know the time the force was applied, e.g., how long the collision took and usually we do not have that data. As we shall see the concept of momentum allows us to circumvent this problem.[55]

1.7.3 Equations of Motion

In one-dimensional systems, the displacement of a moving object is directly proportional to both velocity and time. If the absolute value of the velocity increases, the absolute value of the displacement increases, i.e., the distance travelled by the object increases. If it moves for a longer time, it will also go farther. Acceleration complicates this simple situation because velocity and displacement are now both directly proportional to time. If we write this down it sounds faintly ridiculous, i.e., "Displacement is directly proportional to time and directly proportional to velocity, which is directly proportional to time". Time is now a factor twice, making displacement proportional to the square of time. In developing Newton's equations of motion, I will ignore the vector nature of velocity, acceleration, and displacement and will consider them to be scalar quantities, i.e., positive real numbers.[56]

When acceleration is constant, the velocity will change uniformly from its initial value to its final value and the average velocity will lie halfway between the extremes. If an object is at rest (u=0) and is subject to a constant acceleration for time t and its final velocity is v, then the displacement or distance travelled (s) is calculated using the following argument. The average velocity of the object is $\frac{1}{2}v$ (remember constant acceleration and the object was at rest). Therefore, distance travelled is the average velocity multiplied by the time travelled:

$$s=\frac{1}{2}vt \tag{1.44}$$

If the object was moving with velocity $u(\neq 0)$ before subject to a constant acceleration for time t, then average velocity is $\frac{u+v}{2}$ and Equation (1.44) can then be written as:

$$s=\frac{1}{2}(u+v)t \tag{1.45}$$

The following example should clarify matters. If an object moving at $2\ m.s^{-1}$ undergoes a constant acceleration over a period of 5 seconds and its velocity increases to $12\ m.s^{-1}$, then its average velocity during its period of acceleration is $7\ m.s^{-1}\left(\frac{1}{2}(u+v)\right)$. Therefore, the distance travelled during its acceleration is $35\ m\left(\frac{1}{2}(u+v)t\right)$.

I will now consider the calculation of the final velocity (v). If the object is initially at rest, then

$$v=at \tag{1.46}$$

i.e., the final velocity (remember constant acceleration) is the acceleration multiplied by the time of the acceleration (i.e., the time the force was applied to the object). If the object was initially moving with a velocity u, then

$$v=u+at \tag{1.47}$$

An example will again prove useful. If an object at rest undergoes a constant acceleration of $2\ m.s^{-2}$ then after 1 second, its velocity is $2\ m.s^{-1}$ and after 2 seconds its velocity is $4\ m.s^{-1}$. If its initial velocity was $1\ m.s^{-1}$ then after 1 second, its velocity is $3\ m.s^{-1}$ and after 2 seconds its velocity is $5\ m.s^{-1}$.

Substituting u + at for v (from Equation 1.47]) into Equation 1.45 gives

$$s=\frac{1}{2}(u+v)t=\frac{1}{2}(u+u+at)t=\left(\frac{1}{2}u+\frac{1}{2}u+\frac{1}{2}at\right)t=ut+\frac{1}{2}at^2 \tag{1.48}$$

Substituting $\frac{v-u}{a}$ (from Equation 1.47) for t into Equation 1.45:

$$s=\frac{1}{2}(u+v)\left(\frac{v-u}{a}\right)=\frac{1}{2}\left(\frac{v^2-u^2}{a}\right) \quad (1.49)$$

Further simplification and rearrangement yields:

$$v^2=u^2+2as \quad (1.50)$$

Equations(1.45), (1.48) and (1.50) are known as Newton's equations of motion (as opposed to Newton's laws of motion (see Section 1.9):

$$s=\frac{1}{2}(u+v)t$$

$$s=ut+\frac{1}{2}at^2$$

$$v^2=u^2+2as$$

I now want to consider graphical representations of velocity and acceleration that should reinforce what has been discussed so far in this section.

1.7.4 Graphical Representations of Velocity and Acceleration

Assume an object is moving with a velocity of 1.5 $m.s^{-1}$. At time t = 0 the distance travelled s = 0. When t = 1 second then s = 1 × 1.5 = 1.5 m, and so on. Figure 1.20 is a graphical representation – note velocity is constant and acceleration is zero. Now consider a physical object on Earth in free fall, and we will, for convenience consider the acceleration due to gravity is 10 $m.s^{-2}$. Figure 1.21 shows the distance v. time plot (calculated using Equation 1.48 with u=0). In contrast to Figure 1.20 this plot is not a straight line because velocity is not constant.

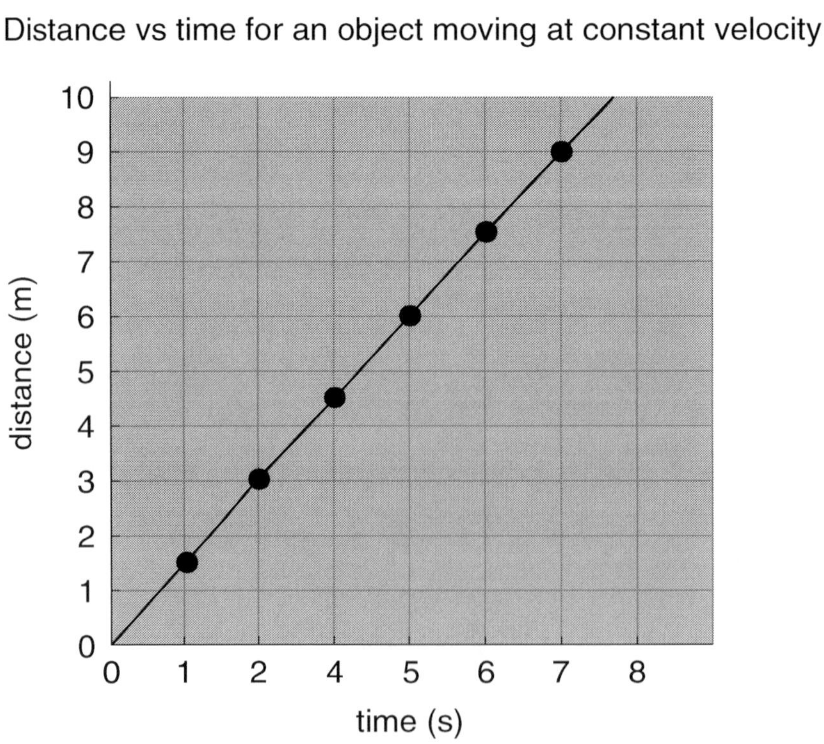

Figure 1.20 Distance v. time plot for an object moving at constant velocity of 1.5 $m.s^{-1}$.

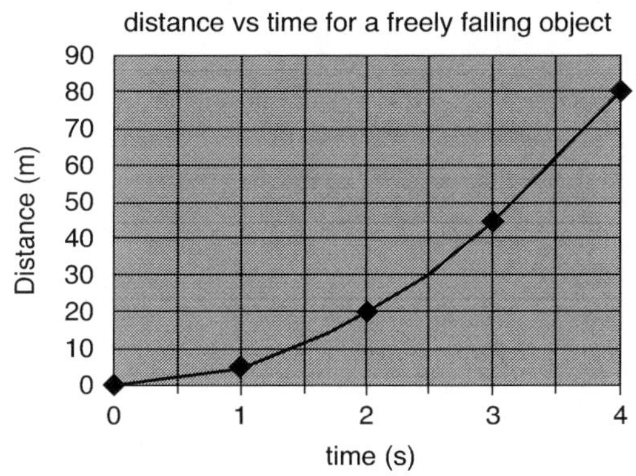

Figure 1.21 Distance v. time plot for an object in freefall on Earth.

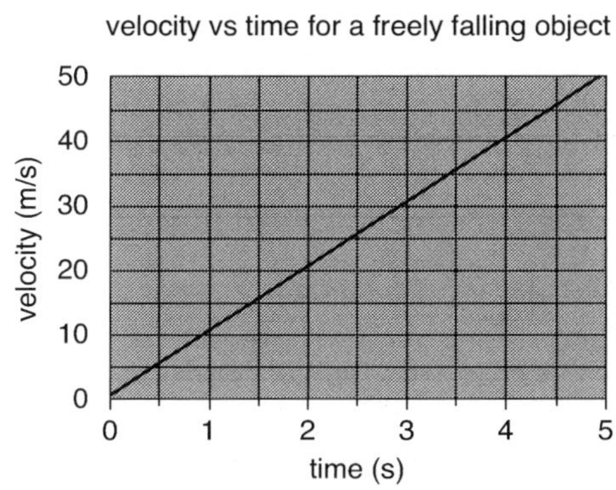

Figure 1.22 Velocity v. time plot for an object in freefall on Earth.

If we plot velocity v time for a freely falling object, we get Figure 1.22 (assuming velocity is zero at t = 0), this is a straight line because acceleration is constant. The slope or gradient of the straight line in Figure 1.22 is 10 $m.s^{-1}$. This is the acceleration of the object. If we now consider the area under the graph when t = 4 seconds, this is equal to $\frac{1}{2} \times 40 \times 4 = 80$ m (remember area of triangle is half the base multiplied by the height). This is the distance travelled in 4 seconds by the physical object in free fall. In other words, the area under the velocity v. time plot gives us the distance travelled. Using the appropriate equation of motion $\left(s = ut + \frac{1}{2}at^2\right)$ when u = 0, t = 4 and a = 10, we, unsurprisingly, get the same result.

The area under an acceleration versus time graph is the velocity of the object at that time. In this book, we will only be concerned with constant acceleration, so acceleration v. time plots are horizontal lines (Figure 1.23) and the area under the graph at time t is the velocity at that time. For example, the area under the graph at t = 8 is 80 $m.s^{-1}$ and this is the velocity of the object after 8 seconds freefall.

An interesting question is what is the physical significance/meaning of the area under a distance v. time graph? The word for this area is absement, a portmanteau of the words "absent" and "displacement", which makes it no easier to understand. Although this a straightforward question there is no straightforward answer and if you Google "absement" you get very few hits. One website describes absement in this way[57]:

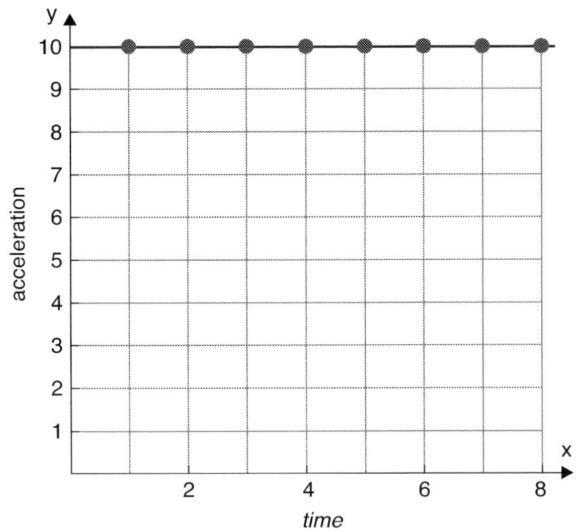

Figure 1.23 Acceleration v. time plot for an object in freefall on Earth.

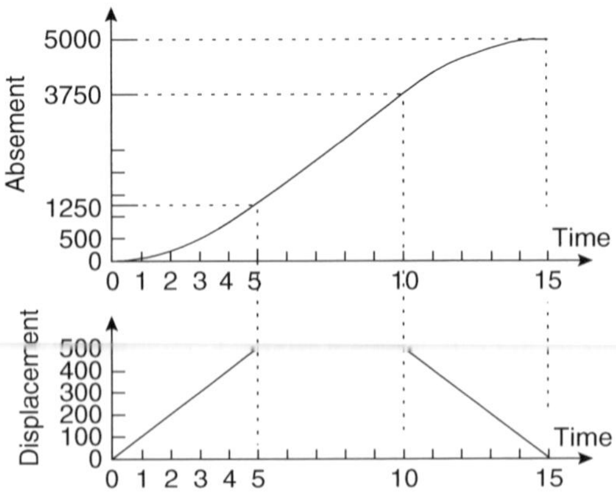

Figure 1.24 Displacement and absement v. time plots.

> Consider a 5-hour train ride that takes you 500 miles directly away from your home, in a straight line, to another destination where you stay for 5 hours and then return. Suppose you want to stay wirelessly logged into your home computer at a 'roaming' communications cost of \$1/mile/hour. For simplicity, assume a linear long-distance rate, i.e. \$1/hour when you're 1 mile away, \$2/hour when you're 2 miles away, \$3/hour when you're 3 miles away, etc. The total cost of your online communications is \$5000, since the absement (time-integral of displacement) is 5000 mile hours (1250 mile hours on the way to your destination, plus 500 miles × 5 hours stay = 2500 mile hours, plus 1250 mile hours of absement during the return trip).

This is shown graphically in Figure 1.24. The bottom plot shows the displacement v. time plot. The first 5 hours are spent in the train going at velocity 100 mph (miles per hour) away from home. The area under this triangular part is one-half of five times 500 mile.hours, which is ½ times 2500 mile.hours, i.e. 1250 mile.hours. The next 5 hours are spent at your destination, 500 miles from your home, the area is five times 500 = 2500 mile.hours. Staying online during your return trip results in another 1250 mile.hours. This gives a cumulative 5000 mile hours. The cost per mile hour is \$1 therefore the total cumulative running cost of staying logged into the computer is \$5000.

I want now to consider Newton's Laws of Motion (as opposed to his Equations of Motion). To introduce this important subject, I will briefly discuss an experiment carried out by Galileo

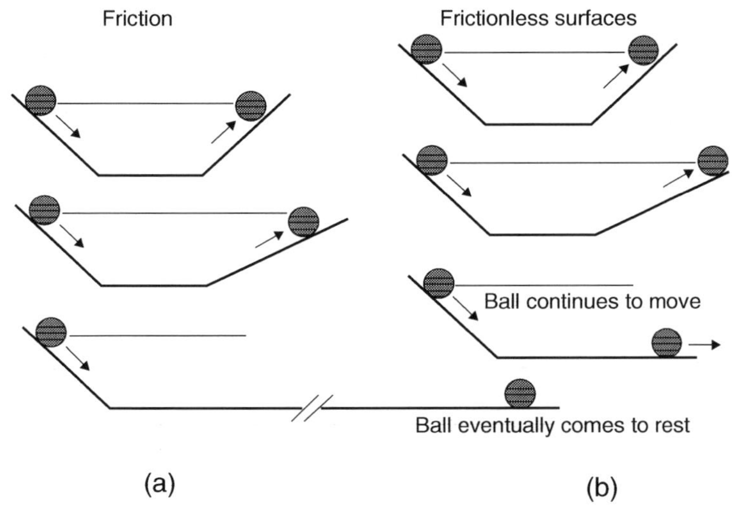

Figure 1.25 Galileo's experiments on the motion of a ball on a pair of inclined planes. (a) Friction surfaces (b) Frictionless surfaces.

1.8 GALILEO'S EXPERIMENT

Arguably what we now call modern science began with Galileo who performed many series of experiments on the motion of physical objects. In this section, I want to consider one sequence of experiments he carried out in the late 1630s. Using a pair of inclined planes facing each other, Galileo observed that a ball would roll down one plane and up the opposite plane to nearly the same height (see Figure 1.25a). If smoother planes were used, the ball would roll up the opposite plane even closer to the original height. Galileo reasoned that any difference between initial and final heights was due to the presence of, what we now call, friction. Friction occurs when physical objects, in contact, move relative to each other, e.g., two touching solid surfaces moving relative to each other, or when a solid object moves through a liquid or gas, or when liquid or gas volumes move relative to each other (this relative movement of gas or liquid volumes is known as turbulence). Friction/turbulence resists the relative movement between the contacted physical objects in question. In Galileo's experiment friction between the surface of the ball and the surface of the wooden plane friction resists the lateral motion of the ball (see Section 1.9.3 for further considerations on friction).

Galileo postulated that if friction could be eliminated (Figure 1.25b), then the ball would reach the same height. This effect was observed regardless of the angle at which the planes were oriented. If the slope of the opposite incline were reduced, then the ball would roll a further distance to reach that original height. He then reasoned that if the opposite incline were elevated at a 0° angle, then, in a frictionless environment, the ball would roll forever, at constant velocity, to try to reach its original height. Galileo said, using Aristotle's terminology (see Section 1.5), that this is the *natural motion* of the object. This conclusion was in direct contradiction to that of Aristotle who considered objects remained stationary and only moved if they were given a 'push' or a 'pull' and, when not subject to a push or pull, they stopped moving. Galileo had shown that in the absence of friction if you give an object a push to get it going then it moves at a constant velocity forever, provided it is not acted on by another force. This conclusion also removed the need to invoke a "prime mover" and begins to explain the flight of thrown spear (see Section 1.5).

1.9 NEWTON'S LAWS OF MOTION

1.9.1 Introduction

Newton (in the 1680s) built on Galileo's thoughts about motion (and devised the necessary mathematics – "the calculus"[58] – (see Section 1.6) to allow him to develop the necessary mathematical equations). He also built upon the published work of Rene Descartes who published a set of laws of motion in 1644, two years after Newton was born.

Newton proposed three laws of motion known, unsurprisingly, as the First, Second, and Third Laws. These laws were published in the *Principia* in 1687 in which he formalised how physical objects move under the influence of external forces. Newton's First Law describes how a physical object moves when it is not subject to an external force (Descartes' laws of motion are very similar to Newton's first law of motion). His Second Law describes what happens when the object is subjected to an external force, and his Third Law describes when two physical objects interact. As we shall see it is the Third Law that is the most difficult to understand and the one which many sources get wrong (one reason for this is that many people think they understand the Third Law when they do not).

1.9.2 Newton's First Law of Motion

Newton's First Law of Motion states that *physical objects at rest will stay at rest, and objects in motion will stay in motion in a straight line unless acted upon by an external force.*

We already know what is meant by a physical object and a force. Let me define rest and motion:

1. *Rest* is the state of a physical object when it is not changing position in relation to its surroundings. If you are sitting in a seat, you can be said to be at rest. However, "rest", is relative because your seat may be on a moving bus. The important point is that you are not moving in relation to your immediate surroundings (the bus) because if rest were defined as a total absence of motion, it would not exist in nature. Even if you were sitting in a chair at home, you would still be moving because your chair is sitting on the surface of a spinning planet that is orbiting a star, and the star is moving through a rotating galaxy that is, itself, moving through the Universe. In other words, while sitting "still", you are, in fact, travelling at a speed of hundreds of kilometres per second in relation to the Universe.
2. *Motion* means a physical object changing position in relation to its immediate surroundings. When you are sitting in your seat on a bus you are at rest; if you get up and walk down the aisle, then you are in motion.

Now that the three major terms of Newton's First Law of Motion have been explained (rest, motion and force), it is possible to restate this law:

> If a physical object is at rest, it takes an applied force to make it move. If the object is already moving, it takes an applied force to stop it or to change its direction or speed.

Newton's First Law of Motion declares that a force (a push or a pull) is not needed to keep an object in motion.[59] The reason Aristotle and everyone else before Galileo and Newton got it so wrong was that they ignored a ubiquitous force that acts on all objects in the real world. This force is friction and always acts in opposition to the movement (or potential movement) of the physical object and slows (decelerates) the object often bringing it to rest. Before considering Newton's Second and Third Laws of Motion I want to briefly consider friction in a bit more detail.

1.9.3 Friction

There are four fundamental forces in nature:

1. Gravitational – attractive force related to the mass of objects.
2. Electromagnetic – attractive or repulsive forces between electrically charged sub-atomic particles such as electrons and protons.
3. Strong nuclear force – holds the nucleus of atoms together.
4. Weak nuclear force – involved in radioactive decay.

Without these fundamental forces, all matter, stuff, physical objects, etc. in the Universe would fall apart into elementary particles.[60] The two nuclear forces will not be mentioned again in this book. Gravitational force has already been mentioned, and the principal reason the electromagnetic force is mentioned in this book is that it is the force responsible for friction (as well as the force that holds "stuff" together).

Friction is a force between two surfaces that are moving, or trying to move, across each other. For example, when you try to push an object across a table, friction makes this difficult. Friction always works in the direction opposite to the direction in which the object is moving, or trying to move, i.e., friction is always a force that resists movement. Friction is an incredibly complex area and all I want to do here is to give you a flavour.

One model of friction invokes surface "bumps". Imagine two smooth surfaces such as a smooth tabletop and a smooth block. If the block, resting on the table, is given an initial push, it will slide for a bit and then stop. The smoother we make the surfaces, the farther it will slide. But even the smoothest blocks and table surfaces, when viewed under a microscope, have what appear to be hills and valleys. Those bumps get stuck on each other a bit, causing the block to slow (Figure 1.26).

Ultimately, however, friction is the result of electromagnetic forces between the two moving surfaces. Even with perfectly smooth surfaces at the atomic or molecular particle level, there is still a frictional force because of the electromagnetic attractions between the constituent particles (Figure 1.27).

Figure 1.28 shows the forces acting on a physical object of mass m on an inclined plane at angle θ to the horizontal. The weight of the physical object is mg and it acts vertically downward (towards the centre of the Earth) – remember weight is a force, and force has direction, i.e., it is a vector. By relatively simple Euclidean geometry[61] and trigonometry, this force can be resolved into two components one acting downward and parallel to the inclined plane $(mg\sin\theta)$ and the other, also acting downward, but at right angles to the plane $(mg\cos\theta)$. We know that weight of the physical object is represented by the length of the vertical line which is the hypotenuse (h). We can show that this is so using the Pythagorean theorem again together with the trigonometric theorem proved in Section 1.6.9:

$$h^2 = (mg\sin\theta)^2 + (mg\cos\theta)^2 = (mg)^2(\sin^2\theta + \cos^2\theta) \tag{1.51}$$

Knowing that $\sin^2\theta + \cos^2\theta = 1$, we can see that $h^2 = (mg)^2$ and $h = mg$.

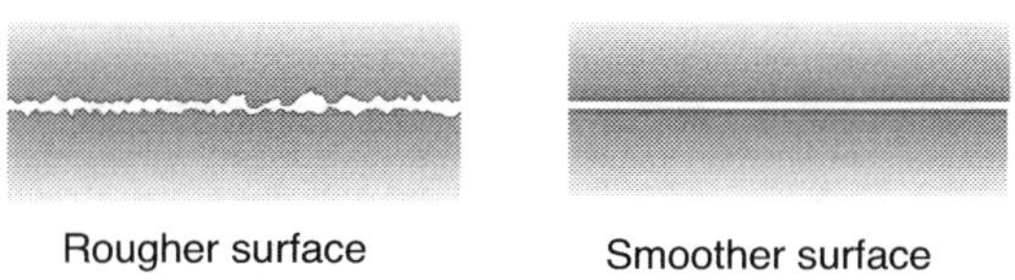

Figure 1.26 The "bump" model of friction.

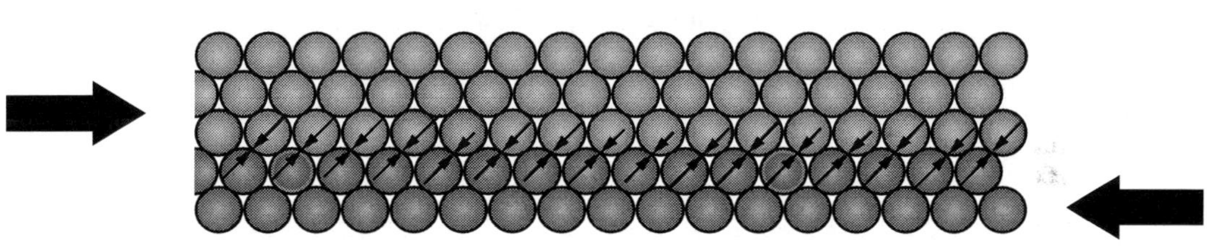

Figure 1.27 Schematic diagram showing attractive electromagnetic forces between constituent atoms or molecules that resist the sliding movement of two "perfectly" smooth surfaces.

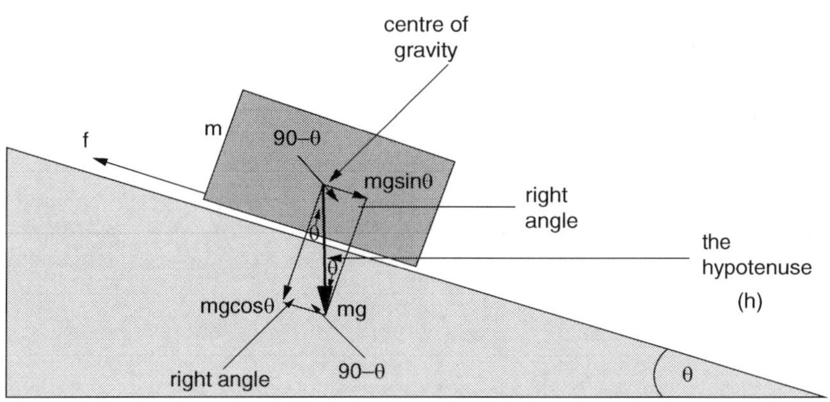

Figure 1.28 Forces acting on a physical object on an inclined plane.

The force that opposes $mg\sin\theta$ (movement down the inclined plane), is the frictional force (f). Now we know from experience that at some point (say θ_f) as θ increases from zero the physical object will start to move. At this point $mg\sin\theta_f$ is equal to the frictional force f. But what is the situation when $\theta < \theta_f$, i.e., when $mg\sin\theta < mg\sin\theta_f < f$? In this case, common sense suggests that the block simply remains at rest (it certainly does not accelerate up the inclined plane). Therefore, f (the frictional force) is the maximum force that friction can generate in order to impede the motion of the physical object. If the applied force $(mg\sin\theta)$ is less than this maximum value then the applied force is cancelled out by an equal and opposite frictional force, and the block remains stationary. This is static friction (f_{static}).

Once the physical object starts to slide you do not need to apply quite as much force to keep the object sliding as you needed to originally break free of static friction (although you still need to apply a force to keep the object sliding) – this is kinetic friction $(f_{kinetic})$. In our example when the physical object starts to move when $\theta = \theta_f$, it will continue to move when θ is reduced. In theory (but probably not in practice) the angle at which the object just keeps moving could be determined (θ_k) and this would allow kinetic frictional force $(f_{kinetic})$ to be determined.

To summarise:

- Static friction is friction between two or more solid objects that are not moving relative to each other, e.g., static friction can prevent an object from sliding down a sloped surface.
- Kinetic or dynamic friction occurs when two objects are moving relative to each other and rub together. Static friction opposes the relative motion of the objects and is usually greater than dynamic friction.

What I want you to take away from this discussion is not only the complexity of frictional forces, but more importantly, that the force of friction opposes the force that accelerates a physical body. Friction arises from interatomic and intermolecular electromagnetic forces between the two contacting surfaces. The complexity of these interactions makes the calculation of friction from first principles impractical (impossible?) and necessitates the use of empirical methods[62] for analysis and the development of theory.

Although in this book I will often tend to ignore all effects due to friction to make things easier, it should be realised that friction is ubiquitous in the real world[63] and I will return to it a number of times in this book. It should be emphasised that the effects of friction do not contravene Newton's First Law of Motion because friction is an applied force on the physical object that acts in the direction opposite to the actual or potential movement of the physical object.[64]

1.9.4 Newton's Second Law of Motion

Newton's Second Law of Motion tells us what happens when a force is applied to a physical object, it states:

1. If a physical object is subject to a force, it will accelerate (change its velocity), and it will change its velocity in the direction of the force (remember force is a vector), i.e., a physical object accelerates or decelerates in the direction it is pushed or pulled.
2. The acceleration is directly proportional to the force. For example, if you are pushing on an object, causing it to accelerate, and then you push, say, three times harder, the acceleration will be three times greater.
3. This acceleration is inversely proportional to the mass of the object. For example, if you are pushing equally on two objects, and one of the objects has five times more mass than the other, it will accelerate at one-fifth the acceleration of the other.

In summary, Newton's Second Law of Motion can be stated as "the acceleration of a body is directly proportional to, and in the same direction as, the force acting on the body, and inversely proportional to its mass". The mathematical statement of the second law is one of the most famous scientific equations:

$$F = ma \tag{1.52}$$

1.9.5 Newton's Third Law of Motion

Newton's Third Law of Motion describes the interaction of physical objects, and it is important to be aware that a force can only exist because of an interaction between physical objects.

Something must push or pull a physical object and the only thing that can do that is interaction with another physical object. Classical mechanics defines two kinds of forces: non-contact forces and contact forces. A contact force is a force applied to an object by another object that is in direct contact with it (such as when a hand pushes a door open); while a non-contact force is a force applied to an object by another object that is not in direct contact with it; the origin of all contact forces can be traced to electromagnetic interaction.[65] The four fundamental forces of nature – strong nuclear, weak nuclear, electromagnetic, and gravitational – are non-contact forces.

In Section 1.3 (on gravitational force), we have already seen if one object is much, much more massive than the other, virtually all of the acceleration is imparted to the second object, and the acceleration of the first object can be safely ignored. Consider Equation(1.2):

$$F = G\frac{m_1 m_2}{r^2} \tag{1.2}$$

Although the equation only considers one force, we know a force pair is involved because mass m_1 is attracting mass m_2 and mass m_2 is attracting mass m_1. We also know that forces do not have independent existences and that they are inextricably linked: you cannot have one without the other. This is what the third law is saying: it is not possible for m_1 to attract m_2 without m_2 attracting m_1. The Third Law represents a certain symmetry in nature because forces always occur in pairs.

Unfortunately, the Third Law is shrouded in confusion and wrong interpretations. It starts when the student is taught that the statement of the Third Law is "for every action, there is an equal and opposite reaction". This statement is deceptively simple the student (and lecturer) think they have understood the Third Law, and many examples of the Third Law given in scientific texts are just plain wrong. The following extract is revealing[66]:

> If I could only change one thing about physics education, it would be the phrasing of Newton's 3rd law. …the 'To every action there is always opposed an equal reaction...' phrasing is Newton's. And it's been causing confusion ever since.

As the above quote indicates the source of this problem rests with Newton himself (Figure 1.29). Translated from Latin, Sir Isaac Newton wrote[67]:

> To every action there is always opposed an equal reaction: or the mutual actions of two bodies upon each other are always equal, and directed to contrary parts.

If the words "action" and "reaction" are replaced by force we have – "To every *force* there is always opposed an equal *force*: or the mutual actions of two bodies upon each other are always equal, and directed to contrary parts". As we shall see in Section 1.11 this statement is a better definition of the Third Law than Newton's.

The apparently simple "to every reaction there is an equal and opposite reaction" leads to a common difficulty as the following two examples indicate:

> If A pushes B, then B pushes A with an equal and opposite force. If these forces are equal and opposite, they cancel, producing a net force of zero. This means that neither object can accelerate, which means that Newton's Laws predict that nothing can ever move.[68]

LAW III.

To every action, there is always opposed an equal reaction: or the mutual actions of two bodies upon each other are always equal, and directed to contrary parts.

Figure 1.29 English translation of Newton's original Latin.

> A horse is harnessed to a cart. If the horse tries to pull the cart, the horse must exert a force on the cart. By Newton's third law the cart must then exert an equal and opposite force on the horse. Newton's second law tells us that acceleration is equal to the net force divided by the mass of the system. ($F = ma$, so $a = \frac{F}{m}$.) Since the two forces are equal and opposite, they must add to zero, so Newton's second law tells us that the acceleration of the system must be zero. If it doesn't accelerate, and it started it rest, it must remain at rest (by the definition of acceleration), and therefore no matter how hard the horse pulls, it can never move the cart.[69]

Section 1.11 addresses these issues together with other aspects of the Third Law, but before that I want to consider absolute values and scientific equations.

1.10 ABSOLUTE VALUES AND SCIENTIFIC EQUATIONS

In this book I will only consider the mathematics of vectors in one-dimensional systems,[70] this allows the use of a simple sign convention such as a vector acting to the right is positive and acting to the left is negative. In other words, vectors can take either a positive or negative real number value depending on our agreed sign convention. This type of sign convention (the use of positive and negative) numbers will also be used in thermodynamics when considering energy flows. The inconsistent application of these conventions in scientific equations is a major source of confusion not only in thermodynamics but also in other areas of science (including applications of Newton's Third Law). This is a subject that does not appear to be covered in the scientific literature but is one that I consider to have a major impact on the rigorousness and understanding of science.

Consider the simple mathematical function $y = ax$ and ask the following straightforward question – *what do* a, x, *and* y *represent*? The answer is, of course, that a, x, and y represent real numbers which can be positive, negative, or zero. Now consider a scientific equation such as Newton's famous equation $F = ma$ and ask the same question. We know that F and a are vector quantities and m is a scalar quantity. In the real world of experimental science mass cannot be negative.[71] Therefore, we can immediately see that (i) m does not represent a real number and (ii) a scientific equation linking physical quantities is not equivalent to a mathematical function. In the scientific equations used in this book some physical quantities are always positive real numbers, e.g., absolute pressure (as opposed to gauge pressure), volume, absolute temperature (see Chapter 3), volume, mass etc. are always positive. These are all scalar quantities. Are all scalar physical quantities positive? The short answer is no, e.g., gauge pressure can be negative and temperature scales such as Celsius and Fahrenheit can have negative values. But, if we ignore these "arbitrary" scales then, as far as we are concerned in this book, all "absolute" scalar physical quantities are positive real numbers. The word "absolute" is often used in science to denote this, e.g., absolute temperature is often used where zero is the coldest possible temperature. The term absolute pressure is less often used but can be rigorously defined[72]:

> The definition of absolute pressure is the pressure of having no matter inside a space, or a perfect vacuum. Measurements taken in absolute pressure use this absolute zero as their reference point.

Absolute volume is more difficult as it normally refers to the volume a granular solid occupies in water. But length, area, and volume are always expressed with positive numbers: in the real world (as opposed to the mathematical world) there is no such thing as negative length, area, or volume. I do not want to pursue this any further because (having Googled "negative volume") I will, undoubtedly, stray into areas outside of my knowledge.

This leads to another question – *are vectors real numbers?* I have already mentioned that, in this book, I will only consider the mathematics of vectors in one-dimensional systems.[73] This allows the use of simple sign conventions such as a vector acting to the right is positive and acting to the left is negative, or vector acting downward is positive and a vector acting downward is negative. If you carry out a Google search of the above question you find yourself in the world of advanced mathematics. However, if we revisit the definition of a real number (see Section 1.6.2) – "a real number is a value of a continuous quantity that can represent a distance along a line", then it is apparent that the magnitude of a one-dimensional vector using a +/– sign convention is a real number. This applies to the vector quantities of displacement, velocity, acceleration, force etc. In addition, differences in absolute scalar physical quantities ($\Delta V, \Delta P, \Delta T, dV, dP, dT$ etc.) are real numbers. Volume, pressure, temperature, length, area etc. can increase (+) or decrease (–) and I will call these "vector-like" quantities.

In thermodynamics we need to carry out energy accountancy and therefore, we must have an agreed sign convention, e.g., energy flowing into a system or object is assigned a positive number and that flowing out a negative number. Bank accounts use exactly the same convention a deposit is assigned a positive number and withdrawals a negative number.[74] These flows of money can be considered to act like one-dimensional vectors and these "vector-like" flow quantities can take either a positive or negative real number value. They can flow in, or they can flow out, or they can flow or act in the upwards direction, or they can flow or act in the downwards direction (and only in these two directions).

In the scientific equations in this book, I will use bold font to represent these one-dimensional vectors and vector-like quantities or flows that can be positive or negative numbers.[75] The magnitude of all other physical quantities are positive numbers. The following examples should clarify matters.

Newton's First Law of Motion is written as $\mathbf{s} = \frac{1}{2}(\mathbf{u} + \mathbf{v})t$. Consider the case when t = 1, $|\mathbf{u}| = 2\ \mathrm{m.s^{-1}}$ and $|\mathbf{v}| = 10\ \mathrm{m.s^{-1}}$ we have four combinations (dropping the units): (i) $\mathbf{u} = +2$ and $\mathbf{v} = +10$ where $\mathbf{s} = +6$; (ii) $\mathbf{u} = -2$ and $\mathbf{v} = -10$ where $\mathbf{s} = -6$; (iii) $\mathbf{u} = +2$ and $\mathbf{v} = -10$ where $\mathbf{s} = -4$; and (iv) $\mathbf{u} = -2$ and $\mathbf{v} = +10$ where $\mathbf{s} = +4$. These results are shown against a number line in Figure 1.30.

1.11 NEWTON'S LAW'S OF MOTION IN MORE DETAIL

1.11.1 Introduction

The teaching of Newton's Third Law is a mess. I considered setting down the erroneous and confusing information out there on the Third Law, but decided it was not a productive way forward. However, I must take issue with categorising force pairs into those that the Third Law applies to and to those that the Third Law does not. Consider the following:

> It is possible to confuse a pair of forces. For example, with a book on a table there is the weight of the book and the contact reaction force of the table. These forces are not a Newton's third law pair – they are not the same type of force, and they act on the same object. When you take away the table the weight of the book remains. The third law pair could be, for example, the gravitational pull of the Earth on the book and the gravitational pull of the book on the Earth.[76]
>
> On entry to university, high achieving physics students from all across Australia struggle to identify Newton's Third Law force pairs. In particular, less than one in ten can correctly identify the Newton's Third Law reaction pair to the weight of (gravitational force acting on) an object. Most students incorrectly identify the normal force[77] on the object as the Third Law force-pair to weight, rather than the gravitational force of the object on the Earth.[78]

Unfortunately, both of these extracts are wrong. Newton's Third Law applies to all force pairs as I will demonstrate in the rest of this section.

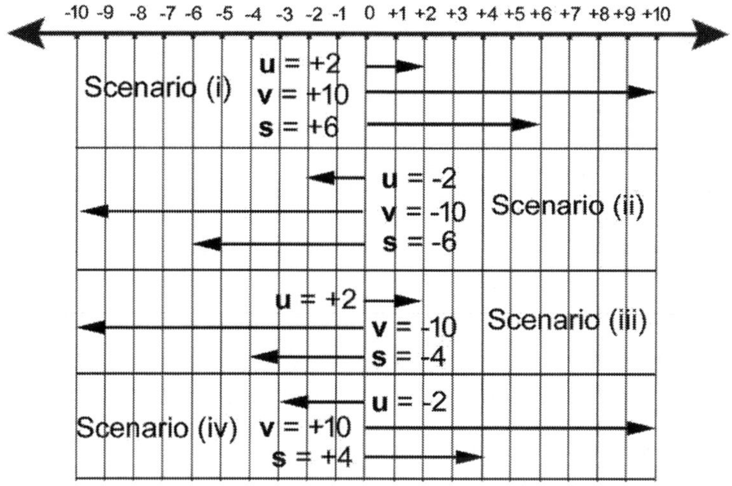

Figure 1.30 An example of the use of bold fonts in scientific formulae.

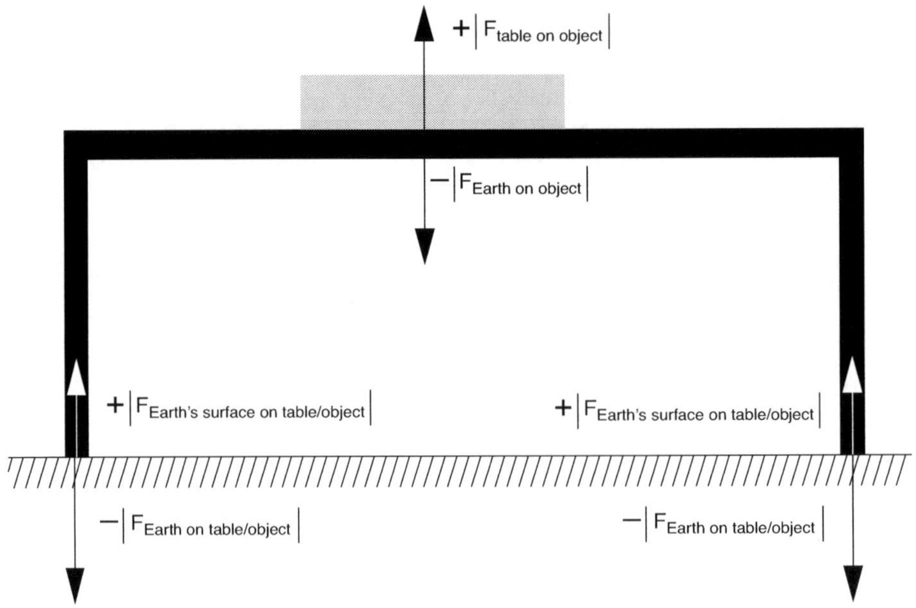

Figure 1.31 Equal and opposite force pairs.

1.11.2 Stationary Object on Earth

Consider an object resting on a table (see Figure 1.31) and we will use the sign convention that a downward force is negative, and an upward force is positive. The object pushes down on the table with a force $-|\mathbf{F}_{\text{Earth on object}}|$, while the table pushes up on the object with an equal and opposite force $+|\mathbf{F}_{\text{table on object}}|$. We know $|\mathbf{F}_{\text{Earth on object}}|$ is the object's weight. In this case the other part of the force pair is $|\mathbf{F}_{\text{table on object}}|$, a good way to imagine this is to consider that the placing of the object on the table causes the table to bow and the restoring force pushes up on the object. The Third Law tells us that $|\mathbf{F}_{\text{Earth on object}}| = |\mathbf{F}_{\text{table on object}}|$. We know both of these forces apply to the object and applying Newton's Second Law we have:

$$m_{\text{object}} \times |\mathbf{a}_{\text{downwards}}| = m_{\text{object}} \times |\mathbf{a}_{\text{upwards}}| \tag{1.53}$$

Therefore, the downwards acceleration is equal to the upwards acceleration: $|\mathbf{a}_{\text{downwards}}| = |\mathbf{a}_{\text{upwards}}|$. If we add the acceleration vectors to get the overall acceleration of the object, we have:

$$\left(-|\mathbf{a}_{\text{downwards}}|\right) + \left(+|\mathbf{a}_{\text{upwards}}|\right) = 0 \tag{1.54}$$

Consequently, the object does not move. The same argument applies to the $|\mathbf{F}_{\text{Earth's surface on table/object}}| / |\mathbf{F}_{\text{Earth on table/object}}|$ force pair. Although we intuitively knew this, the application of Newton's Laws in this example is instructive (note $|\mathbf{a}_{\text{downwards}}| = |\mathbf{a}_{\text{upwards}}| = |\mathbf{g}|$).

The upwards force $\mathbf{F}_{\text{table on object}}$ is often known as the normal force which is normally written $\mathbf{F}_N$ (see the second extract in the previous section). Many texts (including the two in the previous section) incorrectly state that the Third Law does not apply to these types of force pair. They are wrong, and in this, I am supported by Newton who stated in the English translation of the Principia (originally published in Latin)[79]:

> Whatever draws or presses another is as much drawn or pressed by that other. If you press a stone with your finger, the finger is also pressed by the stone.

The stone (or the person) does not have to move for the Third Law to apply.

If you think about it the above is relatively straightforward so, why is there so much confusion? One reason is the student is incorrectly told that the normal (upwards) force is always $\left(|\mathbf{F}_{\text{object on Earth}}|\right)$ – "Most students incorrectly identify the normal force[80] on the object as the Third Law force-pair to weight, rather than the gravitational force of the object on the Earth". (Section 1.11.1). This is why some sources say that the force pairs considered in this section "are not a

Newton's third law pair". This is wrong, consider Figure 1.32 the force pair on the stationary object must be $\left|\mathbf{F}_{\text{Earth's surface on object}}\right| / \left|\mathbf{F}_{\text{Earth on object}}\right|$. If one of the force pair was $\left|\mathbf{F}_{\text{object on Earth}}\right|$, then the Earth's surface (compare the table in the above example) has been "magically" removed and as the second extract in Section 1.11.1 suggests, the object would not be stationary it would be falling to the centre of the Earth, which is patently wrong. This is demonstrated in the following section.

1.11.3 Falling Object on Earth

Let us now make the table, but not the mass, in Figure 1.31 magically disappear. This is shown in Figure 1.33a and note the force $+\left|\mathbf{F}_{\text{table on object}}\right|$ has also disappeared because the table is not present to produce this force. This leaves an unpaired force which contravenes the requirement that forces always come in pairs. Remember this from Section 1.9.5:

> It is important to be aware that a force can only exist because of an interaction between physical objects. Something must push or pull a physical object and the only thing that can do that is another physical object.

Also consider Equation (1.2) from Section 1.3 and notice there is only one force produced by this equation, but this "one force" acts in two opposite directions (see Figure 1.2). Therefore, there are two different forces: m_1 attracting m_2 and m_2 attracting m_1, although these forces do not have independent existences because it is not possible for m_1 to attract m_2 without m_1 being caught up in the very same force of attraction between the two particles. This "unpaired force" situation when the table is removed is resolved by including the (gravitational) force of the object "pulling" on the Earth, this $+\left|F_{\text{object on Earth}}\right|$ and is shown in Figure 1.33b. Note the situation in Figure 1.33b is the same as shown in Figure 1.2 when we were considering Newton's Law of Universal Gravitation and Figure 1.32 when the Earth's surface has been removed.

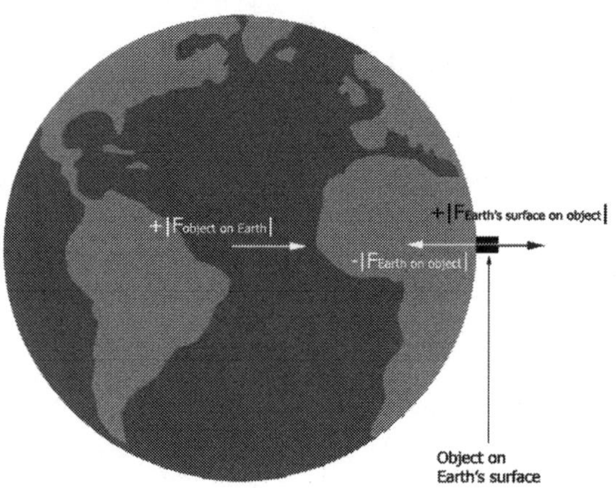

Figure 1.32 Object on Earth's surface.

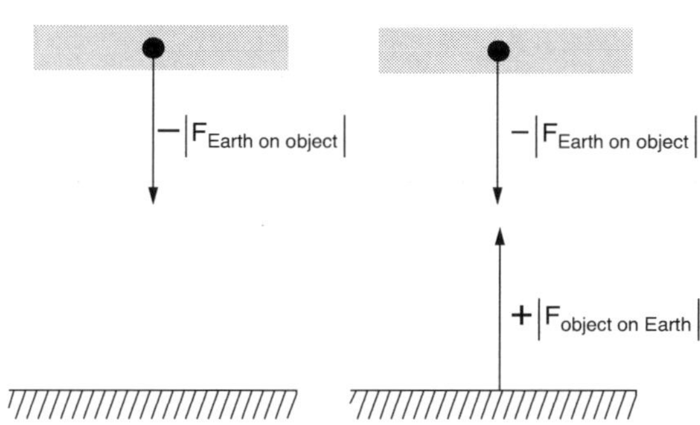

Figure 1.33 Object before and after table was removed.

In analysing the situation in Figure 1.33b, we will use the same sign convention as before, i.e., a downward force is negative, and an upward force is positive. The gravitational force on the object $\left(\left|\mathbf{F}_{\text{Earth on object}}\right|\right)$ is in the downward direction and using our sign convention we can represent this as the vector $-\left|\mathbf{F}_{\text{Earth on object}}\right|$. We also know that gravitational force on the Earth as a result of the object is acting upwards and we can represent this vector as $+\left|\mathbf{F}_{\text{object on Earth}}\right|$. The Third Law tells us that $\left|\mathbf{F}_{\text{Earth on object}}\right| = \left|\mathbf{F}_{\text{table on object}}\right|$ and applying Newton's Second and Third Laws we know:

$$m_{\text{object}} \times \left|\mathbf{a}_{\text{downwards}}\right| = m_{\text{Earth}} \times \left|\mathbf{a}_{\text{upwards}}\right| \tag{1.55}$$

In contrast to the previous example (object on a table) the force pair does not apply to the same object, but to two objects with different mass. In this situation, the downwards acceleration is not equal to the upwards acceleration, i.e., $\left|\mathbf{a}_{\text{downwards}}\right| \neq \left|\mathbf{a}_{\text{upwards}}\right|$.

We know m_{object} is very much less than m_{Earth} (which is, to say the least, an understatement). Therefore, $\left|\mathbf{a}_{\text{downwards}}\right|$ must be very much greater than $\left|\mathbf{a}_{\text{upwards}}\right|$. In fact, we know from our analysis of Newton' Law of Gravitation (Section 1.4) that for anything other than heavenly bodies we can safely assume $\left|\mathbf{a}_{\text{upwards}}\right| = 0$. In other words, as $\dfrac{m_{\text{object}}}{m_{\text{Earth}}} \to 0$ then $\dfrac{a_{\text{upwards}}}{a_{\text{downwards}}} \to 0$, i.e., in other words, as $\dfrac{m_{\text{object}}}{m_{\text{Earth}}} \to 0$ then $a_{\text{downwards}} \to g$ (the acceleration due to gravity).

Before moving on I thought it worthwhile to set down the following extract from a website discussing Newton's Third Law[81]:

> Sadly, much of the help for Newton's Third Law that is available for students on the internet is poorly explained or just plain wrong. The confused student is very likely to seek help where help is supposed to be. Why wouldn't a student struggling with Newton's Third Law not simply pop "newton's laws of motion" into a search bar? If they do, they will find the world stacked against them.
>
> The first hit [searching on "Newton's Laws of Motion"] on just such a Google search results in a college site (must be vetted, no?) that gives the standard unhelpful "For every action there is an equal and opposite reaction" followed by a single example that talks about motion (not forces, not changes in motion, but motion). This example does nothing to help explain the Third Law. So the student keeps looking. "Hey, next is rice.edu? Mr. Hammond [82] went to Rice, this should be good." The same unhelpful phrase is given followed by an example of a rocket. However the example misidentifies the appropriate force-pair and seems to indicate a rocket cannot accelerate unless it has the ground to push off of… confusing. Go on to the third hit, which leads to Wikipedia. Lots of symbols, calculus right off the bat, Latin, blah, blah, blah, nothing helpful to the beginning student. The fourth hit: NASA. Now we're making progress! But the worksheets from NASA all include the misconception that both forces described by the Third Law are exerted on the same object! That's right, the NASA site is dead wrong! Fifth site: Discovery Channel! Yay, TV! Yet the physics is all wrong, making the same mistake as the NASA site. Sixth site: physics4kids.com, a terrible website full of incorrect physics, including an incorrect explanation of the Third Law (same mistake as the NASA site made). Finally, on the seventh and eighth hits, we get some explanations that are correct and might be useful to the beginning student. How many students are going to get past the unhelpful and incorrect explanations (which probably coincide with their current misconceptions)?
>
> Oh, you can go look at Khan Academy. At least they changed their wrong explanation after some physics teacher complained (who was that?[83]).

I will now set down some examples where the situations are analysed by Newton's Third Law. I will start with traditional examples of firing a gun and rockets.

1.11.4 Firing a Gun

When a gun or cannon is fired, an explosion propels a projectile out of the open end of the barrel, and the gun is pushed backward, i.e., it undergoes a recoil or 'kick' (Figure 1.34). All we need to know here is that we have two objects, *viz* the projectile and the cannon, and something produces a force **F** (in this case the explosion of gunpowder). I will use the sign convention that a vector acting in the right direction is positive and *vice versa*. Let the force on the projectile be the vector $+\left|\mathbf{F}_{\text{projectile}}\right|$ and the force on the cannon be the vector $-\left|\mathbf{F}_{\text{cannon}}\right|$.

Figure 1.34 Firing a cannon.

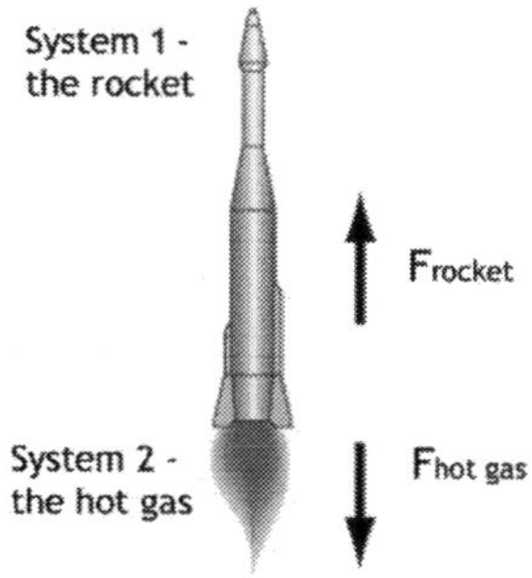

Figure 1.35 The upward push on the rocket equals the downward push on the exhaust gases.

From the Second Law, we can write:

$$+|\mathbf{F}_{\mathbf{projectile}}| = \mathrm{m}_{\mathrm{projectile}} \times (+|\mathbf{a}_{\mathbf{projectile}}|) \tag{1.56}$$

$$-|\mathbf{F}_{\mathbf{cannon}}| = \mathrm{m}_{\mathrm{cannon}} \times (-|\mathbf{a}_{\mathbf{cannon}}|) \tag{1.57}$$

We know from the Third law that $|\mathbf{F}_{\mathbf{projectile}}| = |\mathbf{F}_{\mathbf{cannon}}|$. Therefore, we can write:

$$\mathrm{m}_{\mathrm{projectile}} \times |\mathbf{a}_{\mathbf{projectile}}| = \mathrm{m}_{\mathrm{cannon}} \times |\mathbf{a}_{\mathbf{cannon}}| \tag{1.58}$$

The analysis then follows that set down in Section 1.11.3. Because $\mathrm{m}_{\mathrm{cannon}} \gg \mathrm{m}_{\mathrm{projectile}}$ then $\mathrm{a}_{\mathrm{projectile}} \gg \mathrm{a}_{\mathrm{cannon}}$. Therefore, as the cannonball has a relatively small mass, its acceleration is very much greater than the cannon which has a much larger mass. To reduce the recoil of the cannon and still maintain the same projectile acceleration, we could increase the mass of the cannon (and if the cannon is fixed in the ground its mass is essentially that of the Earth).

1.11.5 Movement of a Rocket

Rockets move forward by burning fuel and this controlled explosion taking place inside the rocket's engine results in an acceleration of hot gas out of the rockets nozzle (Figure 1.35).

I will use the same sign convention as before, i.e., a downward force is negative, and an upward force is positive. The downward of acceleration of the gas results in a negative vector $(-|\mathbf{F}_{\mathbf{expelled\ gas}}|)$ and the paired force is represented by the upwards vector acting on the rocket $(+|\mathbf{F}_{\mathbf{rocket}}|)$. From the Second Law, we can write:

$$+|\mathbf{F}_{\mathbf{rocket}}| = \mathrm{m}_{\mathrm{rocket}} \times (+|\mathbf{a}_{\mathbf{rocket}}|) \tag{1.59}$$

$$-|\mathbf{F}_{\mathbf{expelled\ gas}}| = \mathrm{m}_{\mathrm{expelled\ gas}} \times (-|\mathbf{a}_{\mathbf{expelled\ gas}}|) \tag{1.60}$$

We know from the Third law that $|\mathbf{F}_{\mathbf{rocket}}| = |\mathbf{F}_{\mathbf{expelled\ gas}}|$. Therefore, we can write:

$$\mathrm{m}_{\mathrm{rocket}} \times |\mathbf{a}_{\mathbf{rocket}}| = \mathrm{m}_{\mathrm{expelled\ gas}} \times |\mathbf{a}_{\mathbf{expelled\ gas}}| \tag{1.61}$$

We can see from Equation (1.61) that to increase $|\mathbf{F}_{\mathbf{rocket}}|$ (known as "thrust") requires an increase in the mass of expelled gas and/or its acceleration. Consider the following[84]:

> A rocket engine uses a nozzle to accelerate hot exhaust to produce thrust as described by Newton's third law of motion. The amount of thrust produced by the engine depends on the mass flow rate through the engine, the exit velocity of the flow, and the pressure at the exit of the engine. The value of these three flow variables are all determined by the rocket nozzle design.

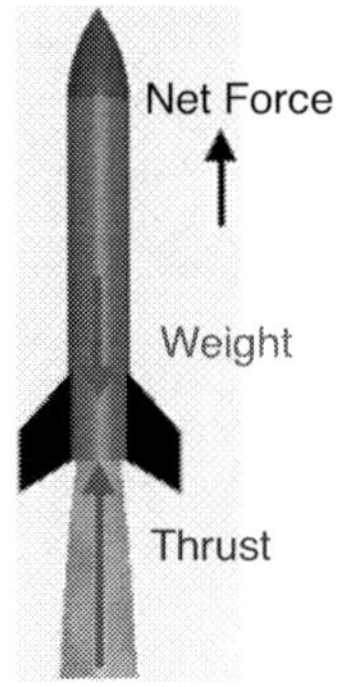

Figure 1.36 From a NASA website.

There are phenomena that occur with rockets that do not happen with firing a gun. When a gun is fired the thrust lasts for just a moment, whereas the thrust for the rocket continues as long as its engines are firing. Furthermore, the mass of the rocket changes during flight: the largest part of the rocket's mass is its propellants and that reduces as the engines fire. This means that the rocket's mass gets smaller during flight, and if the rate of the expelled mass of gas and its acceleration remains constant then we can see that the rocket's acceleration will increase.

Be aware of explanations that are wrong or confusing. Consider Figure 1.36 from the NASA web site. Using Newton's Third Law we can write $|\mathbf{F}_{weight}| = |\mathbf{F}_{thrust}|$ and as these appear to be acting on the same object (the rocket) then there is no movement (see Section 1.11.2). I assume that this is the type of diagram that Hammond (Section 1.11.3) is referring to when he wrote:

> The fourth hit: NASA. Now we're making progress! But the worksheets from NASA all include the misconception that both forces described by the Third Law are exerted on the same object! That's right, the NASA site is dead wrong!

I assume the NASA diagram is telling us that when upward thrust force overcomes the downward weight the rocket moves upward. But if you consider the blue and red arrows represent a force pair then it is "dead wrong". I will be generous and say it is, at best, confusing. However, the following extract is most certainly "dead wrong":

> The rocket's action is to push down on the ground with the force of its powerful engines, and the reaction is that the ground pushes the rocket upwards with an equal force. [85]

If the above extract is correct, then it begs the question how does a rocket move when there is no ground to push it? I think the author is probably aware of Newton's Laws of Motion and walking (see Section 1.11.7), has not really understood it and applied it to rockets. It is a common misconception that rockets propel themselves by pushing on the ground or on the air behind them. In fact, they work more efficiently in a vacuum, where they can more readily expel the exhaust gases.

1.11.6 A Hose Pipe

We have all seen photographs of fighters holding a hose steady (Figure 1.37). What causes such a large backward force that it requires two, three or even four fire fighters to hold the hose steady? I have chosen this example because I came across so much rubbish on the internet about it.

You know that the pressure of water in the hose is greater than atmospheric pressure (try stopping the flow by placing your thumb against a garden hose). As the water issues from the hose, the pressure decreases, and the water undergoes an acceleration in the forward direction. This is an example of what is known as the Bernoulli Principle which states that for a fluid flow an increase in speed occurs with a decrease in pressure.[86] Newton's Third Law tells us there must be a backward force that is equal and opposite to this forward force that causes the water to accelerate (Figure 1.38). It is this backward force that the fire fighters are steadying in Figure 1.37. Comparison of Figures 1.35 and 1.38 shows the similarities of a rocket and a hose, the only difference is the mechanism of producing the "exhaust" fluid.

The Bernoulli Principle tells us why windows tend to explode, rather than implode in hurricanes: the very high speed of the air just outside the window causes the pressure just outside to be

Figure 1.37 Firefighters in action.

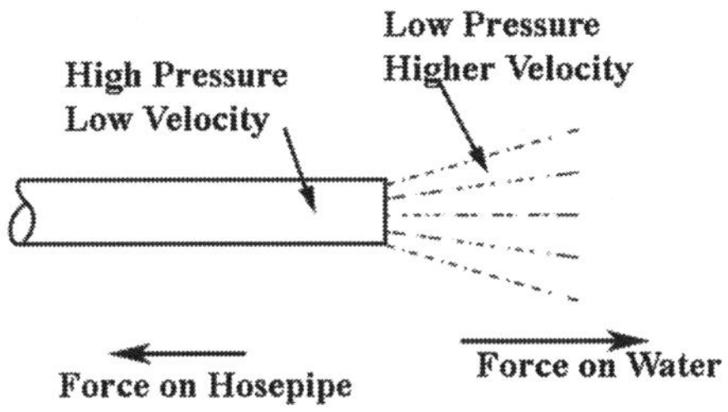

Figure 1.38 Forces on a squirting hosepipe.

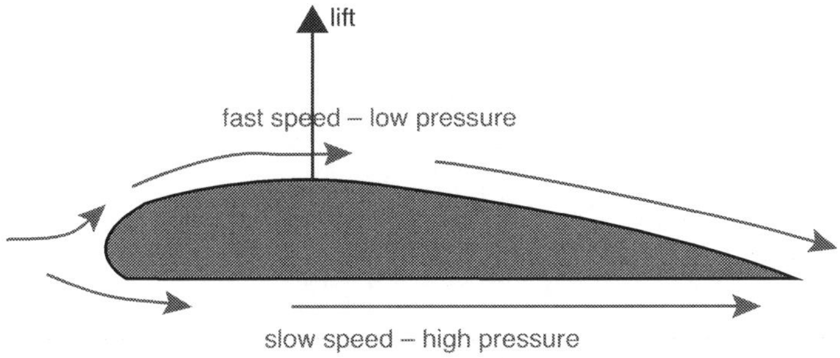

Figure 1.39 Aircraft wing and the Bernouilli Principle.

much less than the pressure inside, where the air is stationary (or essentially so). The difference in force pushes the windows outward. Another example of Bernoulli's principle at work is in the lift of aircraft wings (Figure 1.39), where the design is such as to create a speed differential of the flowing air past the object on the top and the bottom. Such a speed differential leads to a pressure

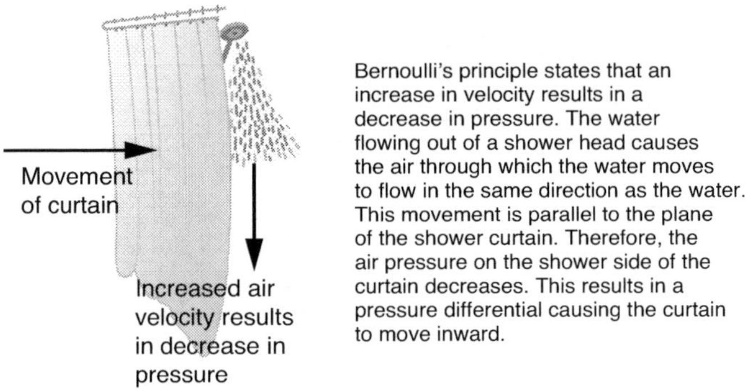

Figure 1.40 The "Shower-curtain Effect".

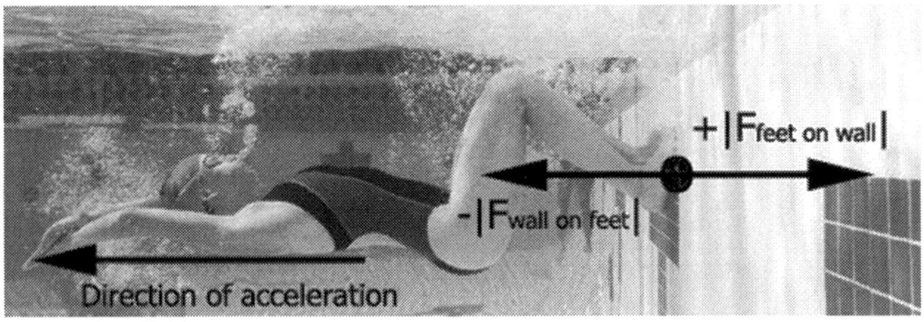

Figure 1.41 Swimmer pushing off from wall.

difference between the top and bottom of the object, resulting in a net force being exerted, either upwards or downwards.

Finally, and on a more a more quixotic note I want to consider the "shower-curtain effect". This describes the phenomenon that a shower-curtain gets blown inward with a running shower. This was featured in Scientific American magazine, with several theories given to explain the phenomenon but no definite conclusion. The most popular theory is based on the Bernoulli Principle (Figure 1.40). Other proposed theories include the chimney effect, horizontal vortex hypothesis, and condensation effect.[87]

1.11.7 A Swimmer Pushing Off from the Side of a Pool[88]

Consider a swimmer pushing off from the side of a pool, as illustrated in Figure 1.41. She pushes against the pool wall with her feet and accelerates in the direction opposite to that of her push. We have two separate objects – the swimmer and the wall of the swimming pool (aka the Earth). Let us use the sign convention that a vector acting to the right is positive. The wall has exerted an equal and opposite force back on the swimmer. There are two objects: the swimmer and the wall. As the swimmer pushes against the side of the swimming pool, she exerts a force represented by the vector $+|\mathbf{F}_{\text{wall on feet}}|$ on the wall (Earth) and the wall exerts a force a force represented by the vector $-|\mathbf{F}_{\text{feet on wall}}|$.

We know from Newton's Third Law that $|\mathbf{F}_{\text{wall on feet}}| = |\mathbf{F}_{\text{feet on wall}}|$ and applying Newton's Second Law:

$$-|\mathbf{F}_{\text{wall on feet}}| = m_{\text{swimmer}} \times \left(-|\mathbf{a}_{\text{swimmer}}|\right) \tag{1.62}$$

$$+|\mathbf{F}_{\text{feet on wall}}| = m_{\text{wall}} \times \left(+|\mathbf{a}_{\text{wall}}|\right) \tag{1.63}$$

Therefore, we have:

$$m_{\text{swimmer}} \times |\mathbf{a}_{\text{swimmer}}| = m_{\text{wall}} \times |\mathbf{a}_{\text{wall}}| \tag{1.64}$$

As $m_{\text{wall}} = m_{\text{Earth}}$ and $m_{\text{Earth}} \ggg m_{\text{swimmer}}$ we are in essentially the same situation as in Section 1.11.3 where $|\mathbf{a}_{\text{swimmer}}| \ggg |\mathbf{a}_{\text{wall}}|$, i.e., the swimmer accelerates to the left.

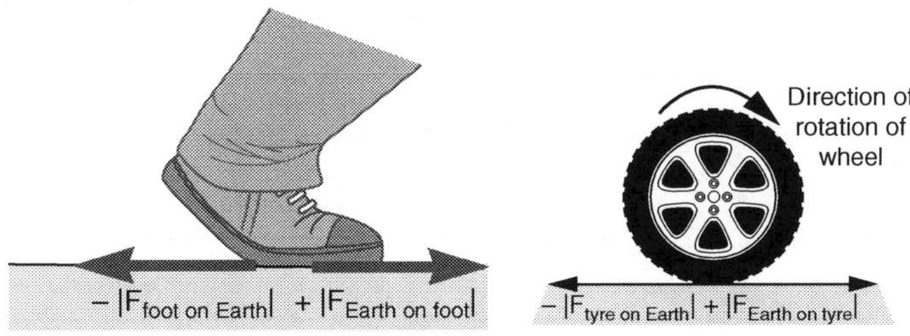

Figure 1.42 Walking and driving.

The mass of the Earth is about 6×10^{24}kg and let us assume the mass of the swimmer is 50 kg and that her acceleration, when she pushes of from the wall, is 1 m.s^{-2}.[89] Using Newton's Third Law we can calculate the acceleration of the Earth caused by the swimmer to be $\frac{50}{6\times10^{24}}\times1\cong8\times10^{-24}$ m.s^{-2}. This is, to say the least, a negligible quantity.

1.11.8 Walking and Driving

Consider a person standing still and then beginning to walk. We know the person has accelerated, and we know that there must be some force pushing the person in the direction of acceleration. This force cannot be an internal force within the person because an object can only accelerate if acted on by an external force. What is this external force? We can rule out the non-contact force of gravity because that is only keeping the person on the surface of the Earth. Therefore, we are left with a contact force. The only two things the person is in contact with while walking are the ground and the surrounding atmosphere. We do not use the air to move (you can walk in a vacuum if you have a space suit on) and so, by a process of elimination, we know that it is the ground pushing the person forwards. In other words, in order to move forward the person needs something to push on and this is the Earth. We know the contact force involved in walking is friction because if there was no friction between the person's foot and the ground the person could not walk. I will use the same sign convention as before.

In the act of walking to the right (Figure 1.42) the person is exerting a force $-|\mathbf{F}_{\text{foot on Earth}}|$ leftwards on the Earth with his or her leg. Newton's Third Law tells us there must be an equal and opposite force on the ground (aka the Earth), this is $+|\mathbf{F}_{\text{Earth on foot}}|$. Accordingly, using the above analysis, the person moves to right and the Earth moves to the left (by an infinitesimally small distance). The movement of a car (Figure 1.42) uses exactly the same arguments where $-|\mathbf{F}_{\text{tyre on Earth}}|$ is the force exerted by the car tyre (turning in a clockwise direction) on the road (the Earth). The car moves to the right in the opposite direction to the force exerted by the tyre on the Earth.

Another extract should reinforce the message given in the previous section regarding the movement of the Earth in these types of processes[90]:

> So why doesn't the road accelerate? The car accelerates, so shouldn't the road accelerate in the opposite direction? Believe it or not, it does; Newton's law is in full effect. Your car pushes the Earth, affecting the motion of the Earth in just the tiniest amount. Given the fact that the Earth is about 6,000,000,000,000,000,000,000 times as massive as your car, however, any effects aren't too noticeable!
>
> Similarly, when a hockey player slaps a puck, the puck accelerates away from the spot of contact, and so does the hockey player. If hockey pucks weighed 1,000 pounds[91] ... you'd notice this effect much more; in fact, the puck wouldn't move much at all, but the player would hurtle off in the opposite direction after striking it.

1.11.9 Discussion

Another quote may help to hammer the message home[92]:

> Unfortunately, the frequency with which I hear students spout "for every action there is an equal and opposite reaction!" whenever they see any two forces pointing in opposite directions with equal magnitudes tells me that this kindergarten version of the Third Law is omnipresent. The students are presented with this version in middle school or on TV science programs, and they have probably had at least one confused adult give an incorrect explanation of what it means. For such a widely memorized tidbit, there sure is a lot of

misunderstanding about what it means! To make things worse, many kids come into physics thinking that this is the ONLY bit of physics they already understand!

In many school settings, the Third Law is stated simply as "for every action there is an equal and opposite reaction." This might be fine if the explanation that followed developed action and reaction as changes in momentum.[93] But nearly universally, this version of Newton's Third Law is followed by a lot of talk about "action forces" and "reaction forces" and very little talk about what objects are exerting forces and what objects are being subjected to forces. Most middle school textbook explanations I've seen are simply impossible to understand.

Although we may think nothing could be as simple as walking. It is the interplay of Newton's laws of motion and properties of frictional forces that allows us to understand our ability to walk. There are many examples of applications of the third law in the scientific literature. Unfortunately, many are wrong and/or confusing. If NASA cannot get it right about rockets, then what hope is there that Newton's Third Law will be taught in a rigorous manner. Newton's Third Law is one of the most fundamental laws in science and, it is a shame that students receive so much erroneous information.

On reading the literature, you will come across the following terms: "internal force pairs", "external force pairs", "balanced forces", "unbalanced forces". It is my experience that these terms may add to the already existing confusion and should be used with care. All I can offer is the following guidance – identify whether a force pair acts on the same object (internal force) or on different objects (external force) and take it from there. Consider Figure 1.43a where the person's feet and the (immovable) wall are immersed in block of concrete are all considered to be part of the object known as the Earth. The force pair (vectors equal in magnitude but opposite in direction) act on the same object (the Earth) and cancel each other out, another way to state this is:

> A system cannot 'bootstrap' itself into motion with purely internal forces – to achieve a net force and an acceleration, it must interact with an object external to itself.

The force pair in Figure 1.43a are internal because they act on the same object and are balanced.

In contrast, if the man is standing on a friction-free surface (Figure 1.42b) he is not now a part of the physical object known as Earth, i.e., force pair is now acting on two objects – the man and the Earth. In this situation motion occurs and the man accelerates to the left (compare this to the swimmer in Section 1.11.7).

Part of the problem when considering human (or horse) muscles is that they must constantly expend energy at a molecular level in order to stay contracted. Therefore, it is easy to confuse force exertion with expenditure of energy.[94] In addition, because humans have reasoning and carry out work, it feels wrong when statements similar to the following are made – "The person pushes on the object and the object pushes on the person" feels wrong because the person is expending energy and the object is not". Intuitively this does not make sense because you know that you are the one doing the pushing, i.e., the person initiates the push and the inanimate object initiates nothing. What is really happening is that you create a repulsive force between the object and yourself.[95] If the object is fixed to Earth and you are not, the repulsive force manifests itself as you push yourself away from the object. However, if you are fixed to the ground, by friction or some

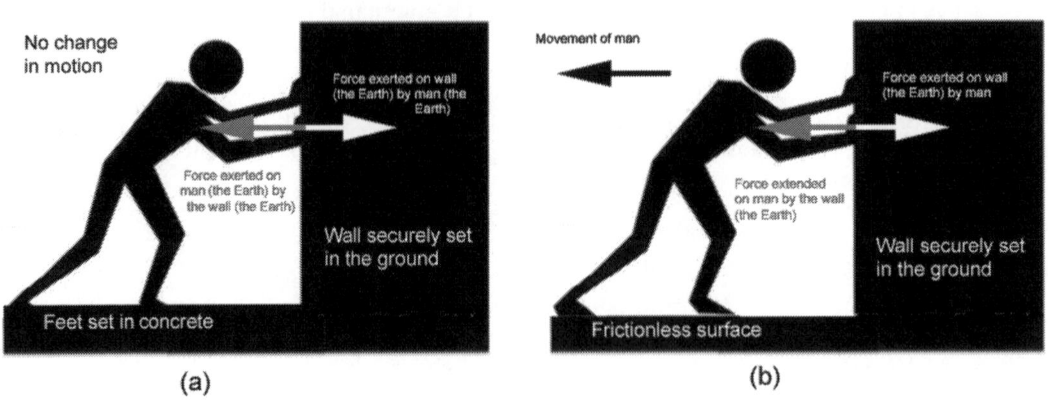

Figure 1.43 Man pushing wall.

other means, then you and the fixed object are part of the Earth and as far as classical mechanics is concerned nothing happens (but you do expend energy doing this pushing).

Finally, I want to return to the horse and cart problem mentioned in Section 1.9.5. The following correspondence is from the 1888 issues of Nature.

The Definition of Force and Newton's Third Law.

PERHAPS your correspondents now engaged in discussing the value of dynamic terms could extend the range of their controversy a little, and deal with a subject of great importance which no text-books touch. It seems to me that the definition of force as that which causes or alters motion is not reconcilable with Newton's law which asserts that every force is always opposed by an equal and opposite force. How can a force opposed by an equal and opposite force perform work, or affect the motion of anything? We have here either a fallacy or an indefiniteness, and the matter is worth clearing up because it incessantly worries students who think.

Nemo[96]

Force, and Newton's Third Law

THE point mentioned by "Nemo" in your issue of March 29. (p. 511) is undoubtedly one that troubles most students at some stage or other, but there is no room for discussion about it; the matter lies in a nutshell: a body does not exert force upon itself. Think, for instance, of a horse and cart. The horse pulls the cart, and the cart pulls back the horse equally; how, then, can the cart move? The only puzzle lies in the false implication that the cart's pull-back is exerted upon the cart. Directly it is perceived that there is only one force acting on the cart, viz. the pull of the horse, no difficulty is felt as to why it moves. The "action" of A is not exerted upon A, but upon B. The "reaction" of B is not exerted upon B, but upon A. …

Oliver J. Lodge[97]

There are a number of scientific texts concerning this supposed horse and cart paradox some of which get rather involved. All I want to do here is to demonstrate there is no paradox and Newton's Laws of Motion do permit the horse to pull the cart.

Consider Figure 1.44. The first point to note is that the horse and cart is a single object and $\left(+\left|\mathbf{F}_{\text{horse on cart}}\right|\right)/\left(-\left|\mathbf{F}_{\text{cart on horse}}\right|\right)$ are internal forces and they contribute nothing to the acceleration of the horse-cart system; they cancel and can be neglected.[98] To move across the ground, there must be an interaction between the horse-cart system and the ground. If we assume there is no friction between the cartwheels and the ground $\left(\left|\mathbf{F}_{\text{cart on Earth}}\right| = \left|\mathbf{F}_{\text{Earth on cart}}\right| = 0\right)$ then the situation is essentially the same as a person walking (Section 1.11.8) and the system (horse and cart) move to the right. What force pushes the horse forward? It is the force exerted by the Earth $\left(+\left|\mathbf{F}_{\text{Earth on horse}}\right|\right)$. The horse pushes backward on the ground $\left(\left|\mathbf{F}_{\text{horse on Earth}}\right|\right)$, so the Earth pushes forward with an equal force. If there is friction between the cart and the Earth then the horse needs to exert a larger force on the Earth to get the physical object (aka the horse and cart) moving.

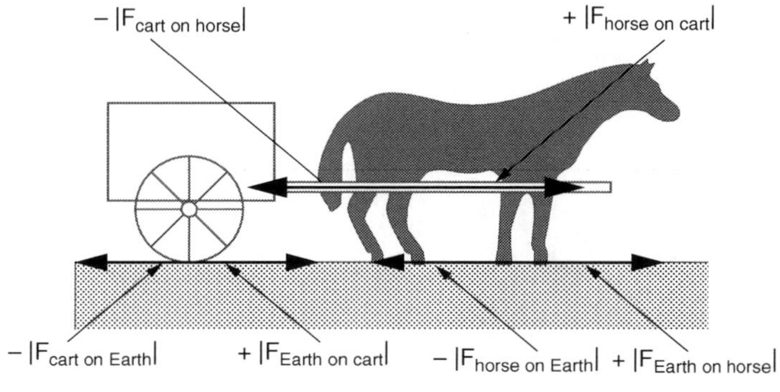

Figure 1.44 Horse and cart.

1.12 MOMENTUM

1.12.1 Introduction

The conservation of momentum is a fundamental concept of classical mechanics along with the conservation of mass (Section 1.3). The Law of Conservation of Momentum (more rigorously in this context the Law of Conservation of Linear Momentum) states that when you have a system closed to the transfer of mass and not acted upon by external forces, *the initial total momentum of objects before a collision equals the final total momentum of the objects after the collision*. This law plays a central role in understanding the classical mechanics of systems of multiple physical objects. The principal reason for introducing the concept of momentum in this book is that it is used in the derivation of the ideal gas equation in Chapter 3.

Unfortunately, the common symbol for momentum is lower case "p",[99] "p" is also used as a symbol for pressure. In this book, I will always use upper case "P" for pressure and lower case "p" for momentum. The linear momentum **p** of a body is defined as the product of its mass m and its velocity **v**:

$$\mathbf{p} = m\mathbf{v} \tag{1.65}$$

Where **p** and **v** are positive or negative, real numbers and m is a positive real number. Like velocity and force, linear momentum is a vector quantity, possessing a direction as well as a magnitude. Because we are only concerned in this book with one-dimensional systems, we can use the simple plus/minus sign convention (see Section 1.2.2).

Using simple differential calculus and Newton's Second Law of Motion it can be shown that force is the rate of change of momentum (remember $\frac{\mathbf{dv}}{dt}$ is acceleration):

$$\frac{\mathbf{dp}}{dt} = \frac{\mathbf{d(mv)}}{dt} = m\left(\frac{\mathbf{dv}}{dt}\right) = m\mathbf{a} = \mathbf{F} \tag{1.66}$$

In other words, force is the rate at which momentum changes with respect to time.

1.12.2 Derivation of the Law of Conservation of Momentum

Conservation of momentum follows logically from Newton's Second and Third Laws of Motion. Suppose two objects (object 1 and object 2) of mass m_1 and mass m_2 respectively undergo a collision (Figure 1.45). The objects move with the velocities $\mathbf{u}_1$and $\mathbf{u}_2$ before the collision. During the collision, there is an unknown period of contact between the objects, when object 1 exerts an average force $\mathbf{F}_2$ on object 2, and object 2 exerts an average force $\mathbf{F}_1$ on object 1. It is important to note that these forces ($\mathbf{F}_1$ and $\mathbf{F}_2$) are the only forces the objects experience, e.g., there is no friction.

From Newton's Third Law, we know that $\mathbf{F}_1$ and F_2 are equal and opposite, i.e., $|\mathbf{F}_1| = |\mathbf{F}_2|$ or $\mathbf{F}_1 = -\mathbf{F}_2$. These forces cause changes in the velocities of objects 1 and 2, i.e., they result in objects 1 and 2 undergoing acceleration. Suppose after the collision, the objects move apart and have the velocities $\mathbf{v}_1$ and $\mathbf{v}_2$. Applying Newton's Second Law to object 1, we have:

$$\mathbf{F}_1 = m_1 \times \mathbf{a}_1 = m_1 \times \left(\frac{\mathbf{v}_1 - \mathbf{u}_1}{t}\right) \tag{1.67}$$

where t is the (unknown) time of contact, and $\mathbf{F}_1$, $\mathbf{u}_1$ and $\mathbf{v}_1$ are real numbers which can be positive or negative depending on our agreed sign convention (movement or application to the right is positive and to the right is negative). A similar relation applies for object 2:

$$\mathbf{F}_2 = m_2 \times \mathbf{a}_2 = m_2 \times \left(\frac{\mathbf{v}_2 - \mathbf{u}_2}{t}\right) \tag{1.68}$$

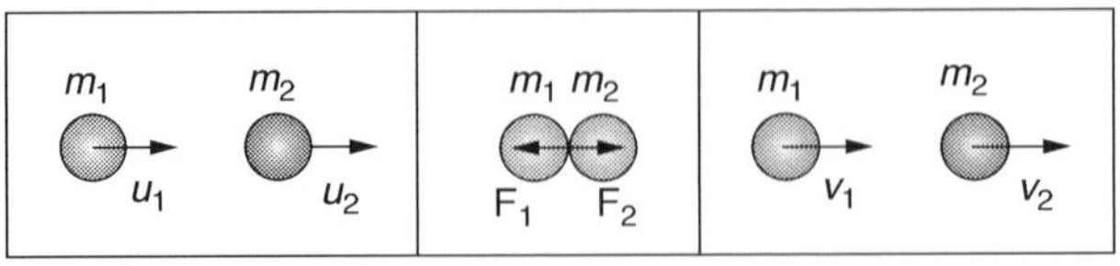

Figure 1.45 Collision between two objects – conservation of momentum.

Adding Equations (1.67) and (1.68) knowing that $\mathbf{F}_1 + \mathbf{F}_2 = 0$ or $\left[\left(-|\mathbf{F}_1|\right) + \left(+|\mathbf{F}_2|\right) = 0\right]$ gives:

$$\left[\left(\frac{m_1}{t}\right) \times (\mathbf{v}_1 - \mathbf{u}_1)\right] + \left[\left(\frac{m_2}{t}\right) \times (\mathbf{v}_2 - \mathbf{u}_2)\right] = 0 \tag{1.69}$$

Multiplying through by t and rearrangement gives:

$$m_1\mathbf{u}_1 + m_2\mathbf{u}_2 = m_1\mathbf{v}_2 + m_2\mathbf{v}_2 \tag{1.70}$$

The Law of Conservation of Momentum (Equation1.70) has been proved using Newton's Second and Third Laws of Motion. The states that the sum of the momenta of the objects before the collision is equal to the sum of momenta after the collision, i.e., the total momentum of the system is unchanged or conserved. The Law of Conservation of Momentum is usually stated as: the momentum of a system is constant if there are no external forces acting on the system.

1.12.3 Examples of the Conservation of Momentum

These examples should allow an appreciation of the Law of Conservation of Momentum. I want to consider an infinite horizontal frictionless one-dimensional surface on Earth.[100] with two spherical physical objects whose only interactions are collisions with each other. There is no horizontal force acting on these two balls because the force of gravity is acting at 90° to the frictionless horizontal surface (more rigorously a line), i.e., there is no horizontal force acting on these balls until a collision occurs. As there are (by definition) no external forces acting on the system, the Law of Conservation of Momentum applies. The following examples will use the sign convention that the left to right direction is positive and that right to left direction is negative). I will use absolute numbers when I think their use will aid clarification.

Example 1.1 – Two Objects Moving in the Same Direction Collide (Figure 1.46)

The Law of Conservation of momentum tells us:

$$\mathbf{p}_1^{\text{before}} + \mathbf{p}_2^{\text{before}} = \mathbf{p}_1^{\text{after}} + \mathbf{p}_2^{\text{after}} = m_1\mathbf{u}_1 + m_2\mathbf{u}_2 = m_1\mathbf{v}_1 + m_2\mathbf{v}_2 \tag{1.71}$$

Using absolute numbers Equation 1.71 can be written:

$$\mathbf{p}_1^{\text{before}} + \mathbf{p}_2^{\text{before}} = \mathbf{p}_1^{\text{after}} + \mathbf{p}_2^{\text{after}} = m_1\left(+|\mathbf{u}_1|\right) + m_2\left(+|\mathbf{u}_2|\right) = m_1\left(+|\mathbf{v}_1|\right) + m_2\left(+|\mathbf{v}_2|\right) \tag{1.72}$$

For m_1 to collide with m_2, we know that $+|\mathbf{u}_1| > +|\mathbf{u}_2|$ and that in the resultant system $+|\mathbf{v}_2| > +|\mathbf{v}_1|$. To illustrate (using arbitrary units for the velocities), suppose $m = m_1 = m_2$, $|\mathbf{u}_1| = 2$ and $|\mathbf{u}_2| = 1$. Firstly, let us suppose m_1 gives m_2 a push and increases its velocity from 2 to 3. Knowing momentum is conserved we know:

$$m \times \left(+|3|\right) + m \times \left(+|2|\right) = m \times \left(+|v_1|\right) + m \times \left(+|3|\right) \tag{1.73}$$

and therefore $|\mathbf{v}_1| = +2$.

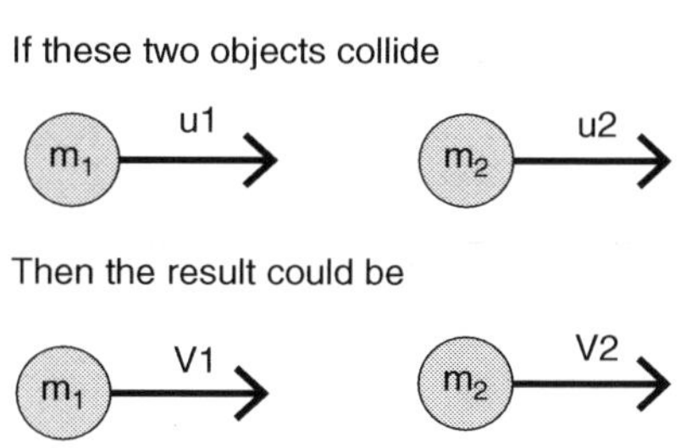

Figure 1.46 Conservation of momentum – Example 1.1.

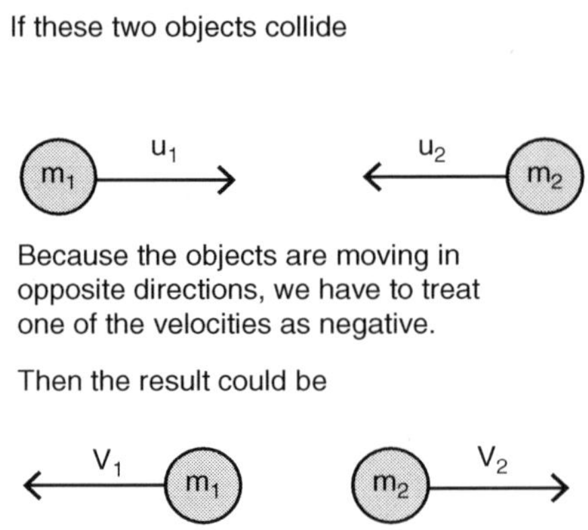

Figure 1.47 Conservation of momentum – Example 1.2.

Example 1.2 – Two Objects Moving in Opposite Directions Collide (Figure 1.47)

The Law of Conservation of Momentum yields:

$$m_1(+|\mathbf{u}_1|)+m_2(-|\mathbf{u}_2|)=m_1(-|\mathbf{v}_1|)+m_2(+|\mathbf{v}_2|) \tag{1.74}$$

I will consider two scenarios:

SCENARIO I

Two objects having the same speed collide and move in opposite directions at the same speed they had before the collision (Figure 1.48).

In this scenario, both balls have the same mass ($m = m_1 = m_2$) and the same speed after the collision, but their velocities have opposite signs $|\mathbf{u}|=|\mathbf{v}|$. It should, by now be apparent that the initial and final momenta of the system are zero:

$$m(+|\mathbf{u}|)+m(-|\mathbf{u}|)=m(-|\mathbf{v}|)+m(+|\mathbf{v}|)=0 \tag{1.75}$$

This type of collision is known as an elastic collision. Snooker balls or ball bearings colliding are very close to undergoing elastic collisions.

It can be seen that any value of **u** and **v** meet the conservation of momentum requirement. For example, if $|\mathbf{u}|=1$ then conservation of momentum is met if $|\mathbf{v}|=100$. Most people would be very surprised if the two balls, initially travelling at 1 $m.s^{-1}$ did rebound at 100 $m.s^{-1}$ and would consider that $|\mathbf{u}|$ and $|\mathbf{v}|$ would be equal in this situation. To fully understand the term "elastic" when used in this context I need to introduce the concept of energy (which I do in Chapter 2). The concept of energy will also explain why $|\mathbf{u}|<|\mathbf{v}|$ is not a valid outcome. However, let us stay with momentum for now and consider Scenario II.

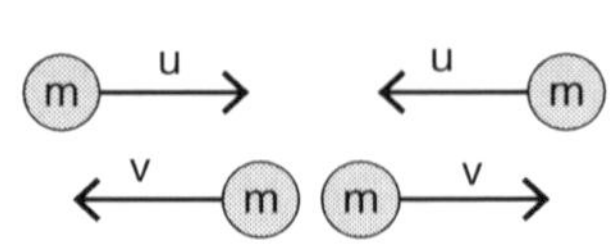

Figure 1.48 Conservation of momentum – Scenario I.

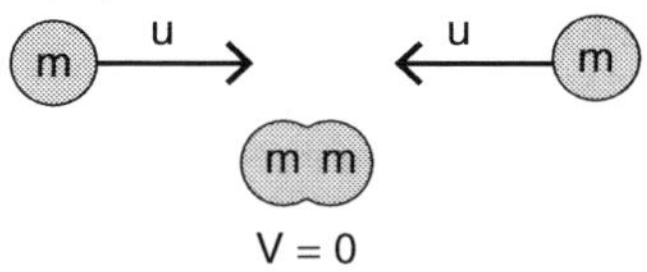

Figure 1.49 Conservation of momentum – Scenario II.

SCENARIO II

Two objects collide and stick together (Figure 1.49).

Now consider the objects are made of a material such as putty or clay such that the two bodies stick together when they collide. As before the initial total momentum of the system is zero. If the objects stick together then $2\,m|\mathbf{v}| = 0$ because of the Law of Conservation of Momentum. We know that m is not equal to zero and is a positive number, therefore, **v** must be equal to zero. This example is an extreme example of what is known as an inelastic collision, and this example demonstrates that the Law of Conservation of Momentum applies whether a collision is elastic or inelastic.

Example 1.3 – Object Bouncing Off a Massive Wall (Figure 1.50)

Consider the situation of an elastic collision of a moving object of mass m with a stationary wall of mass M (where $M \ggg m$), i.e., M is that cliché – an immovable object. All we need to know here is that in an elastic collision at 90° with an immovable wall the initial and final velocity have the same magnitude but opposite directions (i.e., opposite signs).

The momentum of the system before collision is $m(+|\mathbf{u}|) + M \times 0 = m(+|\mathbf{u}|)$, and the momentum of the system after the collision is $m(-|\mathbf{u}|) + M \times 0 = m(-|\mathbf{u}|)$. Clearly, momentum is not conserved because $m(+|\mathbf{u}|) \neq m(-|\mathbf{u}|)$. So, what is going on? The problem seems to be related to the requirement of a stationary wall. Let us consider that after the collision the velocity of the massive wall is not zero. Consider the situation in Figure 1.51.

Assuming the Law of Conservation of Momentum holds in this case we have:

$$m(+|\mathbf{u}_1|) + M \times (0) = m(-|\mathbf{u}_2|) + M(+|\mathbf{v}_2|) \tag{1.76}$$

Therefore:

$$m(+|\mathbf{u}_1|) - m(-|\mathbf{u}_2|) = m(+|\mathbf{u}_1|) + m(+|\mathbf{u}_2|) = M(+|\mathbf{v}_2|) \tag{1.77}$$

Equation (1.77) can be rewritten as:

$$\frac{m}{M} = \frac{(+|\mathbf{v}_2|)}{\left[(+|\mathbf{u}_1|) + (+|\mathbf{u}_2|)\right]} \tag{1.78}$$

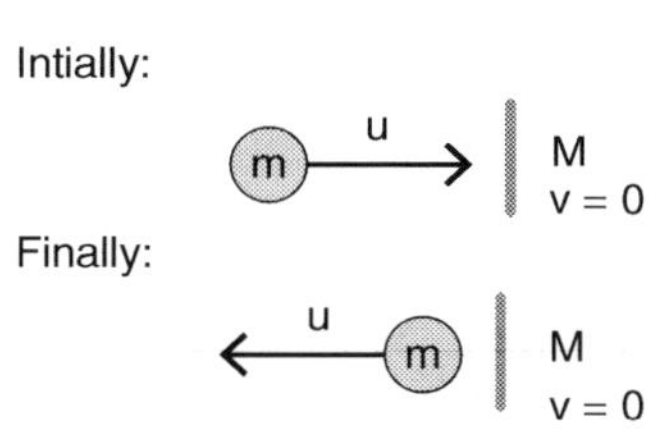

Figure 1.50 Perfectly elastic collision with a massive wall.

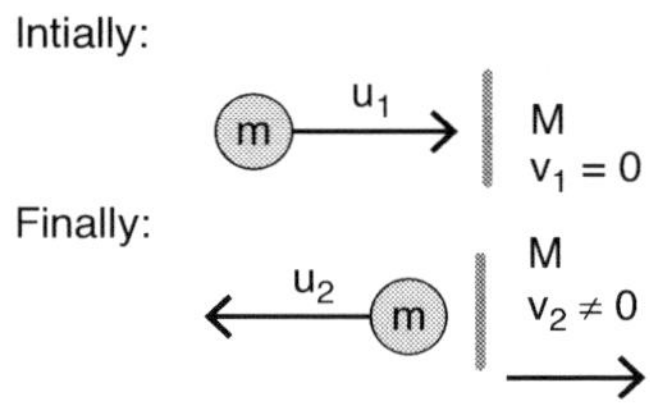

Figure 1.51 Final velocity of massive wall has changed.

Let us assume that M is so much larger than m that the quotient $\frac{m}{M} \to 0$ and we know that $\left[\left(+|\mathbf{u}_1|\right)+\left(+|\mathbf{u}_2|\right)\right]$ cannot tend to infinity therefore it must be the case that $|\mathbf{v}_2| \to 0$. Consequently, the reason the Law of Conservation of Momentum does not appear to hold when considering an object bouncing off massive wall is that the change in the velocity of wall, although not zero, is small enough to be considered zero (as in calculus when we state "in the limit"). It is not right to say that the massive wall has no momentum after the collision. Letting the mass ratio $\frac{M}{m}$ tend towards infinity means that the infinitely massive wall (after the collision) will indeed have zero velocity (in the limit) but, nevertheless, it will still carry the missing momentum $\left[m\left(+|\mathbf{u}_1|\right)+m\left(+|\mathbf{u}_2|\right)\right]$. If we return to Figure 1.51 we know that $|\mathbf{u}_1| = |\mathbf{u}_2| = |\mathbf{u}|$ and change in momentum of the moving object equals its final momentum $m\left(-|\mathbf{u}|\right)$ minus its initial momentum $m\left(+|\mathbf{u}|\right)$:

$$\Delta\mathbf{p} = m\left(-|\mathbf{u}|\right) - m\left(+|\mathbf{u}|\right) = \left(-m\left(+|\mathbf{u}|\right)\right) - \left(-m\left(+|\mathbf{u}|\right)\right) = -2m\left(+|\mathbf{u}|\right) \tag{1.79}$$

Where $\mathbf{\Delta p}$ is the change in momentum of the colliding object after collision with the massive wall – note that the negative value of the momentum change is a direct result of our sign convention. Consider the following abstract[101]:

… momentum is only conserved for a system in the absence of any external force. [I] f you treat the ball as your system, the force exerted by the wall on the ball constitutes an external force for this system, and thus conservation of momentum for the ball simply does not apply.

What if you consider the ball plus wall as the system? Then the forces are internal, and conservation of momentum applies. However, you must also assume that the wall recoils, however slightly, from the collision with the ball. …

The simplest method is to assume a collision between a ball, mass m_b and wall, mass m_w. Allow them to collide elastically, and conserve … linear momentum. Keep track of the positive and negative signs on the velocities. Finally, see what happens to the final velocity of the ball, as the mass of the wall tends to infinity.

In other words, in Figure 1.50 the wall is not part of the system and in Figure 1.51 it is part of the system. If some of the content of this section seems familiar, that is because it is. We have covered the same ground when discussing Newton's Third Law of Motion in Section 1.11, e.g., see Section 1.11.7 (a swimmer pushing off from the side of a pool). The example covered in this section (object bouncing off a massive wall) is used in Chapter 3 in the proof of the ideal gas law.

1.12.4 Discussion

The concept of momentum is a consequence of Newton's Third Law of Motion: when two physical objects collide, the two objects exert equal and opposite forces on each other. The Law of Conservation of Momentum is a direct consequence of Newton's Second Law of Motion.

In Chapter Section 1.12.1 I wrote – "The Law of Conservation of Momentum … states that when you have a system closed to the transfer of mass and not acted upon by external forces ….". I want to explain what is meant by the phrase "not acted upon by external forces"? Consider a horizontal frictionless surface, on Earth, of essentially infinite dimensions containing balls moving at different velocities – this is our system. The balls remain on the table by the force of gravity, but the force of gravity (a vector acting vertically) has no component in the horizontal component. If in doubt see Figure 1.28 in Section 1.9.3, when $\theta = 0$ (i.e., the "inclined" plane is now horizontal) the

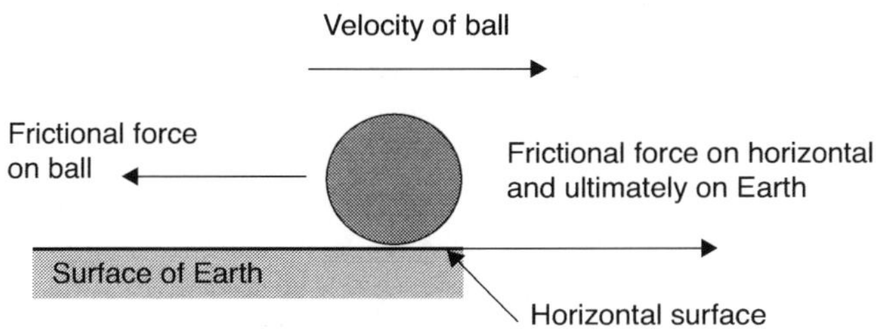

Figure 1.52 Forces on a ball moving on a horizontal surface on the Earth.

horizontal component of the force due to gravity is equal to $mg\sin 0 = 0$. Therefore, if the movement of the ball on the table is frictionless, then the Law of Conservation of Momentum holds for this system irrespective whether collisions are elastic or not. If we then allow frictional effects, our experience tells us that the Law of Conservation of Momentum does not apply to this system. We know this because a ball or balls travelling at certain velocities will eventually come to a halt even if all the collisions between the balls are perfectly elastic. In this case friction is an external force and our system is acted upon by this external force. Please note that it is not the fact that the balls come to a rest, it is the way they do so. Consider if the collisions between our set of moving balls are inelastic then the balls may eventually come to a stop even in a friction-free system, but in this process, momentum is conserved.

We can still define a system in which momentum is conserved but it must be more than just our near-infinite horizontal surface. In fact, we must include Earth (to which the surface is attached to) in the system to eliminate external forces and the Law of Conservation of Momentum applies to this rather large system (Figure 1.52).

As the ball, moving to the right, slows down (i.e., it undergoes a negative acceleration[102]) it experiences a frictional force whose direction is horizontal left. We know from Newton's Third Law that the Earth experiences the same force but acting in the opposite direction (horizontal right). Let us consider a snooker ball weighing 100 g (0.1 kg) having an initial velocity of 0.1 $m.s^{-1}$ that eventually comes to rest, i.e., it has a final velocity of 0 $m.s^{-1}$. The ball loses $0.1 \times 0.1 = 1 \times 10^{-2}$ $kg.m.s^{-1}$ momentum. The Earth's momentum therefore changes by the same amount in the opposite direction, but since the Earth's mass is around 6×10^{24} kg, this corresponds to a velocity change of $\frac{10^{-2}}{6 \times 10^{24}} \approx 10^{-27}$ $m.s^{-1}$, which is immeasurably small. This example is similar to the arguments set out in the previous section on why the Law of Conservation of Momentum applies to a system consisting of an object bouncing off a massive and (essentially) stationary wall and some of the examples given in Chapter Section 1.11 on Newton's Third Law of Motion.

1.13 MASS

1.13.1 Introduction

No matter where you are in the Universe a physical object has the same mass. On Earth, mass is typically measured using a balance. Old-fashioned pan balances (sometimes called scales) involve balancing two scale pans with a known mass in one pan and the item you want to weigh in the other pan (Figure 1.53). This is a straightforward technique based on Equation1.4.

$$F = mg \tag{1.4}$$

If the object of unknown mass, placed on one of the pans, is balanced by a set of known standard masses in the other pan, then we know the masses in the two pans are equal. The beauty of this method is that no matter what planet you are on, the equal masses will always balance out because the gravitational acceleration on each object will be the same. In other words, scales will give the same result for the mass of an object irrespective of whether they are used on Earth, the Moon, Mars, etc. This is because, although the weight of an object will be different in these locations, the change will be proportionately the same on both the object and the standard masses because of the relationship between weight and mass given by Equation (1.4) where the degree of proportionality is g (the acceleration due to gravity). The mass measured using scales is known as "gravitational mass" because its measurement method is based on Newtons Law of Gravitation

Figure 1.53 An old-fashioned analytical balance.

(see Section 1.3). But how do you determine mass when there is no gravitational effect in zero gravity or, more rigorously a "weightless" environment?[103] Before I answer that question, I need to consider the concept of "inertia".

1.13.2 Inertia

We know that an object will not change its velocity on its own, and an external force is necessary to change an object's velocity. Newton's Second Law ($F = ma$) states that the greater mass of an object the greater the force required to change its acceleration. This tendency of an object to resist a change in its state of motion is known as inertia, i.e., inertia is that property of matter which opposes changes in velocity. Newton's Second Law tells us that the inertia of a physical object increases with its mass, in other words, a more massive object has a greater tendency to resist changes in its state of motion than a less massive object, i.e., mass is not just a measure of how much "stuff" is present in an object, it is also a measure of its inertia. Suppose that there are two seemingly identical bricks at rest on a frictionless table. But one brick is made of expanded polystyrene. Without lifting the bricks, you could tell which brick the real brick was, by giving both bricks a push (i.e., subject them to a force) to change their state of motion. The brick that offers the least resistance to the same push is the brick with the least inertia and the least mass. A more massive object has a greater tendency to resist changes in its state of motion. We experience this all the time in life because we know that stopping a car moving at 30 kilometres per hour requires more effort than, say, stopping a tennis ball moving at the same speed. Inertia is the resistance of any physical object to any change in its state of motion, i.e., to acceleration.

Mass, as related to the inertia of a body is defined by two (related) equations $F = ma$ and $p = mv$, i.e., the greater its mass, the less a body accelerates under given force and the greater mass the greater the momentum of a body at a given velocity. Masses defined by these two equations are equal because one can be derived from the other.[104] The "mass" as defined by these equations is known as the "inertial mass", and it is a quantitative measure of that body's inertia which is a measure of how difficult it is to change the velocity of that body. The following quote is from Wikipedia[105]:

> Inertia is the resistance of any physical object to any change in its state of motion; this includes changes to its speed, direction, or state of rest. It is the tendency of objects to keep moving in a straight line at constant velocity. The principle of inertia is one of the fundamental principles of classical physics that are used to describe the motion of objects and how they are affected by applied forces.

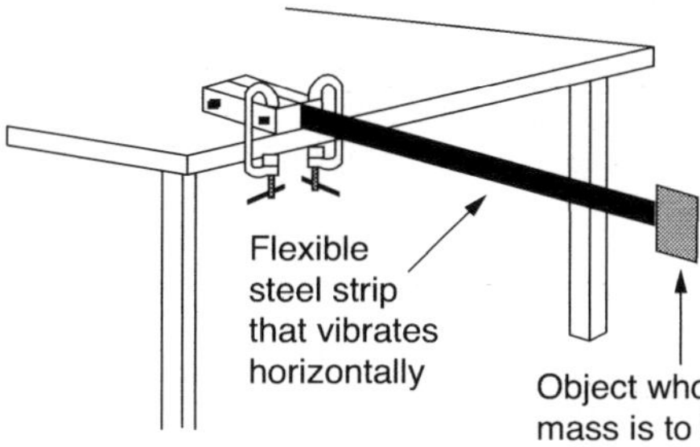

Figure 1.54 An inertial balance allows the determination of inertial mass of an object by measuring the back-and-forth motion of the object. Calibration masses are used to create a graph of T versus the square root of the mass or T^2 versus the mass.

When a bug hits the windscreen of a car, they both experience the same magnitude of force but in opposite directions (Newton's Third Law). But the inertia of the fly is very much less than that of the car, so the fly experiences a high negative acceleration (resulting in it being splattered) and the car experiences a negligible change in motion.

1.13.3 Measuring Mass in Zero Gravity or Weightlessness

In a weightless environment, we cannot use balances to measure mass. I have already indicated a possible method in Section 1.13.2 when discussing bricks of identical appearance but different masses. You could place the object whose mass is to be measured on a near frictionless surface, apply a known force and measure its acceleration. Because of the practicalities of designing the necessary measuring equipment this technique is not used. In practice, an instrument called an inertial balance is used to measure mass in weightless environments. The principle of operation is based on a vibrating spring-mass system: Figure 1.54 shows a representation of the instrument. Since the spring vibrates horizontally, the measured inertial mass is independent of gravity which operates in the vertical direction, i.e., an inertial balance could be used in a zero g environment.

The duration of one cycle (T) is[106]:

$$T = 2\pi\sqrt{\frac{m}{k}} \tag{1.80}$$

where m is the inertial mass of the physical object and k is the spring constant. The inertial balance is calibrated using standard masses and this allows measurement of an unknown mass in weightless environment:

$$\frac{m_{standard}}{m_{unknown}} = \frac{T^2_{standard}}{T^2_{unknown}} \tag{1.81}$$

1.13.4 Inertial Mass and Gravitational Mass

A subtlety emerges when we compare the Law of Gravitation with Newton's Second Law of Motion. The Law of Gravitation provides a definition of gravitational mass (https://www.physicsclassroom.com/class/circles/lesson-3/newton-s-law-of-universal-gravitation) as the property of matter that creates and responds to gravitational forces. In contrast, Newton's Second Law of Motion, F = ma, describes how any force, gravitational or not, changes the motion of an object. In other words, Newton's Second Law provides a definition of inertial mass (https://www.physicsclassroom.com/class/newtlaws/lesson-3/newton-s-second-law) as the property of matter that resists changes in motion, i.e., mass is a measure of an object's inertia.

Let me state this another way. I have set out two ways of measuring mass based on two equations (F = ma and F = mg). Although mass appears in both equations, they have been independently derived. The ("gravitational") mass is defined by Newton's Law of Gravitation which states that there is a gravitational force between any pair of objects, which is given by Equation (1.2):

$$F = G\frac{m_1 \times m_2}{r^2} \tag{1.2}$$

where F is the gravitational force, G is the universal gravitational constant, m_1 and m_2 are the masses of the two objects, and r is the distance between them. This defines the gravitational mass of an object. When one of the objects is a planet, i.e., one of the masses is enormous compared to the other then Equation (1.2) can be rewritten as Equation (1.4) or to be more specific:

$$F_{gravitational} = m_{gravitational} \times g \tag{1.82}$$

In contrast, the inertial mass is defined by Newton's Second Law of Motion, which states that when a force F is applied to an object, it will accelerate proportionally, and that constant of proportion is the inertial mass of that object. This equation can be written more specifically as:

$$F_{inertial} = m_{inertial} \times a \tag{1.83}$$

Gravitational mass is measured with a balance and can only be measured in a gravitational field when the object is in a state of rest. In contrast, inertial mass can be measured (in theory) by giving an object a push and measuring its acceleration (see Section 1.13.3). Notice that the two concepts of mass are completely different: one (inertial) relates interactions (causes) and motion (effects), while the other (gravitation) is a coupling constant of a very particular type of interaction, the gravitational one, i.e., there are two masses – a gravitational mass and an inertial mass. Now we have two possibilities:

1. Gravitational mass and an inertial mass are equal or directly proportional to each other.
2. They are not equal or directly proportional to each other.

This question troubled Galileo, Newton, and many others since. Experiments have shown that all materials fall at the same rate in a uniform gravitational field, providing gravity is the only acting force, i.e., all other forces (friction or air resistance) must be absent or at least negligible. For example, if an iron ball and a feather are dropped from the same height in a vacuum on Earth, the iron ball and the feather will hit the ground at the same time. The popular myth (but probably not true) is that Galileo dropped balls of different mass from the Leaning Tower of Pisa and demonstrated that the time of their descent was independent of mass. These experiments demonstrate, what is now known as, the "Universality of Freefall" [UFF] or the "Equivalence Principle".

As I will now explain UFF implies that gravitational mass and inertial mass are equal or directly proportional to one another. We are now beginning to enter a complex area of physics and my reading of the literature indicates the following argument is not completely rigorous, but I do think it gives a good foundation to anyone who wants to take this subject further.

We can measure the weight of an object on Earth by suspending it from a spring balance. Earth's gravity pulls the object down with a force (weight) of $m_{gravitational} \times g$, where g is the acceleration due to gravity and $m_{gravitational}$ the gravitational mass of the object. The weight, i.e., gravity's pull on the object is balanced by the upward force provided by the stretched spring. When different objects stretch identical springs by identical amounts, we say they have the same gravitational mass. When the spring is cut the object falls. At this stage, the object is in free fall acceleration and does not experience a gravitational field. This is a quote from Einstein[107]:

> Thus, for an observer in free fall from the roof of a house there exists, during his fall, no gravitational field—at least not in his immediate vicinity.

However, in free fall the object experiences an acceleration, and we know that experiments (such as Galileo's probably fictitious experiment on the Leaning Tower of Pisa) find the acceleration, a, to be the same for all objects falling from the same point in the absence of air friction, e.g., if we drop two balls with different weights (different $m_{gravitational}$) in a vacuum both balls will reach the ground at the same time. This must imply acceleration is the same for the two balls, just as we know g is the same for both balls.

Einstein concluded that free fall and inertial motion were physically equivalent by considering that an observer in a windowless room cannot distinguish between being on the surface of the Earth and being in a spaceship in deep space accelerating at one g.

From Newton's Second Law, we can write:

$$m_{inertial} \times a = m_{gravitational} \times g \tag{1.84}$$

Rearranging gives:

$$\frac{m_{gravitational}}{m_{inertial}} = \frac{g}{a} \tag{1.85}$$

Therefore, $\frac{g}{a}$ is the same for all objects and therefore $\frac{m_{inertial}}{m_{gravitational}}$ is the same for all objects.
Traditionally the ratio is written as K:

$$K = \frac{g}{a} \tag{1.86}$$

The ratio of gravitational to inertial mass of any object is equal to some constant K, if and only if all objects fall at the same rate in a gravitational field. Einstein later showed that K = 1 in Equation (1.86) and made it the foundation of his theory of general relativity[108]:

> The general theory of relativity owes its existence in the first place to the empirical fact of the numerical equality of the inertial and gravitational mass of bodies.

Crudely, the Equivalence Principle with K = 1 states that an object accelerated in the absence of any gravitational attraction (inertial mass acceleration), will experience events indistinguishable from that produced by an equivalent magnitude gravitational field (gravitational mass acceleration). No experiments to date have disproved UFF,[109] but examination of the literature indicates significant controversy abounds in this area.

1.13.5 Discussion

In Section 1.5 I mentioned that Aristotle's theories of free fall and projectile motion were only questioned (by Galileo) in the late 16th century. I possibly should have stated "seriously" questioned. Aristotle's theories of free fall and projectile motion were criticised in the 6th century, e.g., by John Philoponus (c. 490 – 570 AD), who anticipated the modern empirical approach, he wrote[110]:

> But this [Aristotle's view] is completely erroneous, and our view may be corroborated by actual observation more effectively than by any sort of verbal argument. For if you let fall from the same height two weights, one many times heavier than the other you will see that the ratio of the times required for the motion does not depend on the ratio of the weights, but that the difference in time is a very small one.

History tells us nobody listened until Galileo and Newton.[111]

Now let me give a simple explanation for why it is natural that UFF comes about. Suppose we have two half-sphere masses. If we drop them separately (in a vacuum) they will both take time T to fall to the ground. On the other hand, if we attach them together, UFF, tells us they will take the same length of time to fall to the ground (remember this is in a frictionless environment). If the two halves of the sphere were dropped together, then they would fall together. This is no different from screwing them together and dropping them together, i.e., there will not be any force on the screws (Figure 1.55). Therefore, the combined (or screwed together) sphere must fall at the same rate as the split sphere.

Finally, another extract that may help[112]:

> Approximately 450 years ago, Galileo, as some have reported, dropped cannonballs of different sizes from the Leaning Tower of Pisa to prove that they would hit the ground at the same time. 350 years later, Einstein published his Theory of General Relativity, which included the equivalence principle, and as Galileo demonstrated, stated that objects with different masses fall at the same rate due to gravity. In this new effort [recreating Galileo's free-falling objects experiment in a satellite], the researchers have performed roughly the same experiment as Galileo, except they have done it in space and measured the results with far more precision.

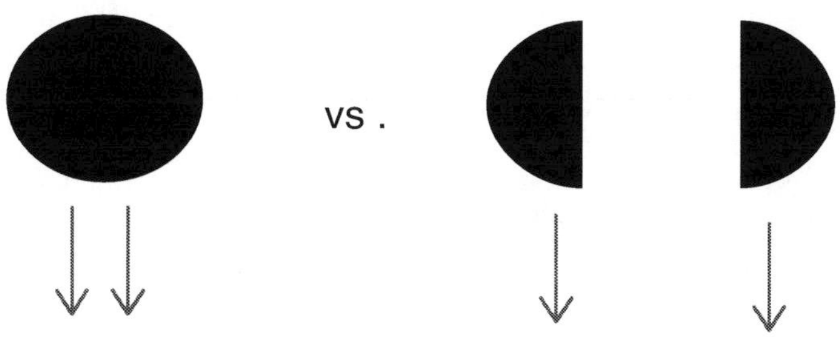

Figure 1.55 Falling of two half-sphere masses when joined together and separately.

> Scientists continue to test the equivalence principle because they believe that if they can find a violation, it could reconcile the impasse regarding quantum theory and general relativity.

It is worth noting that at the end of the Apollo 15 moon walk, Commander David Scott held out a geologic hammer and a feather and dropped them at the same time. Because they were essentially in a vacuum. The following is from the Apollo 15 Preliminary Science Report[113]:

> During the final minutes of the third extravehicular activity, a short demonstration experiment was conducted. A heavy object (a 1.32-kg aluminum geological hammer) and a light object (a 0.03-kg falcon feather) were released simultaneously from approximately the same height (approximately 1.6 m) and were allowed to fall to the surface. Within the accuracy of the simultaneous release, the objects were observed to undergo the same acceleration and strike the lunar surface simultaneously, which was a result predicted by well-established theory, …

1.14 SUMMARY

I started this chapter by considering what is meant by the term "physical object" (*consists of matter, occupies space, endures through time, and can move about in space*). Unfortunately, matter cannot be rigorously defined in science because of wave-particle duality, but it is a useful concept and is used both in this book and throughout science (probably because we have an intuitive feel for the word). I defined physical quantities as (i) things that can be measured and (ii) things that can be divided into:

- Scalar quantities (have magnitude only) and vectors (have magnitude and direction).
- Extensive quantities (dependent on the extent of the system) and intensive quantities (independent of the extent of the system).

In Section 1.3 I considered mass, gravity, and force where I introduced the first of many conservation laws – The Law of Conservation of Mass. A mass (i.e., a physical object) exerts a pull on another mass, this is known as gravitation and was quantified by Newton in his Law of Gravitation. Despite the above-mentioned problem with the definition of matter I moved on to define the mass of an object as the amount of matter it contains. An inherent property of mass is that it exerts a force on any other mass ("gravitation") and Isaac Newton derived what is now known as Newton's law of Universal Gravitation:

$$F = G\frac{m_1 \times m_2}{r^2} \quad (1.2)$$

I then considered the force (F) on a mass m on the surface of the Earth using Equation 1.2. This led to Equation 1.4.

$$F = mg \quad (1.4)$$

where F is known as the weight of the object on Earth of mass m where g is the acceleration due to gravity. In Section 1.4 I considered acceleration due to gravity and weight.

In Section 1.5, I set out a brief introduction to classical mechanics mainly considering how a dogmatic belief in Aristotelian principles held back the advent of the modern science of Galileo and Newton for about 1,500 years.

Section 1.6 contains an introduction to the mathematics needed for this book, concentrating on the mathematics of change – calculus. Differentials appeared with the invention of calculus in the second half of the seventeenth century. The differential dx was understood as a small increment of the variable x. If another variable y depends on the independent variable x, then the resulting increment dy of y is its differential. The quotient of these two differentials, $\frac{dx}{dy}$, was interpreted geometrically by Leibniz[114] as the slope of the tangent at the point x. Differential calculus may have its genesis in solving geometric problems such as that of drawing tangents or of locating the maxima and minima of curves. Its origins are found within the domain of analytic geometry where geometry is studied by means of a coordinate system. A curve drawn in a plane is represented in analytic geometry by an equation involving two algebraic variables, the abscissa, and the ordinate,[115] which in Cartesian geometry are understood as the distances of a point on the curve from two orthogonal axes. In his first publication on the differential calculus of 1684, Leibniz denoted the differential of a variable x by dx, the same notation that is used today. It was interpreted as a small increment in the abscissa x, which implies an increment, or a differential, dy in the ordinate y. Differential calculus was by its origin a natural method for solving geometric problems, the results being expressed in terms of algebraic expressions. In this sense, it was a method to solve problems in analytic geometry. This changed with Euler,[116] who made the subject a theory of functions without the need for a connection with geometry.

Section 1.7 developed equations of motion, usually known as Newton's equations of motion (as opposed to his laws of motion) (see below):

$$s = \frac{1}{2}(u + v)t$$

$$s = ut + \frac{1}{2}at^2$$

$$v^2 = u^2 + 2as$$

Graphical representations of the above equations allowed further insights into calculus.

Section 1.8 briefly considered one of Galileo's experiments (balls rolling down inclined planes) and showed how he showed Aristotle's view of motion was flawed. Newton built upon Galileo's (and others) work and developed his laws of motion (as opposed to his equations of motion) and this is the subject of Sections 1.9 and 1.11. Newton's three laws are:

1. The First Law: an object at rest will remain at rest and an object in motion will remain in motion at constant velocity unless acted upon by an external force.
2. The Second Law: the external force acting on an object is equal to its mass times its acceleration (F = ma).
3. The Third Law: is unhelpfully and confusedly stated as – for every action, there is an equal and opposite reaction.

To analyse the consequences of these laws, Isaac Newton (who independently, along with Leibniz developed calculus) developed a body of mathematical techniques known as the calculus and applied it to develop and analyse his laws of motion.

Section 1.10 considered the differences between a mathematical equation such as y = ax and a scientific equation such as F = ma. In the mathematical equation, y, a, and x represent real numbers (which can be positive or negative). Whereas scientific equations link scalars and vectors. Because this book is restricted to the mathematical consideration of one-dimensional systems, we can use a simple + / – sign convention for vectors, these vectors can take on both positive and negative values in scientific equations. I will represent this by using a bold font, i.e., **F** and **a** represent vectors in one-dimensional systems that can be positive or negative real numbers. Additionally, I will use bold font when considering the differences between two positive variables such as pressure, i.e., **ΔP** or **dP** ("vector-like" quantities). Absolute scalar quantities such as mass are always positive real numbers and are represented by the use of a regular font.[117] I consider this convention together with the use of absolute values (introduced in Section 1.6.8) will assist in clarifying many of aspects covered in this book and, when considering the vector or vector-like nature of physical quantities in scientific equations I will use this convention, e.g., $\mathbf{F} = m\mathbf{a}$ and $-|\mathbf{F}| = m(-|\mathbf{a}|)$.

Although in today's world of quantum mechanics and relativity Newton's laws are no longer considered to be exactly correct, the conservation of momentum (derived from Newton's laws) remains true and is considered in Section 1.12.[118] The Law of Conservation of Momentum is one of the most important laws in classical mechanics which can be stated as follows:

> For a collision occurring between object 1 and object 2 in a closed mechanical system,[119] the total momentum of the two objects before the collision is equal to the total momentum of the two objects after the collision. That is, the momentum lost/gained by object 1 is equal to the momentum gained/lost by object 2.

The above statement tells us that the total momentum of a collection of objects (a system) is conserved – that is, the total amount of momentum is a constant or unchanging value. We have seen inelastic collisions do not invalidate the Law of Conservation of Momentum. In addition, frictional forces do not invalidate the conservation of momentum provided the forces are internal (between two objects considered part of the system). For example, if an object slides on a table the momentum of the object is not conserved as the object comes to rest because of frictional forces. However, if the system is the object and the table (and the table does not move), then the momentum of the object and momentum are conserved. Forces external to the system may change the total momentum when their sum is not zero, but internal forces, regardless of the nature of the forces, will not

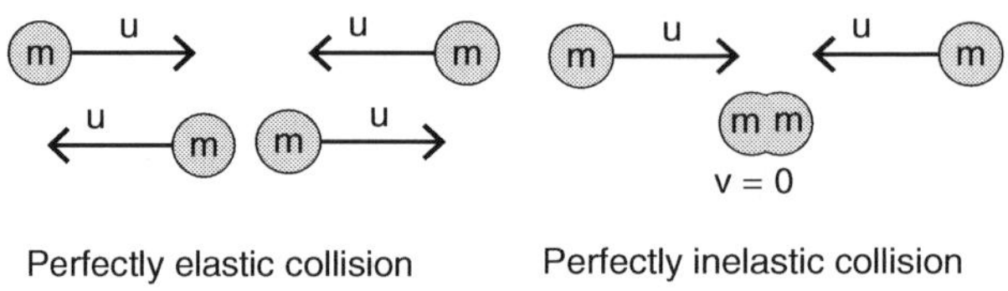

Figure 1.56 Elastic and inelastic collisions.

contribute to the change in the total momentum. To analyse a mechanical system, it is important to recognise which forces are internal and which are external. Once a mechanical system is clearly defined, it is not hard to understand what part should be considered external.

- External forces: forces caused by external agent outside of the system.
- Internal forces: forces exchanged by the particles in the system.

In Section 1.13, I considered the differences between gravitational mass and inertial mass and showed that they are equivalent and how this equivalence is related to the development of the theory of relativity.

Finally, I want to consider the following question (see Figure 1.56):

Although momentum is conserved what are the differences between a perfectly elastic and a non-elastic collision?

At this point, I cannot answer this question because I need to introduce two concepts. One of these is to consider what physical objects are made of this is considered in Chapter 3. The other concept (energy) which has been mentioned a few times in this chapter is introduced in the next chapter.

NOTES

1 Isaac Newton (1643–1727) was an English physicist and mathematician who developed the principles of modern physics. In 1687, he published his most acclaimed work, *Philosophiae Naturalis Principia Mathematica (Mathematical Principles of Natural Philosophy)*, which has been called the single most influential book on physics.

2 One of the consequences of Einstein's special theory of relativity (1905) is that the mass of a physical object increases with its velocity relative to the observer. When an object is at rest (relative to the observer), it is the mass (often called 'rest mass') that we are all familiar with (see Section 1.3).

3 Wave–particle duality is the concept that every elementary particle may be partly described in terms not only of particles but also of waves. It expresses the inability of the classical concepts "particle" or "wave" to fully describe the behaviour of quantum-scale objects. In physics, the term "light" refers to electromagnetic radiation of any wavelength, whether visible or not. Electromagnetic radiation consists of electromagnetic waves, which are synchronized oscillations of electric and magnetic fields.

4 What is meant by the term "non-physical objects"? Good examples are found in mathematics such as the definition of a line in geometry which requires it to have no width. This "line" does not exist in the physical world but is an essential concept in geometry. A mental picture of a tree is non-physical object, but a photograph of a tree is a physical object. The "ideal gas" (considered in Chapter Three), like the mathematicians' line, does not exist in the physical world but it is an essential concept in science: in this book, I consider the ideal gas to be a physical object.

5 An interesting article on this subject is "Physical Object", Ned Markosian, in "A Companion to Metaphysics" 2nd Ed [2009].

6 https://www.physicsclassroom.com/class/1DKin/Lesson-1/Distance-and-Displacement.

7 Pythagoras' theorem states that for a right-angled triangle the square of the hypotenuse (the side opposite the right angle) is equal to the sum of the squares of the other two sides.

8 Pressure, volume, and temperature of an ideal gas are covered in some detail in Chapter 3.

9 Rasmus Persson in https://www.quora.com/How-is-pressure-an-intensive-property.

10 The International System of Units is the modern form of the metric system and is the world's most widely used system of measurement, used in both everyday commerce and science. The abbreviation "SI" is used in all languages – from the French "**Système International d'Unités**".

11 The other fundamental quantity is luminous intensity. These are the definitions before 2019, some of these definitions have subsequently changed.

12 Since 1889 the magnitude of the kilogram has been defined as the mass of an object called the international prototype kilogram which is a cylinder of a platinum/iridium alloy. In 2019 it is now defined in terms of Planck's constant, h.

13 When the property of an object remains unchanged, it is said to be conserved. Certain physical quantities do not change for isolated systems (the Universe is an isolated system).

14 The Law of Conservation of Mass was first outlined by Mikhail Lomonosov (1711–1765) in 1748. He proved it by experiments - though this is sometimes challenged. Antoine Lavoisier (1743–1794) had expressed these ideas in 1774. Others whose ideas pre-dated the work of Lavoisier include Joseph Black (1728–1799), Henry Cavendish (1731–1810), and Jean Rey (1583–1645). (From Wikipedia - https://en.wikipedia.org/wiki/Conservation_of_mass).

15 In this chapter, a system is the portion of the Universe that is being studied – the concepts of system, surroundings and Universe are of paramount importance in understanding the main subject of this book. I will return to these concepts many times and will set out formal definitions in Chapter 4.

16 "Energy" is considered later in Chapter 2.

17 In this context and throughout this chapter the word Universe means everything. "The Universe is everything we can touch, feel, sense, measure or detect. It includes living things, planets, stars, galaxies, dust clouds, light, and even time. Before the birth of the Universe, time, space and matter did not exist." https://www.esa.int/kids/en/learn/Our_Universe/Story_of_the_Universe/The_Universe.

In thermodynamics the term "universe" has a related but subtly different meaning (in this book I will spell the "thermodynamic" universe with a lower case 'u').

18 The full stops used in these abbreviations is optional, i.e., $m.s^{-1}$ and $m.s^{-1}$ are both valid. I will use full stops when I think it aids clarity.

19 Remember (see Prologue) inductive reasoning makes broad generalisations from specific observations.

20 Also known as Newton's Law of Universal Gravitation.

21 Pronounced "pringkipia".

22 Torsion occurs when a physical object, such as a wire is twisted. The twisting force or torsional force acting on the wire is known as the torque (which is the tendency of a force to rotate an object about an axis or pivot). To put it another way, torsion is the twisting of the wire due to an applied torque.

23 Density of an object = mass of object divided by the volume of object.

24 Virtually all textbooks and articles take this as read, but the mathematical proof is complex.

25 I must emphasise here that this does not rigorously apply to anything other than spherically symmetrical physical objects, but if the distance between the physical objects is very much greater than their dimensions the use of point masses is a useful approximation.

26 https://en.wikipedia.org/wiki/Earth_mass.

27 The Earth's radius falls between the polar minimum of about 6,357 km and the equatorial maximum of about 6,378 km. This indicates the Earth deviates from a perfect sphere by only a 0.3%: this is sufficiently close to treat it as a sphere in most contexts. Several different ways of modelling the Earth as a sphere each yield a mean radius of 6,371 km (6.371×10^6 m).

28 https://en.wikipedia.org/wiki/Moon.

29 We have all heard of "weightlessness" usually when applied to astronauts in a "zero g" or "zero gravity" environment. However, the definition and use of "weightlessness" are difficult because the sensation of "weight" in everyday terrestrial experience results not from gravitation acting alone (which is not felt), but instead by the mechanical forces that resist gravity. A physical object in a straight free fall, or in a more complex inertial trajectory of free fall (such as within a reduced gravity aircraft or inside a space station), all experience weightlessness, since they do not experience the mechanical forces that cause the sensation of weight. A common misconception about spacecraft orbiting the Earth is that they are operating in a gravity free environment. According to classical mechanics, this is technically inaccurate. Spacecraft are held in orbit by the gravity of the planet which they are orbiting. In classical mechanics, the sensation of weightlessness experienced by astronauts is not the result of there being zero gravitational acceleration (as seen from the Earth), but of there being no gravitational force that an astronaut can feel because of the free-fall condition, and also there being zero difference between the acceleration of the spacecraft and the acceleration of the astronaut. I will stop at this point with a remark I make all the time – "nothing is simple".

30 For example, see http://www.physics.usu.edu/coburn/Measurement%20Projects/Diameter%20of%20the%20Earth.html.

31 https://www.wired.com/2012/02/a-modern-measurement-of-the-radius-of-the-earth/.

32 To put this in perspective "modern" science which (arguably) started with Galileo is about 400 years old.

33 The unmoved mover or prime mover moves other things, but is not itself moved by any prior action, i.e., the unmoved mover would have initiated movement within the Universe, but and more importantly, the unmoved mover would not have been set in motion by another thing. If you think about it, the unmoved mover must be some form of god. Although I do not intend to get into a religious debate, I do want to make the following point – if, whenever a difficulty arises in a scientific explanation, a god is invoked then we are not considering science but magic.

34 Galileo complained that some of the philosophers who opposed his astronomical discoveries (e.g., the phases of Venus which demonstrated that Venus must circle the sun and not the Earth) had refused even to look through a telescope. Galileo did not name the philosophers concerned, but Galilean scholars have identified two of them as Cesare Cremonini and Giulio Libri.

35 The web sites I used were: https://www.mathsisfun.com/index.htm, also https://sydney.edu.au/stuserv/documents/maths_learning_centre/integration_part1.pdf and https://sydney.edu.au/stuserv/documents/maths_learning_centre/integration_part2.pdf; https://www.

khanacademy.org/math/calculus-1; https://www.wiley.com/legacy/Australia/Landing_Pages/c14Anti-DifferentiationAndIntroductionToIntegralCalculus_web.pdf; and Chapter 10 in "Mathematics for the Million", Lancelot Hogben, WW Norton & Company, 1983.

36 The definition of a real number is given below.

37 https://www.mathsisfun.com/calculus/introduction.html.

38 Division by zero is not defined in arithmetic, i.e., 0/0 is indeterminate (it could be any value) – see https://www.mathsisfun.com/numbers/dividing-by-zero.html for a simple explanation why this is so.

39 The formula is a simpler version of the equation relating distance objects fall under the influence of Earth's gravity with time: this equation is $d = \frac{1}{2}gt^2$ where g is rounded up from 9.81 to $10\,m.s^{-2}$. See Section 1.7.3 for derivation of this equation.

40 One way of viewing calculus is that it is a method that circumvents the "forbidden" division by zero.

41 Why? There are a number of ways to answer this and here is one of them. When we raise a number to the nth power, that means we multiply that number by itself n times, e.g., $2^3=2\times2\times2$. Therefore, raising a number to the zeroth power means we multiply the number by itself zero times - but that means we are multiplying no numbers at all. Now consider when we add no numbers at all, we would expect to get zero. However, zero is a special number in addition: it is called the additive identity, because it is the only number which you can add to any other number and leave the other number the same, i.e., zero is the only number such that for any number x, $x+0=x$. Therefore, it makes sense that if adding no numbers at all gives back the additive identity, then multiplying no numbers at all should give back the multiplicative identity. The multiplicative identity is the number 1, because for any other number x, $1\times x=x$. Therefore, the reason that any number to the zero power is one is because any number to the zero power is just the product of no numbers at all, which is the multiplicative identity, 1.

42 ln (also written as $\log_e$) is the natural logarithm, i.e., log to the base "e" where e is the unique number whose natural logarithm is equal to one (i.e., ln(e)=1). It is approximately equal to 2.71828 and is the limit of $\left(1+\frac{1}{n}\right)^n$ as n tends to infinity, e.g., when n=1000 then $\left(1+\frac{1}{1000}\right)^{1000} = 2.71692$. See Section 1.6.9 for more on logarithms. Like π, e is an irrational real number (e is also known as Euler's number).

43 If you want to check this or carry out other similar spreadsheet calculations, then $A_0(x) = \sum_{r=0}^{r=n-1} y_r \Delta x$ and $A_1(x) = \sum_{r=1}^{r=n} y_r \Delta x$.

44 This is known as the Fundamental Theorem of Calculus which relates differentiation and integration, showing that these two operations are inverses of one another. Before the discovery of this theorem, it was not recognized that these two operations were related. Ancient Greek mathematicians knew how to compute area via infinitesimals, an operation that we would now call integration. The origins of differentiation likewise predate the Fundamental Theorem of Calculus by hundreds of years; for example, in the fourteenth century, the notions of continuity of functions and motion were studied by the Oxford Calculators and other scholars.

45 Division by zero – if we use the rule $\frac{x^{n+1}}{n+1} + C$ we find the indefinite integral of $x^{-1} = \frac{x^{-1+1}}{-1+1}$. This involves the "forbidden" division by zero. However, note the third rule of differentiation in Section 1.6.4, i.e., if $f(x) = \ln(x)$ then $f'(x) = \frac{1}{x}$. Therefore, the indefinite integral of $\frac{1}{x}$ is $\ln(x) + c$.

46 It is customary in maths to use F(x) to represent any indefinite integral of f(x).

47 Based on content from the following websites:
Algebra - Absolute Value Equations (lamar.edu).
https://tutorial.math.lamar.edu/classes/alg/SolveAbsValueEqns.aspx.
https://www.mathsisfun.com/algebra/absolute-value-solving.html.

48 https://en.wikipedia.org/wiki/Inexact_differential.

49 See Section 4.6 for a consideration of state and path functions.

50 M Oliveira, "Exact and inexact differentials in the early development of mechanics and thermodynamics", *Revista Brasileira de Ensino de Física* 42, e20190192, 2020.

51 https://chemistry.stackexchange.com/questions/22171/why-is-du-an-exact-differential-and-dq-an-inexact-differential.

52 https://www.khanacademy.org/math/multivariable-calculus/multivariable-derivatives/partial-derivative-and-gradient-articles/a/introduction-to-partial-derivatives.

53 See Section 1.6.4 Endnote 41.

54 Our first was in Section 1.2.2 when considering vector addition.

55 This is from Wikipedia, https://en.wikipedia.org/wiki/Impulse_(physics).

"In classical mechanics, impulse … is the integral of a force, F, over the time interval, t, for which it acts. Since force is a vector quantity, impulse is also a vector quantity. Impulse applied to an object produces an equivalent vector change in its linear momentum, also in the same direction."

…

"A resultant force causes acceleration and a change in the velocity of the body for as long as it acts. A resultant force applied over a longer time therefore produces a bigger change in linear momentum than the same force applied briefly: the change in momentum is equal to the product of the average force and duration. Conversely, a small force applied for a long time produces the same change in momentum—the same impulse—as a larger force applied briefly."

56 Time is usually considered to be a scalar quantity in classical mechanics; mathematically time is vector (it can go forwards or backwards). Confusing isn't it. See Section 1.2.2 and Section 3.7.

57 http://wearcam.org/absement/examples.htm.

58 Liebniz independently developed "the calculus" at or around the same time as Newton did.

59 Rather more rigorously - an object either is at rest or moves at a constant velocity unless acted upon by an external force.

60 In particle physics, an elementary particle is a subatomic particle with no substructure, i.e., it is not composed of other particles. Quarks, leptons, and the Higgs boson are examples of elementary particles.

61 Not covered in this book. The relevant theorem is "If two parallel lines are cut by a transversal, then each pair of alternate interior angles are congruent (equal)" and its corollaries. See for example https://math.libretexts.org/Courses/Mount_Royal_University/Mathematical_Reasoning/4%3A_Basic_Concepts_of_Euclidean_Geometry/4.1%3A_Euclidean_geometry#:~:text=Euclidean%20geometry%2C%20sometimes%20called%20parabolic,is%20three%2Ddimensional%20Euclidean%20geometry.

62 Empirical methods are ways of gaining knowledge by means of direct and indirect observation or experience. Empiricism is the theory that all knowledge is based on experience derived from the senses. Empirical – "relying on experience or observation alone/originating in or based on observation or experience".

63 As opposed to the unreal, hypothetical, or imaginary world of thought experiments.

64 Newton's First Law of Motion - If a physical object is at rest, it takes an applied force to make it move. If the object is already moving, it takes an applied force to stop it or to change its direction or speed.

65 Modern science considers that physical objects interact at the electron level at or near the objects' surfaces. Contact force is a simplification used in applications of classical mechanics.

66 https://physics.stackexchange.com/questions/45653/with-newtons-third-law-why-are-things-capable-of-moving.

67 https://archive.org/details/bub_gb_Tm0FAAAAQAAJ/page/n63/mode/2up.

68 https://byjus.com/physics/newtons-second-law-of-motion-and-momentum/.

69 https://www.lockhaven.edu/~dsimanek/physics/horsecart.htm.

70 Apart from in Section 1.9.3.

71 "Hypothetically, matter can have negative mass in the same sense that an electric charge can be either negative or positive. People rarely think in these terms, and our everyday world sees only the positive aspects of Isaac Newton's Second Law of Motion, in which a force is equal to the mass of an object times its acceleration, or $F=ma$. In other words, if you push an object, it will accelerate in the direction you're pushing it. Mass will accelerate in the direction of the force." Physicists create 'negative mass': Experimental technique can help probe phenomena in astrophysics and cosmology – ScienceDaily (https://en.wikipedia.org/wiki/Negative_mass#:~:text=In%20theoretical%20physics%2C%20negative%20mass,for%20an%20applied%20force%20orientation).

72 https://www.quora.com/Is-a-negative-volume-theoretically-possible.

73 Apart from in Section 1.9.3 when I briefly consider forces acting on a physical object on an inclined plane.

74 The analogy is not perfect because a bank account is considered to in the "black" if the total quantity of money in it is positive or, is in the "red" (or overdrawn) if the total quantity of money is negative. In thermodynamics a system cannot have "negative" energy (the concept of negative energy is used in theoretical physics to explain the nature of certain fields and various quantum effects).

75 Time is considered a scalar quantity – see Section 3.7.

76 Newton' third law - Forces, acceleration and Newton's laws of motion - Eduqas - GCSE Combined Science Revision - Eduqas - BBC Bitesize.

77 I will explain what a normal force is later in the section.

78 http://unsworks.unsw.edu.au/fapi/datastream/unsworks:45221/bince2fe389-6859-43fa-8974-0bc68ef1461f?view=true.

79 "Newton's Principia. The Mathematical Principles of Natural Philosophy", Sir Isaac Newton, translated into English by Andrew Motte, page 83. https://ia802706.us.archive.org/0/items/newtonspmathema00newtrich/newtonspmathema00newtrich.pdf.

80 I will explain what a normal force is later in the section.

81 https://physicsparsimony.wordpress.com/2011/12/22/confusion-regarding-newtons-third-law-of-motion/.

82 Mr. Hammond is the author of this extract.

83 Mr. Hammond.

84 https://www1.grc.nasa.gov/beginners-guide-to-aeronautics/nozzle-design/.

85 This is from https://spaceplace.nasa.gov/launching-into-space/en/#:~:text=The%20exhaust%20pushes%20out%20of,That's%20the%20reaction%20force.

86 A fluid's velocity must increase as it passes through a constriction. Mathematically, this can be written as $V=A_1\times v_1=A_2\times v_2$ where V=volumetric flow rate, A=the cross sectional area of flow and v=the mean velocity. If A decreases then velocity of the fluid must increase. Therefore, any gain in kinetic energy a fluid may accrue due to its increased velocity through a constriction is balanced by a fall in pressure. In other words, when a fluid is accelerated, its pressure drops. This can be thought of in terms of energy. Pressure is a measure of potential energy and velocity is a measure of the kinetic energy of the fluid flow. When you have more kinetic energy (velocity), you then have less potential energy (pressure). [I will be considering kinetic energy in Chapter 2.] The Khan Academy website is worth a visit if you want to know more. https://www.khanacademy.org/science/physics/fluids/fluid-dynamics/a/what-is-bernoullis-equation.

87 See, for example, https://en.wikipedia.org/wiki/Shower-curtain_effect.

88 Example from https://courses.lumenlearning.com/physics/chapter/4-4-newtons-third-law-of-motion-symmetry-in-forces/.

89 Acceleration figure taken from https://physicinswimming.weebly.com/newtons-laws-of-motion.html.

90 https://www.dummies.com/education/science/physics/newtons-third-law-action-and-reaction/.

91 As opposed to about 6oz (about 150 g).

92 https://physicsparsimony.wordpress.com/2011/12/22/confusion-regarding-newtons-third-law-of-motion/.

93 Momentum is covered in the next section.

94 Energy is the subject of Chapter 2.

95 At a molecular level, this is electromagnetic repulsion.

96 *Nature*, 37, 511, 1888. https://www.nature.com/articles/037511a0#Abs1.

97 *Nature*, 37, 558, 1888. https://www.nature.com/articles/037558b0. Sir Oliver Joseph Lodge, (1851–1940) was a British physicist and writer involved in the development of radio.

98 Remember "A system cannot "bootstrap" itself into motion with purely internal forces - to achieve a net force and an acceleration, it must interact with an object external to itself."

99 'p' is used because the word "impetus" comes from the Latin, "petere" - to go towards or rush upon – and 'impetus' is a precursor to the concepts of inertia, momentum and acceleration in classical mechanics.

100 Yes, I know you cannot have a one-dimensional surface and neither can you have one-dimensional spherical objects. This is one of many examples in this book of systems that cannot exist. The best way to describe them is as "thought experiments". This is from the Stanford Encyclopedia of Philosophy - *Thought experiments are devices of the imagination used to investigate the nature of things. They are used for diverse reasons in a variety of areas, including economics, history, mathematics, philosophy, and the sciences, especially physics.*

101 https://physics.stackexchange.com/questions/167804/does-the-conservation-of-linear-momentum-for-a-ball-hitting-a-wall-and-changing.

102 In normal speech negative acceleration is usually called "deceleration".

103 See Section 1.4, Endnote "29" – I will use the more rigorous term "weightless environment".

104 Providing mass does not depend on time and velocity, i.e., ignoring relativistic effects.

105 https://en.wikipedia.org/wiki/Inertia.

106 The derivation of this equation is given in most texts.

107 Excerpt from essay by Einstein on 'Happiest Thought' in His Life" from an unpublished essay written by Albert Einstein in longhand about 1919. See https://www.nytimes.com/1972/03/28/archives/excerpt-from-essay-by-einstein-on-happiest-thought-in-his-life.html.

108 Einstein, Lecture at King's College, London, 1921. Einstein combined the equivalence principle with special relativity to predict that clocks run at different rates in a gravitational potential, and light rays bend in a gravitational field, even before he developed the concept of curved space time.

109 On 25 April 2016, France's Microscope satellite was launched into space to test UFF principle. The following is from https://phys.org/news/2016-04-satellite-universality-freefall.html. "At the core of the satellite there are two pairs of nested cylinders. In one pair, both cylinders are made of the same material, an alloy of the heavy metals platinum and rhodium. In the other pair, the inner cylinder is again platinum–rhodium, while the outer one is an alloy of light metals: titanium, aluminium and vanadium."

"Microscope is orbiting in Earth's gravitational pull. According to the equivalence principle, all four cylinders should feel the same force regardless of the density of the metals, and therefore they should move along the same orbit."

"If, however, the experiment were [sic] to measure tiny differences in the way that the two pairs of cylinders move, this could point to a breakdown of the principle and call for an revision of our current theory of gravity."

110 https://en.wikipedia.org/wiki/John_Philoponus#cite_note-7.

111 Philoponus's critique of Aristotle in the Physics commentary was a major influence on Galileo Galilei, who cited Philoponus substantially in his works. [Willam A. Wallace, Prelude to Galileo: Essays on Medieval and Sixteenth Century Sources of Galileo's Thought (Dordrecht, 1981).]

112 https://phys.org/news/2017-12-galileo-free-falling-space-equivalence-principle.html - see Endnote 108.

113 https://nssdc.gsfc.nasa.gov/planetary/lunar/apollo_15_feather_drop.html. The 0.03 kg value for the weight of the falcon feather is way too high, typical feathers have masses between 0.0003 and 0.003 kg. This may be due to a typo in the report.

114 Gottfried Leibniz (1646–1716) was a German philosopher, physicist, and mathematician. He developed calculus independently of Newton. He became one of the most prolific inventors in the field of mechanical calculators and invented the first mass-produced mechanical calculator. He also refined the binary number system, which is used in virtually all digital computers.

115 Abscissa - the distance of a point from the y-axis on a graph in the Cartesian coordinate system. It is measured parallel to the x-axis. For example, a point having coordinates (2,3) has 2 as its abscissa.

Ordinate - the distance of a point from the x-axis on a graph in the Cartesian coordinate system. It is measured parallel to the y-axis. For example, a point having coordinates (2,3) has 3 as its abscissa.

116 Leonhard Euler (1707–1783) was a Swiss mathematician, physicist, astronomer, geographer, logician, and engineer who founded the studies of graph theory and topology and made pioneering and influential discoveries in many other branches of mathematics such as analytic number theory, complex analysis, and infinitesimal calculus.

117 Remember in this book time is considered a scalar quantity – see Section 3.7.

118 This is an active area of theoretical physics where there is much debate. The following quote is from The Feynmann Lectures under the heading "Relativistic Mass" – 'The mass of the object which is formed when two equal objects collide must be twice the mass of the objects which come together. You might say, "Yes, of course, that is the conservation of mass." But not "Yes, of course", so easily, because these masses have been enhanced over the masses that they would be if they were standing still, yet they still contribute, to the total M, not the mass they have when standing still, but more. Astonishing as that may seem, in order for the conservation of momentum to work when two objects come together, the mass that they form must be greater than the rest masses of the objects, even though the objects are at rest after the collision!' http://www.feynmanlectures.caltech.edu/I_16.html.

119 In other words, a system closed to all transfers of matter and energy, as we shall see a closed mechanical system is equivalent 0to an isolated thermodynamic system.

2 Energy

2.1 INTRODUCTION

Let me start this chapter as I ended Chapter 1 with another question. Objects A, B, and C of identical mass are suspended by string of negligible mass at different heights above the Earth's surface (Figure 2.1). The question is – *what are the differences between the three situations?* The trivial answer is $h_A > h_B > h_C$, but there is a more fundamental difference that I want to explore in this chapter.

If the suspension strings are cut, then objects A and B start to fall, and object C remains stationary (assuming the Earth's surface is horizontal). We know from Newton's equations of motion (Section 1.7.3) that ($v^2 = u^2 + 2as$ and, in this situation $a = g$ (where g is the acceleration due to gravity). Therefore, $v_A = \sqrt{2gh_A}$ and $v_B = \sqrt{2gh_B}$, where v_A and v_B are the respective velocities of objects A and B when they hit the Earth's surface. In other words, when the strings are cut, object A hits the Earth's surface at a greater velocity than object B, the reason for this is that (by definition) $h_A > h_B$. However, that is not the question I asked because the suspension strings are not cut in Figure 2.1. I could argue that this is a "potential" difference between the three situations, i.e., $v_A > v_B > v_C = 0$. This is about as far as I can go with the physics covered in Chapter 1. To proceed further, I need to introduce the concept of mechanical work, and, as we shall see, this leads directly to the concept of energy.

2.2 WORK

In classical mechanics, if the point of application of a force moves in the direction the force is acting, the force is said to do work (remember force is a vector). However, be aware that for a force to qualify as having done work, there must be a displacement of the force in the direction the force is acting, i.e., force is the agent of change and work is a measure of the change. In Figure 2.1, when the suspension string on object A is cut, there is a movement of F_A (the weight of object A) in its direction of its application (vertically downwards), i.e., the force has done work. The work done by a constant force of magnitude F acting on an object that moves a displacement s is the product of the force (**F**) and the displacement (**s**) both of which are vector quantities. Remember that displacement is a vector that is the shortest distance from the initial to the final position of a physical object, i.e., it is a vector quantity that refers to the object's overall change in position. In other words, displacement quantifies both the distance and direction of an imaginary motion along a straight line from the initial position to the final position of the point.[1] It is important to be aware that the product of vector multiplication is a scalar quantity. I do not want to go into details as it soon becomes complicated, and I only give a brief explanation here. In mathematics,

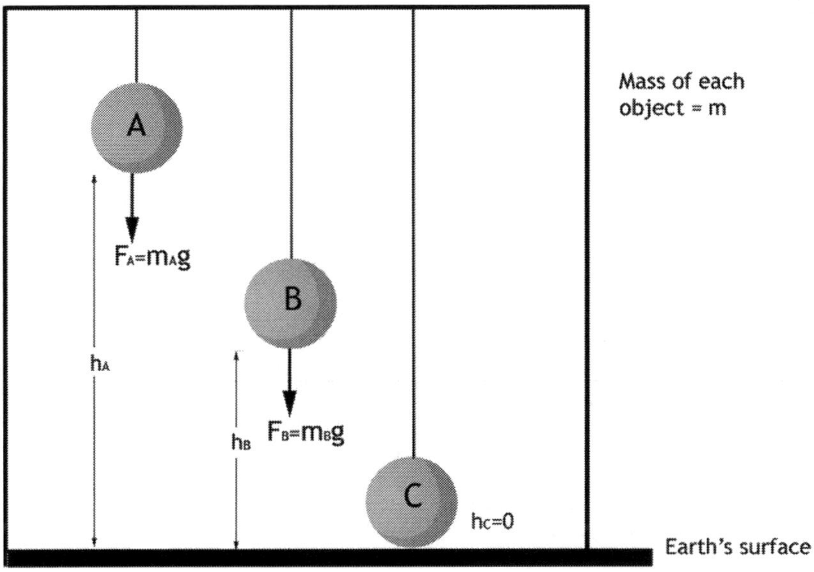

Figure 2.1 Three objects suspended at different heights above the Earth's surface.

DOI: 10.1201/9781003608400-2

the multiplication of vectors is known as the dot product (also known as the scalar product). My maths is not good enough to explain why the result of a dot product is a scalar quantity. The common interpretation is "geometric projection". I came across this explanation which may help[2]:

> The dot product tells you what amount of one vector goes in the direction of another. For instance, if you pulled a box 10 meters at an inclined angle, there is a horizontal component and a vertical component to your force vector. So the dot product in this case would give you the amount of force going in the direction of the displacement, or in the direction that the box moved. This is important because work is defined to be force multiplied by displacement, but the force here is defined to be the force in the direction of the displacement.

In the following discussion, I will only be considering forces acting in one-dimension.[3] In one-dimensional systems, using a + / – sign convention used in Chapter 1, we can calculate work using Equation (2.1):

$$w = |\mathbf{F}| \times |\mathbf{s}| \qquad \text{or} \qquad w = -|\mathbf{F}| \times -|\mathbf{s}| \tag{2.1}$$

Note that if **F** and **s** have opposite signs (+ and –, or – and +), then movement of the force is not in its in its direction of application (see above definition of work). Equation (2.1) shows us that work is a scalar quantity and is represented by a positive real number. The SI unit of work is the joule (abbreviated J). One joule is equal to the work done by a force of one newton when its point of application moves one metre in the direction of action of the force. In terms of SI units:

$$1\text{ joule} = 1\text{ newton} \times 1\text{ metre}, \quad \text{or} \quad 1\text{J} = 1 \times \text{n} \times \text{m} \tag{2.2}$$

The units of the joule are $kg.m^2.s^{-2}$ (Figure 2.2).

The following example should clarify the concept of work. An object is suspended in the Earth's gravitational field and is then released (Figure 2.3).[4] We know the moving force is doing work because the point of application of a force (i.e., the weight of the object) is displaced in the direction the force is acting (i.e., towards the centre of the Earth). After time t, its velocity is v which is not equal to zero. In classical mechanics, a force is said to do work if there is a displacement of

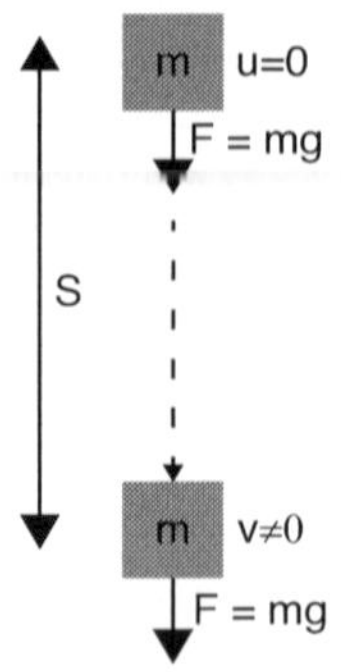

Figure 2.2 Object falling in the Earth's gravitational field.

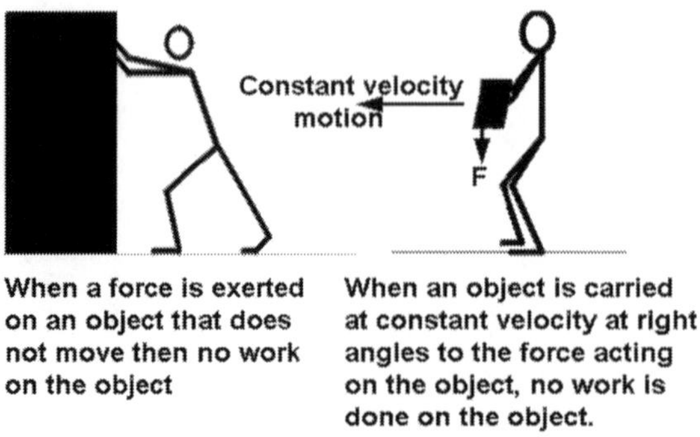

Figure 2.3 Examples of when no work is done.

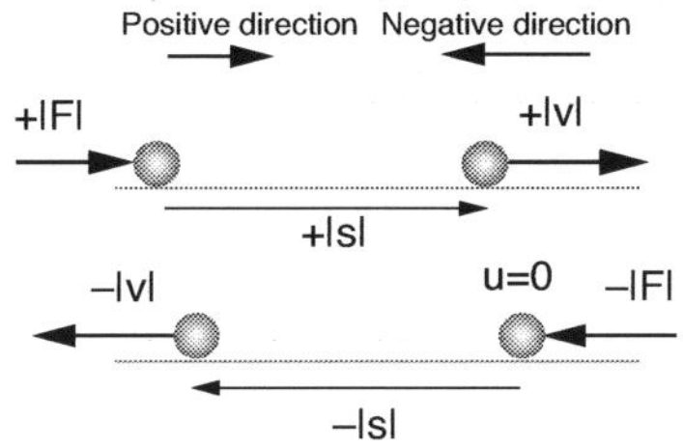

Figure 2.4 A force does work on a stationary particle and particle gains kinetic energy.

the point of application of the force in the direction of the force. This is the case with the example of the falling object where the point of application of the force (i.e., weight of object) has been displaced s (metres) in the vertical down direction.[5]

Returning to my original question in Section 1, we can now say that object A in Figure 2.1 has the potential to do more work than object B, and object C has no potential to do any work (assuming the objects cannot fall lower than the Earth's surface, e.g., by digging a hole).

If we consider the work done on an object that does not move or moves in a direction at right angles to the force of gravity (see Figure 2.4), then this is zero, either because object does not move ($s = 0$) or the object is moving at right angle to the force vector ($\cos(90^\circ) = 0$ – see Section 1.9.3).[6] This contrasts with the use of the term "work" in everyday speech; scientifically, no work is done if a weight undergoes a horizontal displacement. This is important because, in science, work has a very specific meaning, i.e., a force whose point of application does not change (there is no motion) and a force whose point of application moves perpendicularly to its vector does no work.

But how can we say that no work is done when the person pushing the crate which does not move or carrying the object horizontally certainly feels that he or she has done "work"? However, based on the classical mechanics definition of work, if the crate does not move, then the person has done no work on it (since $w = F \times s$ and $s = 0$), and if the object is carried horizontally, then $\cos\theta = \cos(90^\circ) = 0$. So why then does the person feel fatigued if he/she has done no work? When exerting a force, muscle fibres inside the body contract and release and that involves force and motion, so the person does accomplish internal work in his or her body. However, no work has been done on the crate.

2.3 ENERGY

Although the concept of energy is vital to the understanding of modern science, it is a relative newcomer in the development of classical mechanics, e.g., Newton never considered the concept, and it was only with the development of heat engines that drove the Industrial Revolution (see Chapter 4) that the concept of energy was developed. The first to use the word "energy" appears to have been d'Alembert (1717–1783).[7] In the French Encyclopaedia of 1785 d'Alembert wrote: "There is in a body in movement an effort or energie, which is not in a body at rest".[8] However, Wikipedia states that Thomas Young (1773–1829) was the first to define the term "energy" in the modern sense.[9] It is worth noting that 1785 is about 60 years after the death of Isaac Newton and that the accepted Industrial Revolution period is from around 1760 to around 1840. In 1881, William Thomson (later Lord Kelvin) in his presidential address to the British association for the Advancement of Science in 1881 stated[10]:

> The very name energy, though first used in its present sense by Dr Thomas Young about the beginning of this century, has only come into use practically after the doctrine which defines it had ... been raised from mere formula of mathematical dynamics to the position it now holds of a principle pervading all nature and guiding the investigator in the field of science.

During a 1961 lecture for undergraduate students at the California Institute of Technology, the physicist Richard Feynman (1918–1988)[11] and Nobel Laureate said this about energy[12]:

> It is important to realize that in physics today, we have no knowledge of what energy is. We do not have a picture that energy comes in little blobs of a definite amount. It is not that way. However, there are formulas for calculating some numerical quantity, and when we add it all together it gives … always the same number. It is an abstract thing in that it does not tell us the mechanism or the reasons for the various formulas.

To address the question "what is energy", I cannot do any better than the following extract[13]:

> The reason energy is so hard to define is because it's an abstract notion. In physics, the concept of "energy" is really just a kind of shorthand, a tool to help balance the books. Energy is always conserved (or converted into mass) so is incredibly useful in working out the results of any kind of physical or chemical process.
>
> There is no physical "essence" of energy, and no such thing as "pure energy". Energy is always carried by something, usually in the form of movement....
>
> As the Russian physicist Lev Okun said, 'The more basic is a physical notion, the more difficult to define it in words.' For energy, the best we can do is say it's the capacity to cause movement.[14]

In 1898 Mach (1838 – 1916) stated:

> ... "energy" being the term which has gradually come into use for that "indestructible something" of which the measure is mechanical work. [15]

Although energy is one of the most important concepts in science, it is an intangible concept and can only be given meaning through calculation.[16] It is not possible to say what energy is because energy, like time, is a concept so basic that there are no terms available more fundamental to describe it. One important point to make is that energy (like work) is a scalar quantity. If we increase the height of a physical object on the Earth, then we are doing work on the object and we say the object has gained energy. Conversely, if an object falls, it is doing work and we say the object has lost energy. To summarise, work is a process of energy transfer, and energy is an accounting concept.

I now need to introduce one of the most important hypotheses in science. The Law of Conservation of Energy states that the total amount of energy in an isolated system[17] (which does not exchange mass or energy with its surroundings) remains constant over time, i.e., it is said to be conserved over time. This law or principle cannot be proven theoretically – it just is – and after 300 years no one has found a single instance where energy is not conserved. A consequence of the law is that energy can neither be created nor destroyed: it can only be transformed from one form or location, to another form or location. Richard Feynman again[18]:

> There is a fact, or if you wish, a law, governing all natural phenomena that are known to date. There is no known exception to this law - it is exact, so far as we know. The law is called the conservation of energy. It states that there is a certain quantity, which we call energy, that does not change in manifold changes which nature undergoes. That is a most abstract idea, because it is a mathematical principle; it says that there is a numerical quantity, which does not change when something happens. It is not a description of a mechanism, or anything concrete; it is just a strange fact that we can calculate some number, and when we finish watching nature go through her tricks and calculate the number again, it is the same.

Arguably, the most famous/well-known scientific equation is $E = mc^2$, where E, m, and c represent energy, mass, and the velocity of light in a vacuum, respectively. In other words, in modern physics, energy is the substance from which all things in the Universe are made up, and everything that we observe in the world around us represents one of the many manifestations of energy. The one experience we all have of the world is change and to explain and predict change science uses the concept of energy as an accountancy tool because whenever anything happens or changes there is an energy change (form and/or location) but the total quantity of energy in an isolated system remains constant.

Is conservation of energy just an accident? The answer is no, and the reason is due to the following innocent looking statement:

> The laws of physics do not change with time.

From this very simple assumption, the Law of Conservation of Energy can be shown to hold. The first person to fully appreciate this was the mathematician, Emmy Noether (1882–1935)[19]:

> Noether's theorem, which she formulated in 1915, says that symmetries in the universe give rise to mathematical conservation laws. This statement is a crucial underpinning of physical laws, from those that govern the rotation of a wheel or the orbits of planets around stars, to the intricate mathematical frameworks of general relativity, quantum physics and particle physics.

That the invariance of the laws of physics with respect to time lead to something as concrete and useful as conservation of energy is one of the most profound ideas in modern physics. This invariance leads to many other laws of physics as well. For example, conservation of momentum is a consequence of the fact that the laws of physics do not vary from place to place. Noether's theorem is an expression of the fundamental connection between the underlying geometry of the universe and the behaviour of its mass and energy. I will say no more as we soon enter the arcane world of complex mathematics.[20]

It is important to realise that the concept of energy conservation can never be disproved because we can always make up new forms of energy to make it true. However, if we find that we are continually making up new forms of energy for every new system, then we would discard the law of conservation of energy as not being useful. This is from Lehrman in an article entitled "Energy is Not the Ability to do Work"[21]:

> With new discoveries (the neutrino, for instance) it became necessary to invent new terms in order to make the energy equation balance. Up to the present, it has always been possible to find the terms necessary to complete the statement that the total energy at the end of any process is the same as when the process started. We assume that the reason we have always been able to find such terms is that they describe a general property of natural systems. Conservation of energy seems to be a natural law: the description of this law by means of a set of algebraic formulas is a human invention.

New terms invented to make the energy equation balance are the concepts of dark matter and dark energy.[22]

Of course, there is the mass/energy equivalence (remember Einstein's famous equation $E = mc^2$). The point I want to emphasise here is that because "energy" is a word used on a regular basis in today's world, we think we know what it is – we do not – it is a concept that together with the hypothesis (or law) of energy conservation allows us to do simple arithmetic, i.e., in any change a scientist believes that the one thing that remains constant is the energy content of the universe.

In 1686, Gottfried Wilhelm Leibniz proposed that there was a finite quantity of "something" in the universe and that a moving object or a stationary object by nature of its position or deformation possessed an amount of this "something". Christiaan Huygens proposed that this "something" (often called at that time "vis viva" – living force) is conserved in collisions, and Leibniz suggested that the dimensions of this "something" is the product of its mass and the square of its velocity. Conservation of energy became an accepted law, although it only applied to mechanical non-dissipative[23] mechanical systems[24]:

> Mathematical formulation of the energy principle resulted from the needs of the engineers of the early industrial revolution. They needed a way to compare the efficacies of water wheels, draft animals, steam engines and electric motors. It became accepted engineering practice to measure the output of an engine as the product of a weight lifted by the distance it is raised, and this quantity was dubbed work. It was already well known that if a non-dissipative machine[25] could be made, work (so defined) would be equal at the input and output ends. By 1820, the use of algebra in physics had advanced to the point which it was possible to equate the quantity called work with the old idea of vis viva to arrive explicitly at the statement that
>
> $$\int F ds = \frac{1}{2} mv^2$$

The above quote has taken us full circle from the concept of work as energy to the concept of energy as work and leads us neatly to kinetic energy (KE).

2.4 KINETIC ENERGY

In physics, the KE of an object is the energy that it possesses due to its motion. It is defined as the work needed to accelerate a body of a given mass from rest to its stated velocity. If we want to accelerate an object, then we must apply a force. Applying a force requires us to do work. After work has been done, energy has been transferred to the object, and the object will be moving with a new constant speed. The energy transferred is known as kinetic energy, and it depends on the mass and speed achieved.

In Section 1.7.1, I stated: "In the interests of simplicity, I will only consider the mathematics of one-dimensional systems, and, in addition, I will also assume acceleration is constant (such as the acceleration due to gravity at a particular location on Earth)". The restriction to one-dimensional systems (at least when doing the maths) allows a simple derivation of the formula for KE (most texts do not assume constant acceleration and use a calculus-based derivation).

Let us first start with a single-point particle[26] in a frictionless environment. When it is stationary, its KE is zero. To accelerate this stationary article to a velocity v requires the application of a force F for a period of time and over this period of time the displacement of the object is s (Figure 2.4). We know (see Section 2.2) that if a force F acting on an object moves it point of application by s, then the work is done by the force is F × s (providing the vectors **F** and **s** act in the same direction). The relation between the force and the acceleration of the object is given by F = ma (Newton's Second Law), and the object's displacement s is given by $s = \frac{v^2 - u^2}{2a}$.[27] In our example, we know u is zero because the object is initially at rest, therefore:

$$s = \frac{v^2 - 0^2}{2a} \tag{2.3}$$

We know the work done by the force in accelerating the object (w) is by definition F × s, and we have:

$$w = Fs = mas = ma\left(\frac{v^2 - 0^2}{2a}\right) = \frac{mv^2}{2} = KE \tag{2.4}$$

Therefore, the energy of the single-point particle of mass m moving with velocity v is $\frac{1}{2}mv^2$. The work done on a stationary by a force F equals the final KE of the particle. In other words, work transfers energy from one place to another or, as we shall see, from one form to another.

For completeness, let us now consider a single-point particle moving with velocity u and the application of a force for a period of time accelerates the object to a velocity v (Figure 2.5). The relation between the force and the acceleration of the object is given by:

$$s = \frac{v^2 - u^2}{2a} \tag{2.5}$$

We know the work done by the force in accelerating the object (w) is by definition given by F × s, and we have:

$$w = Fs = mas = ma\left(\frac{v^2 - u^2}{2a}\right) = \frac{mv^2}{2} - \frac{mu^2}{2} = \Delta KE \tag{2.6}$$

In this case, the work done by the force on the particle equals the change in the particle's KE $\left(\Delta KE = \frac{mv^2}{2} - \frac{mu^2}{2}\right)$. The KE of the object originally moving with velocity u is $\frac{mu^2}{2}$ and the KE of the object originally moving with velocity v is $\frac{mv^2}{2}$. The relationship set out in Equations (2.4) and (2.6) is contained in Newton's Laws, but it is a restatement of Newton's Laws in terms of energy, which makes many problems much easier to solve.

Now I want to consider friction. In Section 1.9 I stated:

> Friction arises from interatomic and intermolecular electromagnetic forces between the two contacting surfaces. The complexity of these interactions makes the calculation of friction from first principles impractical and necessitates the use of empirical methods for analysis and the development of theory.

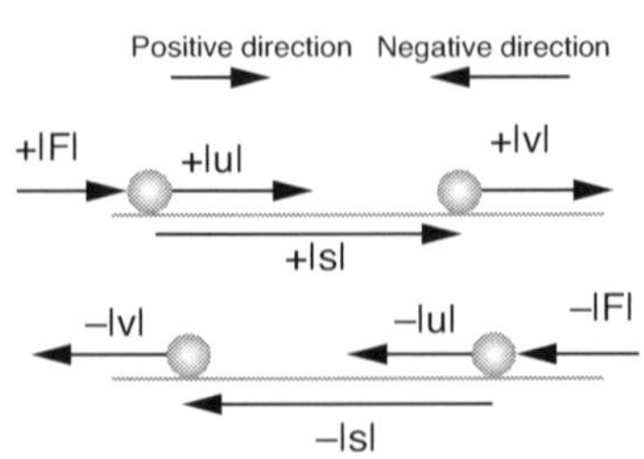

Figure 2.5 A force does work on a particle moving with velocity u and particle gains kinetic energy.

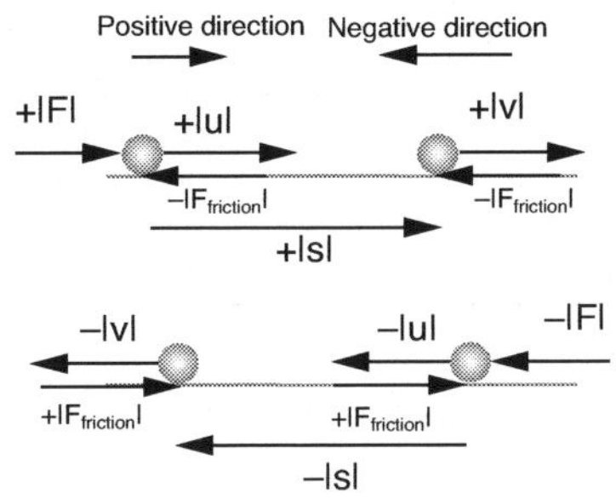

Figure 2.6 Frictional effects added to Figure 2.5.

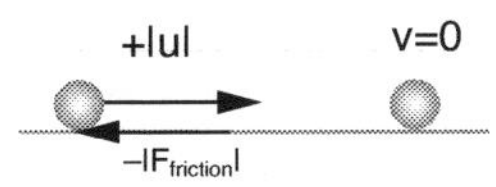

Figure 2.7 Effect of friction on a moving particle.

In the above examples, the force F is the only force acting on the particle, but in the real world there are always frictional affects. The force of friction always opposes the motion of the particle, and the velocity of the particle is reduced until the particle is brought to a halt. Although I have labelled all frictional forces in Figure 2.6 $|\mathbf{F}_{\mathbf{friction}}|$ they will vary with the velocity of the particle. This situation is often stated as energy lost due to friction. We know that energy is conserved, so what has happened to the KE when the velocity moving ball has been reduced to zero by the effect of friction, i.e., its KE has been reduced to zero? This question has been asked in a different manner at the end of Chapter 1 when I asked (but did not answer) what are the differences between a perfectly elastic and a non-elastic collision? It will be asked again (but not answered) in Section 2.8.

In summary:

- If a force acts on a stationary object it accelerates the object in the direction in which the force acts and the displacement of the object is in the same direction as the force, and the object gains KE.
- If a force acts on a moving object and the force and displacement of the object are in the same direction it accelerates the object in the direction in which the force acts, and the object gains KE.
- If a force acts on a moving object and the force and displacement of the object are in opposite directions it accelerates the object in the direction in which the force acts, which is in the opposite direction to the movement of the object and the direction, and the object loses KE.

Analysing systems with frictional effects is complex because $|\mathbf{F}_{\mathbf{friction}}|$ can vary with velocity of the object, and there are differences between static and kinetic friction.[28] However, it can be seen (Figure 2.7) that the presence of frictional forces will eventually bring the moving particle to rest, i.e., reduce its KE to zero.

There is a useful way of understanding what is going on here and that is by considering energy flows. When the force and displacement vectors are acting in the same direction (e.g., $+|\mathbf{F}|$ and $+|\mathbf{s}|$) energy is flowing into the particle and when they act in opposite direction ((e.g., $-|\mathbf{F}_{\mathbf{friction}}|$ and $+|\mathbf{s}|$)s energy is flowing out of the particle. I will return to this use of energy flows later in this book.

Finally, it is worth noting that a single-point particle can only have translational[29] KE (as opposed to say a rod which can also have rotational KE and possibly vibrational KE (Figure 2.8). In this book, I will only be considering single-point particles and, therefore will only be concerned with translational KE.

I now want to briefly return to the Law of Conservation of Momentum which I covered in Section 1.12 where I stated that it applies to systems "closed to the transfer of mass and not acted upon by external forces". This is equivalent of saying that the law only applies to isolated

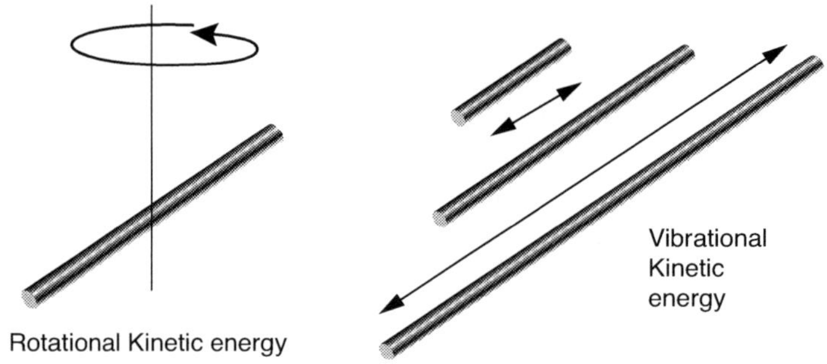

Figure 2.8 Illustrative description of rotational and vibrational energy modes of a rod.

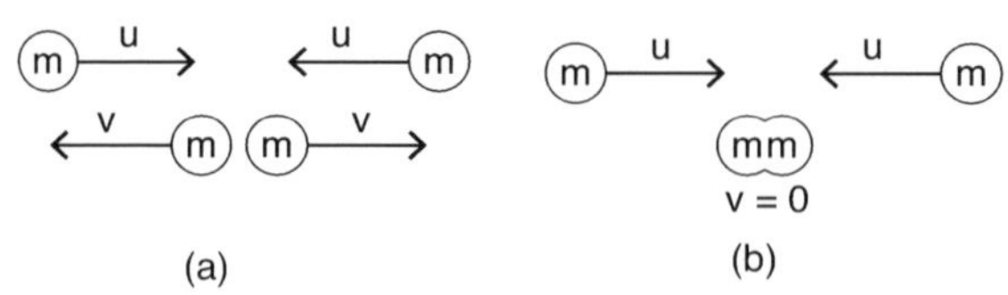

Figure 2.9 Figures 1.48 and 1.49 from Chapter 1.

systems,[30] i.e., ones that do not exchange mass or energy[31] with their surroundings and are not subject to any force whose source is external to the system. I want to consider Figures 1.48 and 1.49 from Section 1.12.3 Example 1.2 (see Figure 2.9).

As I stated, any value of u and v would appear to meet the conservation of momentum requirement, but we would be surprised if the objects collide at $u = 1$ and rebounded at $v = 100$ (Figure 2.9a). Now the concept of energy and energy conservation has been introduced we can now see that requirement that the Principle of Conservation of Momentum only applies to isolated systems, i.e., systems that do not exchange energy with its surroundings, and this precludes $|\mathbf{u}| < |\mathbf{v}|$. Using the above numbers, the initial energy of the two-particle system is $2 \times \frac{1}{2} m \times 1^2 = m$ arbitrary energy units and the final energy is $2 \times \frac{1}{2} m \times 100^2 = 10{,}000 \times m$ arbitrary energy units. If the system is isolated (as it must be for momentum to be conserved) where has the initial energy gone?[32] The same question can be asked of the situation in Figure 2.9b where the initial KE of the system $\left(mu^2\right)$ has decreased to zero. This second time I have asked this question in this chapter (see Section 2.4) which I will consider again in Section 2.8.

2.5 POTENTIAL ENERGY

Before considering potential energy, I need to introduce the concept of conservative forces. I will start by considering a force field which describes a non-contact force acting on a particle at various positions in space; examples are a gravitational field, electric field, or a magnetic field, i.e., at each point in the field, there is a force acting on the particle. Because force is a vector, i.e., it has a magnitude and direction, this is why a force field is usually known as a vector field in physics. A conservative force is one in which work done by a force depends only on the starting and ending points of a motion and not on the path taken (in other words, the work done depends only on the displacement of the force). A conservative force should be contrasted with a non-conservative force such as friction which does depend on the path taken. If an object is pushed along a surface with friction from point A to point B along the two paths shown in Figure 2.10, then the work done along path 2 will be greater than path 1. However, if this was work done in a conservative force field (in a frictionless environment), then the work done along the two paths would only depend on the starting and ending points. This leads to the concept of potential energy, e.g., gravitational potential energy, magnetic potential energy, and electric potential energy.

Most elementary (and not so elementary) courses introduce potential energy using something like the following approach. Consider a point particle of mass m released from rest $(u = 0)$ in a

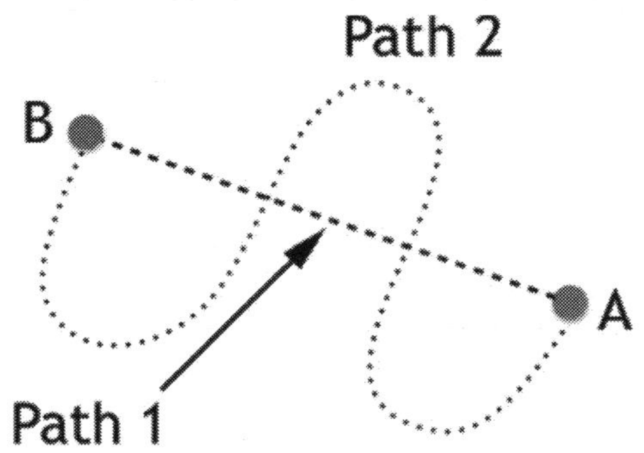

Figure 2.10 Path dependency.

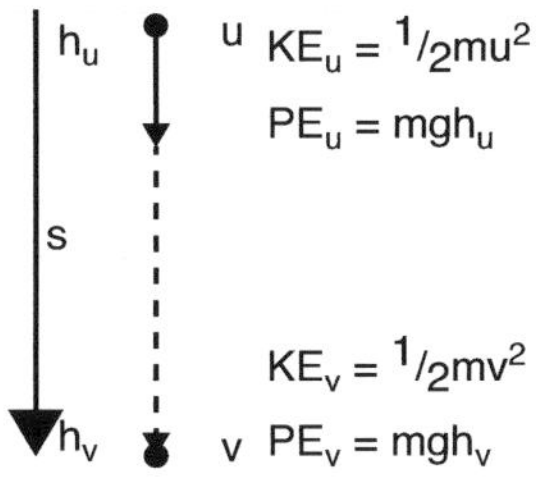

Figure 2.11 Mechanically closed system of a point particle falling in the Earth's gravitational field.

vacuum (no friction) in the Earth's gravitational field (a conservative field) that falls a distance $s = h_v - h_u$ (Figure 2.11). Multiplying both sides of $v^2 = u^2 + 2as$ by m, substituting g for a, $(h_v - h_u)$ for s, dividing through by 2 and rearrange gives:

$$\left(\frac{1}{2}mv^2\right) - \left(\frac{1}{2}mu^2\right) = (mgh_v) - (mgh_u) \tag{2.7}$$

Because v > u, we can write:

$$\left(\frac{1}{2}mv^2\right) - \left(\frac{1}{2}mu^2\right) = +|\mathbf{\Delta KE}| \tag{2.8}$$

We also know $h_u > h_v$:

$$(mgh_v) - (mgh_u) = -|\mathbf{\Delta PE}| \tag{2.9}$$

Therefore, using Equation (2.7), we can write the following absolute value equation:

$$+|\mathbf{\Delta KE}| = -|\mathbf{\Delta PE}| \tag{2.10}$$

Before moving on, it is worthwhile carrying out a numerical example. Let us continue to use the sign convention that the downwards vertical direction is positive. Consider an object of mass 1 kg at height $h_u = 20\,m$ above the Earth's surface moving with a velocity $\mathbf{u} = +10\,m.s^{-2}$ to a position $h_v = 5\,m$ above the Earth's surface, i.e., $\mathbf{s} = +15\,m$. For simplicity, let us assume the acceleration due to gravity is $\mathbf{g} = +10\,m.s^{-2}$. We can calculate the velocity of the object at 5m above the Earth's surface from $v^2 = u^2 + 2as$ as $+\sqrt{400} = 20\,m.s^{-1}$, i.e., $\mathbf{v} = +20\,m.s^{-1}$. The initial KE of the object is $\frac{mu^2}{2} = 50\,J$, and the final KE of the object is $\frac{mv^2}{2} = 200\,J$. The change in KE $\mathbf{\Delta KE} = 200 - 50 = +150\,J$. The initial potential energy of the object is $mgh_u = 200\,J$, and the final potential energy of the object is $mgh_v = 50\,J$. The change in KE $\Delta PE = 50 - 200 = -150\,J$. This agrees with Equation (2.9), which can be rewritten as:

$$\mathbf{\Delta KE} + \mathbf{\Delta PE} = 0 \tag{2.11}$$

The following abstract should prove helpful[33]:

> Forces derivable from a potential are also called conservative forces. The work done by a conservative force is
>
> $$w = -\Delta PE$$
>
> where ΔPE is the change in the potential energy associated with the force. The negative sign provides the convention that work done against a force field increases potential energy, while work done by the force field decreases potential energy. …
>
> Potential energy is the energy by virtue of an object's position relative to other objects. Potential energy is often associated with restoring forces such as a spring or the force of gravity. The action of stretching a spring or lifting a mass is performed by an external force that works against the force field of the potential. This work is stored in the force field, which is said to be stored as potential energy. If the external force is removed the force field acts on the body to perform the work as it moves the body back to the initial position, reducing the stretch of the spring or causing a body to fall.

I will consider the concept of energy stored in a force field in Section 2.7.

Finally, in this section, I want to consider a simple pendulum. There are two reasons for this one is that the action of a simple pendulum in the Earth's gravitational field is a good illustration of the interconversion of gravitational potential energy to KE and vice versa. The other reason is that when we consider a pendulum affected by frictional forces, we come up against the same questions I have asked about the fate of energy in non-elastic collisions and when friction is considered.

Consider the isolated system of a frictionless pendulum in the Earth's gravitational (Figure 2.12), i.e., there are no external forces other than gravitation affecting the movement of the pendulum. When the pendulum bob of mass m is at its highest point, its KE is zero and its potential energy is at its maximum, and when it is at its lowest point, its potential energy is at its minimum. The change in the potential energy of the bob in moving from its highest point to its lowest point is:

$$\Delta PE = PE_2 - PE_1 = mgh_2 - mgh_1 = mg(h_2 - h_1) \tag{2.12}$$

We also know that KE of the bob at its highest point is zero. At its lowest point, its KE is at its maximum. The change in the KE of the bob in moving from its highest point to its lowest point is:

$$\Delta KE = KE_2 - KE_1 = \frac{1}{2}mv_{max}^2 - 0 \tag{2.13}$$

We know that $h_2 < h_1$. Therefore, using absolute values, we know the potential energy of the bob moving from its highest point to its lowest point has decreased, i.e., $-|\mathbf{\Delta PE}| = mg(h_2 - h_1)$. We also know that KE of the bob moving from its highest point to its lowest point has increased, i.e., we know $+|\mathbf{\Delta KE}| = \frac{1}{2}mv_{max}^2 - 0 = \frac{1}{2}mv_{max}^2$. Let us assume the system is frictionless. We know from Equation (2.10) that:

$$mg(h_2 - h_1) = \frac{1}{2}mv_{max}^2 \tag{2.14}$$

Therefore, we can write

$$g(h_1 - h_2) = \frac{1}{2}v_{max}^2 \tag{2.15}$$

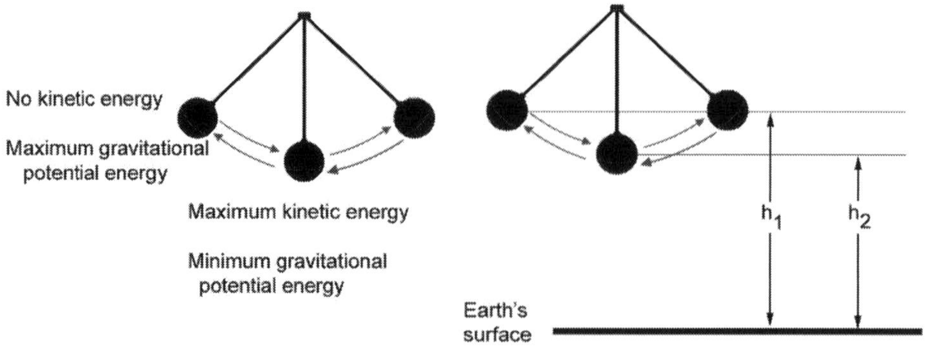

Figure 2.12 Simple pendulum.

Equations (2.14) and (2.15) will apply forever, and the pendulum will keep swinging forever.

In reality, the pendulum experiences frictional forces which oppose the motion of the pendulum. Let $v_{max}^{first\ swing}$ equal the maximum speed reached by bob of the friction affected pendulum on its first swing. We know that $v_{max}^{first\ swing} < v_{max}^{second\ swing} < v_{max}^{third\ swing}$, etc. We also know whether or not the system is or is not affected by friction that:

1. The system is isolated, i.e., its energy content is constant.
2. h_2 is independent of time.

Therefore, for Equation (2.15) to remain valid (as it must because of the Law of the Energy Conservation), then the maximum height pendulum bob achieves after its first oscillation $h_1^{first\ swing} > h_1^{first\ swing} > h_1^{second\ swing} > h_1^{third\ swing}$, etc. Until after n oscillations (where n is less than infinity) $h_1^{n\ swings} = h_2$, i.e., the pendulum eventually comes to less. In other words, the KE of the pendulum bob eventually reduces to zero. By definition, the energy content of the system is fixed so, where has the energy gone? It would appear that the mechanical energy of the pendulum must have been transformed into another form as the pendulum stops swinging. This is the third time I have asked this or a similar question in this chapter; I will return to it in Section 2.8.

2.6 MECHANICAL EQUILIBRIUM

Another way of considering mechanical potential energy is to use the concept of mechanical equilibrium, i.e., an object possesses mechanical energy if it is at a position other than the equilibrium position.

I will first consider static equilibrium. If the size and direction of the forces acting on an object are exactly balanced, then there is no net force acting on the object and the object is said to be in equilibrium, i.e., equilibrium arises if the vector sum of all forces acting upon the object is zero. Any object in a state of equilibrium is either at rest or in motion, and, unless an external force disturbs it, the object will continue in that condition indefinitely (Newton's First Law of Motion). Equilibrium is said to be stable if small, externally induced displacements/disturbances from that state produce forces that tend to oppose the displacement and return the object to the equilibrium state, e.g., a coin lying on a level surface. An equilibrium is unstable if the least departure produces forces that tend to increase the displacement, e.g., a coin standing on its edge.

Consider box resting on a table (Figure 2.13). In both positions A and B, the centre of gravity[34] of the box is lower when compared to the centre of gravity of the box when it is slightly tilted either to the left or to the right (i.e., an unstable equilibrium); it will return to its original equilibrium position. In other words, the gravitational potential energy of the box in positions A or C is at a minimum. However, it is apparent that position A has a lower potential energy than position B, and we say system A is more stable than system B. System B is said to be in a metastable equilibrium. System C is in a state of unstable equilibrium and no additional energy (above that required to tilt the box) is needed to send the box into either equilibrium state A or B. These relations can be presented in a graphical form by plotting the potential energy of the system as a function of the horizontal position of the centre of gravity. Positions of stable or metastable equilibria are seen to be minima in the curve, and the position of highest potential energy (unstable equilibrium) is represented by a maximum (Figure 2.14).

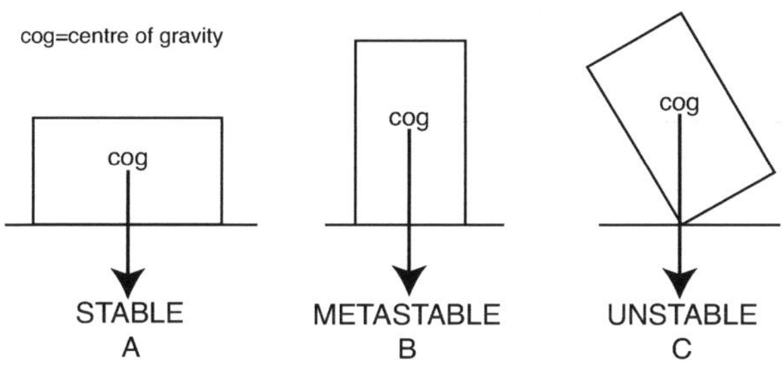

Figure 2.13 Static equilibrium.

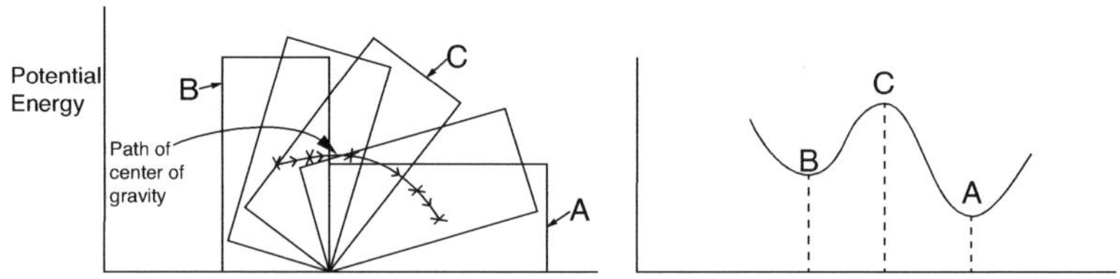

Figure 2.14 An illustration of mechanical static equilibrium.

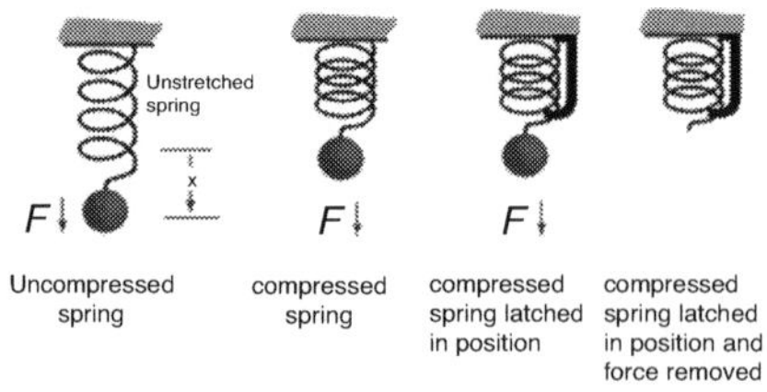

Figure 2.15 Thought experiment that leads to the concept of a conservative field.

I now briefly consider motion equilibrium. The definition is relatively straightforward:

> Motion mechanical equilibrium is when all the forces acting on an object are balanced but the object is moving.

The motion of an object can only be described relative to something else, e.g., other objects, observers, or a set of space-time coordinates. These are called frames of reference. If we chose a frame reference that is not accelerating, i.e., velocity is constant then motion mechanical equilibrium is equivalent to static equilibrium. However, if we choose a frame of reference that is accelerating, then Newton's laws of motion appear to be violated. For example, if your frame of reference is a merry-go-round travelling in a clockwise direction and you throw a ball vertically upwards from the centre, you will see that it tends to curve to the left. To an observer who is not rotating, the merry-go-round turns clockwise and the ball travels in a vertical straight line. In other words, if you refer everything to the frame of the merry-go-round, then you need to invent other fictitious forces which make moving objects turn and "extra forces" must be invoked to retain the accuracy of Newton's laws, e.g., centrifugal forces and Coriolis forces.[35] This all becomes complex and forms the foundation of relativity theories and I plan to go no further.

The concept I want you to take away from this discussion on equilibrium is that there is a relationship between equilibrium and the energy of the system: when a system is at equilibrium, an energy input is required to change the system, and when a system is unstable, the system can change to a lower energy configuration with an energy output.

2.7 POTENTIAL ENERGY REVISITED

Now I want to consider conservative fields and forces as well as potential and kinetic energies in a metaphysical way.[36] To set the scene, I first want to consider an object on a spring and carry out the following thought experiment. If the spring is compressed, then the potential energy of the object has increased. Let us now latch the object and spring in place (see Figure 2.15). If the latch is released, the potential energy appears as KE of the object and, to a lesser extent, of the extending spring. If we had the right design of latch and spring, we could detach the mass from the compressed spring leaving the spring compressed. Now the energy is not associated with the mass at all but is stored in the latched compressed spring.[37] In other words, the potential energy of the object is not an intrinsic property of the object (an intrinsic property is a property that an object or a system has of itself, independently of other things, including its context). This contrasts with its KE which <u>is</u> an intrinsic property of the object.

Let us now consider an object in a gravitational field, we know from the thought experiment that the gravitational potential energy of an object is not an intrinsic property of that object (unlike its KE). In fact, the gravitational potential energy of a given object is stored in the gravitational field (just like a compressed spring). When the object's height in a gravitational field increases, it is usually said that its potential energy increases by an amount ΔPE. However, it is the energy stored in the gravitational field that increases by this amount. Of course (conservation of energy), this increase in the energy stored in the gravitational field is offset by a corresponding decrease in the object's KE. Therefore, when we speak of a body's KE being transformed into potential energy, we are talking about a flow of energy from the object to the surrounding gravitational field. A gravitational field, in fact any conservative force field, stores energy without loss. If the height of the object increases by a certain amount, the object gives up a certain amount of its (kinetic) energy to the surrounding gravitational field, then the field will return this energy to the mass – without loss – if the mass falls by the same distance.

Taking this any further leads us into areas that can only be dealt with by complex mathematics. However, as far as we are concerned, there is one property of a conservative field that is important, and that is the work done in moving an object in a conservative field is path-independent. To clarify, let me restate this in another way – the flow of energy required to move an object from point A to point B in conservative field is exactly the negative of the flow of energy required to move it from point B back to point A, for any given points A and B and completely independent of whatever path you choose.

Let me now rigorously define potential energy:

> The potential energy of an object at position r is equal to the work you must do to move an object from the PE=0 reference point to the position r in a conservative field. The reference point at which you assign the value PE=0 is arbitrary, so may be chosen for convenience.

Now, if you think about it, we can only define potential energy in a system in which the energy of an object is only dependent upon its position and not upon the path by which it reached that position. The corollary is that the concepts of "potential energy" and "conservative field" are interdependent.

2.8 DISCUSSION

Work (a scalar quantity) is the product of force and displacement and is the energy transferred to or from an object via the application of the force along its displacement. A force is said to do work if (when applied) it has a component in the direction of the displacement of the point of application, e.g., a weight falling under the influence of gravity. Work is always a positive quantity because the direction of the force and of the displacement it causes must be the same, i.e., $w = (+|\mathbf{F}|) \times (+|\mathbf{s}|)$ or $w = (-|\mathbf{F}|) \times (-|\mathbf{s}|)$. Let us consider an isolated one-dimensional system where a particle of mass m is moving with a velocity v. If there are no frictional effects (or any other external forces), the particle continues moves with velocity v and the energy content of the isolated system is $\frac{1}{2}mv^2$. Now consider the scenario where the moving particle is affected by friction. In this situation, the velocity of the object is gradually reduced to zero over a period of time, and the energy of the isolated system reduces from $\frac{1}{2}mv^2$ to zero. By definition, the energy content of an isolated system is fixed so where is the original $\frac{1}{2}mv^2$ of energy hiding or moved to? Before I can answer this question, we need to understand what physical objects (stuff or matter) consists of: this is the subject of Chapter 3.

Energy, like time, is a fundamental concept[38] and is undefinable. This is why Richard Feynman said, "we have no knowledge of what energy is". Despite this we know (or think we know):

- *Energy cannot be created or destroyed.*
- *During any change, energy is converted from one form to another or move from one location to another.*
- *The law of conservation of energy states the energy of an isolated system remains constant with time.*

The first form of energy that we encountered was KE. Kinetic energy is the energy associated with the motion of an object with mass m: it is a positive scalar quantity and is equal to $\frac{1}{2}mv^2$.

The second form of energy we encountered is potential energy (PE). Potential energy is the energy associated with an object due to its position in a conservative force field or in a system due to its configuration. It is worth noting that, although there is only one form of KE, there are different forms of mechanical potential energy (gravitational, electric, magnetic, elastic, etc.). However, they are all energy stored by virtue of the position of a physical object in a conservative force field.

Many textbooks make the point that potential energy is "stored" energy, but if you think about it, energy can be stored in motion just as well as it can be stored in position. Implicit in the concept of potential energy is requirement of some "zero" reference point, where gravitational potential is concerned, and this is usually some agreed measure of the Earth's surface.[39]

Energy is always conserved in energy transformations (location and form). This change can be characterised by an initial state that transforms into a final state. There are two points of paramount importance (which will become clearer as you proceed through this book):

1. Only change in energy has meaning.
2. We must account for all the ways energy can change, and if we observe a process, and the changes in energy do not add up to zero, then something is wrong – either the laws for energy transformations are wrong, or there is a new form of energy, or the chosen system is not a closed mechanical system. In other words, energy is always conserved, but sometimes we may need to redefine our system to ensure it is an isolated system.[40]

The concept of energy helps us describe many processes in the world around us:

- A waterfall converts gravitational potential energy into KE, which could be used to drive turbines to generate electrical energy (as well as many other useful activities – there are flour mills paper mills, sawmills, rolling mills, etc.)
- Living organisms transform the stored (i.e., potential) chemical energy of food and/or the energy of sunlight into various forms providing the energy necessary for their existence.
- Car engines convert the stored chemical energy of petrol to drive the pistons of the engine, which can be converted into the movement of the car.
- Stretching or compressing a spring stores elastic potential energy that can be released as KE.

At the end of Chapter 1, I asked the following question: "… what is the difference between an elastic and an inelastic (or non-elastic) collision?" I am now able to answer this question.

> An elastic collision is an encounter between two bodies in which the total kinetic energy of the two bodies after the encounter is equal to their total kinetic energy before the encounter. An inelastic collision is an encounter between two bodies in which the kinetic energy of the two bodies after the encounter is less than their total kinetic energy before the encounter.

Consider Figure 2.15. Experience tells us that the height of a bouncing ball decreases after each bounce. This indicates that in an inelastic collision the KE of the ball decreases after each bounce. Now let us consider an isolated system that consists of a ball and the Earth, and the ball is bouncing on the Earth's surface. When the collisions are elastic, the maximum height the ball reaches (h) is constant. In contrast, h decreases after each bounce when the collisions are inelastic. If h decreases, then kinetic and potential energy decrease, but we know that in an isolated system, the energy is constant. We have already come across a number of similar situations where there is an apparent reduction of energy in isolated systems:

- In Sections 1.12.3 and 1.14 where we considered inelastic collisions;
- In Section 2.4 and the beginning of this section when considering a moving ball slowing down as a result of friction (see Figure 2.6).
- Twice in Section 2.5 (non-elastic collisions – Figure 2.10 – and a pendulum – Figure 2.11).

The following extract is interesting and enlightening, but may make more sense after reading Chapter 3)[41]:

> … if we take two identical lumps of clay and throw them together at high speed, the total system initially has energy according to our provisional definition, but after the collision it has none, because the lumps of clay stick together and the combined lump has zero speed. Therefore, this macroscopic definition of energy does not give a conserved quantity. This definition of energy is essentially equivalent to Leibniz's vis viva, which literally translated means "living force", but using the word "force" to signify what we today would call energy. When Samuel Clarke pointed out that this quantity is not conserved in such collisions, Leibniz replied

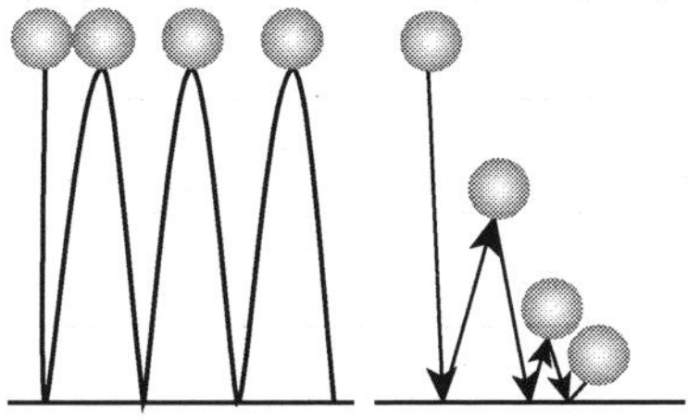

Figure 2.16 Bouncing ball undergoing elastic and inelastic collisions.

> The author [Clarke] objects that two soft or un-elastic bodies meeting together lose some of their energy. I answer no. 'Tis true, their whole lose it with respect to their total motion, but their parts receive it, being shaken by the energy of the collision. And therefore that loss of [energy] is only in appearance. The energy is not destroyed, but scattered among the small parts.
>
> Here we recognize that in order for energy to be conserved we must consider not only the macroscopic kinetic energies of aggregate bodies, but also the microscopic kinetic energies of their constituent particles. The latter is usually regarded as the [internal energy[42]] of the aggregate body. Hence our concept of energy - if energy is to be conserved - must include not only the mechanical kinetic energies of aggregate bodies but also the [internal energies[43]] of those bodies.

In other words, to understand what is going on here, we need to understand what a physical object is or consists of. In Chapter 3, I introduce the simplest model of a physical object – the ideal gas which consists of elementary objects[44] that I call microparticles.

Finally, I want to make an important point. Consider filming a bouncing ball undergoing elastic collisions and one undergoing inelastic collisions (Figure 2.16). Playing the film of the ball undergoing elastic collisions in reverse mode would be indistinguishable from watching it in forward mode. However, you would immediately be able to identify the difference between forward and reverse modes of the film of the bouncing ball undergoing inelastic collisions. This is the first time we have come across an asymmetry in our consideration of science in Chapters 1 and 2. Although this point may be totally underwhelming at this point, it indicates that by the end of this chapter we have reached concepts outside the scope of traditional Newtonian or classical mechanics. This asymmetry is a continuing theme as we proceed through this book.

NOTES

1 As opposed to distance which is a scalar quantity that refers to "how much ground an object has covered" during its motion. See Section 1.2.2.

2 https://math.stackexchange.com/questions/805954/what-does-the-dot-product-of-two-vectors-represent.

3 See Section 1.7.2.

4 Air resistance (a frictional force) is ignored, e.g., the object is suspended and then falls in a vacuum.

5 Also note that the objects in Figures 2.1 and 2.3 cannot move upward without the application of another force acting in the opposite direction to **mg** vector.

6 $\cos 90° = 0$ and $w = F \times \cos\theta \times s$. An example is a moving object on a frictionless horizontal surface.

7 Jean d'Alembert, French mathematician, philosopher, and writer and scientist who was the co-editor (until 1759) of, and contributor to, the *Encyclopédie, ou dictionnaire raisonné des sciences, des arts et des métiers* (Encyclopaedia, or a Systematic Dictionary of the Sciences, Arts, and Crafts). It was a general encyclopaedia published in France between 1751 and 1772, with later supplements, revised editions, and translations. It had many writers, known as the Encyclopédistes.

8 W.J. Moore, *Physical Chemistry*, 5th Edition, Longman, p. 6, 1972.

9 https://en.wikipedia.org/wiki/Thomas_Young_(scientist). Wikipedia reference Fechner, Gustav Theodor (1878). Ueber den Ausgangswerth der kleinsten Absweichungssumme. S. Hirzel, E. Mach in his Popular Lectures states *Although Thomas Young First Called it by the Name of "energy."* - Ernst Mach, Popular Scientific Lectures, Translated by Thomas J. McCormack, The Open Court Publishing Company, p. 173, 1898.

10 https://zapatopi.net/kelvin/papers/on_the_sources_of_energy_in_nature.html.

11 From www.britannica .com – "Richard Feynman, in full Richard Phillips Feynman, (born May 11, 1918, New York, New York, U.S. – died February 15, 1988, Los Angeles, California), American theoretical physicist who was widely regarded as the most brilliant, influential, and iconoclastic figure in his field in the post-World War II era."

12 The Feynman Lectures on Physics: http://www.feynmanlectures.caltech.edu/I_04.html.

13 https://cosmosmagazine.com/physics/what-is-energy.

14 My personal choice would be "energy is the capacity to cause change".

15 E. Mach, *Popular Scientific Lectures*, Translated by T.J. McCormack, The Open Court Publishing Company, p. 139, 1898.

16 An aside – do we know what time is? No, but we know what time **it** is? Like energy, time is a fundamental concept that cannot be defined so paraphrasing Feynman we could say "we have no knowledge of what time is". Chambers Dictionary has the following entries under "time" – "a concept arising from change experienced, expressed by past, present and future, and often measured by the amount of turning of the earth on its axis". I will return to the concept of time in due course.

17 A closed mechanical in classical mechanics is equivalent to an isolated system in thermodynamics. In this book we shall be concentrating on thermodynamics rather than classical mechanics, so I will use the term "isolated system" in this chapter to ensure there is minimum confusion. When I start to consider thermodynamics in Chapter 4, I will return to this subject.

18 The Feynman Lectures on Physics: http://www.feynmanlectures.caltech.edu/I_04.html.

19 https://www.newscientist.com/people/emmy-noether/.

20 What Noether's theorem says, in essence, is that wherever you find some sort of symmetry in nature, some predictability or homogeneity of parts, you will find a corresponding conservation of momentum, electric charge, energy or the like. If a bicycle wheel is radially symmetric, if you can spin it on its axis and it still looks the same in all directions, well, then, that symmetric translation must yield a corresponding conservation – in this case the conservation of angular momentum.

21 R.L. Lehrman, "Energy is not the ability to do work", *Phys. Teach.* 11, 15–18, 1973.

22 "Galaxies in our universe seem to be achieving an impossible feat. They are rotating with such speed that the gravity generated by their observable matter could not possibly hold them together; they should have torn themselves apart long ago. The same is true of galaxies in clusters, which leads scientists to believe that something we cannot see is at work. They think something we have yet to detect directly is giving these galaxies extra mass, generating the extra gravity they need to stay intact. This strange and unknown matter was called "dark matter" since it is not visible".

"Dark energy makes up approximately 68% of the universe and appears to be associated with the vacuum in space. It is distributed evenly throughout the universe, not only in space but also in time – in other words, its effect is not diluted as the universe expands. The even distribution means that dark energy does not have any local gravitational effects, but rather a global effect on the universe as a whole. This leads to a repulsive force, which tends to accelerate the expansion of the universe."

Both extracts from https://home.cern/science/physics/dark-matter.

23 A dissipative process is one in which when energy transfer occurs the capacity of the final form of the system to do mechanical work is less than that of the initial form, as we shall see, all real processes are dissipative. The only dissipative process considered in this book is the process of heat (see Chapter 4).

24 R.L. Lehrman, "Energy is not the ability to do work", *Phys. Teach.* 11, 15–18, 1973.

25 As all real processes are dissipative (see previous, but one, footnote) then "a non-dissipative machine" must be hypothetical.

26 A point object is an object whose dimensions are neglected or ignored compared to its motion. It is a hypothetical physical object used to simplify the calculations.

27 One of Newton's equations of motion.

28 See Section 1.9.3.

29 Translation - the process of something moving from one place to another.

30 In classical mechanics isolated systems are confusingly known as closed systems.

31 If a system is subject to external forces it can exchange energy with its surroundings. A good illustration of why this is the case is to consider a system, consisting of an object, in the Earth's gravity field (external force).

32 The second time I have asked this question (see Section 2.4).

33 https://en.wikipedia.org/wiki/Potential_energy I have changed the symbols in the equations to match the one I have used in this book.

34 Centre of gravity is the point at which all of the weight of an object appears to be concentrated, e.g., if an object is suspended so that it can move freely, its centre of gravity is always directly below the point of suspension. An object can be balanced on a sharp point placed directly beneath its centre of gravity.

35 Centrifugal and Coriolis forces are often referred to as "fictitious" or "apparent" forces. Consider this extract from https://physics.stackexchange.com/questions/109500/does-centrifugal-force-exist - "Newton's laws work in so-called "inertial frames of reference". However, a point on the surface of the Earth is not really an inertial frame of reference because it is spinning around the center of the Earth. (So you can think of it as a rotating

coordinate system.) So Newton's mechanics don't apply if you want to describe motion and use a reference point on the Earth. This is quite inconvenient, because we mostly want to engineer things that work on the Earth.

"Fortunately, there is a trick: you can use a point on the surface of the Earth as your reference and pretend that it's an inertial frame of reference, if you also pretend that some external "imaginary" (fictious) forces exist in addition to the real ones. These are the centrifugal force and the Coriolis force."

36 Metaphysics is a traditional branch of philosophy and it attempts to clarify the fundamental notions by which people understand the world, e.g., existence, objects and their properties, space and time, cause and effect, and possibility.

37 Rigorously I think I should say "associated with" not "stored in".

38 Fundamental concepts in any field are the foundations on which the whole structure of knowledge for that field is built upon.

39 Often it is "mean sea level" (MSL) – a measure of average level of the surface of one or more of the Earth's oceans. In the UK, the Ordnance Datum (the 0 metres height on UK maps) is the MSL measured at Newlyn in Cornwall between 1915 and 1921. In theory, we could have an absolute PE_{Earth} where zero energy is at the centre of the Earth, but (one of many problems) with this is the fact the acceleration due to gravity varies with height.

40 The concept of the term "system" and its associate "closed system" is considered further in Chapter 4.

41 https://mathpages.com/home/kmath184/kmath184.htm.

42 The author wrote "heat content"

43 The author wrote "internal heats"

44 I will consider the term elementary objects in Chapter 3.

3 Ideal Gas

3.1 INTRODUCTION

In the real world, collisions between macroscopic[1] physical objects are never perfectly elastic. The harder an object is, the closer the collisions approximate truly elastic collisions. A steel ball dropped on a steel plate will be very close to an elastic collision. However, none of these objects undergoes perfect (100%) elastic collisions and a ball will bounce several to many times, but each time it bounces, some kinetic energy is lost. When it stops bouncing, its kinetic energy is zero, so what has happened to the original gravitational potential energy of the bouncing ball? The same question can be asked when a falling object hits a soft surface (say a sandy beach) when there is no bounce at all. The kinetic energy of the system is again zero and so is the potential energy (when using the Earth's surface as the zero-reference level). We also know that a pendulum on its own does not swing for ever, frictional forces (mainly air resistance in a well-designed pendulum) result, eventually, in the pendulum coming to rest, but even in a vacuum, a pendulum will come to rest because of other frictional forces. A more fundamental example is when two objects collide head on and the collision is 100% inelastic, momentum is conserved but kinetic energy is not. To understand how the Law of Conservation of Energy is obeyed in these situations, we need to consider what physical objects are made of.

In Chapter 1, I introduced the concept that physical objects consist of matter – although I was honest enough to say that the definition of matter and mass is circular ("a physical object consists of matter" and "matter makes up all physical objects"). In Chapter 2, the concept of energy was introduced, and this allowed us to understand the difference between elastic and inelastic collisions. However, if we assume the Law of Conservation of Energy holds (as we must in modern science), then classical mechanics, as set out in Chapters 1 and 2, is not able to explain the fate of energy when inelastic collisions occur or frictional forces are at play. In order to do this, I first need to consider what a physical object consists of.

In this chapter, I consider the simplest scientific model of matter – the ideal gas. Although the ideal gas does not exist, many gases including air (mainly a mixture of nitrogen and oxygen) around room temperature and atmospheric pressure behave like an ideal gas. The concept of an ideal gas will allow us to investigate thermodynamics using relatively simple mathematics (especially with regard to calculus). I will begin with a consideration of the kinetic theory of matter.

3.2 KINETIC THEORY OF MATTER

Around 450 B.C., the Greek philosopher Democritus considered the following question – can you keep cutting a piece of matter in half *ad infinitum*? Logically, he said that at some point, you must reach the smallest possible bit of matter. He called these smallest bits of matter atoms from the Greek adjective "atomos" or "atomon" meaning indivisible. Over 2000 years later in the 1800s, an English chemist, John Dalton,[2] performed experiments that indicated matter was composed of elementary chemically indivisible[3] particles he called atoms[4], but it was not until the early 20th century that models forming the basis of today's atomic theory were developed. I only briefly consider the structure of atoms in this book (Section 3.6.2) and will only consider the basic tenet of atomic theory which is that all matter is made up of microscopic particles (known in this book as "microparticles"). Particles are considered microscopic if they are roughly of atomic/molecular dimensions (1×10^{-9} m to 1×10^{-11} m).

A macroscopic object or system is usually regarded as one we can see or (in the case of an invisible gas) occupies a volume whose dimensions are visible to the naked eye (considered to be 0.1 mm). As we shall see later, a gas whose volume is 0.1 mm^3, at atmospheric pressure and room temperature, typically contains around 10^{17} microscopic particles, i.e., sufficient particles to apply statistics with high precision. What do I mean by "apply statistics with high precision"? Consider a container divided into two equal volumes by a partition that does not prevent (or hinder) the passage of gas microparticles and assume the gas microparticles are moving at sufficiently high speeds that an equilibrium situation is quickly reached. If we have two gas particles in this container, then, at equilibrium,[5] there is a 50% chance that each half contains one gas particle. If we wish to keep the statistical error below one percentage point, then the system would have to contain about 10,000 microparticles. As the number of gas particles increases to many billions of particles, the situation where both halves contain essentially equal numbers of gas particles is effectively a certainty, i.e., a statistical error of zero[6] (I will return to a consideration of statistics in this context in Chapter 8).

DOI: 10.1201/9781003608400-3

The kinetic theory of matter considers that all matter is composed of microscopic particles – "microparticles" – all in constant random motion and have space[7] between them. This means that irrespective of what phase matter is in, it consists of separate, moving microparticles (this can be translational, and/or rotational and/or vibrational movement). There are three states of matter, or phases; gas, liquid, and solid.[8] Kinetic theory allows an explanation of the properties of the three different states of matter (gas, liquid, and solid) by assuming different arrangements of their constituent microparticles. To explain these different arrangements, we need to postulate a force of attraction between the constituent microparticles. This force of attraction is not gravity (it is not strong enough to explain the observed phenomena), but is a result of another force known as the electromagnetic force – this is the same force that is the source of friction.[9] As we shall see, the ideal gas model assumes no inter-particle interaction.

Before I set out how kinetic theory explains the differences in the structure and properties of gases, liquids, and solids, I need to make the following statement concerning the three phases of matter and energy:

> If we consider a given mass of the same matter, the gas contains more energy than the liquid which contains more energy than the solid – this energy is known as "internal energy".[10]

The concept of internal energy is important and is covered extensively in the chapter. Gases, liquids, and solids of the same type of matter have different internal energy contents, which are reflections of how the microparticles are arranged in the different phases. This determines different physical properties of the three phases (as explained below). Electromagnetic force, unlike gravitational force, can be attractive and repulsive; this repulsive behaviour is the cause of the virtually perfect elastic collisions between microparticles (see later) and, conversely, it is the attractive behaviour that stands as the reason why liquids and solids exist.

I now want to consider how kinetic theory explains the differences between gases, liquids, and solids (see Figure 3.1). A gas, when sufficient energy has been removed from it, condenses into a liquid. This is because, as the energy content of the gas falls, the kinetic energy of the microparticles falls, which is reflected in a decrease in the velocity of the microparticles, and the electromagnetic forces of attraction between the microparticles become more apparent. Because the forces of attraction between the microparticles are significantly greater than those in gases, these attractive forces confine the material to a specific volume, i.e., the gas liquefaction. The microparticles can slide past each other and can move around in any direction. Although the microparticles are not fixed in position, their movement is constrained by the volume of the liquid.

Although the average distance travelled by a moving microparticle between successive collisions is relatively small in liquids (and relatively large in gases), the speeds with which individual microparticles are moving will, at any instant, vary considerably. Therefore, there will be a few microparticles possessing kinetic energy greater than the average, and if such a particle approaches the surface, it may have sufficient energy to escape from the attractive electromagnetic forces in the liquid and travel through the surface into the volume outside the liquid. This effect is termed evaporation.

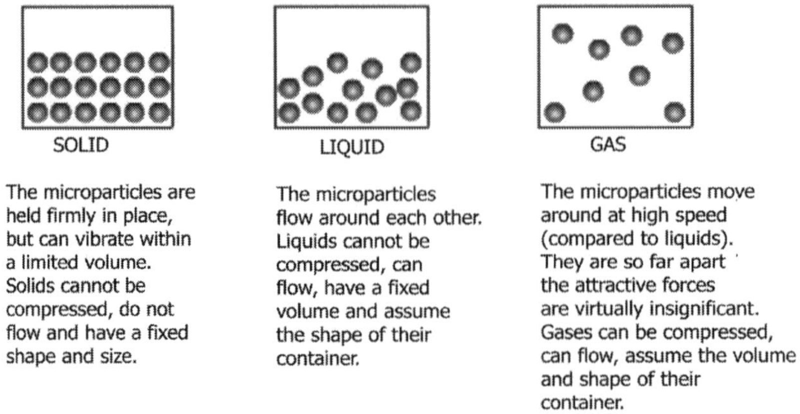

Figure 3.1 Summary of the kinetic theory of the three phases of matter.

A liquid solidifies when sufficient energy has been removed from it. In solids, the microparticles are packed very close together in a regular array and can only vibrate about their average position. These strong forces of electromagnetic attraction confine the material to a specific shape because the microparticles are packed so tightly that the solid is rigid and its volume is fixed. Liquids and gases are fluid because the microparticles are arranged at random, rather than in the regular patterns found in solids where the microparticles are packed closely together and are in fixed positions. Melting and sublimation[11] of a solid are explained in a similar manner to evaporation of a liquid, i.e., microparticles with sufficient energy can escape the forces of attraction between the microparticles in a solid. It is worthwhile noting that because evaporation from a liquid and sublimation from a solid result in the loss of microparticles with relatively high kinetic energies, this results in a reduction in the mean kinetic energy of the liquid microparticles, i.e., a liquid undergoing evaporation and solid undergoing sublimation cools down. The same effect occurs when solid melts (but is less noticeable as the solid and liquid remain in (essentially) the same volume).

Earlier in this section, I stated "[for] a given mass of the same matter, the gas contains more energy than the liquid which contains more energy than the solid". Non-scientists know that for a given material its gaseous form is at higher temperature than its liquid form, which, in turn, is at higher temperature than its solid form. If this is combined with kinetic theory, then we may conclude that temperature is related to the internal energy content of a given material.

Most of us know that temperature is measured by observing its effects on some temperature-dependent variables such as the change in the volume of a liquid (alcohol or mercury are often used) or the change in electrical resistance of a solid (usually a platinum wire). These materials are used over a temperature range when there is a linear relationship between the change in volume or resistance change and temperature. A scale is required to express a temperature numerically, and this is marked off in uniform increments called degrees. The Celsius[12] scale uses 100 equal degree increments between the two reference points defined by the freezing[13] and boiling points of water, i.e., the Celsius scale is a "centigrade scale". The Celsius scale was conceived in 1742 by Anders Celsius. However, Celsius's original scale had water boil at 0°C and freeze at 100°C. Jean-Pierre Christin (1743) and Carolus Linnaeus (1744) independently proposed a temperature scale with 0°C as the freezing point of water and 100°C as the boiling point; with a slight modification (see later), this is the Celsius temperature scale used today. The Fahrenheit scale was proposed in 1724 by Polish physicist Daniel Gabriel Fahrenheit; it has the freezing and boiling points of water at 32°F and 212°F, respectively. Both of these scales are completely arbitrary, and this was of concern to early scientists.

> The definition of temperature was felt to be unsatisfactory because it contained much that was arbitrary: (i) the choice of thermometer substance; (ii) the choice of the quantity X to be measured (X=pressure, volume, relative volume change etc.); (iii) fixing of the function T=f(X).[14]

Concerns about the arbitrary nature of temperature were mitigated to some extent by experiments on gases and the concept of the ideal gas. However, it was as late as 1848 before Lord Kelvin (based on the results of Carnot's thought experiment) finally resolved this concern (see Section 5.9).

The next section considers early experiments on gases and how this led to the gas laws and the concept of an ideal gas.

3.3 EXPERIMENTALLY DETERMINED GAS LAWS

3.3.1 Introduction

The simplest form of matter is gas, and it is not surprising that the earliest scientific investigations concerning matter were on the physical and chemical properties of gases. It is the results of these early studies (between 1650 and 1800 AD) that I want to consider here. Before proceeding, I will first introduce the necessary terms:

- Pressure (P) is the force applied, per unit area, perpendicular to the surface of an object over which that force is distributed. The SI unit for pressure is the pascal (Pa), equal to one newton per square metre ($N.m^{-2}$ or $kg.m^{-1}.s^{-2}$).
- Volume (V) is the quantity of three-dimensional space enclosed by some closed boundary, e.g., the space that matter (solid, liquid, and gas) occupies or (in the case of a gas is contained by). The SI unit for volume is the cubic metre (m^3).
- Temperature (T or θ) is a physical quantity that was mentioned in Section 3.2 for the first time. At this stage, all that I will say is that temperature is a numerical measure of hot and cold as

measured by a thermometer. For reasons that will become clear, I need to differentiate between an arbitrary temperature and a systematic or absolute temperature scale (which is a thermometric scale on which a reading of zero coincides with absolute zero – the lowest temperature that is theoretically possible). To do this, I will use the symbol θ to represent temperature on an arbitrary scale such as the Celsius scale and the symbol T to represent temperature on an absolute scale.

- A permanent gas was believed to be incapable of liquefaction. Most gases in Earth's atmosphere were considered to be permanent gases, e.g., helium, argon, oxygen, nitrogen, and hydrogen. The term is now considered archaic as all gases can be liquefied. All the gases used in the experiments considered in the next two sub-sections were considered, at the time, to be permanent gases.

3.3.2 Boyle's Law

Robert Boyle was one of the first scientists to investigate the physical properties of gases. He examined the effect of pressure on a volume of air trapped in a bent tube filled with mercury (Figure 3.2). The longer end of the tube was open to the atmosphere. Mercury was added to or removed from the open section of the tube to vary the pressure on the trapped volume of air. The volume of the air was calculated from the mercury level read on the graduated scale. The pressure of the air is equal to the atmospheric pressure $\left(P_{atm}\right)$ plus the weight of the column of mercury of height h divided by the area of the column (radius=r). The weight of mercury $\left(mg\right)$ is equal to $\pi r^2 h\rho$ (where πr^2 is the area of the cross section of the column of radius r, h is the height of the mercury column and ρ is the density of mercury[15]). Therefore, the pressure of the trapped air is $P_{atm} + \dfrac{\pi r^2 h\rho g}{\pi r^2} = P_{atm} + h\rho g$ where g is the acceleration due to gravity[16].

Boyle carried out experiments on air at room temperature and he observed[17]

> … tis evident, that as common Air when reduc'd to half Its wonted extent, obtained near about twice as forcible a Spring as it had before; so this thus-comprest Air being further thrust into half this narrow room, obtained thereby a Spring about as strong again as that It last had, and consequently four times as strong as that of the common Air.[18]

To put it simply, Boyle noted that the product of the pressure of the trapped gas and its volume was constant, e.g., the volume of the gas was halved when the pressure was doubled. Boyle published these results in 1662 in a book titled "New Experiments Physico-Mechanical Touching the Spring of Air".

In modern terms, Boyle's Law can be stated as:

The pressure exerted by a given mass of an ideal gas in an isolated system is inversely proportional to the volume it occupies if the temperature remains unchanged, i.e., $P = \dfrac{K}{V}$ (where K is a constant).

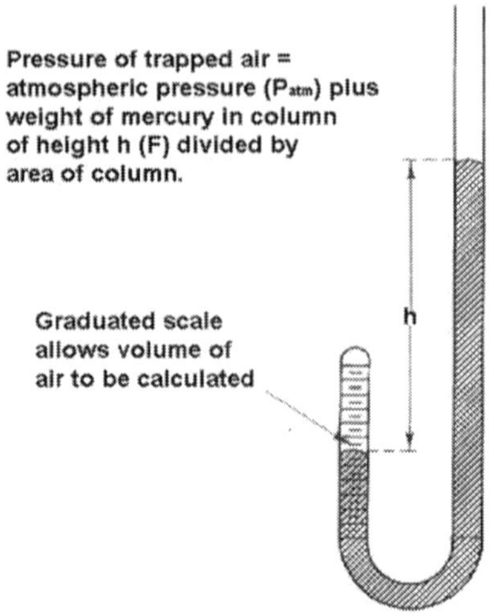

Figure 3.2 A diagram of apparatus used by Boyle.

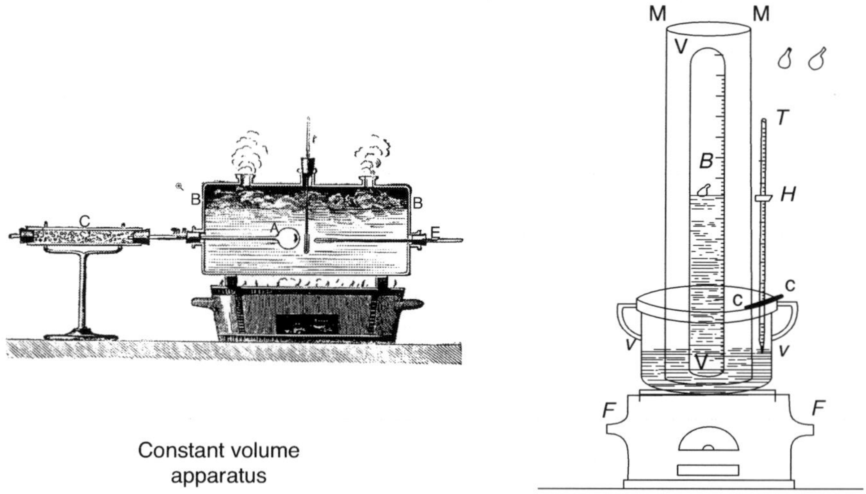

Constant volume apparatus

Constant pressure apparatus

Figure 3.3 Typical apparatus used by Gay-Lussac.

3.3.3 Gay-Lussac's Laws

Joseph Louis Gay-Lussac (in papers published between 1802–1808) observed that:

1. Volumes of given gas samples all showed the same dependence on temperature at constant pressure, i.e., $V = C_1 + C_2 \times \theta$ where C_1 and C_2 are constants and θ is temperature using an arbitrary temperature scale. This is known as the volume-temperature law.
2. Pressures of these same gases showed a similar dependence on temperature at constant volume, i.e., $V = D_1 + D_2 \times \theta$ where D_1 and D_2 are constants and θ is temperature using an arbitrary temperature scale. This is known as the pressure-temperature law.

The constants C and D are dependent on the mass of the sample, the type of gas, and the temperature scale. Figure 3.3 gives some idea of the equipment Gay-Lussac worked with.

When publishing his results on the volume-temperature law, Gay-Lussac conscientiously acknowledged some haphazard experiments of Jacques Charles in 1787, and apart from in France, this law has been mistakenly attributed to Charles ever after. Likewise, the pressure-temperature law is often known as Amonton's Law as he investigated the relationship between pressure and temperature of gases in the late 17th century. However, because he lacked precise thermometers, his results were more qualitative than quantitative. In this book, I will refer to both of these relationships as Gay-Lussac Laws as he was the first to publish convincing quantitative evidence of the relationship between temperature, volume, and pressure.

Using the arbitrary Celsius scale, I will first consider the volume-temperature law. If V_0 is the volume of a given sample of gas at 0°C, we have from Gay-Lussac's work that $C_1 = V_0$:

$$V = V_0 + C_2\theta = V_0\left(1 + \frac{C_2}{V_0}\theta\right) = V_0(1 + \alpha\theta), \tag{3.1}$$

where θ is the temperature in Celsius and α is equal to ss. Equation (3.1) can be rewritten as:

$$\frac{V}{V_0} = (1 + \alpha\theta) \tag{3.2}$$

The left-hand side has no units (is "dimensionless"), therefore the right-hand side has no units. We know that θ has the units of temperature, and therefore α must have the units of reciprocal temperature. Gay-Lussac found α to be constant for permanent gases and to be approximately equal to $\frac{1}{267}$, when the Celsius scale is used; in 1847 Regnault, with improved procedures, obtained $\alpha = \frac{1}{273}$. Equation (3.1) may be written:

$$V = V_0\left(1+\frac{\theta}{273}\right) = V_0 + V_0\frac{\theta}{273} \quad (3.3)$$

This relation is called the Law of Gay-Lussac and it states that a gas expands by $\frac{1}{273}$ of its volume at 0°C for each degree rise in temperature when pressure is constant. The graph of Equation (3.3) is shown on Figure 3.4.

It is now possible to define an "absolute" temperature scale[19] denoted by $T = \theta + 273$ (see Figure 3.5). I will call the unit of temperature on this scale the kelvin (K),[20] (note it is not 0K), and is read as "200 kelvin" or "200 K"). We can immediately see that volume is directly proportional to the absolute temperature, i.e., $V \propto T$. As the slope of the graph in Figure is $\frac{1}{273}$ we can write:

$$V = \frac{T}{273} \text{ or } \frac{V}{T} = \text{constant} \quad (3.4)$$

I will call this the "ideal gas temperature" because as we will see Gay-Lussac's Laws (and Boyle's Laws) are only valid for an ideal gas (which is a remarkably accurate model for "permanent" gases at normal temperatures and pressures, e.g., at atmospheric pressure and ambient temperature – see Section 3.6). This approach is not scientifically or historically accurate because Kelvin developed his absolute temperature scale on Carnot's work (see Section 5.9), but it will do for now.[21]

The Gay-Lussac's pressure-volume relationship can be analysed in a s similar manner:

$$P = P_0 + D_2\theta = P_0\left(1+\frac{D_2}{P_0}\theta\right) = P(1+\beta\theta) \quad (3.5)$$

where is β a constant equal to $\frac{D_2}{P_0}$. Equation (3.5) may be written:

$$P = P_0\left(1+\frac{\theta}{273}\right) \quad (3.6)$$

If we plot P vs θ (Figure 3.6) the point when $P = 0$ is at $\theta = -273$ and the definition of the absolute temperature scale follows as before. Therefore, we can write:

$$P = \frac{T}{273} \text{ or } \frac{P}{T} = \text{constant} \quad (3.7)$$

In other words, the pressure of a given gas sample is directly proportional to its absolute temperature when volume is constant, i.e., $P \propto T$ where T is absolute temperature. When T equal 0 K or −273°C the pressure of the ideal gas is zero.

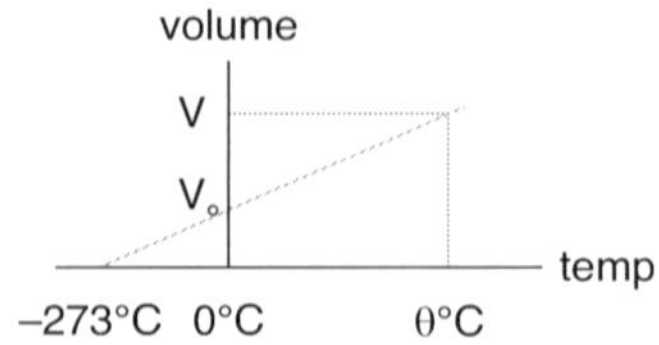

Figure 3.4 Plot V vs θ.

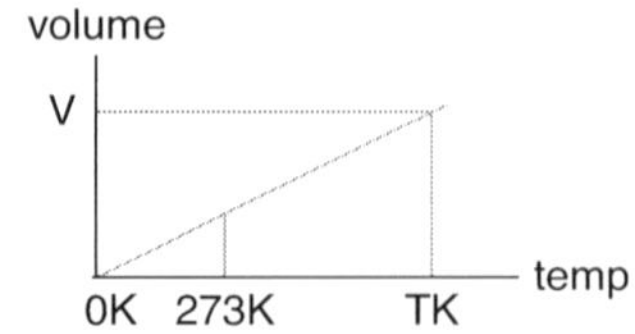

Figure 3.5 Plot V vs T.

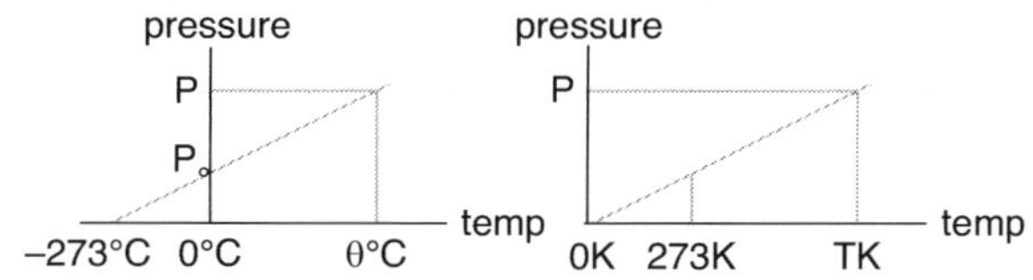

Figure 3.6 Plot P vs θ and P vs T.

This absolute temperature scale is based solely on the behaviour of an ideal gas. Unlike the arbitrary temperature scales only one fixed point is needed. The fixed point used is not the freezing point of water (0°C), but the "triple point" of water, which is the unique pressure and temperature situation where solid, liquid, and vapour all coexist.[22] In other words, the triple point of water where solid, liquid, and vapour are in equilibrium. Water's triple point is specifically chosen because it is relatively convenient to attain in the laboratory.

For reasons too boring to go into in any detail, the temperature of the triple point of water was defined to be exactly 273.15 K in 1954 and 273.16 K in 1990. The historical reason for the choice of 273.16K for the triple point of water is that it gave, to the accuracy of the best measurements then available, 100K units for the difference between the ice point (the freezing temperature at standard pressure[23]) and the steam point (the boiling temperature at the standard atmospheric pressure of water). However, more accurate measurements now give the difference as 99.97K. While the triple point of water completely determines the absolute temperature scale, temperature measurements require approximating the absolute temperature by a "practical" scale that contains other fixed points at which instruments can be calibrated.[24]

The following quote is of interest:[25]

> Temperatures are assigned to these points based on the best scientific estimate of their true [absolute] temperatures, and procedures are specified for interpolating between the fixed points. While previous temperature scales used the atmospheric boiling point of water as a fixed point (assigning it 373.15 K, which is 100°C), the reproducibility of that point is not as good as other choices. The new International Temperature Scale adopted in 1990 (known as ITS-90) covers this region with the solid/liquid equilibrium (melting/freezing) points of gallium (302.9146 K) and indium (429.7485 K). On ITS-90, the atmospheric boiling temperature of water turns out to be approximately 373.124 K (99.974°C).

One final point – why is it "kelvin" and not "degrees kelvin"? The reason is that "degrees" are found in units which are arbitrary such as degrees Fahrenheit or Celsius. The same goes for angle degrees, since that also divides the circle arbitrarily into 360°. When the units are absolute, or measured directly with respect to something else, the "degrees" are dropped from the units. However, it should be noted that degrees are erroneously used with kelvin temperatures in some sources.

3.3.4 Summary

To summarise:

1. Boyle's Law – the product of the pressure (P) and volume (V) of a given mass of a gas is constant at a fixed temperature: $PV = K_1$ (temperature is constant).
2. Gay-Lussac's Volume Law – the volume of a given mass of gas at constant pressure is directly proportional to its temperature: $V = K_2T$ (pressure is constant).
3. Gay-Lussac's Pressure Law – the pressure of a given mass of gas at constant volume is directly proportional to its temperature: $P = K_3T$ (volume is constant).

Combining Boyle's, Charles's, and Gay-Lussac's Laws gives – $PV = K_4T$ which, for the moment, I will call the "gas law".

A relatively simple model known as the "ideal gas" can model all three of the above experimentally determined gas laws. The development and consideration of this model form the basis for the rest of this chapter.

3.4 THE IDEAL GAS

Gas is the simplest form of matter and the simplest model of a gas is known as an "ideal gas". An ideal gas meets the following conditions.

- It consists of microparticles (i.e., microscopically small particles). Their smallness is such that the total volume of the individual microparticles when added up is negligible compared to the volume of the container that holds the gas. This is equivalent to stating that the average distance separating the gas particles is very large compared to their size.
- All the microparticles have the same mass and are perfectly spherical in shape.
- The number of microparticles is so large that a statistical treatment can be applied to high precision. I have mentioned this in Section 3.2, and I ask you at this stage to accept that in a macroscopic volume of an ideal gas, there are at least 10^{17} microparticles. I will return to this when considering Avogadro's number in Section 3.6.3.
- The microparticles are in constant, random, and rapid motion in straight lines.
- The rapidly moving microparticles constantly collide among themselves and with the walls of their container.
- All these collisions are perfectly elastic. Remember in an elastic collision, the total kinetic energy of the two bodies after the encounter is equal to their total kinetic energy before the encounter. This means that the total kinetic energy of an ideal gas system is constant at constant temperature, and there is no net conversion of kinetic energy into potential energy.
- The only time a force is applied to an ideal gas microparticle is when it collides with another ideal gas microparticle or with the walls of the container holding the ideal gas.
- Newton's laws of motion apply to the movement of the microparticles.
- The average kinetic energy of a microparticle is very much larger than its gravitational potential energy, so that the effects of gravity can be considered negligible.

Using these relatively simple assumptions, kinetic theory can be used to theoretically derive the gas law for the ideal gas.

3.5 THEORETICAL DERIVATION OF GAS LAW FOR AN IDEAL GAS

3.5.1 Introduction

Many textbooks do not give the theoretical derivation of the gas law equation, and instead rely on the derivation from the experimentally derived gas laws (as we have already done). However, I consider that an appreciation and understanding of the theoretical derivation is important if you expect to understand the implications of the derived gas law. It is not mathematically demanding (apart from the determination of the distribution of the velocities of the microparticles – which I will skip over), you will need to know what vectors are (covered in Chapter 1) and, most importantly, you do need to think.

Daniel Bernoulli, in 1738, was the first to understand air pressure from a kinetic theory point of view. He considered a cylinder (Figure 3.7), closed at the bottom, with a piston at the top, the piston having a weight on it, both piston and weight being supported by the gas pressure inside the cylinder. He described what went on inside the cylinder as follows:

> let the cavity contain very minute corpuscles, which are driven hither and thither with a very rapid motion; so that these corpuscles, when they strike against the piston and sustain it by their repeated impacts, form an elastic fluid which will expand of itself if the weight is removed or diminished….[26]

In other words, the force exerted by a gas is due to collisions between the microparticles of the gas and the walls of the container. We know pressure is the ratio of force to the area over which that force is distributed.

His hypothesis, although today considered essentially correct, was not widely accepted at the time (nothing new there then). Most scientists believed that the microparticles in a gas stayed more or less in place, repelling each other from a distance, held somehow in the aether.[27] In the 1820s an Englishman, John Herapath, derived the relationship between pressure and speed of microparticles (similar to that given later in this section). The paper was rejected by Humphry Davy

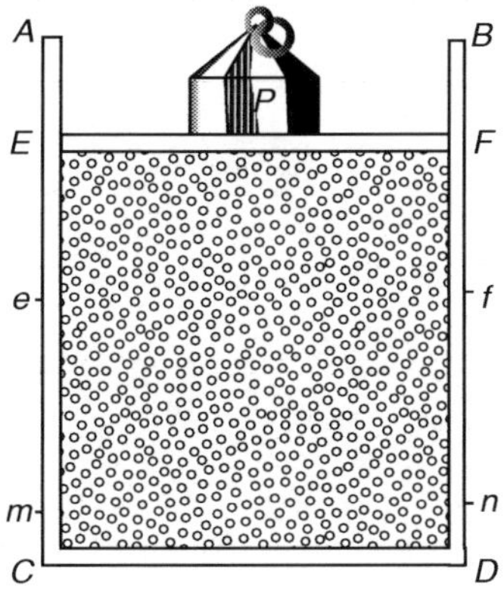

Figure 3.7 Bernoulli's thought experiment concerning microparticles and pressure of a gas.

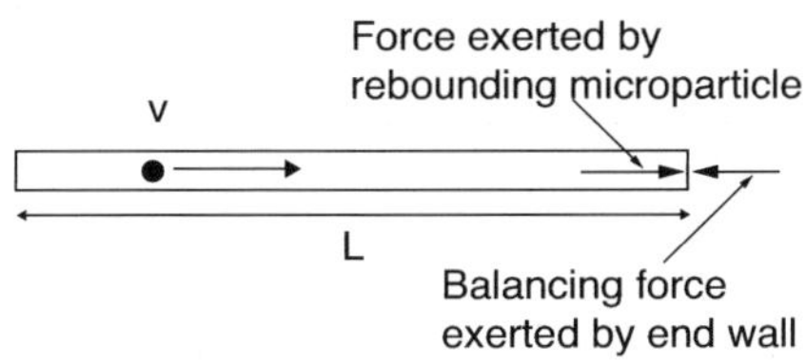

Figure 3.8 One-dimensional gas: a microparticle bounces between the ends of a one-dimensional cylinder.

(president of the Royal Society), who pointed out that equating temperature with motion, implied an absolute zero of temperature, a concept that Davy refused to accept.

3.5.2 One-Dimensional System

The object here is to give you an insight into the approach taken later in this section to (i) theoretically derive Boyle's Law and (ii) to introduce the concept of temperature as a measure of the speed of the microparticles. Let us begin with Bernoulli's premise that the pressure exerted by a gas on the wall of a container is due to the collisions of its constituent microparticles with the wall, let us also assume all collisions are perfectly elastic.

Consider a microparticle, of mass m, bouncing rapidly back and forth at speed v inside a very narrow closed cylinder of length L: the cylinder is so narrow all motion is considered one-dimensional (Figure 3.8). Because all collisions are perfectly elastic the microparticle's kinetic energy must be the same before and after the collision. In this case, the end wall of the cylinder does not feel a smooth continuous force, but a series of equally spaced impacts. However, if the end wall has a much greater mass than the microparticle and the microparticle is moving rapidly, this will have the same effect as a "smooth" force during time scales that are very much longer than the time interval between impacts. I want to calculate the value of this smooth force.

The microparticle's momentum changes by $2\,mv$ each time it hits the end wall (see Section 1.12.3, Equation1.79). The time between hits is $t = \frac{2L}{v}$, so the frequency of hits is $\frac{1}{t} = \frac{v}{2L}$ per second. By the principle of the conservation of momentum, the rebounding microparticle would cause the momentum of the end wall to change by $\left(2\,mv \times \frac{v}{2L}\right)$, i.e. change of momentum every time the microparticle hits (and rebounds from) the piston multiplied by the frequency of the hits.[28] This is the rate of change of momentum $\left(\frac{dp}{dt}\right)$ and we know $\frac{dp}{dt} = F$.[29] We also know the end wall does not move, therefore the balancing force on the end wall is equal to the rate of change of momentum of the microparticle undergoing an elastic collision with it. Therefore, the end wall experiences a force given by Equation (3.8)

$$F = \frac{mv^2}{L} \tag{3.8}$$

Pressure is defined as force per unit area. We know (Newton's Third Law[30]) that the two forces on the end wall are balanced, i.e., acting in opposite directions (see Figure 3.8).

Now, I know that for a one-dimensional system, we cannot have an area but bear with me. If we assume the area of the end wall is A then the pressure on the wall is $P = \frac{F}{A} = \frac{mv^2}{LA}$ and LA is the volume of the vessel (V). Therefore, we know:

$$P = \frac{mv^2}{LA} = \frac{mv^2}{V} \tag{3.9}$$

In other words, $PV = mv^2$. This is essentially Boyle's Law (PV=constant at fixed T) provided temperature is equated with the speed of the microparticles which constitute the gas. Now consider a constant volume and if the velocity of the microparticle is increased then P is increased; or consider a constant pressure and increase the velocity of the particle then V increases. These results, which flow directly from Equation (3.9), have the flavour of Gay-Lussac's Laws if an increase in the velocity of microparticles is equated to an increase in the temperature of the gas.

I must emphasise that extending the results of a one-dimensional analysis to three dimensions is not mathematically or scientifically rigorous. The reason I have considered a one-dimensional system here is that it introduces the concept that the temperature of a gas and the speed of its constituent microparticles are related. Introducing this concept now allows a simpler derivation of the Ideal Gas Law compared to those usually found in text books. But before we can do this we need to consider the distribution of the speeds of microparticles in an ideal gas.

3.5.3 Distribution of Microparticle Speeds in an Ideal Gas

In their constant motion, the microparticles of a gas are constantly in collision with each other and the walls of their container, i.e., the speed[31] of a microparticle is constantly changing. Because of this, it is not valid to ask what the speed of an individual microparticle is. The endless collisions of the microparticles result in a distribution of microparticle speeds and, because of the large number of microparticles in a macroscopic volume of gas, this distribution of microparticle speeds remains constant (when the energy content of the ideal gas system is constant) even though the speeds of the individual microparticles are constantly changing.

The only way we can move forward is to use an average microparticle speed ($< v >$).[32] Most microparticles will have speeds with magnitudes close to the average ($< v >$) and relatively few have magnitudes far above or below the average. Our macroscopic system of an ideal gas contains (by definition) at least 10^{17} microparticles of mass m and allows statistical calculations to be applied to a very high degree of precision. This lets us calculate the energy content of our system $\left(E_{system}\right)$ using Equation (3.10) (where N is the number of microparticles, and m is their mass):

$$E_{system} = \frac{1}{2}\left[Nm(< v >)^2\right] \tag{3.10}$$

In other words, the total energy of the system is the sum of all the particles' kinetic energies. If the mathematical equation of the speed distribution at a fixed energy $\left(E_{system}\right)$ is known, then ($< v >$) can be calculated. The distribution of speeds among microparticles in an ideal gas was partly guessed and partly calculated by James Clerk Maxwell (1831–1879) in 1860. It was later derived, exactly, by Ludwig Boltzmann (1844–1906) in 1872 and is now known as the Maxwell-Boltzmann distribution. The Maxwell-Boltzmann distribution is often represented by the graph shown in Figure 3.9.[33] For an ideal gas whose energy is constant (in other words at constant temperature) the distribution curve remains unchanged with time. Notice that the graph in Figure 3.9 is not symmetrical. There is a longer 'tail' on the higher speed right end of the graph, but to the left, the graph must end at zero (since a microparticle cannot have a speed less than zero). The area under the graph in Figure 3.9 equals the total number of microparticles in the gas (N).

You might think that the speed located directly under the peak of the Maxwell-Boltzmann graph is the average speed of a microparticle in the gas ($< v >$), but that is not the case. The speed located directly under the peak is the most probable speed $\left(v_{prob}\right)$. The average speed ($< v >$) is located a bit to the right of v_{prob}, and this is due to the longer tail on the right side of the curve. There is one other average speed of interest the root-mean square speed, v_{rms}, which is calculated using Equation (3.11):

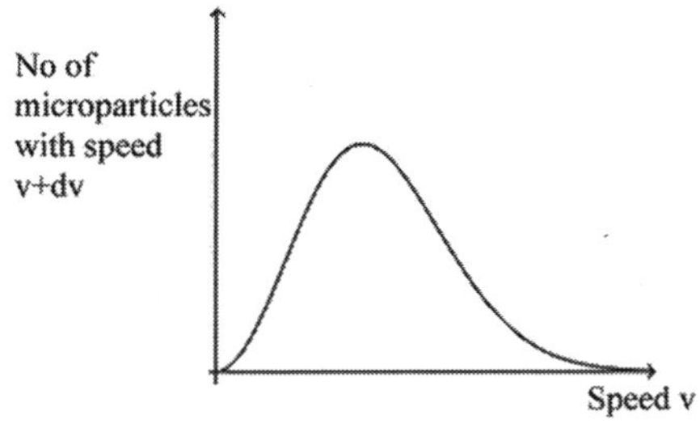

Figure 3.9 Maxwell-Boltzmann distribution.

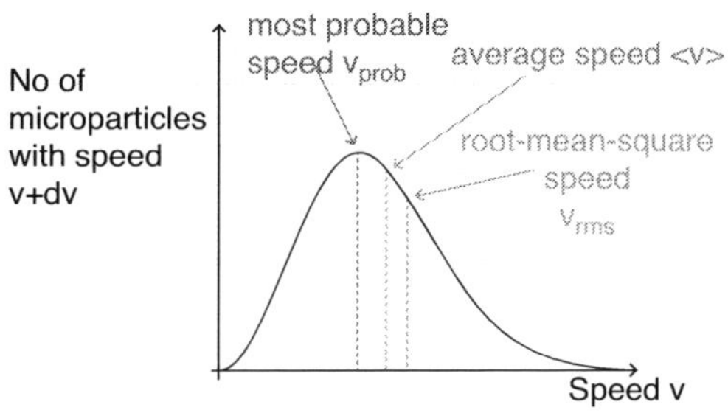

Figure 3.10 The three "average" speeds.

$$v_{rms} = \sqrt{\frac{1}{N} \times \left(v_1^2 + v_2^2 + v_3^2 + \cdots\right)} \tag{3.11}$$

where v_1 is the speed of microparticle 1, v_2 is the speed of microparticle 2, etc., the speed of microparticle n is v_n. This may seem unnecessarily complicated since we squared all the velocities, only to later take a square root. However, root-mean-square speeds are useful to know, because the kinetic energy of a microparticle of mass m moving with a velocity v is $\frac{1}{2}mv^2$, and to find the average kinetic energy, KE_{avg}, of the microparticles we must take the average of the squares of the velocities:

$$KE_{avg} = \frac{1}{N} \times (KE_1 + KE_2 + KE_3 + \cdots) \tag{3.12}$$

where KE_n is the kinetic energy of microparticle n.

$$KE_{avg} = \frac{1}{N} \times \left(\frac{1}{2}mv_1^2 + \frac{1}{2}mv_2^2 + \frac{1}{2}mv_3^2 + \cdots\right) \tag{3.13}$$

$$KE_{avg} = \frac{1}{2}m \times \left[\frac{1}{N}\left(v_1^2 + v_2^2 + v_3^2 + \cdots\right)\right] \tag{3.14}$$

$$KE_{avg} = \frac{1}{2}mv_{rms}^2 \tag{3.15}$$

In other words, the average kinetic energy of a microparticle is directly proportional to the square of the root-mean-square (rms) speed. Figure 3.10 shows these three "average" speeds.

The derivation of the Maxwell-Boltzmann speed distribution function is mathematically complex, and to add to the confusion, there appear to be many versions of the function in the literature. Equation (3.16) is a version of that given on a number of web sites:[34]

$$f(v) = \left(\frac{m}{2\pi k_B T}\right)^{\frac{3}{2}} e^{\left[-\frac{mv^2}{2k_B T}\right]} \tag{3.16}$$

It is often written as:

$$f(v)=\left(\frac{m}{2\pi k_B T}\right)^{\frac{3}{2}}\exp\left[-\frac{mv^2}{2k_B T}\right] \tag{3.17}$$

where $f(v)$ gives the number of microparticles with a speed $v+dv$, m is the mass of the microparticle, k_B is known as Boltzmann's constant, T is absolute temperature, and $v^2 = v_x^2 + v_y^2 + v_z^2$, where v_x, v_y and v_z are the components of the speed v of the microparticle in the x, y, and z directions. The Maxwell-Boltzmann speed distribution function is derived using the same assumptions we used to define an ideal gas (see Section 3.4). The rigorous mathematical derivation by Boltzmann is considered a major achievement in the mathematics of physics.

By more mathematical jiggery-pokery the three average speeds mentioned above can be calculated from the Maxwell-Boltzmann speed distribution function, again I will not present the mathematical details of these calculations, but the results are:

1. The most probable speed ....$v_{prob}=\left(\frac{2k_B T}{m}\right)^{\frac{1}{2}}$
2. The average speed ....$<v>=\left(\frac{8k_B T}{\pi m}\right)^{\frac{1}{2}}$
3. The root-mean-squared speed ....$v_{rms}=\left(\frac{3k_B T}{m}\right)^{\frac{1}{2}}$

All three characteristic speeds are proportional to $\left(\frac{T}{m}\right)^{\frac{1}{2}}$ and the constants of proportionality of all three speeds include k_B (the Boltzmann constant which has a value of 1.380650×10^{-23} joules per kelvin – $J.K^{-1}$). All you need to take away from this are the "speed equations" given above, especially the root-mean-squared speed because this allows us to calculate the average kinetic energy of a microparticle of mass m at a given temperature T.

Figure 3.11 shows how the distribution curves vary with energy/temperature. As we add further energy to the gas, which is equivalent to raising its temperature, the peak of the graph will shift to the right (since the average speed of the microparticles will increase). As the graph shifts to the right, the height of the graph must decrease to maintain the same total area under the curve (because the number of microparticles in the gas sample has not changed). When gas becomes colder, not only does its peak shift to the left (lower temperature=slower microparticles), but its peak also becomes higher and narrower. At lower temperatures, there is less energy available, so the microparticles tend to move at speeds that do not vary so much from each other. In the limit where an ideal gas is taken to absolute zero, the microparticles would stop moving altogether; therefore, it would all be at the same speed $(0\ m.s^{-1})$, i.e., at zero energy, the speed distribution curve is a vertical straight line of height equal to the number of microparticles present.[35] This analysis indicates that as the energy content/temperature of an ideal gas system increases from zero we move from a certainty regarding the speed of its constituent microparticles (they all have zero speed) to a greater and greater uncertainty.

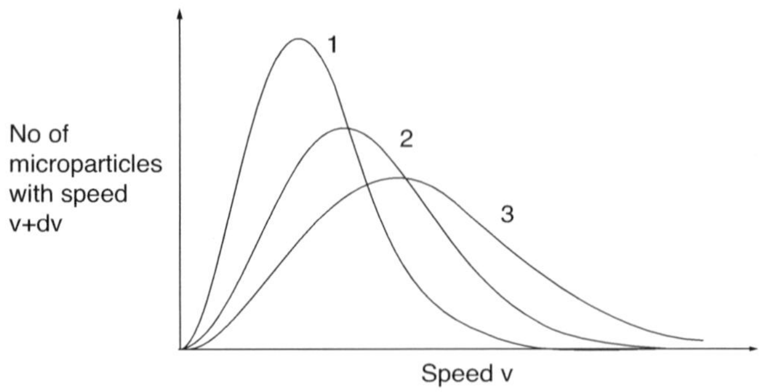

Figure 3.11 Maxwell-Boltzmann distribution curves at different energy contents or temperatures.

I am conscious that I have not given any indication of the magnitude of the mass of the microparticles and their speed at normal temperatures. I will be considering this later in this chapter, but it is worth giving, at this stage, some data relating to a so-called permanent gas and a major constituent of air – nitrogen. The mass of a nitrogen microparticle is about 2.3×10^{-26}kg and the average speed of nitrogen microparticles vary from about 200 m.s^{-1} at 100 K (about –200°C) to about 700 m.s^{-1}, at 1,000 K (about 700°C). These figures demonstrate that gaseous microparticles have very small masses and very large speeds.

We now have the tools to theoretically derive Boyle's Law.

3.5.4 Theoretical Derivation of Boyle's Law

Consider a container holding a macroscopic amount of an ideal gas. We know the pressure of a gas arises from collisions of the gas microparticles with the walls of the container (Section 3.5.2). Since the number of microparticles in the container is large, the number colliding with the wall per unit time is large enough so that fluctuations in the pressure due to the individual collisions are immeasurably small in comparison to the total pressure. Because of the very large number of microparticles in a gas sample, we can use the average speed ($< v >$) of the microparticles in our theoretical derivation of Boyle's Law. We can also convert the speed of any microparticle into v_x, v_y, and v_z components, and finally, because of the large number of microparticles we can state that for the average speeds of the microparticles that $< v_x > = < v_y > = < v_z >$.

First, I will determine the momentum transferred to a container wall in a single collision (I have already done this in Section 3.5.2). Figure 3.12 shows a microparticle of mass m and velocity v colliding with a wall of area A, it also shows the sign convention we will use for the velocity vectors. The wall is in the yz plane and Figure 3.12 shows the collision from two points of view – along the z-axis and along the y-axis. Before the collision, the particle has velocity $+|\mathbf{v}_x|$ and momentum $+|m\mathbf{v}_x|$ along the x direction.[36] After the collision, the particle has velocity $-|\mathbf{v}_x|$ and momentum $-|m\mathbf{v}_x|$ along the x direction. It is important that you note that the components of velocity and momentum along y and z axes remain unchanged. Since the momentum of the particle has changed by $--|2m\mathbf{v}_x|$,[37] the total momentum imparted to the wall must be $+|2m\mathbf{v}_x|$ (remember wall's momentum change is in the opposite direction to the microparticle's momentum change) As the wall is in the yz plane, then the components of the microparticle's velocity in the y and z direct (v_y and v_z) which are parallel to the wall are unchanged. The reason I have used absolute values here is to show that the positive force exerted on the wall by the gas is in the gas-to-wall direction.

The next step is to determine the total number of collisions with the wall in time interval Δt. During this time interval, all microparticles within a distance $< v_x > \Delta t$ of the wall (and travelling towards it) will collide with the wall. Since the area of the wall is A, this means that all microparticles within a volume $A\times < v_x > \Delta t$ will undergo a collision with the wall (Figure 3.13).

We now need to work out how many microparticles will be within this volume and travelling towards the wall. The number density of the microparticles (i.e., the number of microparticles per unit volume) is $\frac{N}{V}$ where N is the number of microparticles in the container of volume V. Our volume of interest is $A\times < v_x > \times \Delta t$. The number of microparticles within our volume of interest is equal to the number density multiplied by our volume of interest (Equation 3.18)

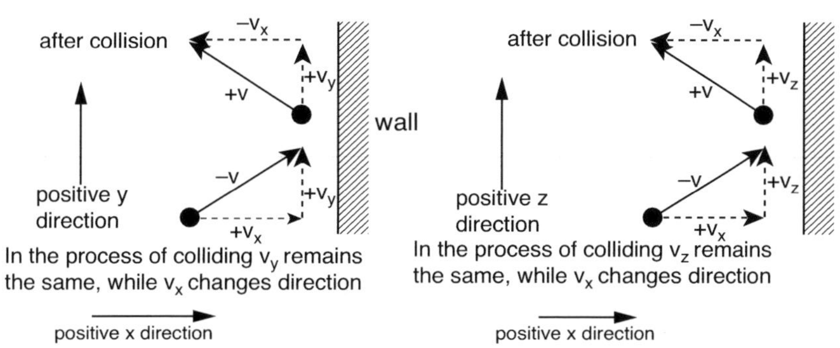

Figure 3.12 Momentum change during elastic collision with container wall.

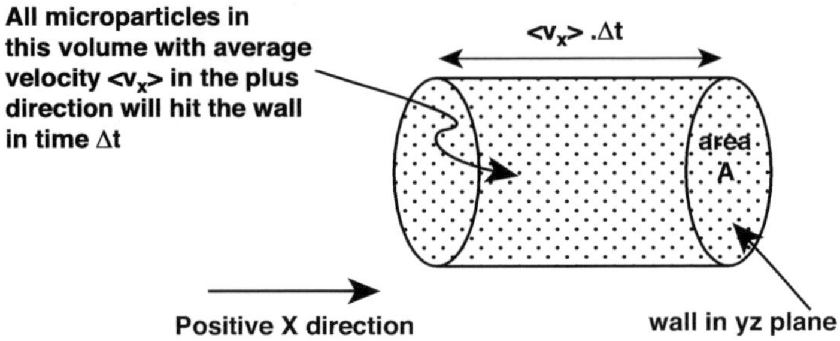

Figure 3.13 Determining the total number of collisions with the wall in time interval Δt.

$$\text{number of microparticles in our volume of interest} = \left(\frac{N}{V}\right) \times \left(A \times <v_x> \times \Delta t\right) \tag{3.18}$$

Since the random velocities of a large number of microparticles mean that on average half of the microparticles in the container will be travelling towards the wall and half away from it, the number of microparticles within our volume travelling towards the wall is half of the above value, i.e., the number of microparticles hitting the yz plane wall in time Δt is given by Equation (3.19):

$$\text{Number of microparticles hitting the yz wall in time } \Delta t = \frac{1}{2}\left(\frac{N}{V}\right) \times \left(A \times <v_x> \times \Delta t\right) \tag{3.19}$$

The total momentum change (Δp_x) imparted to the yz wall in time Δt is the momentum change per collision $(2m <v_x>)$ multiplied by the total number of collisions:

$$\Delta p_x = (2m <v_x>) \times \left[\frac{1}{2}\left(\frac{N}{V}\right) \times \left(A \times <v_x> \times \Delta t\right)\right] = \frac{Nm\left[\left(A \times <v_x>^2 \times \Delta t\right)\right]}{V} \tag{3.20}$$

Pressure is defined as the force per unit area, so we need to convert the momentum change in Equation (3.20) into a force to calculate the pressure. We can do this using Newton's Second Law of Motion:[38]

$$F_x = m <a_x>^2 = \frac{\Delta p_x}{\Delta t} \tag{3.21}$$

where $<a_x>^2$ is the average acceleration of the microparticle along the x-axis.

Substituting for Δp_x from Equation (3.20) yields:

$$F_x = \frac{Nm\left[\left(A \times <v_x>^2\right)\right]}{V} \tag{3.22}$$

The pressure is therefore:

$$P = \frac{F_x}{A} = \frac{Nm\left(<v_x>^2\right)}{V} \tag{3.23}$$

So far, we have based our arguments on a particle with a single average speed $<v_x>$. We have already recognised that the random motion of the microparticles means that the average speed along the x direction is the same as along y and z. This allows us to define a root-mean-square speed:

$$v_{rms} = \left[<v_x>^2 + <v_y>^2 + <v_z>^2\right]^{\frac{1}{2}} = \left[3 <v_x>^2\right]^{\frac{1}{2}} \tag{3.24}$$

Therefore:

$$<v_x>^2 = \frac{1}{3} v_{rms}^2 \tag{3.25}$$

The final expression for the pressure is therefore

$$P = \frac{1}{3}\frac{Nmv_{rms}^2}{V} \quad \text{or} \quad PV = \frac{1}{3}\left(Nmv_{rms}^2\right) \tag{3.26}$$

We know that the average speed of the microparticles of mass m in a gas is constant at constant temperature (remember all three characteristic speeds – average, most probable and rms – of microparticles in an ideal gas are proportional to $\left(\frac{T}{m}\right)^{\frac{1}{2}}$ – see Section 3.5.3). Therefore, we can write PV=constant for a given mass of ideal gas (N and m are constant) at constant temperature (v_{rms}^2 is constant).

Therefore, using kinetic theory, based on basic assumptions about an ideal gas, and relatively straightforward maths (apart from the Maxwell-Boltzmann distribution equation), we have derived the following relationship:

$$P = \frac{K}{V} \qquad \text{where } K = \frac{1}{3}\left(Nmv_{rms}^2\right) \tag{3.27}$$

Equation (3.27) is Boyle's law:

The pressure exerted by a given mass of an ideal gas in an isolated system is inversely proportional to the volume it occupies if the temperature remains unchanged, i.e. $P = \frac{K}{V}$.

I think it is useful to review the assumptions used to derive Boyle's Law together with their implications:

1. The ideal gas consists of an immensely large number of identical microparticles (>10^{17}) which are in high-speed random motion. Although individual microparticles undergo collisions and change direction frequently, the fact that there are so many of them means that at any time there will be equal numbers travelling in, say, the +x and −x directions. This is the crucial assumption of kinetic theory.
2. All collisions, whether between microparticles, or between microparticles and container walls, are perfectly elastic, i.e., a microparticle neither gains nor loses kinetic energy when it collides with another microparticle or the wall. You may be puzzled why perfectly elastic collisions are assumed between microparticles, yet, as we shall see, they never occur in the macroscopic world. Let me quote Richard Feynman again –

   Elastic collisions are especially interesting for systems that have no internal 'gears, wheels, or parts'. Then when there is a collision there is nowhere for the energy to be impounded, because the objects that move apart are in the same condition as when they collided. Therefore, between very elementary objects, the collisions are always elastic or very nearly elastic. For instance, the collisions between atoms or molecules[39] in a gas are said to be perfectly elastic. Although this is an excellent approximation, even such collisions are not perfectly elastic; otherwise one could not understand how energy in the form of light or [infra-red][40] radiation could come out of a gas. Once in a while, in a gas collision, a low-energy infrared ray is emitted, but this occurrence is very rare and the energy emitted is very small. So, for most purposes, collisions of molecules in gases are considered to be perfectly elastic.[41]

3. Even though the received scientific opinion is that quantum mechanics replaces classical mechanics when we consider the very small (such as atoms and molecules) the motion of the microparticles in our ideal gas calculations is governed by Newton's laws of motion and, as we shall see, the derived equations give accurate predictions of many of the properties of real gases.
4. The microparticles experience only contact forces. They interact like hard spheres and this is only when they are in contact with each other. The presence of any long-range forces between the microparticles (such as gravitational and electromagnetic forces) are ignored. These assumptions allow us to consider that, between collisions, the motions of microparticles are along straight lines.
5. The volume occupied by the microparticles themselves is insignificant compared to the volume occupied by the gas, i.e., the diameter of the microparticles is insignificant compared to the average distance between microparticles and the microparticles may be treated as particles with no volume ("point particles").

The use of kinetic theory and the simple assumptions used in the derivation of Equation (3.27) have predicted Boyle's experimental results on the behaviour of real gases: this gives confidence that real gases are collections of essentially very small microparticles that have very little interaction with each other, are very far apart, zoom around at high speed, and all inter-particular collisions and those of microparticles with the container walls are perfectly elastic.

3.5.5 Theoretical Derivation of the Ideal Gas Equation

We have now covered most of the ground in theoretically deriving the "gas law" ($PV = K_4T$).[42] We know the average kinetic energy of a microparticle in an ideal gas is $\frac{1}{2}mv_{rms}^2$ (Equation 3.15), and the root-mean-square speed $v_{rms} = \left(\frac{3k_BT}{m}\right)^{\frac{1}{2}}$ (see Section 3.5.3). The average kinetic energy ($KE_{average}$) of a microparticle in three dimensions is therefore:

$$KE_{average} = \frac{1}{2}mv_{rms}^2 = \frac{1}{2}m\left(\frac{3k_BT}{mT}\right) = \frac{3}{2}k_BT \tag{3.28}$$

In an ideal gas, the energy of a microparticle is fully described by its translational kinetic energy. This is a direct result of the assumptions made regarding an ideal gas (Section 3.4), namely the microparticles are perfectly spherical in shape and there are no forces (interactions) between the microparticles. This means the only energy an ideal gas microparticle can have is a result of its speed (i.e., translational energy) because there are no other modes for a point particle to store energy.[43] This is an important point, so I will repeat it in a slightly different way:

> The energy content of an ideal gas is in the form of the translational kinetic energy of its constituent microparticles and any change in energy content is accompanied by a change in temperature.

For ideal gases, Equation (3.28) is usually written as:

$$E = \frac{3}{2}k_BT \tag{3.29}$$

Equation (3.29) relates the macroscopic quantity[44], temperature, to the microscopic quantity, the average kinetic energy of the microparticles (E). We also know that:

$$E = \frac{1}{2}mv_{rms}^2 \tag{3.30}$$

Therefore:

$$\frac{1}{2}mv_{rms}^2 = \frac{3}{2}k_BT \tag{3.31}$$

Rearrangement of Equation (3.31) gives:

$$v_{rms}^2 = \frac{3k_BT}{m} \tag{3.32}$$

Equation (3.26) $\left(PV = \frac{1}{3}\left(Nmv_{rms}^2\right)\right)$ is a mathematical version of Boyles Law. Substituting $\frac{3k_BT}{m}$ for v_{rms}^2 gives:

$$PV = \left[\frac{1}{3}(Nm)\times\left(\frac{3k_BT}{m}\right)\right] = Nk_BT \tag{3.33}$$

For a given quantity of gas, Equation (3.33) can be written as $\frac{PV}{T} = \text{constant}$. We have now theoretically derived the ideal gas equation from a few relatively simple assumptions. Remember the gas law $PV = K_4T$ (see Section 3.3.5) was obtained by experiment. From now on this equation will be known as the ideal gas equation:

$$PV = Nk_BT \tag{3.34}$$

where:

P=pressure (force per unit area) – units are $N.m^{-2}$ (known as pascals[45])

V=volume – units are m^3.

N=is the amount of matter measured as the number of microparticles.

T = temperature units are kelvin (K); this is absolute temperature where 0 K (known as 'absolute zero') is –273.16°C.

k_B (Boltzmann's constant) – units are $N.m^{-2}.m^3.K^{-1} = N.m.K^{-1} = J.K^{-1}$ – the value of k_B is 1.3805×10^{-23} $J.K^{-1}$.

3.5.6 Temperature Revisited

Equation (3.31) states that the average kinetic energy of a microparticle is $\frac{3}{2}k_BT$. This means $\frac{3}{2}Nk_BT$ is the sum of the kinetic energies of the microparticles in an ideal gas sample containing N microparticles. This energy is known in science as the "internal energy" (and is usually denoted by the letter "U") and represents the microscopic energy content of the ideal gas. From Equation (3.33) we can write:

$$U = \frac{3}{2}Nk_BT = \frac{3}{2}PV \tag{3.35}$$

The concept of internal energy is of paramount importance in science and will feature throughout the rest of this book. It is the existence of an internal energy that can be varied that makes a system "thermodynamic". In mechanics, the term "degrees of freedom" is used to describe the number of independent movements an object can make. An ideal gas microparticle (a point mass) has three degrees of freedom (translating motion along the x, y, and z axes). This book is only concerned with the motions of point particles, but you should be aware that there are other degrees of freedom in three-dimensional molecules that are other forms of motion that contribute to the internal energy of a physical object. An unrestrained rigid three-dimensional body has six degrees of freedom; the translating motions of the point particle plus three rotary motions around the x, y, and z axes (Figure 3.14). For polyatomic gases,[46] there is also the internal energy associated with vibration (because chemical bonds are not rigid) as well with rotation. In Figure 3.14, I have shown the three fundamental modes of vibration for water.

Using Equation (3.35), we can define the temperature of an ideal gas as:

$$T = \frac{PV}{Nk_B} \tag{3.36}$$

We can write Equation (3.26) as:

$$PV = \frac{2}{3}\left(\frac{1}{2}Nmv_{rms}\right) \tag{3.37}$$

If we substitute $\frac{2}{3}\left(\frac{1}{2}Nmv_{rms}\right)$ for PV in Equation (3.36), we get:

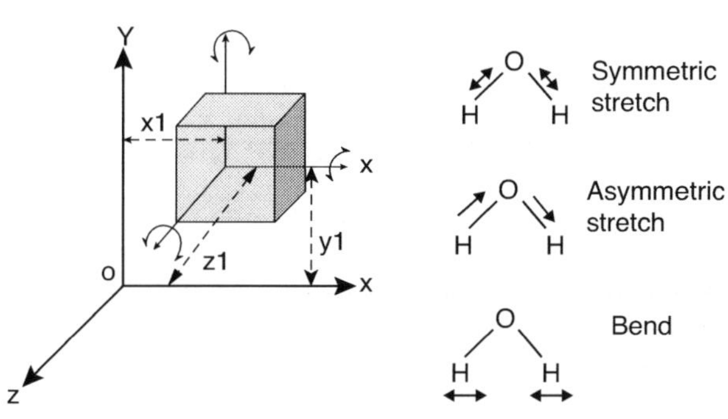

Figure 3.14 The six degrees of freedom of an unrestrained rigid 3-dimensional object and the normal modes of vibration for the water molecule.

$$T = \frac{\left[\frac{2}{3}\left(\frac{1}{2}Nmv_{rms}\right)\right]}{Nk_B} = \frac{1}{3}\left(\frac{mv_{rms}^2}{k_B}\right) \quad (3.38)$$

Equation (3.38) provides a scientifically rigorous definition of temperature based on the ideal gas; when $T = 0$ then $v_{rms}^2 = 0$, and from Equation (3.36) when $T = 0$ then either $P = 0$ and/or $V = 0$. In Section 1.2.2 I stated that physical quantities can be divided into extensive and intensive quantities. We can see from Equation (3.38) that temperature does not depend on the amount of matter present (the "N's" in Equations 3.36 and 3.37 have cancelled out). Therefore, temperature, like pressure, is an intensive quantity. Also, note it meets the requirements given in Section 1.2.2:

> Intensive properties have an 'intensity' in a point in space. Extensive properties are undefined (or zero) in a point in space.

It is instructive to ask the following question. Why can we use a single numerical value (which we call temperature) as a measure of all the speeds of the microparticles that constitute macroscopic ideal gas system? The answer is that a macroscopic amount of an ideal gas contains at least 10^{17} microparticles, and, although we cannot predict the behaviour of any one microparticle with any precision, we can predict the behaviour of large quantities of microparticles with great precision. Think of tossing a coin or throwing a die, we cannot know an individual result but if we do it 10^{17} times we know to a very high precision[47] (provided the coin or die are not loaded) that ½ of the tosses will be heads and the other ½ will be tails, similarly with the die when ⅙ of the throws will be a "1", ⅙ will be a "2" etc. The reason we can use the Maxwell-Boltzmann distribution curve to describe the distribution of the speeds of the microparticles at a given internal energy is because of the ginormous quantities of microparticles present in a macroscopic system.

In Section 3.5.5 I wrote:

> The **energy content** of an ideal gas is in the form of the translational kinetic energy of its constituent microparticles and any change in energy content is accompanied by a change in temperature.

Now we could rewrite this as:

> The **internal energy** of an ideal gas is in the form of the translational kinetic energy of its constituent microparticles and, any change of internal energy is accompanied by a change in temperature.

I now want to consider the role temperature plays in energy transfer.

3.5.7 Temperature and Energy Transfer

Consider a physical object and its surroundings, e.g., a cup of coffee or glass of iced water in a room. If the temperature of the object is higher than the surroundings, the object cools down by transferring some of its internal energy to the surroundings. Conversely, if the temperature of the object is lower than its surroundings the object warms up by transfer of energy from the surroundings. Eventually, the object and the surroundings reach the same temperature and net energy transfer ceases, and an equilibrium is reached in which the transfer of energy from the object to the surroundings is matched by the transfer of energy from the surroundings to the object. It is at this point, that the two objects are said to have reached thermal equilibrium, i.e., at thermal equilibrium all temperatures of interest are unchanging in time and uniform in space.

Mechanical equilibrium (see Section 2.6) is a static equilibrium, but this is not the case with thermal equilibrium. Thermal equilibrium requires that a flow of energy into the object or system of interest is balanced by a flow of energy out of the object or system, i.e., energy flows in both directions but there is no net flow of energy. This type of equilibrium is known as a dynamic equilibrium, i.e., the rates of the forward and reverse processes are equal.

I want now to consider the following situation. Two equal volumes of the same quantity of an ideal gas (i.e., both volumes at the same pressure) at two temperatures $T_H > T_C$ are in a perfectly insulated container[48] separated by a removable insulating wall (Figure 3.15). When the wall is removed the two gases diffuse and the constant collisions of particles will, at equilibrium, result in a different Maxwell-Boltzmann distribution curve and therefore a different final temperature $(T_H > T_{final} > T_C)$.

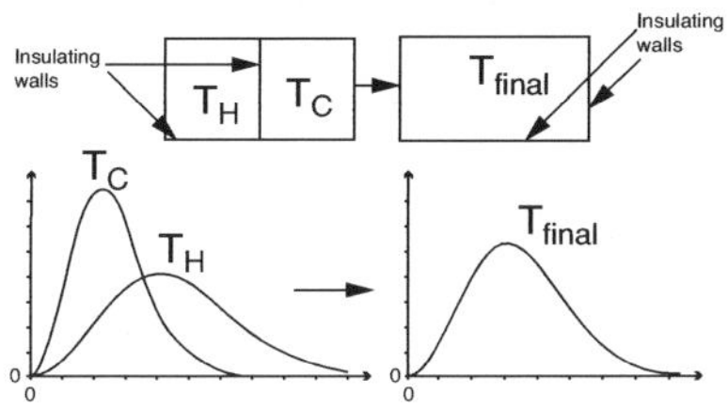

Figure 3.15 Thermal equilibrium attained by the mixing of two ideal gas volumes at different temperatures.

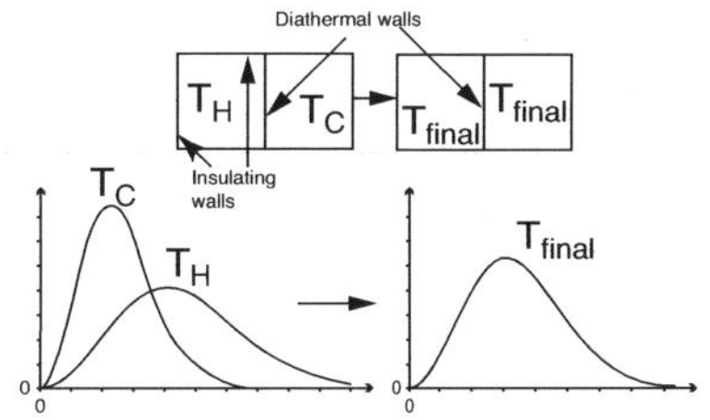

Figure 3.16 Thermal equilibrium obtained by thermal conduction through the diathermal wall.

Now consider a similar experiment but instead of a removable insulating wall, we have a fixed diathermal wall between the two volumes (Figure 3.16). A diathermal wall allows energy to flow between the two volumes but not mass, i.e., it is made from thermally conducting material.[49] We say the two volumes are in thermal contact. Ignoring any energy change in the diathermal wall the result will be the same as before, i.e., both volumes will eventually achieve the same final temperature where again ($T_H > T_{final} > T_C$), although the times taken to reach this equilibrium state will, in all probability, be different.

Neither of the above results is surprising, we know if we mix a hot fluid, at temperature T_H, with a cold fluid, at temperature T_C, the temperature of the resultant mixture is somewhere between T_H and T_C. Similarly, if two insulated objects are in thermal contact, we know both objects will eventually reach the same temperature, T_{final}, which is between T_H and T_C. It is important to note that there is no mention of time, it may take a few seconds or millions of years but, assuming no other factors are involved, both objects will eventually reach the same temperature.

The major point to note is that energy has flowed from a hot object to a colder object, i.e., from a region of high temperature to one of lower temperature; this is an energy transfer process. It is also important to note that even at thermal equilibrium energy is still being transferred across the diathermal wall. However, at thermal equilibrium equal quantities of energy are being transferred in both directions, i.e., we have a dynamic equilibrium.

3.5.8 Discussion and Summary

Once the equation for the distribution of an amount of energy between identical particles is known (the Maxwell-Boltzmann Distribution), then employing simple assumptions from kinetic theory, and, using relatively simple mathematics, an equation can be derived that explains experimentally determined results of how the amount of permanent gas, and its pressure, volume, and temperature are related. In other words, we have a simple model of one form of matter. I consider the theoretical derivation of the experimentally determined gas law is a remarkable result and indicates that gases consist of an immensely large number of microparticles moving at high speed and in constant collision with one another. The resulting distribution of speeds remains unchanged with time providing the internal energy (U) or absolute temperature (T) remains unchanged. The success of this model strongly indicates all macroscopic quantities of matter, whether gas, liquid, or solid, consist of very large numbers of microparticles.

There is one fundamental difference in this model of a physical object and the classical mechanics presented in Chapter 1 – temperature. There are two basic properties of temperature:

1. It is the quantity that ultimately becomes equal when two objects or systems in thermal contact transfer energy, i.e., when no net energy transfer occurs the systems/objects are at the same temperature and are in thermal equilibrium.
2. When the temperatures between two objects or systems in thermal contact are different the net energy transfer is from the system/object at the higher temperature to the system/object at the lower temperature.

3.6 'REAL' IDEAL GASES

3.6.1 Introduction

It was first thought that all "permanent"[50] gases obeyed the ideal gas laws. Later more precise measurements revealed that real gases do not exactly obey the laws of Boyle and Gay-Lussac. The variations for a particular gas are least when a gas is at high temperature and low pressure. Furthermore, the deviations vary from gas to gas, e.g., helium obeys closely whereas oxygen and nitrogen are less obedient, and "non-permanent" gases such as carbon dioxide are very noncompliant. In theory at absolute zero, any quantity of ideal gas would have no pressure and no volume. This could never happen because:

- At these very low temperatures, "real" matter would not exist as a gas.
- The assumption that the microparticles of a real gas have negligible volume is not the case when the gas is near its liquefaction temperature.

3.6.2 Atoms and Molecules

I will not consider atoms and molecules, in any detail in this book, but I do need to briefly mention some relevant details. All matter consists of atoms and/or molecules (groups of atoms). We can consider atoms to have a spherical shape and molecules are groups of spherically shaped atoms held together by chemical bonds. The noble gases helium, neon, argon, krypton, xenon, and radon are monatomic whereas oxygen (O_2), hydrogen (H_2) and nitrogen (N_2) are diatomic. Figure 3.17 indicates the shapes of monatomic (e.g., helium) and a diatomic gas (e.g., hydrogen).

Atoms consist of a nucleus that contains protons and neutrons and is surrounded by electrons. In a neutral atom, the number of protons equals the number of electrons.[51] An element is a substance that cannot be broken down into simpler substances by chemical means and is composed of atoms that have the same number of protons in its nucleus. This number is known as the atomic number, e.g., a helium atom contains two protons in its nucleus and its atomic number is 2, and any atom with two protons in its nucleus is a helium atom. Along with a definitive number of protons all atoms but a "normal" hydrogen atom[52] contain several neutrons in their nuclei, e.g., helium (Figure 3.17) normally has two neutrons in its nucleus. I write normally because there are helium atoms with one neutron in the nucleus. These two types of helium atoms are known as isotopes and are usually called helium-3 and helium-4 atoms. An isotope is one of two or more species of atoms of the same chemical element that have different atomic masses, i.e., the same number of protons (otherwise they would not be the same element), but different numbers of neutrons. The mass of a proton and neutron are essentially the same at 1.66×10^{-24}g and the mass of an electron is 9.11×10^{-28}g.[53]

The noble gases helium, neon, argon, krypton, xenon, and radon are the only monatomic "permanent" gases. Because of the assumption that the microparticles of an ideal gas "are perfectly spherical in shape", the noble gases more closely meet the definition of an ideal gas than the diatomic permanent gases such as oxygen (O_2), hydrogen (H_2) or nitrogen (N_2).

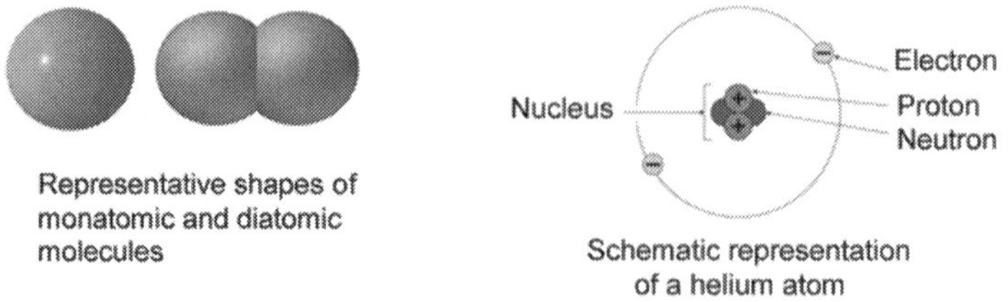

Figure 3.17 Atoms and molecules.

Table 3.1: Atomic Data for First Three Noble Gases

	Helium	Neon	Argon
No of protons = atomic number	2	10	18
No of neutrons	2	10	22
Relative atomic mass or atomic weight	4.003	20.180	39.948
Atomic mass	6.647×10^{-24}g	3.351×10^{-23}g	6.633×10^{-23}g

Finally, I want to introduce, the concept of atomic weight which is now known as the relative atomic mass.[54] I will not be considering atomic weight in this book, but I will set down a definition. The (dimensionless) unit of atomic weight is 1/12 the mass (not weight as some sources state) of a carbon-12 atom (six protons and six neutrons), i.e., the atomic weight of a carbon-12 atom is 12 units. Although the atomic number defines an element, the atomic weight of a particular element may vary slightly, e.g., boron from Turkey has a lower atomic weight than boron from California, because of its different isotopic composition. The "molecular weight" (or relative molecular mass) of a molecule is the sum of its constituent atomic weights. It is worth re-emphasising that, despite containing the words "weight" or "mass", the atomic weight/relative atomic mass is a dimensionless quantity. It is also worth saying that an approximate value of atomic weight (often adequate for many calculations) can be calculated by the addition of the number of neutrons and protons in the nucleus of the atom of interest.

Table 3.1 sets out some atomic data for the first three noble gases. The atomic weights are often not whole numbers primarily because they reflect the typical ratio of the natural abundance of its isotopes on Earth abundances yields, e.g., the atomic weight of helium is 4.003 because on Earth 99.9998% of helium in the atmosphere is He-4 (nucleus contains two protons and two neutrons) and 0.0002% is He-3 (nucleus contains two protons and one neutron).[55]

3.6.3 Avogadro's Number and Moles

I want now to consider the number of microparticles (atoms or molecules) in a mass equivalent to the atomic or molecular weight in grams. I suppose I should be using the SI unit of mass – the kilogram – but everyone uses grams. We already have the data to do this in Table 3.1. We know the mass of helium atoms (or microparticles) are 6.647×10^{-24}g. Therefore, the number of atoms in a mass of helium of 4.003 g is equal to:

$$\text{Number of atoms in } 4.003\text{g helium} = \frac{4.003}{6.647\times10^{-24}} = 6.02227\times10^{23}$$

Similarly, for neon and argon, we have:

$$\text{Number of atoms in } 20.180\text{g neon} = \frac{20.180}{3.351\times10^{-23}} = 6.02208\times10^{23}$$

$$\text{Number of atoms in } 39.948\text{g argon} = \frac{39.948}{6.633\times10^{-23}} = 6.02261\times10^{23}$$

The result of these calculations is a constant (the small differences are due to rounding errors). This constant is equal to the number of microparticles (atoms or molecules) in a mass equal to its atomic or molecular weight in grams. This number is known as Avogadro's number (A)[56] and leads to an important concept in chemistry – the "mole" ("moles" is Latin for a heap or pile). A mole of a substance is defined as

> The mass of substance containing the same number of microparticles (atoms or molecules) as there are atoms in exactly 12.000 g of carbon-12.

The concept of a mole applies not only to atoms but to molecules. Many scientists erroneously consider a "mole" to be short for molecule and consider that Avogadro's number of atoms is therefore an "atom". The mole is widely used in chemistry as of unit of quantity. The abbreviation for a mole is only slightly shorter – "mol.".

Avogadro's number is an experimentally (not mathematically) derived unit[57] and despite its name Avogadro did not determine its value. The reason it is given his name is from Avogadro's Law (1811):

The volume occupied by a gas is directly proportional to the number of microparticles of the gas provided T and P are fixed: $\frac{V_1}{N_1} = \frac{V_2}{N_2}$ (P and T constant).[58]

where N_1 is the number of microparticles in volume V_1 and N_2 is the number of microparticles in volume V_2. Avogadro's Law, like Boyle's and Gay-Lussacs' laws, is only 100% valid for the hypothetical ideal gas. It (Avogadro's Law) be derived from the ideal gas equation $(PV = Nk_BT)$. We know $\frac{V}{N} = \frac{k_BT}{P}$ and if T and P are constant then for any ideal gas $\frac{V}{N}$ is constant. We can now see that at constant T and P one mole of ideal gas has the same volume, i.e., 4.003 g of helium, 20.180 g of neon and 39.498 g of argon all have the same volume at a fixed T and P. This volume (V_{mol})can be calculated for an ideal gas knowing (where A is Avogadro's Number):

$$PV_{mol} = Ak_BT \tag{3.39}$$

Therefore, the molar volume of one mole of ideal gas at standard atmospheric pressure $(101.325 \times 10^3 \text{ Pa})$ and 273.15 K is:

$$V_{mol} = \frac{(6.022 \times 10^{23} \times 1.3805 \times 10^{-23} \times 273.15)}{101.325 \times 10^3} = 0.0224 \tag{3.40}$$

The units of V_{mol} in Equation (3.40) are cubic metres which equal 1,000 L, i.e., molar volume is 22.4 L (or more rigorously 22.4 dm^3).[59]

We can write the ideal gas equation in terms of number of moles (n) instead of the number of microparticles (N). You should be able to see that for n moles of microparticles:

$$PV = nAk_BT = nRT \tag{3.41}$$

where R is known as the ideal gas constant and has the value 8.314 $J.mol^{-1}.K^{-1}$. Equation (3.41) is the form of the ideal gas law most familiar to students and scientists.

3.6.4 Internal Energy of Ideal Gas

I introduced the concept of internal energy in Section 3.2 and defined it in Section 3.5.6 for an ideal gas:

> The **internal energy** of an ideal gas is in the form of the translational kinetic energy of its constituent microparticles and, any change of internal energy is accompanied by a change in temperature.

The internal energy (U) of n moles of ideal gas at constant temperature (T) is:

$$U = nA\left(\frac{1}{2}mv_{rms}^2\right) \tag{3.42}$$

Remember, n moles of ideal gas contain nA microparticles, and the average kinetic energy of each microparticle is $\frac{1}{2}mv_{rms}^2$ (Equation 3.30). We also know (Equation 3.26 where N=nA):

$$PV = \frac{1}{3}\left(nAmv_{rms}^2\right) \tag{3.43}$$

Multiplying both sides of Equation (3.43) by $\frac{1}{2}$ and rearranging yields:

$$\left(\frac{1}{2}mv_{rms}^2\right) = \frac{3}{2}\left(\frac{PV}{nA}\right) \tag{3.44}$$

Substituting into Equation (3.42), we obtain:

$$U = nA\left(\frac{3}{2}\frac{PV}{nA}\right) = \frac{3}{2}PV \tag{3.45}$$

We also know that PV=nRT (Equation 3.41), so for a system of n moles of an ideal gas:

$$U = \frac{3}{2}PV = \frac{3}{2}nRT \tag{3.46}$$

From Chapter 2 we know the unit of energy is the joule and that 1 joule=1 newton×1 metre. Therefore, PV and RT must have the units of energy. The units of pressure are the pascal (Pa) equal to $N.m^{-2}$ and the units of volume are m^3. Therefore, the units of PV are $Nm.mol^{-1}$ (energy per mole) which are the units of energy and R has the units $J.K^{-1}.mol^{-1}$.

3.6.5 Average Speed of Noble Gases

Suppose two gas systems, one containing helium and the other of argon have the same temperature. We know from above that their microparticles must have the same average kinetic energy.[60] But do the helium atoms and the argon atoms have the same average speed? Remember kinetic energy is given by $KE = \frac{1}{2}mv^2$. The mass of a helium atom is 6.647×10^{-24} g and that of argon is 6.633×10^{-23} g (see Table 3.1). If the two systems are at the same temperature, then the average kinetic energy of the atoms is equal. Therefore,

$$6.647 \times 10^{-24} \times v_{He}^2 = 6.633 \times 10^{-23} \times v_A^2 \tag{3.47}$$

This gives,

$$\frac{v_{He}^2}{v_A^2} = \frac{6.633 \times 10^{-23}}{6.647 \times 10^{-24}} = 9.96 \tag{3.48}$$

taking square roots of both sides yields

$$\frac{v_{He}}{v_A} = 3.2 \tag{3.49}$$

This indicates, that at the same temperature, the average speed of the helium atoms is about three times that of argon atoms, i.e., at the same temperature lighter microparticles move faster (on average) than heavier ones.

Figure 3.18 gives some feel of the speed distribution of noble gases at 298 K;[61] you can see speeds of noble gas atoms are typically around a few hundred metres per second. Gases composed of lighter microparticles have microparticles with a broad speed distribution that peak at higher velocities. While gases composed of heavier microparticles have microparticles with a narrower speed distribution that peaks at lower velocities. Figure 3.19 shows speed distributions at different T when m is constant. The trend to higher speeds and flatter curves at high temperatures mirrors the trend seen with the increasing mass of microparticles at constant temperature.

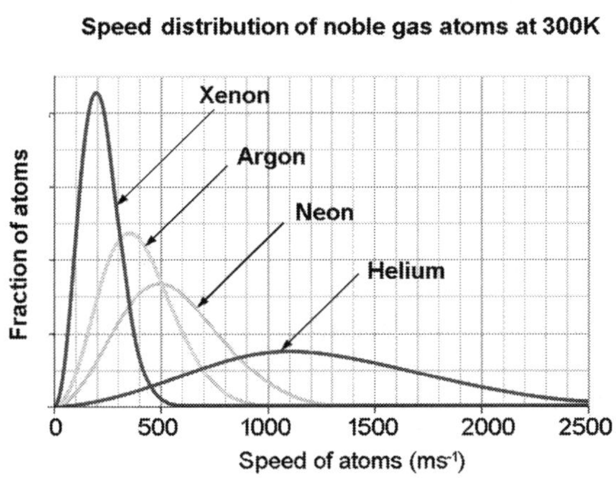

Figure 3.18 The speed distribution curves of noble gases at a temperature of 300 K.

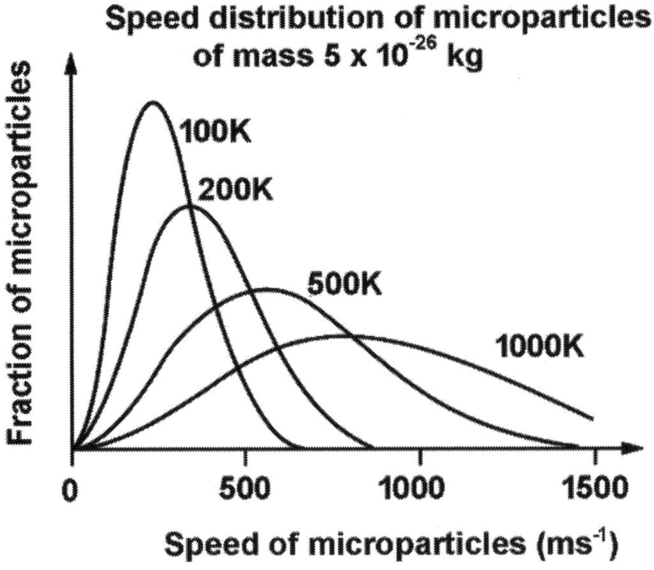

Figure 3.19 Calculated speed distribution curves at different temperatures for an ideal gas with microparticle mass of 5×10^{-23}g.

3.7 DISCUSSION AND SUMMARY

The ideal gas is a model of the simplest form of matter. Although there is no such thing as an ideal gas, it is a conceptual model that allows us to understand how gases respond to changing temperature, volume, or pressure. Experiment has shown that under many conditions, so-called permanent gases, exhibit behaviour that closely approximates that of an ideal gas, and the ideal gas law can be used to predict the behaviour of real gases under conditions of relatively high temperatures and relatively low pressures, e.g., at atmospheric pressure and the typical ambient temperatures on Earth. Experiment has also demonstrated that deviations from ideal behaviour are most commonly observed at low temperatures or high pressures, where the following requirement for an ideal gas starts to fall away (see Section 3.4):

> It consists of microparticles (i.e., microscopically small particles). This smallness is such that the total volume of the individual microparticles when added up is negligible compared to the volume of the container that held the gas. This is equivalent to stating that the average distance separating the gas particles is very large compared to their size.

In Section 3.5.8 I wrote – "I consider the theoretical derivation of the experimentally determined gas law is a remarkable result and indicates that gases consist of an immensely large number of microparticles moving at high speed and in constant collision with one another". We now have a simple model of one form of matter and its success strongly suggests all macroscopic quantities of matter consist of immensely large numbers of microparticles. There are four physical quantities in the Ideal Gas Law – amount of material, pressure, volume, and temperature. I highlighted the fact that, of these four quantities, the one we understand least is temperature. In this chapter we have considered some of the properties of temperature:

1. Temperature is the quantity that becomes equal when two systems in thermal contact transfer energy.
2. Energy is only transferred from a higher temperature to a lower temperature.
3. If two objects are in thermal contact and no net transfer of energy occurs, then they are at the same temperature, i.e., they are in thermal equilibrium.
4. If two objects are in thermal contact and a net transfer of energy occurs, then they are not at the same temperature, i.e., they are not in thermal equilibrium.

5. The temperature of an ideal gas is proportional to the average translational kinetic energy of the microparticles.
6. The mathematic definition of temperature of an ideal gas based on ideal gas laws is (see Equation 3.38): $T = \frac{1}{3}\left(\frac{mv_{rms}^2}{k_B}\right)$.
7. The reason we can use a single numerical value (temperature) as a measure of the translational speeds of the microparticles that constitute macroscopic ideal gas system is a direct consequence of the immensely large number of microparticles in a macroscopic system. This large number means the resulting speed distribution of the microparticles remains unchanged with time at constant temperature, and as a consequence, the root-mean-square speed of the microparticles is known with considerable precision.
8. Temperature is a measure of the internal energy of a given macroscopic ideal gas system.

Consider the following situations:

1. If a system is not in thermal equilibrium then, by definition, the temperature of the system is not uniform.
2. If the pressure of a system is changing system, then, again, the system is not in equilibrium and is not uniform throughout the system.

In both of these cases, the ideal gas law cannot apply because we do not know what temperature or pressure to use in the ideal gas law. This leads to the following statement – "the ideal gas law only applies only when the system is in equilibrium". This is an important statement that is often ignored in many scientific texts. I will be returning to this subject in later chapters.

In Section 3.6.3, I introduced the concepts of the mole which is the quantity of material that contains Avogadro's number (A) of microparticles, i.e., 6.022×10^{23} microparticles. All we need to know in this book is that the atomic or molecular weight (or more properly relative atomic/molecular mass) in grams of an atom or molecule contains Avogadro's number of microparticles, e.g., the atomic weight of helium is 4.003 and 4.003 g of helium contains 6.022×10^{23} microparticles (in this case helium atoms). Similarly, molecular hydrogen (H_2) has a molecular weight of 2.016, therefore 2.016 g of molecular hydrogen contains 6.022×10^{23} microparticles (in this case molecules of hydrogen – H_2).

Schrödinger in his interesting little book "What is Life"[62] asked what he called the "almost ludicrous" question "why are atoms so small?". The following is an extract:

> Why are atoms so small? To begin with, they are very small indeed. Every little piece of matter handled in everyday life contains an enormous number of them. Many examples have been devised to bring this fact home to an audience, none of them more impressive than the one used by Lord Kelvin: Suppose that you could mark the molecules in a glass of water; then pour the contents of the glass into the ocean and stir the latter thoroughly so as to distribute the marked molecules uniformly throughout the seven seas; if then you took a glass of water anywhere out of the ocean, you would find in it about a hundred of your marked molecules. The actual sizes of atoms lie between about 1/5000 and 1/2000 the wave-length of yellow light. The comparison is significant, because the wave-length roughly indicates the dimensions of the smallest grain still recognizable in the microscope. Thus it will be seen that such a grain still contains thousands of millions of atoms. Now, why are atoms so small? Clearly, the question is an evasion. For it is not really aimed at the size of the atoms. It is concerned with the size of organisms, more particularly with the size of our own corporeal selves. Indeed, the atom is small, when referred to our civic unit of length, say the yard or the metre. In atomic physics one is accustomed to use the so-called Angstrom (abbr. A), which is the 10^{10}th part of a metre, or in decimal notation 0.0000000001 m. Atomic diameters range between 1 and 2 A. Now those civic units (in relation to which the atoms are so small) are closely related to the size of our bodies. There is a story tracing the yard back to the humour of an English king whom his councillors asked what unit to adopt - and he stretched out his arm sideways and said: 'Take the distance from the middle of my chest to my fingertips, that will do all right.' True or not, the story is significant for our purpose. The king would naturally indicate a length comparable with that of his own body, knowing that anything else would be very inconvenient. With all his predilection for the Angstrom unit, the physicist prefers to be told

> that his new suit will require six and a half yards of tweed - rather than sixty-five thousand millions of Angstroms of tweed. It thus being settled that our question really aims at the ratio of two lengths -that of our body and that of the atom - with an incontestable priority of independent existence on the side of the atom, the question truly reads: Why must our bodies be so large compared with the atom? I can imagine that many a keen student of physics or chemistry may have deplored the fact that every one of our sense organs, forming a more or less substantial part of our body and hence (in view of the magnitude of the said ratio) being itself composed of innumerable atoms, is much too coarse to be affected by the impact of a single atom. We cannot see or feel or hear the single atoms. Our hypotheses with regard to them differ widely from the immediate findings of our gross sense organs and cannot be put to the test of direct inspection. Must that be so? Is there an intrinsic reason for it? Can we trace back this state of affairs to some kind of first principle, in order to ascertain and to understand why nothing else is compatible with the very laws of Nature? Now this, for once, is a problem which the physicist is able to clear up completely. The answer to all the queries is in the affirmative.
>
> If it were not so, if we were organisms so sensitive that a single atom, or even a few atoms, could make a perceptible impression on our senses - Heavens, what would life be like! To stress one point: an organism of that kind would most certainly not be capable of developing the kind of orderly thought which, after passing through a long sequence of earlier stages, ultimately results in forming, among many other ideas, the idea of an atom. Even though we select this one point, the following considerations would essentially apply also to the functioning of organs other than the brain and the sensorial system.

In Section 3.6.4, Equation (3.46) I demonstrated the internal energy (U) of one mole of ideal gas is equal to $\frac{3}{2}PV = \frac{3}{2}nRT$ where $R = k_B \times A$. I also wrote that for ideal gases internal energy is the sum of the chaotic microscopic translational kinetic energy associated with the system's constituent microparticles. Consider the following thought experiment. Assume that a sample of ideal gas in an unbreakable perfectly insulated container of negligible mass (aka the "system") is suspended a height h metres above the Earth's surface. By convention, the potential energy of the mass is $m \times g \times h$ because the mass on the Earth's surface is considered to have zero potential energy. If the system falls under the influence of gravity when $h = 0$ the falling mass hits the Earth's surface, bounces around and then remains stationary, i.e., the system's kinetic energy is again zero. The total initial energy of the ideal gas system is $(U_{initial} + m \times g \times h)$ and the final energy of the ideal gas is U_{final}. When the container of ideal gas is falling there is velocity vector **v** applied to all the ideal gas microparticles in the container (on top of their chaotic velocity vectors) and after it has hit the ground and stopped moving this is no longer the case. We know $U_{final} > U_{initial}$ because some of the "organised" potential energy (mgh) is converted into the "chaotic" kinetic energy of the ideal gas by the fall (the rest increases the internal energy of the container and the Earth). For an ideal gas, we know all the increase in its internal energy is reflected in an increase in temperature.[63]

We are now in the realms of thermodynamics and not classical mechanics. In the first three chapters of this book, we have moved from considering classical mechanics, via developing a model of the simplest form of matter (the ideal gas), to thermodynamics. In Chapter 4, I introduce the First Law of Thermodynamics. Before doing so I want to consider briefly the subject of perpetual motion. If you Google "perpetual motion" most hits mention "perpetual motion machines" but at this stage, all I want to consider is "perpetual motion" *per se*. The first hit (ignoring Wikipedia's tautological – "Perpetual motion is motion of bodies that continues indefinitely".) that did not mention "perpetual motion machines" was:[64]

> Perpetual motion, the action of a device that, once set in motion, would continue in motion forever, with no additional energy required to maintain it. Such devices are impossible on grounds stated by the first and second laws of thermodynamics.

The second hit was in a similar vein:[65]

> A perpetual motion … device is a device that, barring mechanical breakdown, is capable of running for an arbitrarily long period of time with no outside intervention or energy input. Such a device has never been built since the concept violates the laws of thermodynamics.

Newton's First Law of Motion (a body remains in a state of constant velocity unless acted upon by an external force) suggests that perpetual motion is not only possible but inevitable for any object already in motion. However, "external forces", primarily frictional forces, are everywhere in our universe. A pendulum eventually stops swinging because of friction with the air. The moon orbits the Earth every 28 (and a bit) days and has been doing so at almost constant speed for eons. However, it can be shown that this is not a perpetual motion device. With advanced instruments and careful measurements, we can determine that the moon's motion is changing, and it gets farther away from the Earth on average by about two centimetres each year. Why? Because even in space there are frictional forces. For objects here on the Earth, the frictional forces are relatively large and tend to slow motions down after a short period of time. For objects like the moon, frictional forces are small compared to what would be needed to slow down such a large object, so the changes are very slow. In all of these macroscopic examples, perpetual motion requires an initial force and a sustaining force.[66]

However, we know perpetual motion does occur at the microscopic scale because at $T > 0$ K the microparticles in an ideal gas are in perpetual motion, as are the atoms and molecules of gases (and as it turns out so are the atoms and/or molecules of liquids and solids). What is going on here? What all of the sites I visited should have said was:

1. The laws of physics indicate that perpetual motion will always occur provided there are no external force pairs.
2. In the macroscopic world there are always external force pairs (friction).
3. In the microscopic world there are no external force pairs.

This demonstrates that there are fundamental differences between the microscopic (i.e., at atomic/molecular scale) world and the macroscopic world (i.e., when the number of microparticles in an object is large enough to allow statistical arguments to be applied to a high precision – see Section 3.2). This suggests there is a discontinuity between the microscopic and macroscopic worlds.

In classical mechanics, there is no way of telling whether any given process is moving forward in time or backwards in time. Therefore, in any classical mechanical equation time could be considered a vector quantity or a vector-like quantity because it can take on positive and negative values. Consider the following quote from Tom Stoppard's 1993 play "Arcadia":

> When you stir your rice pudding, Septimus, the spoonful of jam spreads itself round making red trails like the picture of a meteor in my astronomical atlas. But if you stir backwards, the jam will not come together again. Indeed, the pudding does not notice and continues to turn pink just as before. Do you think this is odd?

As we move from the hypothetical world of classical mechanics to the science of the non-relativistic real world, we can see that time moves in only one direction. In this "real world" temperature plays an important role (i.e., we have entered the world of thermodynamics) and we have seen processes involving temperature spontaneously proceed in one direction, i.e., net spontaneous positive energy flows (a vector-like quantity) only occur from a higher temperature to a lower temperature and never the other way round.

In the real world, we cannot reverse processes and go back in time. In other words, in the scientific equations describing this "real" world, time is a positive quantity, and time is considered a scalar quantity. Unlike walking one kilometre north, we cannot walk one hour later or earlier. All I want you to take away from this is that there is a discontinuity on moving from the microscopic world of microparticles (atoms and molecules) to the macroscopic world of macro-physical objects.

In Chapter 4, we start to explore the world of thermodynamics.

NOTES

1 See next section for definition of a macroscopic object.

2 John Dalton (1766–1844) was an English chemist and meteorologist. He kept daily records of the weather for nearly sixty years. His interest in meteorology and the atmosphere led to research on the nature of gases, and this was the foundation on which he built his atomic theory (for which he is best known).

3 The term "chemically indivisible" is used because the definition of an atom is the smallest part of an element that has the same chemical properties of the element (see Section 3.6.2). A chemical property is a characteristic of a particular substance that can be observed in a chemical reaction.

4 Some sources cite John Dalton as the first to use the term atom to refer to what we now think of as atoms (possibly in a now lost 1802 paper presented to the Manchester Literary and Philosophical Society). A contemporary of Dalton's, John Gough, did use the term in his paper read on 27 January 1804 and published in the Memoirs and Proceedings of the Manchester Literary and Philosophical Society, Volume 1, page 405 (1805), and the term atom appears in the index to the same volume, so it appears to have been in common usage by 1805.

5 At this stage, I will not define equilibrium in this context – one dictionary definition is that equilibrium is a state of balance – this should prove sufficient in the current context.

6 A statistical error is the amount by which an observation differs from its expected value.

7 In this context space (often called physical space) is emptiness in which physical objects are located. Please note – this definition, like that of a physical object, is circular.

8 There is a fourth state – plasma. A plasma is an ionised gas consisting of equal numbers of positively charged ions and negatively charged electrons. The characteristics of plasmas are significantly different from those of ordinary neutral (uncharged) gases so that plasmas are considered a distinct "fourth state of matter." Plasma is the most abundant form of ordinary matter in the Universe (ignoring the hypothetical "dark matter" which may or may not be explained by theories of "ordinary" matter), most of which is in the rarefied intergalactic regions. Plasmas will not concern us here and will not be mentioned again.

9 Four fundamental forces (already mentioned in Section 1.9.3) have been identified in the Universe: gravitational, electromagnetic, weak interaction, and strong interaction. The weak interaction and strong interaction are the forces that hold atomic nuclei together and are not mentioned again in this book. Gravitational force was mentioned in Chapter 1. Many everyday experiences such as friction and air resistance are due to the electromagnetic force, as are all forms of chemical phenomena. In this book, the electromagnetic force is only mentioned in passing because it is the force that "binds" microparticles into liquid and solid phases and is behind energy dissipative processes such as friction.

10 See Section 3.5.6.

11 Sublimation is the transition of matter directly from the solid phase to the gas phase without passing through a liquid phase.

12 Celsius and centigrade temperature scales are the same: Celsius has been the recommended term since 1948.

13 More accurately the triple point of water – there is a 0.01°C difference between the triple point or water and its freezing point at standard atmospheric pressure (1 bar). Theoretically, the triple point of any substance is the temperature at which gas, liquid, and solid are in equilibrium.

14 G. Falk, "Entropy, a resurrection of caloric – a look at the history of thermodynamics", *Eur. J. Phys.* 6, 108–115, 1985.

15 The density of an object is found by dividing its mass by its volume.

16 Ensuring all units are the same – if units of pressure are $N.m^{-2}$ or $kg.m^{-1}.s^{-2}$ then h must be in m, g in $m.s^{-2}$ and density in $kg.m^{-3}$.

17 "A defence of the doctrine touching the spring and weight of the air propos'd by Mr. R. Boyle in his new physico-mechanical experiments, against the objections of Franciscus Linus; wherewith the objector's funicular hypothesis is also examin'd, by the author of those experiments." Author: Boyle, Robert, 1627–1691. Found at the following website: https://quod.lib.umich.edu/cgi/t/text/pageviewer-idx?cc=eebo;c=eebo;idno=a28956.0001.001;%20node=A28956.0001.001:4.2.5;page=root;view=text. "Franciscus Linus had objected that the pressure of the atmosphere was insufficient to raise a mercury column by over 70 cm. To answer this, Boyle designed and made a U-shaped glass tube with one long, open limb of almost 2.5 m and a short, sealed limb of 30 cm. He poured mercury into the long limb and measured the compression of air in the short limb. He showed that the compressed air in the short limb was capable of supporting a long column of mercury, proving Linus wrong. Graphs were not in general use in 1662, but when Boyle's data are plotted on a standard xy graph, they strongly support his hypothesis that the volume and pressure of a gas are inversely related." From https://www.chemistryviews.org/details/ezine/1419939/Boyles_Law_350th_Anniversary.html.

18 For 'spring' read pressure.

19 Remember an absolute temperature scale a thermometric scale on which a reading of zero coincides with absolute zero – the lowest temperature that is theoretically possible.

20 After British mathematician, physicist, and engineer Lord Kelvin (1824–1907) formerly (Sir) William Thomson.

21 Lord Kelvin used the Second Law of Thermodynamics (arguably first discovered by Carnot – see later in this book) to define a thermodynamic temperature scale which is completely independent of any thermometric substance. As we shall see the ideal gas temperature scale and Kelvin's thermodynamic temperature are equivalent.

22 Simply put, the triple point of water is the only temperature at which water can exist in all three states of matter; solid (ice), liquid (water), and gas (water vapour). By international agreement (in 1960), the triple point of water has been assigned a value of 273.16 K and partial vapor pressure of 611.66 pascals.

23 Standard atmospheric pressure is the pressure of the Earth's atmosphere under normal conditions at sea level and is defined to be 1.013×105 $N.m^{-2}$. The SI unit of pressure is N/m^2; this unit has been given the name pascal (Pa).

24 This changed in 2019 (but has no impact as far as we are concerned – see https://www.lgcgroup.com/newsroom-and-blog/blogs/counting-down-to-the-si-redefinition-kelvin-and-degree-celsius/.

25 http://www.iapws.org/faq1/temper.html - I have inserted "absolute" temperature rather than "themodynamic" temperature which was used in the article.

26 http://galileo.phys.virginia.edu/classes/252/kinetic_theory.html.

27 Newton showed that Boyle's Law (PV = constant) followed if the repulsion between the microparticles was proportional to the inverse-square of their distance apart. The aether (also spelled ether) was considered to be a space-filling substance, thought to be necessary as a transmission medium for the propagation of electromagnetic or gravitational forces. This early modern aether has little in common with the aether of classical elements from which the name was borrowed. Aether theories have fallen out of use in modern physics and have been replaced by more abstract models.

28 I have ignored (and will ignore) the vector characteristics of force in this discussion because we will only be interested in the magnitude of the momentum change.

29 See Section 1.12.1 Equation 1.66 and knowing that Newton's Second law of Motion is F = ma (Section 1.9.4, Equation 1.52).

30 See Section 1.11.9.

31 Why have I used the term speed instead of velocity? Remember that velocity is a vector and has a direction. The average gas microparticle velocity is zero, since in a one-dimensional system there are just as many gas microparticles going right (+ velocity) as there are going left (– velocity). This argument equally applies to a three-dimensional system.

32 If the average microparticle speed is <v>, then there are equal number of microparticles with speeds less than this average as there are microparticles with speeds greater than this average.

33 Many distribution graphs label the *y*-axis as a probability of finding a microparticle with a speed of v + dv and this is more rigorous. I have labelled it "number of microparticles with speed v + dv" because the area under the graph is then the total number of microparticles in the system of interest (N). The *y*-axis can be converted to probability by division of number of microparticles with speed v + dv by the total number of microparticles N.

34 Including http://www.thermopedia.com/content/942/.

35 Just a thought – how can a vertical line define an area?

36 Remember velocity is a vector.

37 Total momentum change is $\left(-|m\mathbf{v}_x|\right)-\left(+|m\mathbf{v}_x|\right)=-|2m\mathbf{v}_x|$.

38 Remember in the limit $\frac{\Delta p}{\Delta t}\rightarrow\frac{dp}{dt}$ and $\frac{dp}{dt}=F$ (Equation 1.41Section 1.10.1). This approach is not 100% rigorous but it allows an understanding of how the gas law is derived from first principles.

39 Molecules are groups of atoms held together by chemical bonds.

40 Feynman used the word "heat" – a concept that is considered in Chapter 4.

41 http://www.feynmanlectures.caltech.edu/I_10.html.

42 See Section 3.3.4.

43 For example, in real gases, there are electromagnetic forces between the particles (it is the reason a gas liquefies and solidifies at low temperatures). In addition, in gaseous molecules, there can be rotational or vibrational energy. Note – gravitational effects on a microparticle can be neglected (see Section 3.4).

44 A macroscopic physical quantity is a measurement of a macroscopic quantity.

45 1 kilopascal = 1,000 newtons per square metre or 1 kPa = 1,000 $N.m^{-2}$.

46 Polyatomic gases have more than one atom (also see Section 7.2) – polyatomic gases are not considered in any detail in this book.

47 In statistics, precision refers to the dispersion of the observations. High precision denotes a very small spread of results from the "true" probability value which in this case is either ½ (coin) or ⅙ (die).

48 As we shall see a perfectly insulated container, like an ideal gas, is a construct that does not occur in nature but proves incredibly useful in the understanding of science. You will come across many more of these ideal constructs as you proceed through this book.

49 Most texts describe a diathermal wall as one which will allow the flow of heat. First, I have not defined "heat" or "heat flow" yet (see Chapter 4). Second, as we shall see in later chapters, the use of the term "heat" in thermodynamics is a major cause of problems and misunderstandings.

50 See Section 3.3.1.

51 The nucleus of an atom is about 10^{-15} m in size; this means it is about 10^{-5} of the size of the whole atom. A good comparison is that if an atom were the size of the Earth, then the nucleus would be the size of a typical football stadium.

52 There are "abnormal" hydrogen atoms that contain one neutron ("deuterium") and two neutrons ("tritium") – these different types are known as isotopes.

53 More accurately and rigorously (i.e., mass in kg) - neutron = $1.6749286 \times 10^{-27}$ kg; proton = $1.6726231 \times 10^{-27}$ kg; and electron = $9.1093897 \times 10^{-31}$ kg.

54 I was taught to call it atomic weight and I (and many others) still use the term. However, this is another confusing area – atomic weight is not a weight and is still used by people like me because of custom and practice. The use of term "relative atomic mass" is proposed as an alternative to atomic weight. I will use the term atomic weight because (i) I grew up with it and (ii) the use of the more modern "relative mass" has its own problems (e.g., confusion with "atomic mass" which is the term used for the mass of a single atom – now often referred to as "relative isotopic mass). I will continue to use the old-school term – atomic weight – but with no units.

55 For further information see https://www.sciencedirect.com/topics/chemistry/atomic-mass.

56 The standard value is $6.02214076 \times 10^{23}$ (SI definition) – see https://en.wikipedia.org/wiki/Avogadro_constant.

57 See https://www.scientificamerican.com/article/how-was-avogadros-number/ for some information on experimental methods used.

58 Usually written as "equal volumes of all gases, at the same temperature and pressure, have the same number of atoms or molecules"

59 dm is the abbreviation of the decimetre, which is equal to one-tenth of a metre.

60 This must be the case because if they did not have the same average kinetic energy then the object whose microparticles had (on average) a higher kinetic energy would tend to transfer energy to the object whose microparticles had a (on average) a lower kinetic energy. This does not occur because two objects or systems in thermal equilibrium are always at the same temperature. It has never been observed that two objects at the same temperature when placed in thermal contact exchange energy such that a temperature difference exists between them. Therefore, if two objects have the same temperature, then their microparticles must have the same average kinetic energy.

61 I gave an indication of speeds of microparticles in gases in Section 3.5.3.

62 "What is Life?" by E.Schrödinger (First published 1944 and based on lectures delivered under the auspices of the Dublin Institute for Advanced Studies at Trinity College, Dublin, in February 1943) http://www.whatislife.ie/downloads/What-is-Life.pdf.

63 It makes no difference to the overall argument if the system is not an ideal gas – real solids, liquids, and gases have other contributions to their internal energy other than the kinetic energy of the microparticles.

64 https://www.britannica.com/science/perpetual-motion.

65 https://rationalwiki.org/wiki/Perpetual_motion.

66 See https://engineering.mit.edu/engage/ask-an-engineer/is-it-possible-to-construct-a-perpetual-motion-machine/ for a good review of the subject.

4 First Law of Thermodynamics

4.1 INTRODUCTION

As we progressed though Chapters 1, 2, and 3, we moved from classical mechanics to thermodynamics. We also moved from processes with no directionality (classical mechanics) to processes that have directionality. Thermodynamics is universal, and it applies to everything in nature including engineering, living organisms, cosmology, chemistry, physics, etc. Because the development of classical thermodynamics grew out of the desire to design and build efficient heat engines, no other discipline in mathematical science is riddled with so many logical and mathematical inconsistencies, differences in definitions, and ill-defined notations as thermodynamics. The development of the theory of thermodynamics followed two conceptually rather different lines of thought. The first (historically), known as classical thermodynamics, is based on fundamental laws that are assumed as axioms, which in turn are based on experimental evidence. Conclusions are subsequently drawn from them using the notion of a thermodynamic state of a system (known as the macrostate), which includes temperature, volume, and pressure. The second is known as statistical thermodynamics where the state of a thermodynamic system is completely specified by the position and velocity of all the microparticles at an instant time; this is known as a microstate. The rest of this book is devoted to thermodynamics.

In Section 3.5.6, I wrote that it is the existence of variable internal energy that makes a system thermodynamic. Internal energy of an object or system is the energy associated with the random, disordered motion of microparticles. It excludes the kinetic energy of motion of the object/system as a whole, and the potential energy of the object/system as a whole due to external force fields. Though it is a macroscopic quantity, internal energy can be explained in microscopic terms by two components. One is the sum of the kinetic energies associated with system's microparticles (translations, rotations, and vibrations). The other is the potential energy associated with inter-particular interactions as a result of electromagnetic forces (this includes chemical bonds). When considering the ideal gas, we can neglect the potential energy associated with interactions of the microparticles, and the kinetic energy associated with rotation and vibration of the microparticles. In other words, as far as ideal gases are concerned, internal energy is only the chaotic microscopic translational kinetic energy associated with the system's constituent microparticles.

If you Google "thermodynamics", you will get something like the following:

- *Thermodynamics is a branch of physics that deals with the energy and work of a system.*
- *Thermodynamics is a branch of physics concerned with heat and temperature and their relation to energy and work.*
- *Thermodynamics is the branch of physics that deals with the relationships between heat and other forms of energy. It describes how thermal energy*[1] *is converted to and from other forms of energy and how it affects matter.*
- *Thermodynamics is the study of heat,* "thermos", *and work,* "dynamics".
- *Thermodynamics is the study of energy changes accompanying physical and chemical changes.*

Not one mentions internal energy, but three out of the five mentions "heat", a term which I have not yet defined. As modern science has evolved from its beginnings in the 17th century, there have been two principal theories of heat. The first theory was that heat is something akin to a fluid that flows from one location to another. This theory is known as the caloric theory and is hardly ever taught today. However, the caloric theory is very much akin to our intuitive feeling for heat; for example, when we say a hot object contains more heat than a cold object, and that on a cold day heat flows out a house. The other theory of heat is that, like work, it is an energy transfer process. This is known as the mechanical theory of heat,[2] and this approach is almost universally used in modern science and in the teaching of thermodynamics. In this chapter, I will be using the term "heat" in its modern sense as an energy transfer process, but unfortunately in modern science, the term "heat" is used in both as a process and in the caloric sense (sometimes in the same sentence): to add to the confusion, the term "heat" is also used as a synonym for internal energy (when, to add to the confusion, it is often called "thermal energy" – a meaningless term which we will encounter a number of times in this chapter; I will demonstrate why the term "thermal energy" should not be used in Section 4.5). A good definition of heat (in the modern scientific sense) is:

DOI: 10.1201/9781003608400-4

> When energy is exchanged between two systems at different temperatures, in thermal contact, the process of energy transfer is called heat.

The Scottish chemist Joseph Black (1728–1799) was the first person who used the word "heat" in a scientific sense, when he clearly distinguished between the intensive quantity of temperature and the extensive "quantity of heat". He also stated[3]:

> By the use of these instruments [thermometers] we have learned, that if we take 1000, or more, different kinds of matter, such as metals, stones, salts, woods, cork, feathers, wool, water and a variety of other fluids, although they be all at first of different heats, let them be placed together in a room without a fire, and into which the sun does not shine, the heat will be communicated from the hotter of these bodies to the colder, during some hours, perhaps, or the course of a day, at the end of which time, if we apply a thermometer to all of them in succession, it will point to precisely the same degree.

This led Black to conclude that temperature difference is the driving force for a flow of "heat" (the first time this was mentioned in the scientific literature). Note that "Black's heat" behaves like our intuitive sense of heat – the "different kinds of matter" have "different heats", and this "heat" flows from "the hotter of these bodies to the colder". We say nowadays that objects in thermal contact eventually come into thermal equilibrium, which means they finally attain the same temperature, after which no further net transfer of energy[4] takes place (or, using the term in its modern scientific sense – there is no heat).

This chapter introduces the First Law of Thermodynamics, by considering the energy capacity[5] of a fixed quantity of ideal gas at constant volume and constant pressure. Unfortunately, energy capacity is almost universally known as heat capacity, and this is the term I will use in this book.[6] Before considering the heat capacity of an ideal gas, I now need to rigorously define the concepts of system, surroundings, universe, and heat reservoirs.

4.2 SYSTEM, SURROUNDINGS, UNIVERSE AND HEAT RESERVOIRS

One of the fundamental concepts of thermodynamics is that we can divide the universe into a system and its surroundings (Figure 4.1):

1. The universe[7] is the system plus the surroundings.
2. A system is a precisely defined macroscopic region of the universe – in this context, the term "system" is simply anything that we wish to consider. The system is separated from its surroundings by a boundary (sometimes called "thermodynamic walls"), which may be notional or real and may or may not be fixed in shape or size.

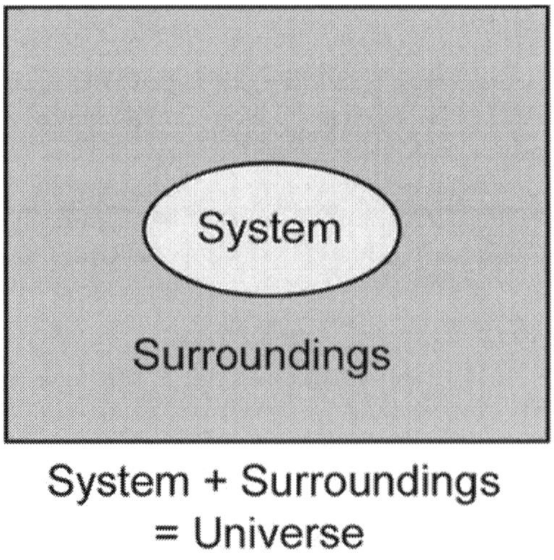

Figure 4.1 System, surroundings and universe.

3. The region that is not the system is known as the surroundings. In theory, the surroundings are everything else that is not the system, i.e., the rest of the universe. While the system is the part of the universe being studied, the surroundings is the rest of the universe that lies outside the boundaries of the system. Depending on the type of system, it may interact with the system by exchanging mass and energy. Pragmatically, the surroundings may be considered to be a subset of the universe providing relevant properties of the surroundings (such as temperature, pressure, and volume) that do not change when the thermodynamic process occurring in the system.[8] When a cup of coffee cools in a room, the temperature of the room, for all intents and purposes, does not change as the cooling process occurs. Therefore, the room can be considered the surroundings, the cup of coffee the system and the universe the cup of coffee and the room.
4. Exchange of energy or mass may or may not take place across the system/surroundings boundary, but what goes out of the system must go into the surroundings and *vice versa*. Although it is rigorously correct to say that the surroundings are everything in the universe apart from the system, the following definition of surroundings is a more practical definition – *the surroundings are the rest of the universe that interacts with the system.*

The system is separated from the surroundings by thermodynamic walls; they can be:

- Restrictive and constrain an extensive system parameter (such as mass or energy).
- Non-restrictive and allow an extensive system parameter to change.

Remember extensive quantities are dependent on the extent of the system, and intensive quantities are independent of the extent of the system (Section 1.2.2). Energy, volume, and mass (or number of particles) are extensive parameters, and temperature and pressure are intensive parameters.

There are three types of thermodynamic systems (see Figure 4.2):

1. Isolated systems – neither matter nor energy can be transferred from these systems to the surroundings.
2. Closed systems – matter cannot be transferred from a closed system, but energy can.
3. Open systems – both mass and energy can move into and out of an open system.

Open systems are difficult systems to study scientifically because there are many variables to track. The Earth, for example, is an open system. Energy comes in from the Sun and is re-radiated back to space. Meteors strike the Earth and cause parts of the Earth to be ejected into space. Another example of an open system is a living organism. The thermodynamic walls in open systems can be real or imaginary. Open systems are not considered, to any extent, in this book.

Isolated systems are surrounded by a restrictive wall that does not allow exchange of energy (by the processes of heat and/or work) or matter with the surroundings. This type of wall is known as an adiabatic[9] wall, which is defined as insulated, rigid, and non-permeable barrier. If you Google "adiabatic wall", you will get something like "an adiabatic wall between two thermodynamic systems does not allow heat or matter to pass across it".[10] Heat in modern science is a process, but in this sense, heat is being used either a synonym for energy or in the caloric sense (and I assume it is the former as the study of caloric is *verboten* in modern science). This is an example of the

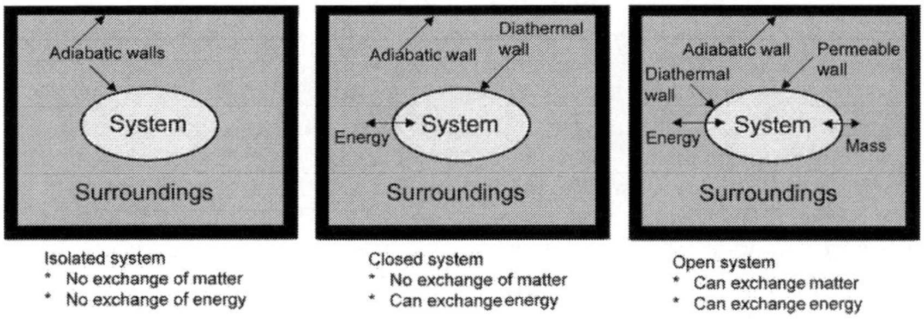

Figure 4.2 The three thermodynamic systems.

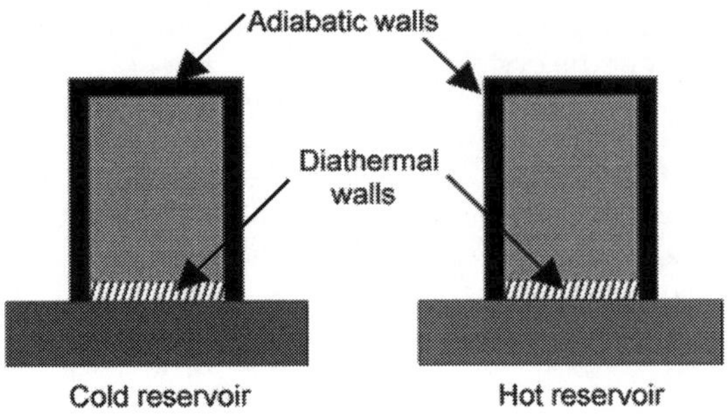

Figure 4.3 Heat reservoirs.

sloppiness which pervades thermodynamics. Although an isolated system can be approached, a truly adiabatic wall does not exist in reality, i.e., there is no such thing as a truly isolated system apart from the universe.

A closed system is surrounding by a wall that allows transfer of energy (but not mass) with the surroundings. A thermodynamic wall that is non-permeable (no mass transfer) but allows the transfer of energy between system and surroundings, by the process of heat, is known as a diathermal wall. A closed system can also transfer energy by the process of work (see Section 4.3).

I now need to introduce the concept of a heat reservoir,[11] which is a hypothetical object that can supply to or receive from another object infinite amounts of energy at a constant temperature, by the process of heat, when the two objects are in thermal contact. A heat reservoir is another concept, like the ideal gas and isolated systems, that, in reality, do not exist but can be closely approached, e.g., lakes, oceans, and rivers, large masses of air often serve as heat reservoirs in geophysical processes, such as the weather. The point to remember is that the temperature of the heat reservoir does not change when energy is added or extracted from it by the process of heat because of its *near infinite energy capacity* compared to the system.[12] A reservoir that supplies energy to a system by the process of heat is called a hot reservoir or source, and one that absorbs energy from a system by the process of heat is called a cold reservoir or sink (see Figure 4.3).[13]

Before moving on, I want to introduce the concept of a "work reservoir". The concept of a work reservoir is not commonly used, but its use does convey some advantages. When I Googled "work reservoir thermodynamics", I only found one scientific source using the term "work reservoir"; this was titled "Work Reservoirs in Thermodynamics".[14] Below are two quotes from the abstract:

> We stress the usefulness of the work reservoir in the formalism of thermodynamics …
> …, in thermodynamics, there is a significant supplementary aspect, that cannot be circumvented, related to the absolute necessity to introduce the concept of work reservoir. This is in close analogy with the need for introducing the concept of heat reservoir now in relation with the heat transfer to or from the system. ….

At this stage, the best way to illustrate what is meant by a work reservoir is by an example from classical mechanics – I will use a pendulum.[15] When the pendulum is stationary, its potential energy is at maximum,[16] and the energy in our hypothetical work reservoir is at minimum. When the weight is at minimum height, the energy in the work reservoir is at a maximum. Unlike the heat reservoir, the work reservoir is not considered to be a source or sink of infinite amount of energy that can be transferred to or from the system by the process of work. A work reservoir, as used in this book, should only be considered as an energy accounting device, e.g., it is not an infinite energy source. The concept will become clearer when heat engines are considered in Chapter 5.

In the following sections, I will consider the change, with temperature, of the internal energy of an ideal gas system under conditions of constant volume and constant pressure. In particular, I will consider the increase in internal energy when the temperature of one mole[17] of ideal gas changes by 1 K. A good, but rather longwinded, phrase to describe this characteristic would be molar internal energy change per kelvin. However, the phrase that is generally used in science is "molar heat capacity", and that is the terminology I will use in this book.

4.3 HEAT CAPACITY OF AN IDEAL GAS

4.3.1 Introduction

I have already mentioned in the previous section that the term "heat" in modern science is, like work, an energy transfer process. If you Google "heat capacity", you will come across:

- "The heat capacity of a defined system is the amount of heat (usually expressed in ... joules) needed to raise the system's temperature by one degree (usually expressed in Celsius or Kelvin). It is expressed in units of thermal energy per degree temperature".[18]
- "Heat capacity or thermal capacity is a measurable physical quantity equal to the ratio of the heat added to (or removed from) an object to the resulting temperature change".[19]

In both of the above bullet points, heat is not a process. I would suggest that most people reading the above definitions would consider heat to be a substance. I do not intend, at this stage, to pursue the problem of modern science (and modern scientists) with "heat". All I want to say here is that this problem leaves students bewildered and confused.[20] I will be again discussing this issue throughout in this book. For now, all I want to say is that in this book I will use the phrase "process of heat" to emphasise that in modern science "heat" is, like work, an energy transfer process. The first of the above quotes also uses the meaningless term "thermal energy" that we first encountered in Section 4.1.

Although I consider "energy capacity" to be a more scientifically rigorous term than "heat capacity", I will, as previously stated above, continue to use the phrase "heat capacity". My definition of heat capacity is:

> The heat capacity of an object or system is the amount energy needed to raise the temperature of the object or system by one Kelvin and this energy is input into the object or system by the process of heat.

I now want to consider how the internal energy of one mole of an ideal gas changes when the temperature is increased by 1 K; this is known as the molar heat capacity. We know, from the gas laws, that increasing the temperature of an ideal gas will result in an increase in pressure if the volume remains constant or an increase in volume if the pressure remains constant. I will start by considering molar heat capacity at constant volume.

4.3.2 Molar Heat Capacity of an Ideal Gas at Constant Volume

Based on Equation (3.46)[21] in Chapter 3, the change in internal energy (ΔU) of one mole of ideal gas by a 1 K increase in temperature is given by:

$$\Delta U = \frac{3}{2}R(T+1) - \frac{3}{2}RT = \frac{3}{2}R = C_V \quad (4.1)$$

where C_V is the molar heat capacity of an ideal gas at constant volume. Therefore, if the surroundings transfer $\frac{3}{2}R$ of energy, by the process of heat, to one mole of ideal gas, its temperature increases by 1 K. Since R is 8.314 $J.mol^{-1}.K^{-1}$ (see Section 3.6.3), the molar heat capacity at constant volume for an ideal gas is 12.48 $J.mol^{-1}.K^{-1}$. This value is independent of temperature, volume, and pressure. Table 4.1 sets out experimentally determined C_V values for the noble gases (helium, neon, and argon[22]), together with those for hydrogen and carbon dioxide at atmospheric pressure (0.1 MPa). It is immediately apparent that Equation (4.1) is only valid for the noble gases. As far as hydrogen and carbon dioxide are concerned not only does Equation (4.1) fail to calculate

Table 4.1: Experimentally Determined CV Data

	Molar Heat Capacity C_V $J.mol^{-1}.K^{-1}$						
Gas	**298 K**	**400 K**	**600 K**	**800 K**	**1,000 K**	**1,500 K**	**2,000 K**
He	12.48	12.48	12.48	12.48	12.48	12.48	12.48
Ne	12.48	12.48	12.48	12.48	12.48	12.48	12.48
H_2	20.52	20.87	21.01	21.30	21.89	23.96	25.89
CO_2	28.81	33.00	39.00	43.11	45.98	40.05	52.02

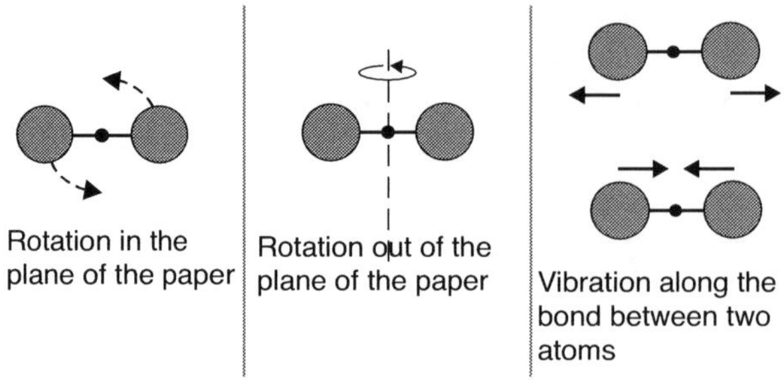

Figure 4.4 Rotational and vibrational contributions to the internal energy of a diatomic molecule.

experimentally measured C_V's, the experimentally determined C_V's exhibit a temperature dependence.

What is it about the noble gases that allows Equation (4.1) to successfully calculate molar heat capacity at constant volume? As we have already mentioned in Section 3.6.2, they are the only chemical elements that form monatomic (single atom) gases at standard temperature and pressure (STP)[23] – remember one of the assumptions of an ideal gas is – "These particles have the same mass and are perfectly spherical in shape".

The reason the molar heat capacities at constant volume of H_2 or CO_2 (and all other polyatomic gases) deviate from the calculated value of an ideal gas is associated with polyatomic nature of these microparticles. This allows energy to be stored in forms other than the kinetic energy of linear movement, i.e., vibration and rotation of the microparticles contribute to the internal energy of the gas (Figure 4.4). Studying these deviations allowed a greater understanding of molecular structure and paved the way to quantum mechanics, but this does not concern us in this book where I will mainly consider ideal gases.

4.3.3 Sign Convention

4.3.3.1 Introduction

In Section 1.10, I considered the difference between mathematical and scientific equations (at least those used in this book): I used the example $F = ma$. If all we are concerned about is the magnitude of the variables, then F, m and a are all positive real numbers, but we know that F and a are vectors, i.e., they have direction as well as magnitude. Now remember in this book I will be principally concerned with one-dimensional vectors (see Section 1.2.2), where a simple ± sign convention will suffice. I will use equations with bold font variables when concerned with direction and magnitude of one-dimensional vectors or vector-like quantities, $\mathbf{F} = m\mathbf{a}$. Using bold font and absolute values it should be apparent that $|\mathbf{F}| = m|\mathbf{a}|$ is identical to $F = ma$. Finally, although in scientific equations mass (m) is always a positive real number, a difference or change in mass (represented in this book by either Δm or dm) can be positive or negative real numbers. Accordingly, when appropriate I will use a bold font to represent this fact ($\mathbf{\Delta m}$ or $\mathbf{dm}$).[24] In other words, physical quantities represented in bold type are vector or "vector-like" and are represented by real numbers (i.e., they can be positive or negative quantities). When I think it helps, I will use equations with signed absolute values, e.g., **F** can be a positive or negative quantity (just like x in an algebraic equation), but if I use $+|\mathbf{F}|$ it is a positive quantity and $-|\mathbf{F}|$ it is a negative quantity. I hope this convention assists the non-mathematical reader.

To carry out energy accountancy we need to have rules or conventions. First let us consider a bank account. The physical quantity money is a positive number, and you cannot hold in your hand negative money. However, in accountancy, we consider money flows into and out of the bank account. The sign convention is the flow of money into the account is considered a positive quantity, and the flow of money out of the account is considered a negative quantity. It is the vector-like flow of money that can be positive or negative, not money *per se* which is always a positive quantity. Summation of these positive and negative numbers gives the balance or the sum of money in the bank account.

Similarly, in thermodynamics, we need to establish a sign convention so we can carry out energy accountancy. Not only that but we need to adhere to it rigorously. Unfortunately, this often does not always happen in the teaching of thermodynamics and is yet another source of confusion among students. In the next section, I am going to set down a sign convention for energy flows and I plan to stick to it throughout this book (unless I state otherwise).

Figure 4.5 Cylinder with piston.

4.3.3.2 The Sign Convention

The energy flows from one location to another, by the processes of heat and work, is represented by **q** and **w**, respectively (many texts use upper case **Q** and **W**). It is important to understand that work and heat are processes, and **q** and **w** represent the energy flows resulting from these processes. The symbol **q** represents the energy that flows, as a result of a temperature difference, between a system and its surroundings (heat reservoirs). For the process of heat to occur the system, its surroundings must be in thermal contact and there must be a temperature difference. Similarly, the symbol **w** represents the energy that flows between a system and its surroundings (work reservoir) that are in mechanical contact. As far as we are concerned in this book, a system is in mechanical contact with its surroundings when the system is a cylinder with a movable gas-tight piston (see Figure 4.5) or can be represented by a cylinder/piston arrangement. It can be seen from Figure 4.5 that if energy is supplied to the ideal gas from a heat reservoir, then the temperature of the gas increases, and its expansion causes the piston to rise. This results in the piston acquiring kinetic energy (which, for example, could be used to raise a weight against the force of gravity), i.e., energy is transferred to the work reservoir.

I will now introduce a sign convention that will be used to keep track of energy changes. We take our point of view as the system and when energy flows out of the system, this is assigned a negative sign, and when energy flows into the system, this is assigned a positive sign.[25] If **q** amount of energy flows into the system by the process of heat, then the flow of energy into the system is $+|\mathbf{q}|$, if the flow of energy is *into* the system *from* the surroundings, and $-|\mathbf{q}|$, if the flow of energy is *from* the system *into* the surroundings. Similarly, if **w** amount of energy flows into the system by the process of work, then the change in the energy content of the system is $+|\mathbf{w}|$, if the flow of energy is into the system from the surroundings, and $-|\mathbf{w}|$, if the flow of energy is from the system into the surroundings. In other words, if the system gains energy from the surroundings, the energy flow is considered positive and if the system loses energy to the surroundings, its energy flow is considered negative, e.g., if 2 joules of energy flow out of the system, then **q** or $w = -2$ J, i.e., -2 joules of energy has flown into the system. A good way of considering this is – "call energy flowing into the system as positive and energy leaving the system negative".[26] It is also worth noting that when the surroundings gain 2 joules of energy then **q** or $w = -2$ J, i.e., 2 joules of energy must flow out of the system to the surroundings.

This convention is similar to the sign convention for vectors in one-dimensional space (see Section 1.7.2). On the face of it, this appears straightforward, but there is an area of confusion, especially for the non-mathematical. The confusion is that many thermodynamic texts consider that q and w represent both energy (a positive scalar quantity) and energy flow (a vector) and interchange between the two without explicitly commenting. This is the principal reason for using, when appropriate, absolute value equations with bold fonts representing one-dimensional vectors or vector-like quantities. If you are confused at this point, the following sub-sections should provide, hopefully, some clarifications.

4.3.3.3 Energy Flow from a Single Reservoir to or from the System

Consider Figure 4.6a showing a closed system in thermal contact with a heat reservoir (aka the surroundings). Using our sign convention ("energy flowing into the system as positive and energy leaving the system negative"), we know **q** is a symbol for a real number whose value can be positive or negative. If energy is flowing from the system to the heat reservoir Figure 4.6b, we know

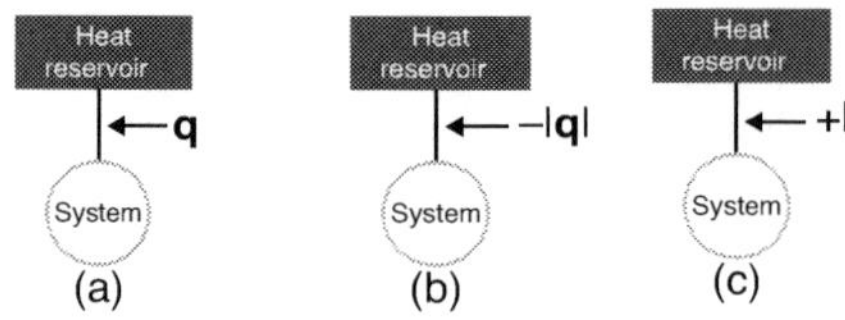

Figure 4.6 Sign convention, vector-like real numbers and absolute values.

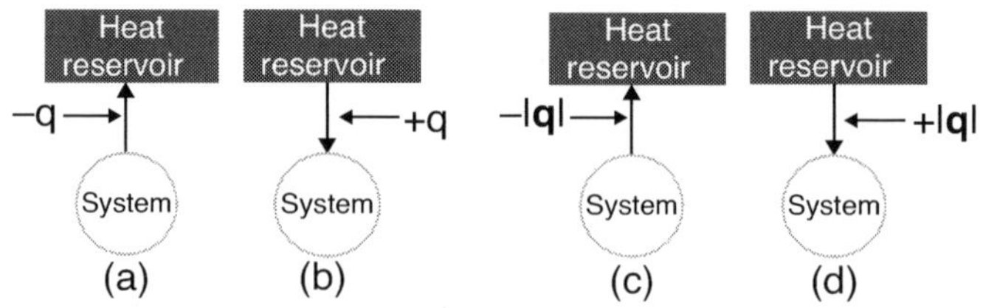

Figure 4.7 Confusing energy flow symbolism used in many thermodynamic texts.

the energy change of the system is $-|\mathbf{q}|$ (a negative real number). If the energy flow is into the system from the heat reservoir Figure 4.6c, then we know the energy change of the system is $-|\mathbf{q}|$.

Figure 4.6b represents a loss of energy from the system, i.e., there is an energy flow from the system to the heat reservoir by the process of heat. While Figure 4.6c represents gain in energy by i.e., there is an energy flow from the heat reservoir to the system. Remember energy, a scalar quantity, is always a positive real number. It is important to realise that Figure 4.6a can represent both the situations shown in Figures 4.6b and c, i.e., **q** represents a real number and can be either positive or negative.

Most thermodynamic texts also use "energy flow arrows" that point in the direction of energy flow and typically, you will come across something like Figures 4.7a and b.[27] In these two figures, q must be a positive scalar quantity or why would you use + or − as prefixes? In both Figure 4.7a and b q must represent energy and what is negative energy (−q)?[28] In Figures 4.6, and 4.7c & d **q** represents an energy flow (a real number) and dependent on the choice of sign convention may take either a positive or negative value. You may think I am splitting hairs but, unfortunately, you will often find that in many thermodynamic texts, the symbols q and w are used in an ambiguous manner – sometimes they represent vectors (flows of energy) and sometimes they represent scalars (energy). When I first started on this book found this ambiguity very confusing and this is the reason I have decided to use the convention set down here of bold font representing vector or vector-like quantities. When I return to this in Section 5, I will demonstrate that this is not a minor issue. I am being pedantic, but this is part of scientific rigorousness. Figures 4.7c and d may contain some redundancy but will be used in this book together with the more general Figure 4.6a. Figures 4.7a and b are at best confusing and at worst just plain wrong.

If the energy change of the system is $-|\mathbf{q}|$ joules, then the energy change of the heat reservoir is $+|\mathbf{q}|$ joules (Figure 4.7c), and the energy change of the universe (heat reservoir and system) is zero. This must be so or else the Principle of Conservation of Energy is contravened (remember the universe is an isolated system.

4.3.3.4 Energy Flows from Multiple Reservoirs to or from the System

As we shall see in Chapter 5, Figure 4.8 depicts the energy flows in a heat engine using our sign and symbol convention. Where $\left(+|\mathbf{q_H}|\right)$ joules of energy flow into the system (the heat engine) by the process of heat, $\left(-|\mathbf{q_C}|\right)$ joules of energy flow into the system by the process of heat and, $\left(-|\mathbf{w}|\right)$ joules of energy flow into the system by the process of work. Alternatively, and slightly more in line with everyday speech we could write $\left(+|\mathbf{q_H}|\right)$ joules of energy flow <u>into</u> the heat engine) by the process of heat, $\left(+|\mathbf{q_C}|\right)$ joules of energy flows <u>out of</u> the heat engine by the process of heat and, $\left(+|\mathbf{w}|\right)$ joules of energy flows <u>out of</u> the system by the process of work. We could also describe the

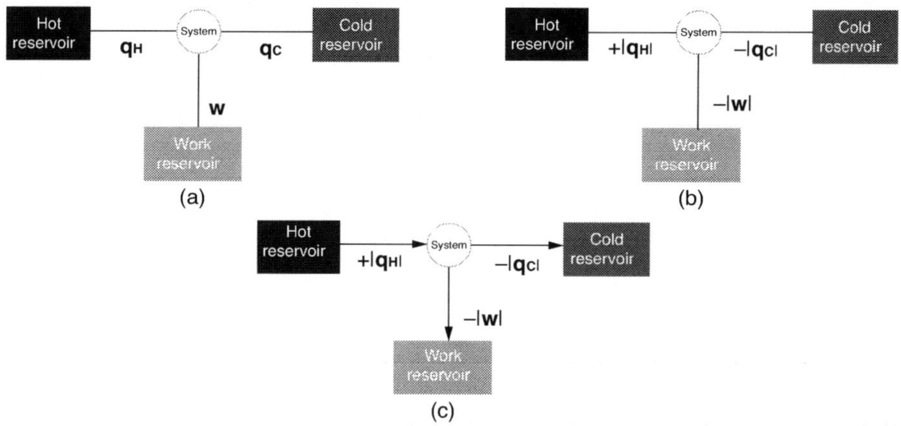

Figure 4.8 Schematic diagrams for a heat engine using our agreed sign and symbol convention.

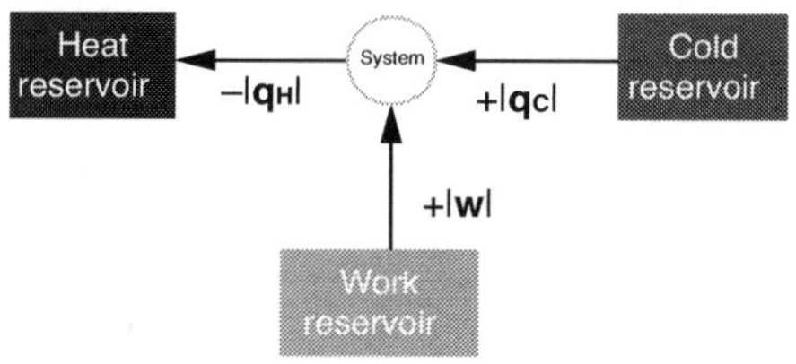

Figure 4.9 Schematic diagram for a heat pump/refrigerator using our agreed sign and symbol convention.

process solely in terms of the surroundings; $(+|\mathbf{q}_H|)$ joules of energy flow out of the hot reservoir by the process of heat, $(+|\mathbf{q}_C|)$ joules of energy flow <u>into</u> the cold reservoir by the process of heat and $(+|\mathbf{w}|)$ joules of energy flows <u>into</u> the work reservoir by the process of work.

We can calculate the change in internal energy of the system by adding $\mathbf{q}_H$, $\mathbf{q}_C$ and $\mathbf{w}$, providing their signs conform to our agreed sign convention. This is because the sign convention considers all energy flows from the system's point of view and $\mathbf{q}_H$, $\mathbf{q}_C$ and $\mathbf{w}$, are real numbers. Therefore, we can write:

$$\Delta\mathbf{U} = \mathbf{q}_H + \mathbf{q}_C + \mathbf{w} \tag{4.2}$$

This indicates an issue with Figure 4.8a since we have no idea of the direction of the energy flows, and although there is nothing wrong with this depiction of a heat engine (or heat pump – see below) it is hardly ever used because of this ambiguity. Because we are dealing with a heat engine, we know $\mathbf{q}_H$ is a positive real number, and $\mathbf{q}_C$ and $\mathbf{w}$ are negative real numbers. Therefore, we can write the following unambiguous equation for the heat engine write signed absolute values:

$$\Delta\mathbf{U} = (+|\mathbf{q}_H|) + (-|\mathbf{q}_C|) + (-|\mathbf{w}|) = |\mathbf{q}_H| - |\mathbf{q}_C| - |\mathbf{w}| \tag{4.3}$$

Although Figure 4.8c contains redundant information, I consider it does offer advantages over Figure 4.8b, and for this reason, I will tend to use this format in this book.

Finally, let us reverse the energy flows in Figure 4.8c: this represents the energy flow of a heat pump or refrigerator (Figure 4.9). As with the heat engine, we can calculate the change in internal energy of the system by using Equation (4.4). However, because we are dealing with a heat pump, we know $\mathbf{q}_H$ is a negative real number, and $\mathbf{q}_C$ and $\mathbf{w}$ are positive real numbers. Using absolute values, we can write:

$$\Delta\mathbf{U} = (-|\mathbf{q}_H|) + (+|\mathbf{q}_C|) + (+|\mathbf{w}|) = |\mathbf{q}_C| + |\mathbf{w}| - |\mathbf{q}_H| \tag{4.4}$$

Equation (4.4) tells us that $|\mathbf{w}|$ of energy flows into the heat pump by the process of work, $|\mathbf{q}_H|$of energy flows out of the heat pump into the hot reservoir by the process of heat, and $|\mathbf{q}_C|$ of energy flows into the heat pump from the cold reservoir by the process of heat.

4.3.3.5 The Universe

The closed system (the heat engine) and the surroundings (heat and work reservoirs) constitute an isolated system known as the universe (Figure 4.10). The Law of Conservation of Energy tells us the universe contains a set amount of energy, i.e., $\Delta \mathbf{U}_{\text{universe}} = 0$. We also know, by definition, that if there is a flow of energy, say $+|\mathbf{energy}_{\text{reservoir}}|$, into the system from a particular reservoir (the system has gained energy), then the energy flow into that particular reservoir is $-|\mathbf{energy}_{\text{reservoir}}|$ (the reservoir has lost energy). Therefore, we can write the following equation for the universe represented by Figure 4.10

$$\Delta \mathbf{U}_{\text{universe}} = \left(+\left|\mathbf{q}_{\mathrm{H}}^{\text{sys}}\right|\right)+\left(-\left|\mathbf{q}_{\mathrm{C}}^{\text{sys}}\right|\right)+\left(-\left|\mathbf{w}^{\text{sys}}\right|\right)+\left(-\left|\mathbf{q}_{\mathrm{H}}^{\text{surr}}\right|\right)+\left(+\left|\mathbf{q}_{\mathrm{C}}^{\text{surr}}\right|\right)+\left(+\left|\mathbf{w}^{\text{surr}}\right|\right)=0 \tag{4.5}$$

where $+\left|\mathbf{q}_{\mathrm{H}}^{\text{sys}}\right|$, $-\left|\mathbf{q}_{\mathrm{C}}^{\text{sys}}\right|$ and $-\left|\mathbf{w}^{\text{sys}}\right|$ are the energy flows into the system and $-\left|\mathbf{q}_{\mathrm{H}}^{\text{surr}}\right|, +\left|\mathbf{q}_{\mathrm{C}}^{\text{surr}}\right|$ and $+\left|\mathbf{w}^{\text{surr}}\right|$ are the energy flows into the surroundings. We know that $\mathbf{q}_{\mathrm{H}}^{\text{sys}}\big| = \left|\mathbf{q}_{\mathrm{H}}^{\text{surr}}\right|, \left|\mathbf{q}_{\mathrm{C}}^{\text{sys}}\right| = \left|\mathbf{q}_{\mathrm{C}}^{\text{surr}}\right|$ and $\left|\mathbf{w}^{\text{sys}}\right| = \left|\mathbf{w}^{\text{surr}}\right|$. Equation (4.5) is evidently correct because the universe is, by definition, an isolated system with a fixed quantity of energy and energy is conserved.

4.3.3.6 Discussion

I struggled with the "energy accounting" mathematics used in scientific texts and it took me some time before I realised that the problem was not mine and it rested with the authors. There is often an ambiguous use of algebraic symbols, sometimes q or w represent energy (scalar) and sometimes they represent energy flows (vectors); often with no explanation given, i.e., there is no consistency or scientific rigorousness. There are occasional texts where attempts are made to deal with this problem but not many. I have come to realisation that most teachers and lecturers (including those who write about the subject) are as confused as their students will turn out to be.

4.3.4 Sign Convention and Molar Heat Capacity at Constant Volume

In Figure 4.11, the closed system contains 1 mole of ideal gas in thermal contact with a heat reservoir (the surroundings). In Figure 4.11a the system and the surroundings (in this case the cold reservoir[29]) are both at the same temperature (T K). They are in dynamic thermal equilibrium and there is no net flow of energy from the cold reservoir to the system, by the process of heat. In Figure 4.11b the cold reservoir at temperature T K is replaced by a hot reservoir at temperature (T+1) K. Energy flows, by the process of heat, from the hot reservoir into the system. Remember a

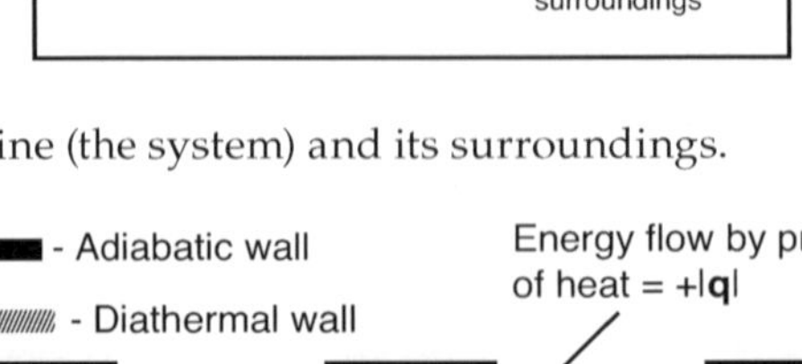

Figure 4.10 The heat engine (the system) and its surroundings.

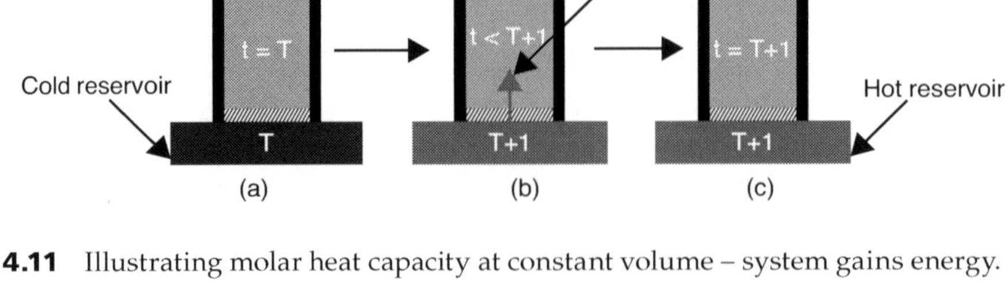

Figure 4.11 Illustrating molar heat capacity at constant volume – system gains energy.

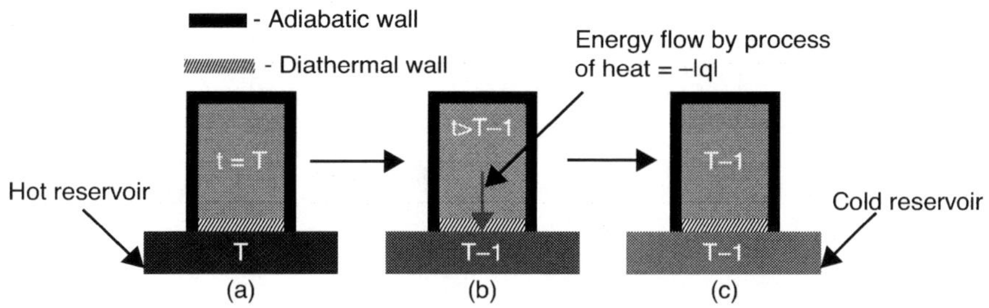

Figure 4.12 Illustrating molar heat capacity at constant volume – system loses energy.

heat reservoir contains, by definition, an infinite quantity of energy, and any energy flowing into (or out of) the system is negligible compared to the quantity of energy in the reservoir and, consequently, the temperature of the heat reservoir remains constant. In Figure 4.11c the system is again in dynamic thermal equilibrium albeit at a higher temperature. Using Equation (4.1) we can write:

$$+\left|\Delta\mathbf{U}_{\mathbf{system}}\right| = +\left|\mathbf{q}\right| = \frac{3}{2}\mathrm{R} = \mathrm{C_V} \tag{4.6}$$

In other words, $+|\mathbf{q}|$ (equal to $\frac{3}{2}$R) joules of energy has flowed into the system of 1 mole of ideal gas at constant temperature and the temperature has risen 1 K: this is the molar heat capacity of an ideal gas at constant volume.

Let us now consider the system cooling down by 1 K (Figure 4.12). In Figure 4.12a the closed system and the hot reservoir are in dynamic thermal equilibrium. In Figure 4.12b the hot reservoir at temperature T K, is replaced by a cold reservoir at temperature $(T-1)$ K. In contrast to Figure 4.11, the energy flows, by the process of heat, from the system into the cold reservoir and the temperature of the system decreases and the temperature of the heat reservoir remains constant. In Figure 4.12c the system is again in dynamic thermal equilibrium at the lower temperature.

We know that in this case **q** is a negative real number and we can write:

$$-\left|\Delta\mathbf{U}_{\mathbf{system}}\right| = -\left|\mathbf{q}\right| = -\frac{3}{2}\mathrm{R} = -\mathrm{C_V} \tag{4.7}$$

4.3.5 Molar Heat Capacity at Constant Pressure

I want now to consider the molar heat capacity of an ideal gas at constant pressure (C_P). In the real world, most processes of interest take place at constant atmospheric pressure, e.g., virtually all life processes, all chemical reactions that are carried out in beakers, flasks open to the atmosphere, most industrial processes, meteorology, etc. The best way to understand the difference between heat capacity at constant volume and constant pressure is to consider an ideal gas in a cylindrical container sealed by a frictionless piston that has no mass (i.e., there are no frictional or gravitational effects on the piston). The closed system, situated in the Earth's gravitational field, is placed in a vacuum, i.e., the pressure of the surroundings (P_{surr}) equals zero. To provide a constant pressure a weight (of mass m) is placed on top of the piston (Figure 4.13).

We know from the gas laws that if pressure is constant an increase in temperature results in an increase in volume. We also know (Section 2.2) that, if the point of application of a force moves it is said to do work. We can see from Figure 4.13 that when the temperature of the gas in the piston/cylinder arrangement changes, the point of application of the force changes, i.e., the weight moves in a vertical direction with or against gravity (dependent on positive or negative temperature change). Therefore, at constant pressure a temperature increase not only increases the internal energy of the ideal gas, it also increases the potential energy of the weight, i.e., there is a flow of energy $+|\mathbf{w}|$ to the surroundings (the work reservoir) from the system by the process of work.[30] This is often stated by saying "the system has done work on the surroundings". In thermodynamics, work is said to be done by a system if the sole effect on the surroundings can be reduced to the raising or lowering of a weight. The weight may not actually be raised, but the net effect external to the system would be the raising of a weight. The definition of work as "weight lifted through a height", was originally defined in 1824 by Carnot in his famous paper Reflections on the Motive Power of Fire (which is considered in the next chapter)[31]:

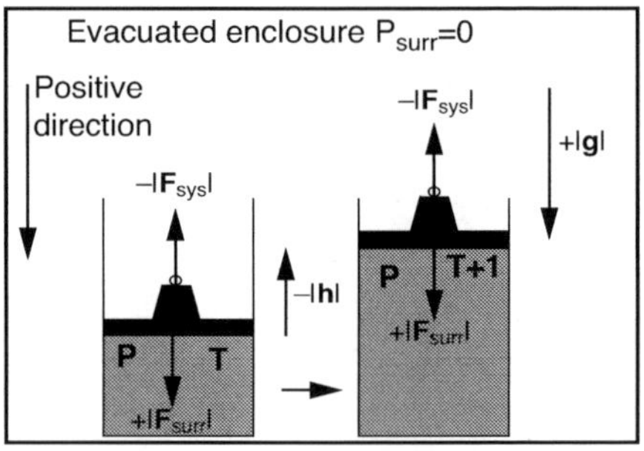

Figure 4.13 Container of an ideal gas sealed by a massless and frictionless piston: the containers are in an evacuated enclosure ($P_{surr}=0$) on Earth and the gas pressure is the result of placing a weight on the piston The temperature of the gas increases from T.K to T+1 K.

> We use here [work][32] to express the useful effect that a motor is capable of producing. This effect can always be likened to the elevation of a weight to a certain height. It has, as we know, as a measure, the product of the weight multiplied by the height to which it is raised.

Let us assume the piston/cylinder arrangement in Figure 4.13 contains one mole of an ideal gas at temperature T.K and the temperature increases by 1 K and the vertical displacement of the weight is $-|\mathbf{h}|$ (note sign convention shown in Figure 4.13). The flow of energy to the system by the process of work is:

$$-|\mathbf{w}| = m(+|\mathbf{g}|)(-|\mathbf{h}|) \tag{4.8}$$

Remember this is a flow of energy from the system to the surroundings because the potential energy of the weight has increased.

At equilibrium, the force exerted on the gas (the system) by the surroundings ($\mathbf{F}_{surr}$)is equal in magnitude, but opposite in sign to the force exerted on the surroundings by the system $(\mathbf{F}_{sys})$, i.e., $|\mathbf{F}_{surr}| = |\mathbf{F}_{sys}|$.[33] Using our agreed sign convention we know $+|\mathbf{F}_{surr}| = m(+|\mathbf{g}|)$ (using our sign convention the vector **g** acting vertically downwards is positive). We know $-|\mathbf{F}_{sys}| = -(A \times P)$ where $\mathbf{F}_{sys}$ is the force acting vertically upwards on the weight (i.e., the surroundings) by the piston, and A is the area of the piston.[34] Note that the $\mathbf{F}_{sys}$ vector is negative as it is acting vertically upwards. Knowing $(+|\mathbf{F}_{surr}| - |\mathbf{F}_{sys}|) = 0$, we can write $m(+|\mathbf{g}|) = A \times P$. Therefore, we can write Equation (4.8) as:

$$-|\mathbf{w}| = A \times P(-|\mathbf{h}|) \tag{4.9}$$

Multiplying both sides of Equation (4.9) by –1 and knowing that $A \times (+|\mathbf{h}|)$ is the change in the volume occupied by the gas $(+|\Delta\mathbf{V}|)$ we can write Equation (4.9) as:

$$+|\mathbf{w}| = P\ (+\Delta V) \tag{4.10}$$

From the ideal gas law, we can calculate $\Delta\mathbf{V}$ when the temperature of one mole gas is raised by 1 K at constant pressure P. Let the volume of the gas at temperature T = V_T and the volume at $T_{T+1} = V_{T+1}$:

$$+|\Delta\mathbf{V}| = V_{T+1} - V_T = \frac{R(T+1)}{P} - \frac{RT}{P} = \frac{R}{P} \tag{4.11}$$

This yields the simple equation for work done when the temperature of one mole of ideal gas is raised by 1°C (or 1 K) at constant pressure:

$$+|\mathbf{w}| = P|+\Delta\mathbf{V}| = \frac{RP}{P} = R \tag{4.12}$$

Unfortunately, we have a problem as I will now explain. When the temperature increases the weight rises, i.e., the energy of the surroundings has increased by the process of work (we have increased the potential energy of the weight) and this must be at the expense of the internal energy of the ideal gas, i.e., energy is flowing out of the ideal gas by the process of work into the surroundings. However, I have just shown in Equation (4.12) that this work is a positive number which is in contravention of our sign convention (all energy flows into the system are positive).

The mathematically rigorous way to deal with this paradox is to change the sign convention such that work done by the system on the surroundings is positive and this is the sign convention used by physicists and mathematicians (I will discuss this further in the next section). The fudge used by chemists to allow us to stick with our sign convention that all energy flows into the system are positive and all energy flows out of the system are negative is to simply redefine **w** such that **w** is positive when $|\Delta \mathbf{V}|$ is negative and *vice versa*. In other words, to simply change the sign of the calculated **w**, i.e., when we calculate $+|\mathbf{w}|$ we change it to $-|\mathbf{w}|$. This "fudge" is mathematically nonsensical because this means we rewrite Equation (4.12) as:

$$-|\mathbf{w}| = \left[P\left(+|\Delta \mathbf{V}|\right)\right] = R \tag{4.13}$$

which is meaningless (a negative quantity cannot equal a positive quantity). A version of Equation (4.13) without the use of absolute values is:

$$\mathbf{w} = -P\Delta\mathbf{V} = -P\left(V_{final} - V_{initial}\right) \tag{4.14}$$

To say the least, not very satisfactory, mathematically flawed and another source of confusion, but there it is, and being chemists, we need to work with it Please view this as a 100% total fudge. I will return this equation in the next chapter (Section 5.6.2).

This type of work is known as "PV work"[35] The temperature of the gas is increasing and the weight (force) is moving upwards against the force of gravity, i.e., the energy flow by the process of work is from the system to the surroundings (work reservoir), i.e., the energy flowing into the system by process of work is $-|\mathbf{w}|$ and it is negative because the system is losing energy to the surroundings by the process of work.

In expansion at constant pressure $\left(+|\Delta \mathbf{V}|\right)$ the potential energy of the weight is increasing; this is at the expense of the internal energy of the ideal gas $\left(-|\mathbf{w}|\right)$. We say the system is doing work on the surroundings. In the case of contraction at constant pressure $\left(-|\Delta \mathbf{V}|\right)$, the potential energy of the weight is decreasing; this results in an energy flow into the system from the surroundings $\left(+|\mathbf{w}|\right)$. We say the surroundings are doing work on the system.

Now I have introduced the concept of PV work let us consider the molar heat capacity at constant pressure (C_P). Standard atmospheric pressure is (P_{atm}) is, by definition, 101.325×10^3 pascals or $N.m^{-2}$. Therefore, if we placed a weight of $101.325 \times 10^3 \times A$ newtons, or a mass of $\frac{\left(101.325 \times 10^3\right)A}{|\mathbf{g}|}$ kg (where $|\mathbf{g}|$ is the acceleration due to gravity ($9.81\ m.s^{-2}$ and A is the area of the piston) on the piston under a vacuum this would be equivalent to the expansion taking place under atmospheric pressure using our frictionless and massless piston (Figure 4.14). As before the 1K temperature rise increases the internal energy of 1 mole of ideal gas by $\frac{3}{2}R = C_V$ (remember for an ideal gas C_V is independent of temperature, pressure, or volume (see Section 4.3.2).

We know that the process of work results in an energy change, in this case (expansion) the system has lost energy (by doing work on the surroundings), and you should now be able to see that it takes more energy to raise the temperature of 1 mole of ideal gas by 1 K at constant pressure compared to the amount of energy required to raise its temperature at constant volume because

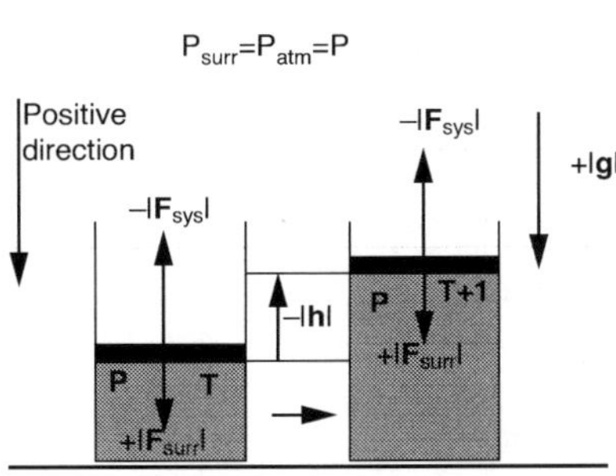

Figure 4.14 Container of an ideal gas sealed by a massless and frictionless piston under atmospheric pressure.

Table 4.2: Experimentally Determined C_P Data

Gas	C_P at 300 K ($J.K^{-1}.mol^{-1}$)
He	20.79
Xe	20.79
H_2	28.9
CO_2	37.0

part of the energy transferred from the surroundings (the heat reservoir) to the system by the process of heat is transferred to the surroundings (work reservoir) by the process of work, i.e., part of the energy transferred to the system by the process of heat does not increase the internal energy of the system (it does work). Whereas at constant volume all the energy transferred to the system increases the internal energy of the system, i.e., all the transferred energy raises the temperature of the system.

Consequently:

$$C_P = C_V + R = \frac{3}{2}R + R = \frac{5}{2}R = 12.471 + 8.314 = 20.785\ J.K^{-1}.mol^{-1} \tag{4.15}$$

Remember this relationship only applies to ideal gases.

Table 4.2 shows some C_P data at 300 K, again the noble gases behave as an ideal gas, whereas polyatomic gases do not. We have already seen that this "non-ideal behaviour" is principal because all "permanent gases" are polyatomic and only noble gases are monatomic.

4.3.6 Summary and Conclusions

We have covered a lot of ground in this section. I defined heat capacity as "the amount energy needed to raise the temperature of the object or system by one Kelvin and this energy is input into the object or system by the process of heat". I defined the molar heat capacity as the energy required, by the process of heat, to raise the temperature of 1 mole of gas by 1 K. There are two types of molar heat capacity of gases:

1. Molar heat capacity at constant volume (C_V).
2. Molar heat capacity at constant pressure (C_P).

Molar heat capacity at constant volume (C_V) is the increase in internal energy of one mole of gas when its temperature increases by 1 K: we know $C_V = \frac{3}{2}R$ (Equation 4.1). Molar heat capacity at constant pressure is equal to the amount energy, transferred by the process of heat, required to raise the temperature of one mole of gas by 1 K: we know $C_P = \frac{5}{2}R$ (Equation 4.15). The reason that $C_P > C_V$ is that some of the energy flow, by the process of heat, into the system, flows back out of the system into the surroundings (the work reservoir) by the process of work, i.e., by raising a weight against gravitational force. In other words, for an ideal gas at constant pressure, it takes more energy to achieve the same temperature increase than it does at constant volume. In introducing the subject, I used a weight on a piston, but the same process of work is used in expanding the gas at, for example, atmospheric pressure. I demonstrated this by showing that a weight of the appropriate magnitude could simulate atmospheric pressure when the piston/cylinder arrangement is housed in a vacuum.

Examination of experimentally determined molar heat capacities of gases indicated that only the monatomic noble gases showed agreement with the theoretically determined ideal gas heat capacities and it was mentioned that deviations from the calculated value of an ideal gas are associated with the polyatomic nature of these microparticles which allows energy in forms other than the kinetic energy of linear movement (vibration and rotation) to contribute to the internal energy of polyatomic gases.

We now know that the change in the internal energy of a closed ideal gas system ($\Delta\mathbf{U}$) is equal to the flow of energy into the system (**q**) as a result of temperature difference, i.e., by the process of heat, plus the flow of energy into the system by the process of work (**w**). We also know that these energy flows ("vector-like quantities") can be positive and negative, and we have an agreed sign convention that decides whether the energy flow is positive or negative ("call energy flowing into the system as positive and energy leaving the system negative".). We also know that for our piston/cylinder closed system if **q** is a positive real number (expansion) then **w** is a negative real number and *vice versa*. The only other procedure by which we can increase the internal energy of any system is by adding or subtracting mass and, by definition, this cannot occur with a closed system and we are not dealing with open systems in this book.

I also showed that our agreed sign convention led to a paradox. For an expansion $\Delta\mathbf{V}(= V_{final} - V_{initial})$ is, by definition a positive real number, as P is always a positive real number so $+|\mathbf{w}| = +P|\Delta\mathbf{V}|$ (Equation 4.12). But using our agreed sign convention, **w** is a negative real number (the system is losing energy by the process of work during an expansion process at constant pressure). The work-around used by chemists is redefined which is to say that the **w** value according to our agreed sign convention is the negative of the calculated value, i.e., if we have calculated a positive value then we convert to a negative value and *vice versa* using the nonsensical equations $-|\mathbf{w}| = +|\Delta V|P$ (for expansion) and $+|\mathbf{w}| = -|\Delta\mathbf{V}|P$ (for compression)). The mathematically rigorous way is not to use the chemists' fudge and change the sign convention and this is how physicists and mathematicians dealt with the issue (Table 4.3).

In the physicists' sign convention if the system loses energy by the process of work then this is considered positive. This is best understood by considering Figure 4.15 which shows both sign conventions.[36] Figure 4.15a is applicable to both sign conventions. Figure 4.15b is the chemists' sign convention and Figure 4.15c is the physicists' sign convention.

Using absolute values and the chemists' sign convention we know from Equation (4.4) that for a heat engine:

$$\Delta\mathbf{U}^{chemists} = (+|\mathbf{q}_H|) + (-|\mathbf{q}_C|) + (-|\mathbf{w}|) = |\mathbf{q}_H| - |\mathbf{q}_C| - |\mathbf{w}| \quad (4.16)$$

Table 4.3: Sign Conventions

Process	Sign Convention	
	Chemistry	Physics
Energy absorbed by the system by process of heat	$+\lvert\mathbf{q}\rvert$	$+\lvert\mathbf{q}\rvert$
Energy evolved by the system by process of heat	$-\lvert\mathbf{q}\rvert$	$-\lvert\mathbf{q}\rvert$
Work done by the system (energy lost by system)	$-\lvert\mathbf{w}\rvert$	$+\lvert\mathbf{w}\rvert$
Work done on the system (energy gained by system)	$+\lvert\mathbf{w}\rvert$	$-\lvert\mathbf{w}\rvert$

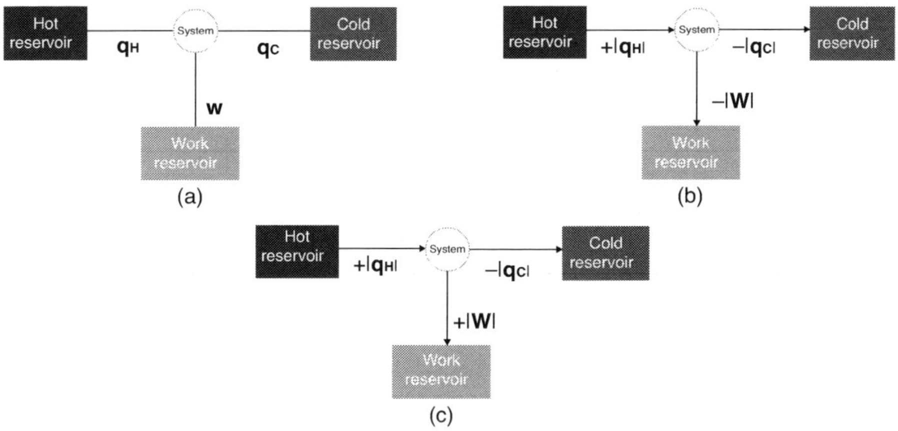

Figure 4.15 Energy flows in a heat engine showing both sign conventions, (b) is the chemists' sign convention, and (c) is the physicists' sign convention.

Using the physicist' sign convention and absolute values Equation (4.16) is written as:

$$\Delta U^{physicists} = \left(+|q_H|\right)+\left(-|q_C|\right)-\left(+|w|\right)=|q_H|-|q_C|-|w| \tag{4.17}$$

The two sign conventions achieve the same result. A little thought should convince you that this must be the case: a sign convention is only an accountancy tool using different conventions does not change the world (as the physicists' sign convention avoids the "chemists' fudge" it must be considered the more rigorous). This then begs the following question - *why not use absolute values and do away with sign conventions?* The answer is simple – manipulating equations with absolute numbers can be as difficult as doing arithmetic with Roman numerals. In addition, we have seen that the chemists' fudge results in nonsensical absolute value equations. In this book the use of absolute values is only to aid clarity and understanding; from now on this book only uses the chemists' sign convention.

4.4 INTERNAL ENERGY

The total energy of an object or system (E) consists of the macroscopic kinetic energy of motion (E_k) of the system, the macroscopic potential energy of the system due to its position in external force fields (E_p),[37] and microscopic energy contained within the object/system which is known as internal energy (U):

$$E = E_k + E_p + U \tag{4.18}$$

Both E_k and E_p depend on the choice of the measurement frame of reference. For example, the macroscopic kinetic energy of a car moving at speed v with respect to a road is $\frac{1}{2}mv^2$. But with respect to a train moving at that same speed parallel to the roadway, the car's macroscopic kinetic energy is zero, since its relative velocity is zero with respect to the train. However, whatever (non-accelerating)[38] reference frame you use, changes of macroscopic kinetic energy will be unaffected by this choice. Likewise, macroscopic potential energy has a value dependent on your choice of measurement frame of reference. A system at one position in a force field may have a different macroscopic potential energy at another position, e.g., an object on Earth resting on a table has zero gravitational potential energy with respect to the table top, but a larger potential energy with respect to the floor below. But whatever "zero" reference frame you choose any changes of macroscopic potential energy will be unaffected by this choice. This is the reason why the (arbitrary) choice of the Earth's surface as the reference frame for zero kinetic and potential energy can be made. Internal energy can be considered as the energy of "being" of the object or system. In contrast to E_k and E_p its magnitude is independent of frames of reference.

Internal energy of a system is the summation of the kinetic and potential energies of all the microparticles in the system. Because the internal energy depends on the amount of mass the system or object contains (i.e., number of microparticles) it is an extensive quantity like mass, energy, length, and volume, but unlike temperature and pressure which are intensive quantities. It is usually not possible to calculate the total internal energy of a real system and a convenient zero reference point is chosen.[39] However, because the internal energy of an ideal gas system is just the translational kinetic energy of the spherical (more rigorously – "point") microparticles, the absolute internal energy and can be calculated from the mass and number of the microparticles in the system, and the temperature, e.g., the absolute internal energy of a system of n moles of an ideal gas at temperature T.K is nC_V or $\frac{3}{2}nR$.

If we are only concerned with changes in energy (and ignore the complexities of frames of reference), then, in thermodynamics, we can ignore E_k and E_p in Equation (4.17) and only consider changes in internal energy (ΔU).

Let us now consider the internal energy changes of the three system types set out in Section 4.2:

1. In an isolated system (no exchange of matter or energy) the internal energy of the system is constant, i.e., $\Delta U = 0$.[40]
2. In a closed system (no exchange of matter) a change in internal energy can only occur if energy is transferred to or from the system by the processes of heat and/or work.
3. In an open system (exchange of matter and energy can occur), the change in internal energy can occur if energy is transferred to or from the system by the processes of heat and/or work. In addition, mass transfer to or from the system can also change the internal energy.

In this book, I will only be considering isolated and closed systems. Once the process of heat and work is over it is not possible by examining the change in internal energy of the system to ascertain whether the change was caused by process of work, the process of heat or a combination of the two. This now leads us to the First Law of Thermodynamics.

4.5 THE FIRST LAW OF THERMODYNAMICS

The First Law of Thermodynamics is:

$$\Delta \mathbf{U}_{\text{closed system}} = \mathbf{q} + \mathbf{w} \tag{4.19}$$

As the First Law of Thermodynamics is an alternative way of stating the Law of Conservation of Energy for a closed system when the two energy transfer processes of heat and work are considered. Consequently, it may not come as a surprise that we have already arrived at Equation (4.19) – see Equations (4.16) and (4.17) in Section 4.3.6. The First Law can be stated, in words, as follows:

> The change in the internal energy of a closed system $(\Delta \mathbf{U}_{\text{closed system}})$ is equal to the flow of energy that has occurred by the process of heat ($\mathbf{q}$) and the flow of energy that has occurred by the process of work ($\mathbf{w}$).

If a closed system has lost energy, then by the Law of Conservation of Energy the surroundings have gained an equal amount of energy, and *vice versa* (remember that by definition the system plus surroundings is an isolated system where $\Delta \mathbf{U} = 0$):

$$\Delta \mathbf{U}_{\text{system}} = -\Delta \mathbf{U}_{\text{surroundings}} \quad \text{or} \quad \left|\Delta \mathbf{U}_{\text{system}}\right| = \left|\Delta \mathbf{U}_{\text{surroundings}}\right| \tag{4.20}$$

where $\Delta \mathbf{U}_{\text{system}}$ is the change in the total internal energy of a system, and $\Delta \mathbf{U}_{\text{surroundings}}$ is the change in the total internal energy of the surroundings. Equation (4.20) unlike Equation (4.19) applies to an open system as well as a closed system.

Note the following important points:

1. Both $\mathbf{w}$ and $\mathbf{q}$ represent the energy flows by the processes of work and heat, respectively; these are vector-like quantities. Whereas the internal energy U_{system} is the quantity of energy that the system contains (a scalar quantity).
2. The total energy of an isolated system is constant (or $\Delta \mathbf{U}_{\text{isolated system}} = 0$).
3. One way of stating the First Law is the total energy of the universe is constant, i.e., $\Delta \mathbf{U}_{\text{universe}} = 0$.
4. Implicit in the First Law of Thermodynamics is the interconversion of energy transferred by the process heat and energy transferred by the process of work. To put it another way, the internal energy of a system cannot be divided into "work energy" and "heat energy" once the energy transfer process(es) is/are complete. In other words, the terms heat energy (usually written as thermal energy) and work energy (admittedly not often seen) are meaningless.

Point 4 should be obvious from the foregoing discussion, but when you read many scientific texts you realise that many scientists do not "get it". When I Googled "thermal energy" I came across the following:

1. "Thermal energy is an example of kinetic energy, as it is due to the motion of particles, with motion being the key. Thermal energy results in an object or a system having a temperature that can be measured. Thermal energy can be transferred from one object or system to another in the form of heat".[41]
2. "Thermal energy refers to the energy contained within a system that is responsible for its temperature. Heat is the flow of thermal energy".[42]
3. "Thermal energy is the energy that comes from heat. This heat is generated by the movement of tiny particles within an object. The faster these particles move, the more heat is generated".[43]
4. "Heat energy is the result of the movement of tiny particles called atoms, molecules or ions in solids, liquids and gases. Heat energy can be transferred from one object to another. The transfer or flow due to the difference in temperature between the two objects is called heat".[44]

I do not intend to spend time commenting on the above extracts: but some are more wrong than others.[45] I strongly advise against the use of ill-defined terms such as "thermal energy" and "heat energy" and use the rigorously defined internal energy of a system. It is also worth comparing the use of the word "heat" in the above examples in light of its modern definition as an energy transfer process. I will return to this subject in a later chapter.

In the next section, I want to consider the processes of heat and work in more general terms. This will lead to the concept of state functions and path functions.

4.6 STATE AND PATH FUNCTIONS

Before considering state and path functions I need to consider the concept of a thermodynamic state. This is from Wikipedia[46]:

> In thermodynamics, a thermodynamic state of a system is its condition at a specific time; that is, fully identified by values of a suitable set of parameters known as state variables, state parameters or thermodynamic variables. Once such a set of values of thermodynamic variables has been specified for a system, the values of all thermodynamic properties of the system are uniquely determined. Usually, by default, a thermodynamic state is taken to be one of thermodynamic equilibrium. This means that the state is not merely the condition of the system at a specific time, but that the condition is the same, unchanging, over an indefinitely long duration of time.

Now consider the following[47]:

> State functions are 'variables' that define the state of a system. When you have a system you need to be able to define the conditions in which it exists before and after a change. We typically referred to these as the initial and final states. By states we mean the system can be described by a set of properties. For example, the state of a system might be that I have 1 mole of argon in a 10 L container with a temperature of 300 K. Here the state of the system is defined by the 'state functions' of volume and temperature as well as the amount of the gas. Likewise, the pressure is also a state function. We'll also see later in thermodynamics that there are a number of variables related to energy that are also state functions.
>
> If I take my 1 mole of gas and do something to it, I might end up in a new state where for example the volume is now 20 L and the temperature is 600 K. For such a process I can look at the change in the state functions. In this case the initial volume, V_i, was 10 L and the final volume, V_f, was 20 L. So my change in volume is given by
>
> $$\Delta V = V_f - V_i = 20\ L - 10\ L = 10\ L$$
>
> The key to this idea is that the change in volume has nothing to do with the particulars of the mystery process that brought me from my initial state to my final state. The difference in volume will always be the same. That is, if I started with a volume of 10 L and ended with a volume of 20 L the difference is always +10 L. This might seem frightfully obvious. However, when we start to think about abstract state functions related to energy it can be more difficult to wrap your head around the ideas. But it is important to know that the concept is exactly the same. If you know the initial state and you know the final state, then you can calculate the change (regardless of the process by which you achieved the change).[48]

A state function is a property that describes a particular state, without depending on the path taken to reach this state. In contrast, functions whose value depends on the path taken to get between two states are called path functions. Another way of defining a state function is that it is a property of a system that depends only on the current state of the system, not on the way in which the system acquired that state. There are three mathematical criteria for state functions[49]:

1. State functions are exact differentials.[50]
2. If the state function is a dependent variable, then the differential of the function can be represented as the partial differentials of the independent state functions on which the function depends.[51]
3. The cyclic integral of state function is zero.

In this book, the only criterion I will use for a state function is the third one. A cyclic process is a process in which the state of the system changes and then returns to its initial state. Using the concept of the cyclic process we can show internal energy is a state function by the following argument. Consider a closed system undergoing a process by which it passes from state A to state B, e.g., by a change in temperature. The change in internal energy will be:

$$\Delta \mathbf{U} = U_B - U_A \tag{4.21}$$

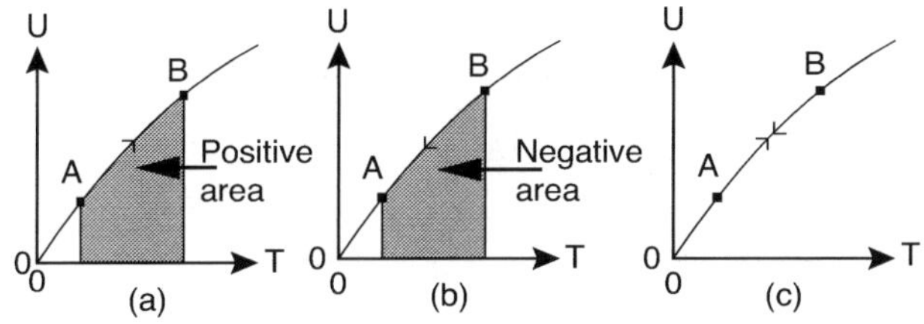

Figure 4.16 Graphical demonstration that $\oint U\,dT = 0$ for the state function U.

where U_A and U_B are internal energies of the system in states A and B respectively. The energy difference ($\Delta\mathbf{U}$)only depends on the initial state (U_A) and the final state (U_B) and in no way depends on the path followed between the initial and final states. We can prove that this is true by assuming the change in internal energy does depend on the path followed (in other words the cyclic process does not return the internal energy of the system back to its original state). We start with a system at state A with an internal energy U_A and there is a path, to state B (with internal energy U_B) requiring an energy input into the system of E_{BA} equal to $U_B - U_A$. Now, if there is a path back to state A in which the energy output from the system is E_{AB}, where $E_{AB} \neq E_{BA}$, then the following argument shows that this violates the Law of Conservation of Energy:

> If, say, $E_{AB} < E_{BA}$, then every time you go around the cycle from state A to state B and then back to state A you create energy, and if you run the system in reverse you destroy energy. In other words, we have either increased the energy of the universe or decreased the energy of the universe. This violates the Law of Conservation of Energy, therefore $E_{AB} = E_{BA}$ and, $(U_A - U_B) = -(U_B - U_A)$.

We have shown above that internal energy is a state function (otherwise we contravene the Law of Conservation of Energy). Now let us consider a graph of the change of internal energy with temperature for any object. First consider the change from U_A to U_B as the temperature increases from T_A to T_B (Figure 4.16a). We know the area under the graph can be calculated by integration, i.e., $\int_{T_A}^{T_B} U\,dT$. Although we do not calculate it, we should be aware that as $T_B > T_A$ this area must be a positive quantity. Now consider the change from U_B to U_A as the temperature decreases from T_B to T_A. We know U is a state function, therefore, this reverse process must follow the same path as the forward process (Figure 4.16b). We know the area under the graph can be calculated by integration, i.e., $\int_{T_B}^{T_A} U\,dT$. This area must be a negative quantity.[52]

Finally, let us consider the cyclic process of T_A to T_B and back to T_A. The area under the graph of the cyclic process is given by $\int_{T_A}^{T_B} U\,dT + \int_{T_B}^{T_A} U\,dT$ and we can immediately see that this area is zero. In mathematics, the integral around a closed curve is known as a cyclic integral and its symbol is $\oint$. In the example, given above $\oint U\,dT = \int_{T_A}^{T_B} U\,dT + \int_{T_B}^{T_A} U\,dT$. For any state function, it can be shown that its cyclic integral is zero, conversely, for a path function its cyclic is not equal to zero. I will return to this subject in Chapter 5.

> Now you are probably sitting there thinking, why on Earth do I care? Why would I ever need to know about state functions? Mathematicians care a lot about state functions and spend a lot of time writing equations about them.
>
> But fortunately for us, we don't really need to understand the math; we just need to know how we can use state functions. We come across state functions a lot in thermochemistry and thermodynamics. Many of the equations involved in these topics include properties that are state functions. What is really cool about state functions is that they make calculations simple and easy.
>
> All we need to know is the state change - the change in value between the final and starting position. We do not need to know how this change happened.[53]

As far as we are concerned in this book the only path functions are distance, speed,[54] heat, and work: all other physical quantities are state functions. Consider the following[55]:

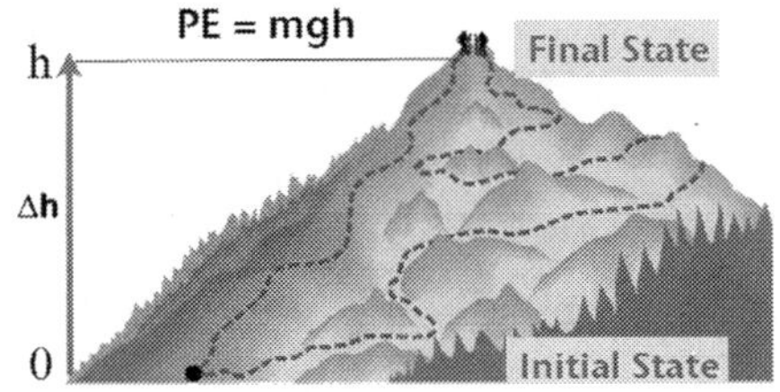

Figure 4.17 The potential energy differences between the two walkers are the same even though they took very different paths to get to the top of the hill: potential energy difference only depends on the initial and final heights. However, the energy (the "walking energy") expended by the two walkers is different.

A state function is a property describes a particular state, without depending on the path taken to reach this state. In contrast, functions whose value depends on the path taken to get between two states are called path functions.

To understand the difference between the two, think of two cities, A and B, that are 100 km away from each other along a straight line. Roads are rarely built in long straight lines, we may have two roads linking A and B, with lengths 150 km and 200 km. So, while the displacement between the two towns is the same for each route (it is a state function), the distance you need to travel to get from A to B depends on which road you take (it is a path function).

To reinforce the concept of state functions, consider two walkers initially at the same location climbing a hill by different routes (Figure 4.17). When they are both at the summit the change in potential energy (PE) for both walkers is the same $(\mathbf{\Delta PE} = \mathrm{mg}\mathbf{\Delta h})$ because both started at the same place. However, the hikers have expended different amounts of energy in getting there. The walkers' bodies have converted the energy in the food they have consumed into "walking energy". The fact that the two hikers have taken different routes suggests they have expended different walking energies to reach the summit, i.e., potential energy is a state function, but "walking energy" is not. The two different routes taken by the hikers involve the expenditure of different amounts of energy to reach the top of the mountain, i.e., "walking energy" is a path function. All "path functions" relate to processes – in this case, the process is the method used to get to the summit (other processes to reach the summit could be to land in a helicopter, parachute from a plane, drive up in a car, etc.).

Let me give another example. Suppose you have a savings account containing a balance of £1000 and there are two deposit/withdrawal processes – "writing cheques" and "bank transfer". Say you withdraw £500: you could do this by writing a cheque for £500 or by a bank transfer of £500 or any of a multitude of combinations of the two. Your bank balance is a state function, and the deposit/withdrawal processes are path functions.[56] Now if we assume that the amount of money in the world (also known as the universe) is constant, then if £500 is paid into your account then the surroundings, i.e., the universe apart from your bank account has lost £500. This analogy to energy and energy transfer processes falls down if your bank account can go into the "red". If we state this is a forbidden practice, then we can take the analogy further. We cannot have a negative bank balance and like energy it is always a positive number (a scalar), but a change in your bank balance can be a positive or negative number (vector-like quantity). Similarly, the deposit/withdrawal processes (flows of money) are vector-like quantities and can be positive or negative (in this case the sign convention is always money flows into an account are positive and out of the account are negative).

The two path functions of interest to us in this book are the processes of work and heat. One consequence of all this is that we can write mathematical symbols like **ΔU** for state functions, but we can never write the path functions **q** or **w** with a Δ in front of them. As we have seen a state function (like U) is independent of path, i.e., **ΔU** depends only on the initial and final states $(\mathbf{\Delta U} = \mathrm{U}_{\mathrm{final}} - \mathrm{U}_{\mathrm{initial}})$. However, **q** and **w** do depend on the path one takes to get from the initial to the final state. State functions can be considered as quantities or properties of a thermodynamic system, while path functions are processes during which the state functions change. The processes of heat or work are flows of energy represented by **q** or **w**. Once the energy transfer processes have ceased then **q** = 0 and **w** = 0, and the energy is now present in the system as internal energy (**U**) and there is no way you can divide internal energy into categories such as "heat energy" or "thermal energy" or "work" energy, i.e., we say there is no heat and no work when a system has reached thermodynamic equilibrium (to distinguish this from mechanical equilibrium I will use the term "thermodynamic equilibrium"). This is because heat and work are processes of energy transfer and once a system has reached equilibrium, these processes (by definition) do not occur. In addition, if there is a flow of energy $+|\mathbf{q}|$ or $+|\mathbf{w}|$ into the system then there is a flow of $-|\mathbf{q}|$ or $-|\mathbf{w}|$ into the surroundings (the heat reservoir or work reservoir respectively. The energy change of the isolated system ("universe") consisting of the heat reservoir, work reservoir, and the system is, as expected, zero $\left[\Delta \mathbf{U}_{\mathbf{universe}} = \left(+|\mathbf{q}|\right) + \left(+|\mathbf{w}|\right) + \left(-|\mathbf{q}|\right) + \left(-|\mathbf{w}|\right) = 0\right]$. When you have understood this then you have grasped one of the important concepts of modern thermodynamics (and will see that a significant number of writers of thermodynamic texts have not).

There is a fundamental difference between a state function (such as temperature or volume) and a path function (such as heat or work): the value of a state function refers to one instant of time; the value of a path function refers to an interval in time. In Chapter 5, I will introduce the concept of the cyclic process. As we shall see the fact that the processes of heat and work are not state functions is vitally important in understanding the concepts of thermodynamics. Heat (**q**) and work (**w**) are not properties of the system; instead, they are quantities of energy transferred across the boundary, the values of **q** and **w** depend in general on the path and are called path functions. Because **q** and **w** are energy flows over a period of time it is incorrect to write $\Delta\mathbf{q}$ or $\Delta\mathbf{w}$.

A system in thermodynamic equilibrium can do no work or do no heat, i.e., there can be no flow of energy by the processes of work and/or heat ($\mathbf{w} = 0$ and $\mathbf{q} = 0$). In everyday English, the phrase "do no heat" sounds ludicrous while the phrase "do no work" sounds reasonable. I would suggest this is because we think we know what the word "heat" means and it is not a process (more on this in later chapters). It is only when thermodynamic equilibrium is disturbed that $\mathbf{w} \neq 0$ and/or $\mathbf{q} \neq 0$ until thermodynamic equilibrium is re-established. When the pressure and temperature of a gas in a cylinder fitted with a piston equals the pressure and temperature of the surroundings, the system is in thermodynamic equilibrium and cannot do any work or heat. The only way this can change is if the temperature and pressure of the surroundings change.

The following extract from Wikipedia[57] may prove helpful:

> Thermodynamic equilibrium is an axiomatic concept of thermodynamics. It is an internal state of a single thermodynamic system, or a relation between several thermodynamic systems connected by more or less permeable or impermeable walls. In thermodynamic equilibrium there are no net macroscopic flows of matter or of energy, either within a system or between systems.

We can now answer the questions posed a number of times in the previous three chapters with regard to the loss of energy in isolated systems. We came across a number of examples:

- Inelastic collisions.
- Moving ball slowing down as a result of friction.
- Pendulum slowing down as a result of friction.

We know that some or all of the macroscopic kinetic energy of the object(s) is transformed into the chaotic microscopic energy of the microparticles constituting the object(s), i.e., it raises the internal energy of the objects. This results in temperature differences and equilibrium is achieved by the process of heat. In other words, the apparent "missing" energy in the isolated system is located as in increase in the internal energy of the system, i.e., the transformation of energy transferred by the process of work into energy transferred by the process of heat. This is usually described in scientific texts as dissipation. The end result of a dissipative process is an increase in the temperature of the system. However, in many cases, this increase in temperature is so small as to be immeasurable, e.g., the cooling down of a cup of coffee has a negligible effect on room temperature.

Consider an isolated system consisting of a ball (mass m) moving with velocity v in the positive direction. The moving ball is affected by a constant frictional force $-|\mathbf{F}|$ which opposes the motion of the ball. Eventually, after time t the ball stops moving, and its displacement is $+|\mathbf{s}|$. We can see that ball has lost $\frac{1}{2}mv^2$ of kinetic energy. During this deceleration process, the ball is continually acted upon by the force $-|\mathbf{F}|$. Unfortunately, the maths is counter intuitive. If the force vector is negative the displacement vector s must also be negative $\left(-|\mathbf{s}|\right)$. This must be the case to ensure the work done by the frictional force is a positive quantity. We know, by definition, that the energy change of the isolated system is zero. Therefore, we can write $\frac{1}{2}mv^2 = Fs$.

We know the acceleration of the ball is:

$$\frac{0-(+|\mathbf{v}|)}{t} = \frac{-|\mathbf{v}|}{t} = -|\mathbf{a}| \tag{4.22}$$

We know the force acting on the moving ball:

$$-|\mathbf{F}| = m(-|\mathbf{a}|) \tag{4.23}$$

Work done by the frictional force:

$$w = (-|\mathbf{F}|) \times (-|\mathbf{s}|) \tag{4.24}$$

Therefore, we can argue that energy by the process of work has been converted into some other form of energy. We know that this is localised in the internal energy of parts of the system, e.g., the surface on which the ball is moving and on the surface of the moving ball. This is reflected in localised temperature increases. To achieve equilibrium, energy must flow by the process of heat down these temperature gradients by the process of heat and at equilibrium the temperature of the system is now everywhere the same and is higher (but often immeasurably so) than the original temperature of the system before the moving ball came to a stop has increased (often immeasurably). Remember this quote from Black in Section 4.1[58]:

> By the use of these instruments [thermometers] we have learned, that if we take 1,000, or more, different kinds of matter, such as metals, stones, salts, woods, cork, feathers, wool, water and a variety of other fluids, although they be all at first of different heats, let them be placed together in a room without a fire, and into which the sun does not shine, the heat will be communicated from the hotter of these bodies to the colder, during some hours, perhaps, or the course of a day, at the end of which time, if we apply a thermometer to all of them in succession, it will point to precisely the same degree.

In other words, the energy transferred by the process of work (the slowing down of the ball) is converted into an increase in the system's internal energy and as far as we are concerned in this book this results in a temperature rise.

To put it another way, the organised macroscopic energy of the system has been converted into the chaotic microscopic energy of the matter in the system.

A shorthand way to describe this is to say work has been converted into heat. Unfortunately, as heat and work are processes this can be seen as confusing. What has happened is organised macroscopic energy has been converted into chaotic microscopic energy.

4.7 SUMMARY

I proved that the formula for the molar heat capacity of an ideal gas at constant volume (C_V) is $\frac{3}{2}R$ and its molar heat capacity at constant pressure (C_P) is $\frac{5}{2}R$. The difference $(C_P > C_V)$ indicates that not all of the energy flowing into the system at constant pressure from the hot reservoir, by the process of heat, is used to increase its internal energy because some of it is being used to do mechanical work on the surroundings. This introduced the concept of "PV" work: because, at constant pressure and with increasing temperature, the system's volume increases and, it carries out mechanical work on the surroundings (i.e., the system does work on the surroundings, or energy flows from the system to the surroundings by the process of work). Conversely, for a temperature decrease, the surroundings do work on the system. I showed that calculated values of C_V and C_P only agreed with experiment for the monatomic noble gases but not for the polyatomic permanent gases. This is even though these gases do obey the ideal gas law $\left(\frac{PV}{T} = \text{constant}\right)$ under certain conditions (for permanent gases like oxygen, nitrogen, hydrogen, etc., these conditions are typically STP/NTP).[59]

In thermodynamics, we focus attention on the system (a defined part of the universe) and its surroundings (the rest of the universe). The First Law of Thermodynamics identifies heat and work as the only processes of energy transfer which can bring about a change in the energy of a closed system. The core of the First Law of Thermodynamics[60] is the introduction of a new state function known as the internal energy and the energy balance equation $(\Delta \mathbf{U} = \mathbf{q} + \mathbf{w})$. In any change, $\mathbf{q}$ represents the flow of energy by the process of heat and **w** represents the flow of energy by the process of work. We speak of the work done on the system by its surroundings $(+|\mathbf{w}|)$, and the work done by the system on its surroundings $(-|\mathbf{w}|)$. After the change has taken place, neither the words work, or heat have any usefulness in describing the final state of the system - we can speak only of the internal energy of the system. In other words, heat and work are processes of energy transfer, and when the process is over, there is no heat and no work ($\mathbf{w} = 0$ and $\mathbf{q} = 0$). The internal energy of the system has either increased, decreased, or remained constant. It is impossible to subdivide the internal energy into the part transferred by the process of heat and the part transferred by the process of work, just as it is impossible to subdivide the water in a lake into parts resulting from (a) groundwater seepage, (b) rain, (c) flow from a stream, (d) condensation, etc. The First Law of Thermodynamics allows us to account for the quantities of energy that are transferred by the processes of work and heat in closed systems. To carry out this energy accounting we need to adopt and strictly adhere to a sign convention. The system we adopted is to consider energy flowing into our system as positive and energy flowing out of our system as negative. Because of our adopted sign convention (the "chemists' sign convention") we have had to introduce a fudge factor to ensure consistency, i.e., $\mathbf{w} = -\mathrm{P}\Delta\mathbf{V}$.

This energy accounting business is not as straightforward as it may first appear, and many thermodynamic texts mess it up by interchangeably considering q and w to represent positive real numbers (energy) or to be vector-like energy flows that can be represented by real numbers (positive and negative values). This is the reason I have spent an inordinate amount of time on sign conventions, absolute values, and energy accountancy mathematics (mainly aimed at the non-mathematical among us) because it is essential to understand the energy accountancy aspects of thermodynamics.

Internal energy (U) is a state function (always a positive real number) whereas **q** and **w** are path functions (positive or negative real numbers dependent on our sign convention). What this means is that if you take the system from one state to another state by two different processes, $\Delta\mathbf{U}$ will be the same independent of the path taken but the energy transferred by the processes of heat (**q**) and work (**w**) to effect that change can be different.

The First Law of Thermodynamics can be written as a differential equation which in this book I will write as:

$$\mathbf{dU} = \mathbf{dq} + \mathbf{dw} \tag{4.25}$$

This is not mathematically rigorous, because **q** and **w** are path functions, and their differentials are "inexact" as opposed to the differentials of state functions which are "exact" and Equation (4.25) should be written as (where 'd' represents an exact differential and 'đ' an inexact differential)[61]:

$$\mathbf{dU} = \mathbf{đq} + \mathbf{đw} \tag{4.26}$$

Is the ability to do work conserved? By now you should know the answer to this one – no it is not. Consider the following extract[62]:

> If the ability to do work were indeed conserved, we could shut down many of our power plants. To run an elevator,[63] for example, it would be necessary to lift it to the top of the shaft only once. As it descended, its ability to do work would decrease, and this ability could be stored in some other system, such as a flywheel or a storage battery. If undiminished, this ability could be used to raise the elevator to the top of the shaft ready for the next descent. Extend this to other systems and it is clear that the problem of diminishing energy resources would disappear.

This lift and associated hardware are a perpetual motion machine, and it was widely accepted in the 18th century, that a perpetual motion machine is impossible. This acceptance was based on experience – no perpetual motion has ever been constructed. Yet, the perpetual motion elevator implies no contradiction with the law of conservation of energy. Experiments have shown the quantitative conversion of work to heat[64] is possible in the real world, but not *vice versa.* As we shall see in the next chapter it is because of this experimental fact that a perpetual motion machine is impossible.

If in an isolated system, energy is constant or conserved, why are we urged to "save energy"? To understand this and directionality (i.e., why processes "spontaneously" occur in one direction only), we need to consider one of the greatest thought experiments in science. Although widely ignored at the time it was this thought experiment that kicked off the science of thermodynamics. This is the subject of Chapter 5.

I want to finish this chapter with three instructive quotes:

The First Law of Thermodynamics is an extension of the principle of the conservation of mechanical energy. ... both work and heat were seen to be entities that described the transfer of energy from one system to another. If a temperature difference exists between two systems in thermal contact, energy can be transferred from one system to another in the form of heat. Heat transfer between open systems can also occur by transport of matter from one system to another. Work is the form in which energy is transferred to system by displacement of parts of the system under the action of external forces. We should not say that heat and work are "forms of energy" since they are entities that have significance only in terms of energy transfers between systems. We cannot speak of the "heat of a system" or "work of system" although we can and do speak of the "energy of a system".[65]

It is correct, then, to say that a system has a large amount of internal energy, but it is not correct to say that a system has a large amount of heat or a large amount of work. Heat is not something that is contained in a system. Rather, it is a measure of energy that flows from one system to another because of a difference in temperature.[66]

This use of the word "heat" takes quite a bit of getting used to. For instance, we should never say that a cup of boiling water "contains lots of heat", because once the energy is in the water, it does not matter whether the energy entered as heat or as work. If you put the water on a hotplate, then energy enters by the process of heat. But you could vigorously shake or stir the water, with the same end result (hot water), accomplished this time by the process of (mechanical) work. Another way would be to put the cup of water into a microwave oven. Since the flow of microwave energy from the magnetron source into the water is not caused by the magnetron being hotter than the water, we should classify this energy transfer as work (in this case, electrical work), not as heat.[67]

NOTES

1 The phrase "thermal energy" is often used in thermodynamic texts. I will not use the term for reasons that I will explain later in this chapter.

2 Also known as the kinetic theory of heat.

3 *Lectures on the Elements of Chemistry Delivered in the University of Edinburgh by the Late Joseph Black, M.D.*, John Robison LLD, page 78, Volume 1 1803.

4 Remember thermal equilibrium is a dynamic equilibrium.

5 Defined as the amount of energy needed to raise the temperature by one energy unit.

6 This is similar to the use of atomic weight instead of relative atomic mass – see Section 3.6.2. Unfortunately, the continued use of the phrase heat capacity perpetuates the confusing use of the word heat in scientific literature.

7 This is thermodynamic definition of "universe" as opposed to the cosmological definition of "Universe" which is "is all of space and time and their contents, including planets, stars, galaxies, and all other forms of matter and energy. The Big Bang theory is the prevailing cosmological description of the development of the universe." This quote is from https://en.wikipedia.org/wiki/Universe.

8 This "constancy" of the surroundings is of paramount importance in understanding thermodynamics. I will return to this subject later in this section when considering heat reservoirs.

9 Adiabatic is derived from the Greek word *adiabatikos*, which means that "heat is not able to go through".

10 https://en.wikipedia.org/wiki/Adiabatic_wall.

11 Also known as "thermal reservoirs", "thermal energy reservoirs" (that dreaded term again), or a "heat baths".

12 This is another concept like the immovable or massive objects used Sections 1.11.9 and 1.12.3.

13 As a result of constant temperature all processes in a thermal reservoir are reversible (see Chapter 5). This reversibility is of paramount importance in the analysis of various thermodynamic systems.

14 Article by *J. Anacleto, Eur. J. Phys.* 31(3), 617, 2010. https://www.researchgate.net/publication/231042729_Work_reservoirs_in_thermodynamics.

15 See Section 2.5.

16 In this example the energy transferred to the work reservoir is equivalent to the energy stored in the gravitational vector field (see Sections 2.5 and 2.7).

17 See Table 1.1 and Section 3.6.3.

18 https://chem.libretexts.org/Core/Physical_and_Theoretical_Chemistry/Thermodynamics/Calorimetry/Heat_Capacity.

19 https://en.wikipedia.org/wiki/Heat_capacity and the use, again, of that dreaded phrase thermal energy.

20 Apart from mentioning the use of thermal energy in the first definition – there is no need to use the adjective thermal because this either causes confusion or, hopefully, prompts the question what is the difference between thermal energy and energy? Answer – there is no difference. Another example of sloppiness.

21 $U = \frac{3}{2}PV = \frac{3}{2}nRT$

22 All the noble gases (helium, neon, argon and krypton, xenon, and radon) have the same molar heat capacity of 12.48 $J.mol^{-1}.K^{-1}$

23 Standard temperature and pressure (STP) - temperature of 273.16 K (0°C, 32°F) and an absolute pressure of 0.1 MPa (14.504 psi, 0.987 atm, 1 bar). Note - normal temperature and pressure (NTP) - is 298 K and 0.1 MPa.

24 I have called these quantities "vector-like" - see Section 1.10.

25 Unfortunately, you may come across another sign convention (this tends to be used by physicists) – I will briefly mention this later in this chapter.

26 Adapted from W.J. Moore, *Physical Chemistry*, 5th Edition, Longman, p. 31, 1972.

27 The positive symbol is often omitted.

28 Alternatively if q represents a real number then +q or −q may take either positive or negative values based on whether the real number q is positive or negative.

29 "Hot" and "cold" in this context are relative terms – I could have used the terms "hotter reservoir" and "colder reservoir", but that can cause intelligibility issues. I have decided to stick with the terms "hot" and "cold" and specifically highlight any ambiguities.

30 Alternatively, we can say there is a flow of energy $-|\mathbf{w}|$ to the system from the surroundings (the work reservoir) by the process of work.

31 Sadi Carnot, "Réflexions sur la puissance motrice du feu" (1824) translated by R.H. Thurston in *Reflections on the Motive Power of Heat*, 2nd Edition, John Wiley and Sons, p. 38, 1897..

32 Carnot used the phrase "motive power".

33 Newton's Third Law of Motion – see Sections 1.9.5 and 1.11.

34 Remember - pressure is force per unit area.

35 More rigorously "PΔV work" because "VΔP work" is zero (constant volume), but it is always known as PV work.

36 Figure 4.15 is based on Figure 4.9.

37 The energy associated with mass $E = mc^2$ is usually ignored as it does not play any role in chemical and (most) physical transformations.

38 This becomes complex as we delve deeper, e.g. if the train now accelerates then the car (which was moving at a fixed velocity in one frame of reference, i.e. when observed from the train moving at a fixed velocity) is now decelerating in another "frame of reference" – this paradox is the reason for the use of "non-accelerating" or inertial frame of reference. It is also worth mentioning that the concept "frame of reference" played a large part in the development Einstein's special theory of relativity (see Endnote 2).

39 Although the assignment of reference states is relatively straightforward for macroscopic physical quantities, the situation with internal energy is more complicated - "The concept of reference states only sounds simple; in reality it can be insidious." - C M Sliepcevich, "Oversights in Energy Reference States." Proceedings of the Oklahoma Academy of Sciences. Vol. 66. 1986. However, we are only concerned in this book with the hypothetical ideal gas whose reference state is "simple".

40 Remember the universe is considered to be an isolated system and consequently we have the conservation of mass and conservation of energy laws (or, more properly, the conservation of mass/energy).

41 From study.com/academy/lesson/what-is-thermal-energy-definition-examples.html – this was the first result when I googled "thermal energy".

42 https://www.khanacademy.org/science/physics/work-and-energy/work-and-energy-tutorial/a/what-is-thermal-energy.

43 http://www.softschools.com/facts/energy/thermal_energy_facts/402/.

44 https://www.sciencelearn.org.nz/resources/750-heat-energy.

45 This is a misquote from George Orwell's Animal Farm ("All animals are equal but some animals are more equal than others.") Another quote comes to mind– "I didn't think it was physically possible, but this both sucks and blows." [Bart Simpson].

46 https://en.wikipedia.org/wiki/Thermodynamic_state.

47 http://ch301.cm.utexas.edu/thermo/#properties/state-functions.html.

48 http://ch301.cm.utexas.edu/thermo/#properties/state-functions.html.

49 https://physics.stackexchange.com/questions/595089/why-is-pressure-a-state-function. There is a fourth criterion whose mathematics have not been covered in this book, namely "The mixed derivatives (second partial derivatives) of the independent variables are equal."

50 See Section 1.6.9.1.

51 See Section 1.6.9.2.

52 See Section 1.6.7 where I stated, "There can be no such thing as negative area, but the negative sign follows directly from the definition of a definite integral since you use the negative values of the function to calculate it."

53 https://study.com/academy/lesson/state-functions-in-thermochemistry.html.

54 See Section 1.2.2 for consideration of distance and displacement and speed and velocity.

55 https://isaacphysics.org/concepts/cc_state_and_path_functions?stage=all.

56 There is another useful analogy here. Once the money is in your bank account you cannot differentiate between "cheque money" and "bank transfer money", compare with "heat or thermal energy" and "work energy".

57 https://en.wikipedia.org/wiki/Thermodynamic_equilibrium.

58 *Lectures on the Elements of Chemistry Delivered in the University of Edinburgh by the Late Joseph Black, M.D.*, John Robison LLD, page 78, Volume 1 1803.

59 STP - Standard Temperature and Pressure - is defined by IUPAC (International Union of Pure and Applied Chemistry) as air at 273.15 K and 10^2 kPa. NTP - Normal Temperature and Pressure - is defined as air at 293.15 K and 1 atmosphere pressure (101.325 kN/m^2, 101.325 kPa).

60 Remember the First Law of Thermodynamics is the Principle of Conservation of Energy applied to thermodynamic systems.

61 See Section 1.6.9.1. This convention originated in the 19th century by German mathematician CG Neumann, indicating that q and w are path-dependent, while U (a state function) is not.

62 R.L. Lehrman, "Energy is not the ability to do work", *Phys. Teach.* 11, 15–18, 1973.

63 US elevator is a British lift.

64 Rigorously – "the quantitative conversion of energy transferred by the process of work to energy that is transferred by the process of heat …"

65 W.J. Moore, *Physical Chemistry*, Longman, p. 38, 1972.

66 P.A. Tippler, *Physics for Scientists and Engineers*, Palgrave, Macmillan, 2003.

67 http://docplayer.net/12389702-Chapter-3-thermal-energy.html.

5 Carnot's Thought Experiment

> Already the steam-engine works our mines, impels our ships, excavates our ports and our rivers, forges iron, fashions wood, grinds grain, spins and weaves our cloths, transports the heaviest burdens, etc. It appears that it must someday serve as a universal motor, and be substituted for animal power, waterfalls, and air currents.[1]
>
> An engine is a machine that turns the energy locked in fuel into force and motion. Coal is no obvious use to anyone: it's dirty, old, rocky stuff buried underground. Burn it in an engine, however, and you can release the energy it contains to power factory machines, cars, boats, or locomotives. The same is true of other fuels such as natural gas, gasoline, wood, and peat. Since engines work by burning fuels to release heat, they're sometimes called heat engines.[2]
>
> Carnot's work is distinguished for his careful, clear analysis of the units and concepts employed and for his use of both an adiabatic working stage and an isothermal stage in which work is consumed.[3]

5.1 INTRODUCTION

A simple thought experiment demonstrates that the First Law of Thermodynamics and the scientific concept of energy that led to its formulation do not give us a total insight into how the world works. Consider a hot object in cooler surroundings. What always happens? You do not have to be a scientific genius to answer – the hot object cools and eventually, the temperatures of the object and the surroundings become equal. You may require a modicum of scientific expertise to answer a subsidiary question – why does not the temperature of the surroundings (e.g., the room that contains a cooling cup of coffee) increase? The answer is that the temperature of the surroundings does increase, but the temperature rise is too small to be immeasurable. Now, there is no contravention of any laws, rules, concepts, principles, or theories that I have covered in this book so far if the temperature of a cup of coffee increases with time. This event would not contravene the Conservation of Energy/First Law of Thermodynamics because energy would flow from the surroundings, by the process of heat, to the cup rather than the other way round (and the temperature of the surroundings would become immeasurably lower instead of immeasurably higher). Therefore, we need *something else* that tells us the directionality of a process. Surprisingly, this "something else" came from investigations into the steam engines that powered the Industrial Revolution, and it was an engineer from a notable French family (Carnot) who started the ball rolling in 1824 when he published a monograph on a thought experiment aimed at improving the efficiency of steam engines. This work, little regarded at the time, provided insight into what this "something else" was and kicked off the science of thermodynamics.

It is worthwhile noting that before the Industrial Revolution, all machines only used the energy transferred by the process of work, e.g., watermills, windmills, falling weights, and pendulums. The Industrial Revolution was heralded by the invention of the steam engine, which used the energy transferred by the process of heat to do work on the surroundings.

Consider the following[4]:

> In the revolutionary years 1791–1797 an army engineer named Lazare Carnot found himself suddenly catapulted into the ruling bodies of France. As a member of the Legislative Assembly, the Committee of Public Safety, and the Directoire, he participated in voting for the execution of Louis XVI, the appointment of Napoleon to his Italian command; and in so many other things that his biography fills two volumes.
>
> But before and after that period, Lazare Carnot published articles on his true lifelong interest, the most general statement of the principles of mechanics and inferring from them the impossibility of a perpetual motion machine.[5] His son Sadi then turned that idea neatly around, inferring a new principle from the assumed impossibility of a perpetual motion machine. (It was Lazare's grandson, another Sadi Carnot, the nephew of our Sadi, who later became President of the French Republic).
>
> An important technical problem of the time was the design of steam engines. How much work can be extracted from a kilo of coal? Can the efficiency be improved by different temperatures or pressures, a different working substance than water; or some different mode of operation than pistons and cylinders?
>
> In the absence of any sound understanding of the nature of heat, misconceptions flourished. In 1818 Petit suggested an air engine would be far more efficient than a steam engine because no heat of vaporisation need be supplied: an erroneous argument that must have

DOI: 10.1201/9781003608400-5

> inspired much wasted effort. Sadi Carnot (1824) resolved this confusion by enunciating a single, intuitively compelling principle that answered all these questions and a hundred others not yet asked. He envisaged a generalised heat engine … which operates by drawing heat[6] q_1 from a source which is at thermal equilibrium at temperature t_1, and delivering useful work W. He saw that, in order to operate continuously, the engine requires also a cold reservoir, $t_2<t_1$, to which some heat q_2 can be discharged. The temperature scale t was any convenient thermometer calibration.
>
> Now Carnot had the happy idea of a reversible engine; one can turn the shaft backwards, delivering the same work W back to the engine, which then delivers the same heat q_1 back to the high-temperature reservoir.

Despite the fact that Carnot is arguably considered to be the founder of thermodynamics, Carnot's work was referenced only twice in scientific papers between 1824 and 1834 (aside from the reviews when it was published in 1824).

> That Carnot's important book Réflexions sur la puissance motrice du feu et sur les machines propres à développer cette puissance should have been neglected at the time of its publication is certainly not because it went unnoticed. It was published on 12 June 1824 and on 26 July of that year Pierre Girard gave a long review of it to the Académie des Sciences in Paris. Pierre Girard's review was a very positive one and was published in the Revue encyclopédique. Perhaps the problem with the review was that although it stated the theorems and the conclusions of Carnot's work fully, it did not comment on the highly original reasoning which Carnot had employed to achieve his results.[7]

In 1834 Émile Clapeyron, a French railroad engineer, quoted Carnot's results, and his work was eventually incorporated into the thermodynamic theories developed by Rudolf Clausius in Germany and William Thomson (later Lord Kelvin) in Britain. It is possible that Carnot's personality played a role in the lack of enthusiasm for his work in his lifetime. He was described as reserved, extremely introverted, and aloof. His work did not immediately lead to more efficient steam engines or any other practical application. It has also been mentioned that his work used the now disregarded caloric theory of heat (see Chapter 6), and this was a factor in it being disregarded; personally, I do not think in 1824 that would have been the case.

I will start this chapter by considering engines and then I will introduce the necessary concepts (reversible isothermal and adiabatic expansion and compression) that will allow us to consider and understand Carnot's brilliant thought experiment.

Before moving on, I want to consider Jayne's use of the word "heat".[8] It appears that he is viewing "heat" as a substance using heat as a noun ("drawing heat q_1 from a source"/ "to which some heat q_2 can be discharged"). However, he is using the word as shorthand for energy transfer by the process of heat as he also does with the process of work ("delivering useful work W"), which rigorously should be written as energy transfer by the process of work. Virtually, all thermodynamics texts use this approach with no explanation given. Unfortunately, some authors use the word "heat" erroneously, and this can be a cause of confusion. I have already considered this (see Chapter 4.2, Sections 1 and 7). In Section 6.1, I will consider an article by Zemansky entitled "The Use and Misuse of the Word "Heat" in Physics Teaching". If the word "heat", in any article, cannot be replaced by "energy transferred by the process of heat", or something similar and still maintains its meaning in the text, then it is probably not being used in its modern sense as a process. Unfortunately, in normal speech, we often use the word "heat" as if it were a substance. In contrast, we often use the word "work" as a process. If I said, "a system contained a lot of heat", then this would make sense to most people. However, if I said, "a system contained a lot of work", then it would sound wrong to someone fluent in English (although both statements are in the modern sense scientifically wrong).

I have covered this subject at some length, both in this chapter and the previous one not just because of pedantry, but because (i) it is a source of confusion in the teaching of thermodynamics, and (ii) "heat as substance" is the basis of the caloric theory of heat which was used by Carnot in his thought experiment.

The real world is incredibly complex and chaotic, and to make sense of it, scientists attempt to reduce its complexity, but still retain its essential aspects. We have already seen how the concept of an ideal gas explains qualitatively and, in some cases quantitatively the macroscopic properties (e.g., pressure, volume, temperature, and heat capacity) of a class of physical objects, the permanent gases,[9] at relatively high temperatures (around 250 K and above) and relatively low pressures (around atmospheric pressure or less). The subject of this chapter is a brilliant thought experiment, which abstracted the essential features of heat engines.

5.2 ENGINES

5.2.1 Introduction

A basic definition of an engine is that is a device or machine designed to convert one form of energy into mechanical energy.[10] Mechanical energy in this context is the sum of the kinetic and potential energy in a macroscopic physical object. By this definition, a stick of dynamite could be considered an engine because when it explodes it converts chemical energy into mechanical energy of macroscopic physical objects (e.g., flying rocks). However, most people would say that explosive material is not a machine, and I would suggest the main reason is that it uncontrollably converts some of its available energy into mechanical energy. Most people would say a rocket is an engine (in fact we say a "rocket engine") mainly because there is a degree of control when a rocket takes off: but what about a firework rocket? I will not answer this question, but it does show that there is no rigorous definition that covers what may or may not be called an engine. The term "engine" was originally a term for any mechanical device that converted force into motion, e.g., pre-industrial weapons such as catapults, trebuchets, and battering rams were called "siege engines". Windmills and watermills are also "engines" by this definition, but the newly invented (in the 18th century) steam engine is a different beast. In order for a steam engine to produce a continual flow of mechanical energy to the surroundings, it needs to operate in a cycle (more on this later).

In Chapter 4, we have seen, by using a piston/cylinder system, how energy transferred from a heat reservoir, by the process of heat, is converted by the system to do mechanical PV work on the surroundings. In 18th century England, engineers discovered that energy released by the burning of coal could be used to expand steam (known as the "working substance") and drive pistons in cylinders. Engines that use the process of heat to convert energy flowing down a temperature gradient into mechanical energy are known as heat engines. The first heat engines were the steam engines that powered the Industrial Revolution in the 18th and 19th centuries: steam engines are a subset of heat engines. Another way of considering it is to say a heat engine converts some of the internal energy of the working substance of the heat engine into the (kinetic or potential) energy of a macroscopic object in the surroundings, i.e., a heat engine converts the chaotic microscopic energy of microparticles into the ordered mechanical energy of macroscopic objects. As we shall see, no engine can convert all of the chaotic internal energy present in the working substance (e.g., steam) into ordered mechanical energy, and it must dump the energy it cannot transfer by the process of work to the engine's surroundings (known as the cold reservoir). This is necessary to complete the engine cycle, i.e., to return the engine to its initial state, so the process of transferring internal energy to mechanical energy can start again.

The first practical steam engine was the "Newcomen Engine" invented by Thomas Newcomen (1664–1729) in 1712 (Figure 5.1).[11] The engine operated by condensing steam drawn into the cylinder, this created a partial vacuum, thereby allowing the atmospheric pressure to push the piston

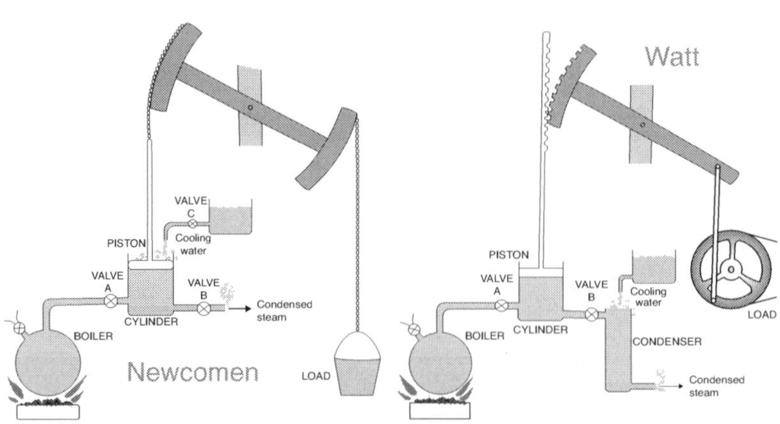

Figure 5.1 Newcomen and watt steam engines.

into the cylinder. It was the first practical device to harness the process of heat to produce mechanical work (i.e., to convert heat into work). Hundreds of Newcomen engines were used throughout Britain and Europe in the 18th century, principally to pump water out of mines. The efficiency[12] of Newcomen's engine was low (about 2%), and the coal was mostly consumed in heating the engine's boiler. The steam cylinder was heated and cooled repeatedly, which wasted energy, and also caused large thermal stresses. Subsequently, James Watt (1736–1819) improved the efficiency up to about 5%–7% by inserting a separate condenser condensation (Figure 5.1). This allowed Watt's steam cylinder to remain hot at all times. The efficiency of the engines gradually continued to be improved, and by the 1820s, it had reached 18%. These steam engines were fundamental in the transition to new manufacturing processes (known as the Industrial Revolution 1760–1840).[13] At first, these steam engines served mainly as pumps for the drainage of coalmines (to allow greater depths to be reached and resulting in increased production) and crushing the extracted ore. Later, steam engines were used to perform a variety of tasks including moving trains along railway tracks. These engines quickly became a strategic industrial resource.

One similarity of the above engines is that they all work on a cycle and deliver a continuous "stream" of mechanical energy to their surroundings by the process of work; provided the engine is provided with energy by the process of heat. Please note that devices such as turbines (i.e., windmills, watermills, jet engines, etc.), electric motors, and rocket engines do not operate in a cycle but are still considered engines.

In this book, the following definition of a heat engine will suffice – *it is a machine that operates in a cycle and converts a continuous input of energy, transferred to the engine by the process of heat, into a continuous output of energy transferred to the surroundings by the process of work.* Or, in our agreed shorthand – *a heat engine operating in a cycle converts heat into work.*

5.2.2 Heat Engines

A basic heat engine consists of a gas confined by a piston in a cylinder (see Figure 5.2). If the gas is heated, it expands, moving the piston. However, this would not be a particularly practical engine because once the gas reaches thermal equilibrium with the heat reservoir, the motion would stop. The great leap forward in the 18th century was the realisation to provide a continuous source of power a heat engine needs to repeatedly go through a cycle and the piston moves back and forth. When the gas is heated, the piston moves out, and when the gas is cooled, the piston moves in. A cycle of heating and cooling moves the piston back and forth. A flywheel provides the inertia to compress the gas and, because (as we shall see) this takes less energy than that generated in the expansion cycle, a heat engine provides a continuous form of mechanical energy (rotation or oscillating movement) from energy provided by the process of heat.

Heat engines are divided into two types:

1. External combustion engines, such as a steam engine, where the fuel burns outside and away from the cylinder where the force and motion are produced.
2. Internal combustion engines, such as diesel and petrol engines, where the fuel burns inside the cylinder.

I will briefly consider the internal combustion engine, which often operates in a cycle of four stokes,[14] i.e., the piston completes four separate strokes before it is back to its initial position, these strokes are intake, compression, power, and exhaust (Figure 5.3). The simplest four-stroke engine

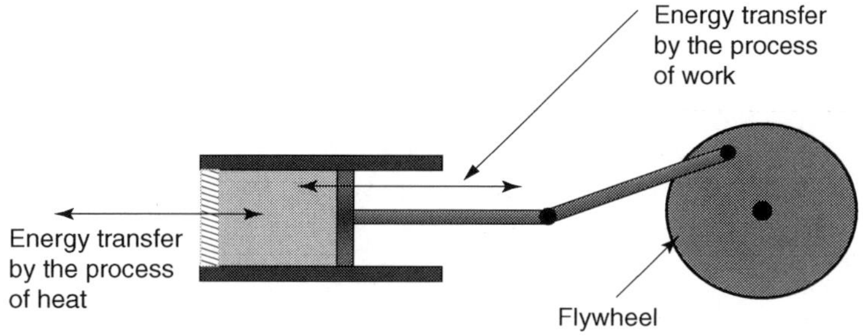

Figure 5.2 Schematic diagram of heat engine.

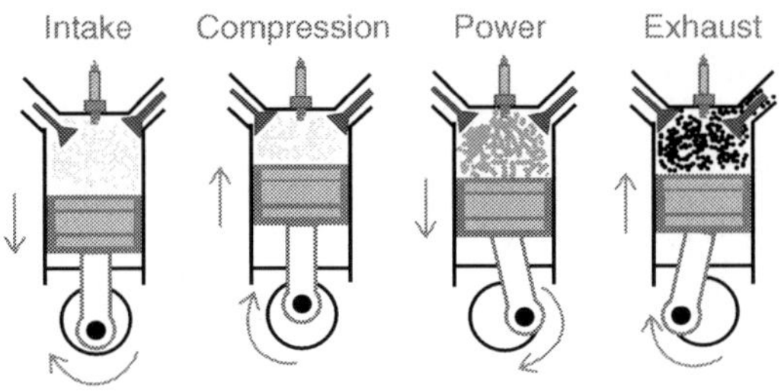

Figure 5.3 The internal combustion engine.

uses four cylinders (pistons) mounted in a straight line. When a piston inside one cylinder moves down, a petrol-air vapour is drawn into the cylinder through a valve ("intake"). This valve then closes, and the piston moves up. It squeezes the vapour into a small space at the top of the cylinder ("compression"). When the piston almost reaches the top of the cylinder, the spark plug produces an electric spark. This spark causes the vapour to combust, and the piston is driven down to the bottom of the cylinder ("power"). As the piston moves back up the cylinder for the second time, it forces the leftover gases out of the exhaust valve ("exhaust"). This ends the cycle, and a new cycle begins. This is called a four-cycle or four-stroke internal combustion engine. The four pistons are attached to a crankshaft which contains a flywheel to provide the necessary inertia for the intake and compression stages and converts the up-and-down motion into a turning motion of the wheels (via a clutch and gearbox). The cycle then repeats itself, i.e., each cycle requires four strokes of the piston – intake, compression, power, and exhaust – and two revolutions of the crankshaft. The hot and cold reservoirs in the internal combustion are not so obvious as with a steam engine. The combustion of the hydrocarbon provides a flow of energy to the engine, by the process of heat (equivalent to the hot reservoir), and the atmosphere is the cold reservoir that the hot gases are exhausted.

Although steam engines were of major economic and industrial importance since the mid-18th century, there was no scientific study of them until the early 19th century, and there was no scientific theory for their operation. One of the most important questions was: *what is the maximum efficiency that can be achieved by a steam engine?* This question was answered brilliantly by the French engineer and scientist Nicolas Leonard Sadi Carnot (1796 – 1832) in his only publication, the 1824 monograph "Réflexions sur la Puissance Motrice du Feu et sur les Machines Propres à Développer cette Puissance" [Reflections on the Motive Power of Fire and on Machines Fitted to Develop that Power]. In this monograph, Carnot gave the first successful theory of the maximum efficiency of heat engines. Carnot's genius was to outline the general principles and laws of the production of mechanical work ("motive power") not only for steam engines but also for all imaginable heat engines, whatever the working substance and whatever operating procedure used. Carnot abstracted the essential features of the steam engine into a conceptual idealised heat engine in what was one of the most impressive thought experiments in science. It was this work that led to the birth of thermodynamics.

The basic idea of a conceptual heat engine is a source of energy (a hot reservoir at temperature T_H) transfers energy to the cylinder/piston system by the process of heat, to increase the temperature of a gas in a cylinder fitted with a piston (see Figure 5.4). In thermodynamic speak, we say $+|\mathbf{q_H}|$ joules of energy flow into the system from the hot reservoir, by the process of heat. The gas expands and drives the piston to the right and the cylinder/piston system does work on the surroundings. In thermodynamic speak, we say $-|\mathbf{w}|$ joules of energy flow into the system from the work reservoir by the process of work. To return the system to its initial state the system transfers energy, by the process of heat, to an energy sink (a cold reservoir at temperature T_C). In thermodynamic speak, we say $-|\mathbf{q_C}|$ joules of energy flow into the system from the cold reservoir. During this process the temperature of the gas decreases and the gas volume contracts driving the piston to the left. The inertia of the revolving flywheel smooths the intermittent reciprocating piston movement to a continuous energy output flow; the energy is stored mechanically in the rotating flywheel as kinetic energy. Notice that the presence of a cold reservoir is an obligatory condition in allowing the heat engine to operate in a continuous cycle. In summary, a heat engine takes in energy from a hot reservoir and, by the process of heat, transfers some of this energy to the surroundings ("work reservoir") by the process of work (i.e., does work on the surroundings)

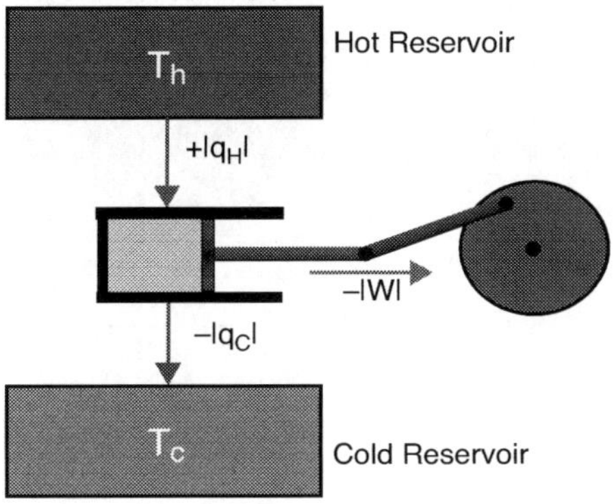

Figure 5.4 Conceptual diagram of heat engine showing the overall energy flows (using our agreed sign convention).

and expels excess energy to a cold reservoir by the process of heat to complete the cycle. Or, more succinctly, heat is converted to work.

In his book, Carnot points out that a cyclic process[15] is the underlying principle for a heat engine and states that all parts of the engine, including the working substance (e.g., steam), must be, at the end of the process, in precisely the same state as at the beginning[16]:

> ... when a body has experienced any changes, and when after a certain number of transformations it returns to precisely its original state, that is, to that state considered in respect to density, to temperature, to mode of aggregation – let us suppose, I say, that this body is found to contain the same quantity of heat that it contained at first, or else that the quantities of heat absorbed or set free in these different transformations are exactly compensated.

Callendar[17] quotes the same passage as the more understandable[18]:

> When a body has undergone any changes and after a certain number of transformations is brought back identically to its original state, considered relatively to density, temperature and mode of aggregation, it must contain the same quantity of heat as it contained originally.

Today we would say that at the end of the cycle, the internal energy of the working substance is the same as it was at the beginning of the cycle. However, note the use of the word "heat" in the above quotes. It cannot be replaced by "energy transferred by the process of heat" (see Section 5.1) and this is because Carnot, like the majority of scientists and engineers of his day, believed in the caloric theory of heat, where heat is considered to be a fluid (the caloric theory of heat is discussed in Chapter 6).

5.3 FRICTION AND TURBULENCE IN EXPANSION AND COMPRESSION PROCESSES

I do not intend to go into turbulence in any detail, but it is important to know what turbulence is and does. This allows us to understand why we need to eliminate turbulence from our models if we want to simplify our systems to the point where we have systems (such as the ideal gas) that can be analysed using relatively simple mathematics. I have abstracted the following from a website (together with the photographs) because I did not feel I could improve on it[19]:

> Let us first define what a flow is: a flow is the continuous movement of a fluid, i.e. either a liquid or a gas, from one place to another. Basically, there exist two types of flows, namely laminar flows, and turbulent flows. Roughly speaking we can say that a laminar flow is a "simple" flow while a turbulent flow is a "complicated" flow. We will illustrate what we mean by "simple" and "complicated" using the following, simple experiment. Go to your kitchen sink and open the faucet. The stream of water that emerges from your faucet is very smooth and very regular. The flow of water is smooth because all the water molecules move, at essentially the same speed, in the same direction. This is called a laminar flow. Furthermore, if you did not open the faucet too much, the water will also flow down the drain in a laminar flow. Now place a cup under the stream of water emerging from the faucet. Although the stream is still laminar, the flow pattern of the water in the sink has become very complicated. This is since now the water molecules tend to move in different directions at different speeds. Such a flow is called turbulent.

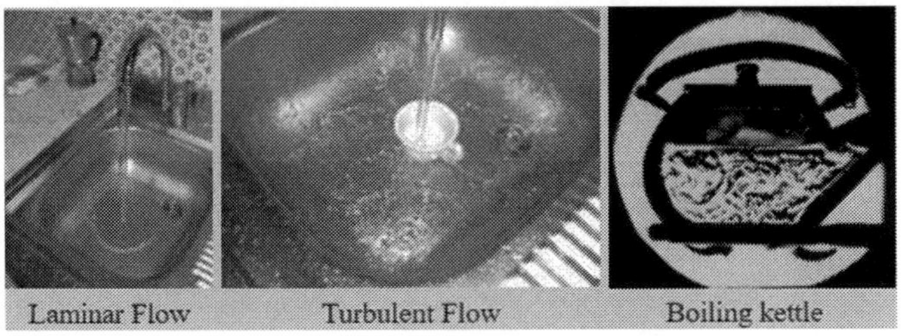

Another example of turbulence you can easily observe at home is the flow of water in a boiling pot of water. Put water in a pot and heat it up on your electric cooker. If you wait for a short while, the water will start to move in a laminar fashion, i.e. in a very regular way. If you wait a bit longer bubbles will start to rise from the bottom to the surface and the motions of the water become very complicated or turbulent. In this case the turbulence is due to convection.[20] To see how complicated, the turbulent motions are look at the picture of the boiling kettle.

Now we can define the terms laminar and turbulent in a more precise fashion. In a laminar flow, all the molecules in the fluid move essentially in the same direction and at the same speed. In a turbulent flow, however, the molecules in a fluid move in many different directions and at many different speeds.

Where does turbulence occur?

Turbulent motions are very common in Nature. Turbulence occurs nearly everywhere: in the oceans, in the atmosphere, in rivers, even in stars and galaxies. In fact, it is easier to find a turbulent flow than a flow that is laminar. Here are a few examples of turbulent flows:

1. The wake of a ship or submarine is turbulent.
2. The swirls and eddies in a fast-flowing river are turbulent.
3. The air currents in the atmosphere are turbulent.
4. Turbulence was observed in the ash plume of the Mt. Spurr volcano as it erupts
5. The outer layer of the Sun, i.e. the convection zone, is highly turbulent.

To put it simply turbulence is the complex, chaotic, random, and unpredictable movement of macroscopic volumes of a fluid. Eddies (movements of water, often circular, running contrary to the main flow) are common in turbulent flow. The relative movement of macroscopic volumes of fluid moving relative to each other is directly analogous to the relative movement of solid objects (see Section 1.9.3), i.e., friction opposes their movement and eventually, the relative movement stops, and the macroscopic kinetic energy of the eddies is converted to the microscopic kinetic energy of the microparticles of the fluid resulting in localised temperature rises within the fluid. To maintain turbulence there must be a continuous external input of energy because the fluctuations of turbulence in a fluid will be damped out over time. You know this – if you stop stirring your cup of coffee the turbulence caused by your stirring will, eventually, stop. This is what is meant by the terms "damping" or "damped out". The end result is that when equilibrium is eventually achieved, the internal energy of the fluid has increased, and this manifests itself in a temperature rise.

I do not want to spend much longer on this, the bottom line is that friction whether it results from the movement of the piston or turbulence results in the generation of temperature differences. In Section 4.3.5 we have seen that a temperature difference can be used to do work. Therefore, we know (from the Principle of Energy Conservation that the creation of a temperature difference requires energy. This immediately leads to the conclusion that frictional effects reduce the amount of energy transferred to the surroundings by the process of work for a given input of energy to the heat engine. All of this tells us what we already know: if you have a rusty piston/cylinder arrangement or increased turbulence in your working substance you are going to have to transfer more energy to the system from the hot reservoir to get the same amount of energy transferred to the surroundings by the process of work. This if often termed "lost work" in thermodynamics.

I want to make the following points:

1. When there are no pressure and temperature gradients, there is no associated turbulence, and for a fixed mass of an ideal gas fixing two of the variables P, V, and T fixes the third because PV = nRT applies. To put it another way, when a system is in equilibrium there can be no turbulence.
2. If there are pressure and temperature gradients then there is turbulence, and under these conditions, the state of the gas is not represented by the two variables P and V. In fact, an enormous number of variables would be required because the system is not at equilibrium: we are now getting into the realms of chaos theory.[21]
3. Pressure and temperature gradients and associated turbulence cause eddies and vortices. These are chaotic forms of macroscopic kinetic energy and as they die away due to the effects of friction (interaction between the microparticles) this macroscopic energy is transferred to the kinetic, rotational, and vibrational energy of the microparticles, i.e., the internal energy content of the system increases.
4. Under turbulent conditions we have a complex system that, without the use of super computers, cannot be analysed to any degree of precision.[22] Richard Feynman has described turbulence as the most important unsolved problem of classical physics.[23] To bring this closer to home, under the non-equilibrium conditions of turbulence we cannot apply the ideal gas law (or variants of it) as there is no T and P or V that applies over the whole system.
5. Turbulence like friction results in "lost work", i.e., if we have turbulent conditions in our gas piston system, then, like friction, you are going to have to transfer more energy to the system from the hot reservoir to get the same amount energy transferred to the surroundings by the process of work.

Before leaving the subject of turbulence I want to return to the frictionless and massless piston/cylinder arrangement operating in a vacuum, and what happens to the oscillations when a weight is placed on the piston (the following is in essence another thought experiment). Laminar flow only occurs when the piston moves very slow, and please note that in these types of systems, laminar flow only occurs in the very small volume adjoining the piston. In laminar flow (Figure 5.5a) the microparticles follow the streamlines[24] exactly and the kinetic energy of the microparticles undergoing laminar flow will increase and this results in an increase in the working fluids temperature. If the direction of the piston is reversed, then the direction of laminar flow reverses and the kinetic energy of the microparticles undergoing laminar flow will decrease (resulting in a decrease in the working fluids temperature. As we shall see, in the next section, these are reversible processes. When the speed of movement of the piston increases then turbulence within the fluid increases (Figure 5.5b), i.e., not all of the microparticles in close proximity to the piston follow the streamlines and eddies are forming (turbulence) which we know increases the kinetic energy of the microparticles as the turbulence is damped out by the opposing force of friction. We now have two processes increasing the temperature of the working substance – the movement of the

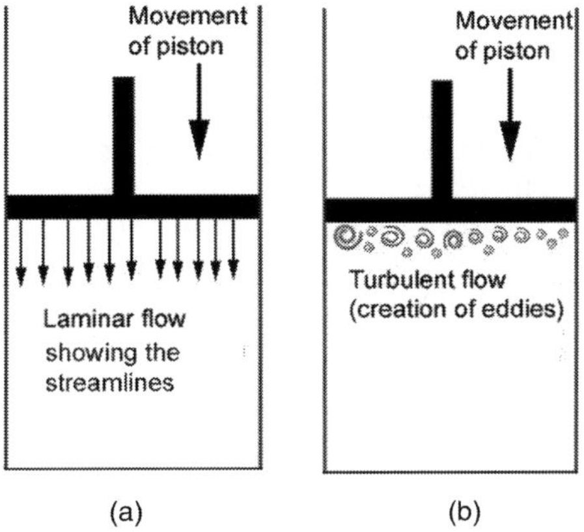

Figure 5.5 Depiction of (a) Laminar and (b) Turbulent flow in a piston.

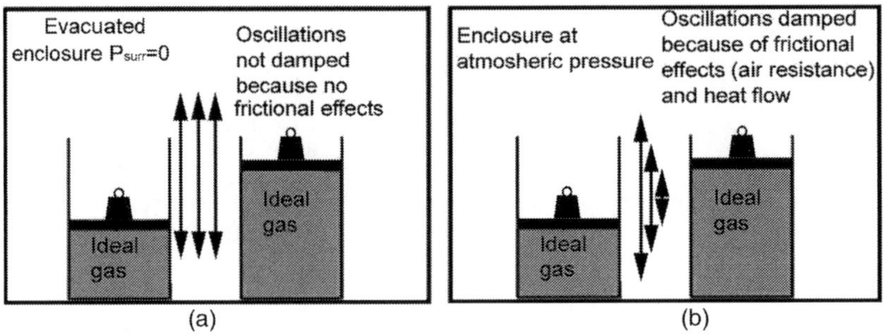

Figure 5.6 Hypothetical scenarios in (a) there are no frictional effects and we have perpetual motion, and although there are no frictional effects within the piston/cylinder in (b) there are frictional effects external to the piston/cylinder and the oscillatory movement of the piston will eventually come to a halt.

piston and friction. If the movement of the piston is reversed turbulence still occurs (and resulting temperature increase), but the outward movement of the piston results in a temperature decrease (Figure 5.6).

The above description only occurs for real gases where there is (electromagnetic) interaction between the particles. By definition there is no interaction between the microparticles of an ideal gas and any turbulence will not be damped out by friction and there will be no temperature increase. Consider the following thought experiment with ideal gas as the working substance. The frictionless piston/cylinder arrangement is housed in a perfect vacuum and a finite weight is placed on the piston which starts to oscillate (please ignore the practicalities). We know turbulence in an ideal gas does not result in temperature differences because there are no frictional effects. We also know there is an oscillation of the temperature of the ideal gas in a boundary layer near the piston, but overall, the temperature of the gas must remain constant, i.e., there is no transformation of energy flow by the process of work into energy flow by the process of heat (no conversion of work into heat). Therefore, the oscillatory motion of the piston is not damped, and we have perpetual motion. Now consider the situation if the cylinder/piston arrangement is contained in an environment containing, say, air at atmospheric temperature. Now there are the frictional effects (drag) on the moving piston that will eventually damp down the oscillatory movement as work is converted to heat.

5.4 REVERSIBLE PROCESSES

If we want to simplify our systems to the point where we have models (such as the ideal gas) that can be analysed using relatively simple mathematics, then we need to eliminate turbulence and friction from our models. Turbulence, like friction, is ubiquitous in nature, however, if the change of force ($\mathbf{dF}$) on the piston is infinitesimally small (i.e., $\mathbf{dF} \rightarrow 0$), then the movement of the piston occurs at an infinitesimally slow rate. If this occurs, then the volume and temperature changes are infinitesimally small, there is no turbulence, and the system is always in equilibrium. When an ideal gas system is in equilibrium, the state functions P and V can be used to describe the system and the ideal gas law is always applicable.[25] These infinitesimally small changes are the characteristics of what is known in thermodynamics as a reversible process. It is called "reversible" because by carrying out infinitesimally small changes in the other direction we can reverse the direction of the change. Reversible thermodynamic processes are analogous to the frictionless laws of motion covered in Chapter 1 and the ideal gas considered in Chapter 2, i.e., they do not exist in the real world, but as a simplifying concept, they give us insights into "real" objects and "real" processes.[26]

A good way to appreciate a reversible process is to consider the frictionless piston systems described in Section 4.3.5. In Figure 5.7a, when the total mass of M kg is placed on the piston in one go, the piston oscillates up and down with an ideal gas; when there is no interaction between the microparticles (and the piston is in a vacuum to eliminate air resistance), this oscillation would carry on for ever (see previous section). For a real gas (interactions between microparticles), this oscillation (and associated gaseous turbulence) dies down over time – this is an irreversible process. In both cases, the gas is not in equilibrium while oscillation continues, and the gas law does not apply.

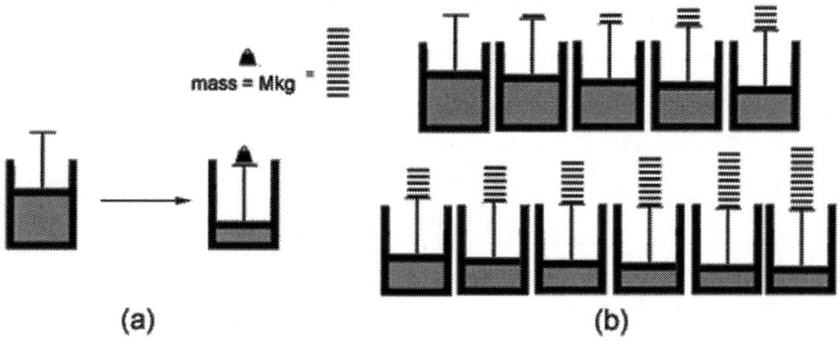

Figure 5.7 Schematic representation of (a) an irreversible process and (b) a reversible process.

However, if we increase the mass on the piston by infinitesimal increments of mass (many texts use the addition of consecutive addition of single grains of sand), then we will have (virtually) eliminated all turbulent effects and the system (either ideal or real gas) is always in equilibrium (Figure 5.7b). Rigorously (think about the theory of calculus) we would have to place an infinite number of infinitely small mass increments onto the piston to have a true reversible process and, in the real world, this can never happen, i.e., a perfectly reversible process is not possible. As with the frictionless and massless piston, and ideal gas, the conditions for reversibility can never be met, but they can be approached, i.e., we are again considering an ideal theoretical system to allow us to simplify the scientific/mathematical analysis. The corollary to this is that all real processes are irreversible, and during any process, the system plus surroundings are never at equilibrium. These important statements will be considered throughout the rest of this book.

One aspect of reversible processes is that they cannot take place in real time, or to put it another way they are independent of time. We would not know, if we saw a film of a reversible process, whether the film was running forwards or backward (note that in the example above your hand placing or removing infinitesimally small masses from the piston is not a reversible process). A simple example should prove useful here. If we watch a film of a ball bouncing in the real world we would know if we were watching the film in reverse because the ball would bounce to a greater height with each bounce.[27] If we now consider an "ideal bouncing ball" where no macroscopic energy is converted to the microscopic internal energy of the ball (the system) and the surroundings, then the ball would continue bouncing to the same height forever[28] and we would not be able to tell whether a film of our ideal bouncing ball was running forwards or backward. The point to take away from this is that the effects of friction, turbulence, and other dissipative processes (see below) all result in processes that are time-dependent, i.e., if we are watching a film of a real process we can tell when the film is running in reverse.[29] We now know that the "lost" kinetic energy in an inelastic collision is transferred to the surroundings by the process of heat. In thermodynamics, these types of processes are known as "dissipative". After a dissipative process has occurred the capacity of a system to do work in its final state is less than that of the initial form. All real processes are dissipative and are, by definition, irreversible. The following is a list of the dissipative processes that can occur in ideal gas systems considered in this book:

1. Transfer of energy, by the process of heat, down a finite temperature gradient.[30]
2. Free expansion of a volume of gas.[31]
3. Friction
4. Diffusion

To this list, we can add turbulence as a dissipative process in all non-ideal gas systems. The corollary is that reversible processes are non-dissipative processes.

Wikipedia states[32]:

> The concept of dissipation was introduced in the field of thermodynamics by William Thomson (Lord Kelvin) in 1852. Lord Kelvin deduced that a subset of … irreversible dissipative processes will occur unless a process is governed by a 'perfect thermodynamic engine'.[33]

The processes that Lord Kelvin identified were friction, diffusion, conduction of heat and the absorption of light.

It was Carnot who first proposed the concept of reversibility. In 1911 Callendar[34] gave his presiden tial address to the Physical Society of London which was titled "The Caloric Theory of Heat and Carnot's Principle". Consider this extract where he first quotes Carnot (the underlined bit)[35]:

'In order to realise the maximum effect, it is necessary that, in the process employed, there should not be any direct interchange of heat between bodies at sensibly different temperatures.' Direct transference of heat between bodies at different temperatures would be equivalent to wasting a difference of temperature which might have been utilised for the production of motive power.[36] This rule excludes all the well recognised thermal sources of waste of power, which practical engineers had already been most successful in reducing. The ideal engine was also assumed to be free from mechanical losses such as friction. Under these conditions there is equilibrium, both mechanical and thermal, at every stage of the process, so that each operation is reversible, and may be supposed to be performed in either direction at will.

Consider the following simple example of a billiard ball moving around a frictionless billiard table with no pockets and all collisions with the cushions are perfectly elastic. At any point in time, the ability of the ball to do work (e.g., in a head-on inelastic collision with a stationary ball) is unchanged. This is a non-dissipative system and if we watched a film of the process, we could not tell whether time is running backwards or forwards. Now, consider, the same situation but the collisions with the cushion are non-elastic. Now we have a dissipative system, the ability of the moving ball to do work reduces with time and the original kinetic energy of the system (i.e., the ball and the billiards table) is converted into temperature differences which then dispersed energy to the system and surroundings by the process of heat. If we watched a film of this process, we would immediately be able to tell whether time was moving forward or running backwards.

As all real processes are dissipative then after any real process the capacity of the universe to do work is less than before the process occurred.[37] The meaning of Carnot's statement that "there should not be any direct interchange of heat between bodies at sensibly different temperatures" will become clear as you proceed through this chapter. In Section 5.1, I wrote, "we need something else that tells us the directionality of a process". This tendency towards a reduced capacity of the universe to do work, after irreversible (real) processes have occurred, may point us to this "something else". This suggests the concept of a dissipative process is vitally important in the understanding of directionality in thermodynamics.

The concept of a reversible process (i.e., a non-dissipative process), like that of an ideal gas, and containers with frictionless and massless pistons, strips away the complexities of the real world and allows a relatively straightforward analysis of certain systems or processes. The results of the analysis can then be tested out in the real world. I now want to consider two reversible ideal gas processes in some detail - reversible isothermal and reversible adiabatic processes. We will then have the necessary background to consider Carnot's thought experiment.

Now that I have introduced the concept of reversibility, I want to revisit heat reservoirs. A heat reservoir is an idealisation (like an ideal gas). The basic idea is that the energy capacity of the heat reservoir is large enough so that any transfer of energy, by the process of heat, does not alter the temperature of the reservoir, i.e., a heat reservoir is a constant temperature heat source or sink (e.g., the room in which a cooling cup of coffee sits can be considered, for all intents and purposes, a cold heat transfer is irreversible when it occurs down a temperature gradient. However, there are no temperature gradients within a heat reservoir. Although this is implicit in the definition of heat reservoirs reservoir), I wanted to explicitly mention it here because many texts do not do so (and it is vitally important). For the energy transfer process to be considered reversible with respect to the universe, the temperature difference between the system and reservoir must be infinitesimal. The exchange of heat between two ideal reservoirs at different temperatures is an irreversible process. The interface between the two reservoirs cannot be at both temperatures at the same time, and the temperature at the interface cannot be discontinuous. We can visualise the interface between the reservoirs as a very thin slab of conductive material (of negligible mass and heat capacity) sandwiched between the two reservoirs. At one surface of the slab we have the hot reservoir temperature and at the other surface we have the cold reservoir temperature, with the temperature varying linearly across the slab. It is at this interface, and only at this interface, that the irreversibility of the energy transfer process occurs.

5.5 REVERSIBLE PROCESSES, ISOTHERMS AND ADIABATS

5.5.1 Introduction

> The problem occupying Carnot was how to design good steam engines. Steam power already had many uses - draining water from mines, excavating ports and rivers, forging iron, grinding grain, and spinning and weaving cloth - but it was inefficient. The import into France of advanced engines after the war with Britain showed Carnot how far French design had fallen behind. It irked him particularly that the British had progressed so far through the genius of a few engineers who lacked formal scientific education. British engineers had also accumulated and published reliable data about the efficiency of many types of engines under actual running conditions; and they vigorously argued the merits of low- and high-pressure engines and of single-cylinder and multi-cylinder engines.[38]

As we have seen real heat engines go through a complex sequence of steps, and Carnot's genius was (i) to reduce its operation to a simple model, and (ii) his consequent "highly original reasoning".[39] He considered two conditions in which the heat engine operated. One condition, known as an adiabatic expansion or compression, is one in which the cylinder/piston system is assumed to be perfectly insulated so that no energy can flow into or out of the cylinder by the process of heat (Figure 5.8a). In an adiabatic process, energy is transferred to or from the system only by the process of work, and the gas cools as it expands, or warms as it is compressed, i.e., $\mathbf{q} = 0$ and $\mathbf{w} \neq 0$. Because the system is either transferring or receiving energy to or from the work reservoir and, by the First Law of Thermodynamics (i.e., Principle of Energy Conservation), this must result, respectively, either in a decrease or in an increase in the internal energy of the system, which for an ideal gas means a decrease or increase of the temperature of the system. As we shall see, in a reversible adiabatic process, the magnitude of the energy change of the system by the process of work is equal to the magnitude of the internal energy change of the system, i.e., $\left|\Delta\mathbf{U}^{\text{reversible}}_{\text{adiabatic}}\right| = \left|\mathbf{w}^{\text{reversible}}_{\text{adiabatic}}\right|$ (from the First Law of Thermodynamics – $\Delta\mathbf{U} = \mathbf{q} + \mathbf{w}$).

The second condition, known as an isothermal expansion or compression, involves keeping the gas at a constant temperature (Figure 5.8b). As the engine (the system) transfers energy to or from the work reservoir by the process of work, the engine must absorb or dump energy from or into a heat reservoir, via a diathermal wall, in order to maintain a constant temperature: otherwise, it would cool down as it expands or, conversely, warm up as it is compressed. In a reversible isothermal energy process, $\Delta\mathbf{U}^{\text{reversible}}_{\text{isothermal}} = 0$ (because, by definition, $\Delta\mathbf{T} = 0$) and, as we shall see,

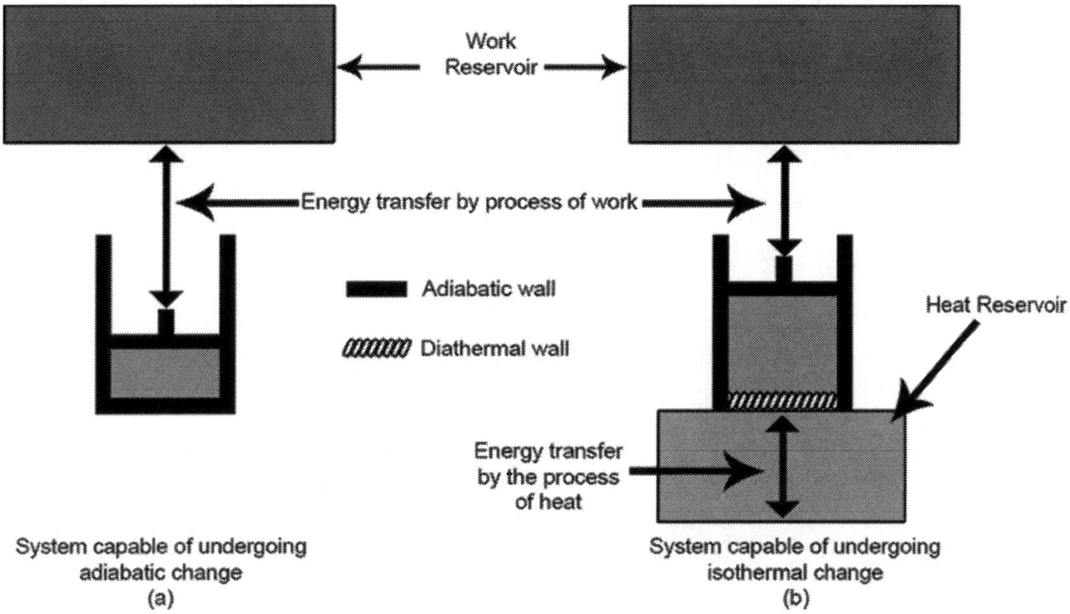

Figure 5.8 Systems capable of undergoing adiabatic and isothermal changes (a) adiabatic and (b) isothermal changes.

$\left|\mathbf{q}_{\text{isothermal}}^{\text{reversible}}\right| = \left|\mathbf{w}_{\text{isothermal}}^{\text{reversible}}\right|$, i.e., in reversible isothermal processes the energy absorbed from or dumped into a heat reservoir, by the process of heat, is transferred to or from a work reservoir with 100% efficiency., i.e., heat is converted to work with 100% efficiency. The concept of reversibility is fundamental in Carnot's analysis and, in modern thermodynamics.

I will now cover isothermal and adiabatic expansion/compression processes in detail.

5.5.2 Reversible Isothermal Expansion or Compression - Isotherms

For an ideal gas if the temperature remains constant then it follows $\Delta\mathbf{U} = 0$.[40] From the First Law of Thermodynamics (Equation 5.1) we know that for an isothermal process occurring in a closed system $\mathbf{q} = -\mathbf{w}$ where $\mathbf{q}$ and $\mathbf{w}$ represent the energy flows into the system and these flows can be positive or negative depending on whether the system gains energy (positive) or loses energy (positive). Alternatively, using absolute values, we can write $\mathbf{q} = \mathbf{w}$ (see Equation 5.2):

$$\Delta\mathbf{U}_{\text{closed system}} = \mathbf{q} + \mathbf{w} = 0 \tag{5.1}$$

$$+\Delta\mathbf{U}_{\text{closed system}} = +\mathbf{q} - \mathbf{w} = 0 \quad \text{or} \quad -\Delta\mathbf{U}_{\text{closed system}} = -\mathbf{q} + \mathbf{w} = 0 \tag{5.2}$$

Consider a fixed quantity (n moles) of ideal gas at temperature T in our hypothetical frictionless and massless piston/cylinder system that has a diathermal wall in contact with a heat reservoir. Placed on the piston is a mass = m_{initial} made up of an infinite number of infinitely small objects. The system (the ideal gas) is kept at constant temperature T by allowing energy to flow back and forth, by the process of heat, from a heat reservoir at the same temperature as the system (see Figure 5.9). Although we know (Section 3.5.7) that you cannot have a net energy flow, by the process of heat, if the temperatures are the same, this requirement of equal temperatures is necessary because energy transfer by the process of heat between two objects at different temperatures is an irreversible process.

The best way to imagine what is going on in reversible isothermal processes is to consider infinitesimally small changes, e.g., the mass on the piston decreases by an infinitesimal amount $\left(-|\mathbf{dm}|\right)$, the volume of the system increases infinitesimally $\left(+|\mathbf{dV}|\right)$ resulting in an infinitesimal decrease in temperature $\left(-|\mathbf{dT}|\right)$ and an infinitesimal quantity of energy $\left(+|\mathbf{dq}|\right)$ flows into the heat reservoir, by the process of heat, into the system via the diathermal container wall to maintain constant temperature (Figure 5.9). After a sufficient number of infinitely small mass increments have been sequentially removed and the mass on the piston is m_{final} the pressure has decreased from P_{initial} to P_{final}, the volume has increased from V_{initial} to V_{final}, the energy transferred from the heat reservoir to the system is $\sum_{\text{initial}}^{\text{final}} \mathbf{dq} = \mathbf{q}$, and the temperature has remained constant. It is important to remember for a reversible process every intermediate state is at equilibrium and this state is described by the ideal gas equation PV = nRT, and, at a fixed temperature, PV = constant.

An isotherm is a curve on a P-V diagram for an isothermal process and there are an infinite number of isotherms (Figure 5.10). The equation of an ideal gas isotherm immediately falls out of

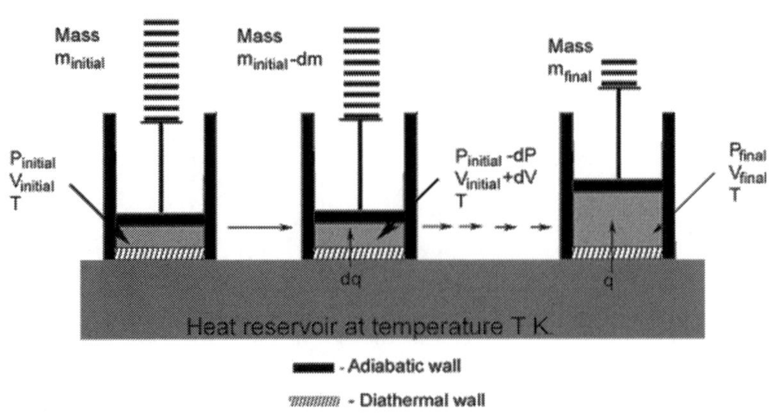

Figure 5.9 Reversible isothermal expansion.

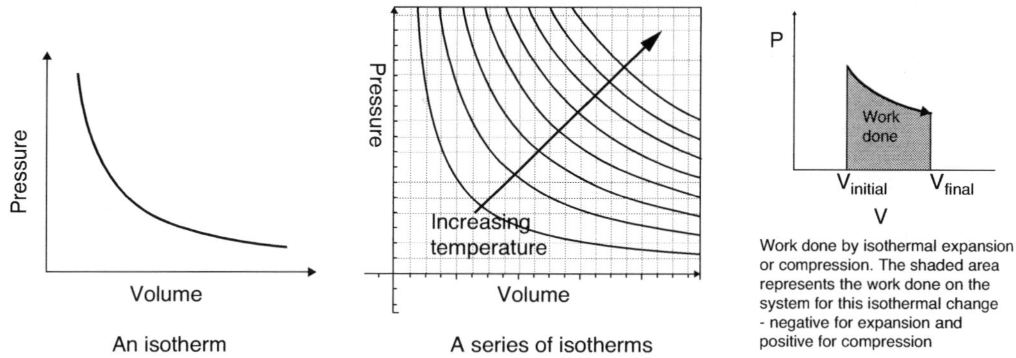

Figure 5.10 Isotherms.

the ideal gas law: PV = constant where the value of the constant varies with the amount of ideal gas and its temperature, where the constant is nRT. It is important to remember that this relationship only holds when the system and surroundings are always in equilibrium, e.g., when a process is reversible.

Now I want to calculate the PV work done on the system by the surroundings during an isothermal expansion/compression at constant T. Consider an isothermal expansion from $V_{initial}$ to V_{final} (i.e., $V_{final} > V_{initial}$), the work done on the system by the surroundings is the area under the isotherm (Figure 5.10) and we know from our agreed sign convention that for isothermal expansion the flow of energy by the process of work is a negative quantity. The reversible isothermal expansion of the gas takes place in a finite number of infinitesimally small intermediate steps. The work done on the system by the surroundings by one infinitesimally small step is[41]:

$$\mathbf{dw} = -\mathrm{P}\mathbf{dV} \tag{5.3}$$

If the gas expands from $V_{initial}$ to V_{final} during isothermal reversible expansion, then the total amount of work done on the system is:

$$\mathbf{w} = -\int_{V_{initial}}^{V_{final}} \mathrm{P}\mathbf{dV} \tag{5.4}$$

The process occurs reversibly, and by definition, the system is always in equilibrium, this means that the ideal gas equation (PV = nRT) holds throughout the process. Therefore, we can substitute $\frac{nRT}{V}$ for P:

$$\mathbf{w} = -\int_{V_{initial}}^{V_{final}} \frac{nRT}{V}\mathbf{dV} \tag{5.5}$$

The system is closed so n is a constant, and, as the process occurs under isothermal conditions, T is also a constant, therefore Equation (5.5) simplifies to:

$$\mathbf{w} = -nRT\int_{V_{initial}}^{V_{final}} \frac{\mathbf{dV}}{V} \tag{5.6}$$

This integral may be solved to give[42]:

$$\mathbf{w} = -nRT\ln\frac{\mathbf{V_{final}}}{\mathbf{V_{initial}}} \tag{5.7}$$

When $V_{final} > V_{initial}$ (expansion) then $\ln\frac{V_{final}}{V_{initial}}$ is a positive number and **w** is a negative number, and if $V_{final} < V_{initial}$ (compression) then $\ln\frac{V_{final}}{V_{initial}}$ is a negative number and **w** is a positive number. This agrees with our sign convention because we know:

1. if the system expands reversibly and isothermally at constant pressure the system has gained energy by the process of heat $(+|\mathbf{q}|)$ the system has lost energy to the surroundings by the process of work $(-|\mathbf{w}|)$,
2. if the system contracts reversibly and isothermally at constant pressure the system has lost energy by the process of heat $(-|\mathbf{q}|)$ the system has gained energy from the surroundings by the process of work $(+|\mathbf{w}|)$.
3. for a reversible isothermal process $|\mathbf{q}| = |\mathbf{w}|$, therefore, $\Delta \mathbf{U}_{\text{closed system}} = 0$.

To clarify using absolute values:

$$-\mathbf{w}^{\text{expansion}} = -\text{nRT}\left(+\left|\ln \frac{\mathbf{V}_{\text{final}}}{\mathbf{V}_{\text{initial}}}\right|\right) \tag{5.8}$$

$$+\mathbf{w}^{\text{compression}} = -\text{nRT}\left(-\left|\ln \frac{\mathbf{V}_{\text{final}}}{\mathbf{V}_{\text{initial}}}\right|\right) \tag{5.9}$$

To reinforce our understanding of isotherms let us consider how kinetic theory explains what is going on. In expansion, the force (pressure/area) on the piston is reduced (by the removal of an infinitesimally small mass). This means that as microparticles of the ideal gas collide with piston they lose some of their kinetic energy to the moving the piston (i.e., by doing work on the surroundings). This results in an infinitesimal decrease in temperature.[43] The heat reservoir supplies an infinitesimally small amount of energy, by the process of heat, to the ideal gas system to "make up" for the loss in energy that has occurred because the system has done work on the surroundings (the system "must" remain at constant temperature). In compression, the force (pressure/area) on the piston is increased (by the addition of an infinitesimally small mass). This means that as microparticles of the ideal gas collide with piston they gain some of their kinetic energy from the moving piston (i.e., the surroundings are doing work on the system). This results in an infinitesimal *increase* in temperature. The heat reservoir gains an infinitesimally small amount of energy, by the process of heat, from the ideal gas system to "make up" for the gain in energy that has occurred because the surroundings have done work on the system.

5.5.3 Reversible Adiabatic Expansion and Compression - Adiabats

When our frictionless and massless piston system is thermally isolated from its surroundings, no energy can flow into or out of the system by the process of heat (in other words $\mathbf{q} = 0$). The only process that can change the internal energy of the ideal gas is PV work ($\mathbf{w} \neq 0$). The plot of P v. V for an adiabatic process is known as an "adiabat". To understand how an adiabat differs from an isotherm, imagine beginning at some point "A" on the lower temperature T_C isotherm (Figure 5.11) and increasing the gas pressure by compressing the gas, i.e., the surroundings are transferring energy to the system by the process of work. This process increases the internal energy of the gas, and because the system is thermally insulated from its surroundings, its temperature increases. Consequently, the PV plot (i.e., the adiabat) does not follow the isotherm but moves towards the

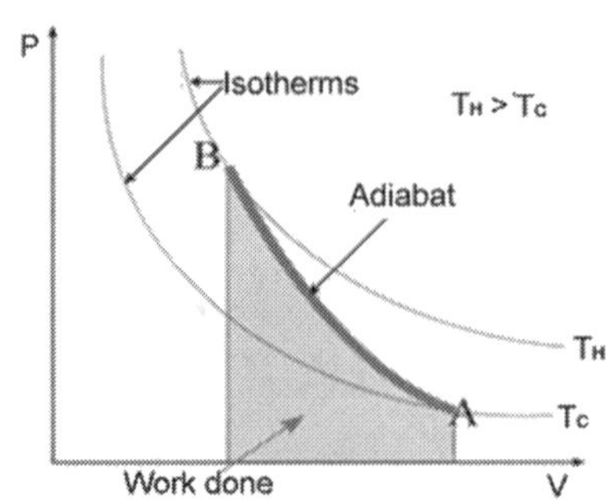

Figure 5.11 An adiabat compared to two isotherms.

higher temperature T_H isotherm at "B" and then (if compression continues) above it. This demonstrates the slopes of adiabats are always greater than those of isotherms. The derivation of the equation of the adiabat, set out below, is more complex than that of the isotherm. The derivation given here is based on that given by Fermi in his book "Thermodynamics".[44]

An adiabatic process is one in which no energy enters the system by the process of heat, i.e., $\mathbf{q} = 0$. The First Law of Thermodynamics takes the form[45]:

$$\mathbf{dU} = \mathbf{dw} = -\mathrm{P}\mathbf{dV} \tag{5.10}$$

In adiabatic expansion **dV** is a positive real number and, **dU** and **dw** are negative real numbers (the system is transferring energy to the surroundings by the process of work), and in adiabatic compression **dV** is a negative real number and, **dU** and **dw** are positive real numbers (the surroundings are transferring energy to the system by the process of work).

The internal energy of an ideal gas depends only on the temperature and is independent of the volume. For n moles of ideal gas, the change in internal energy during adiabatic expansion/compression is given by Equation (5.11):

$$\mathbf{dU} = \mathbf{dw} = \mathrm{nC_V}\mathbf{dT} \tag{5.11}$$

In adiabatic expansion **dT** is a negative real number and, **dU** and **dw** are negative real numbers (the system is transferring energy to the surroundings by the process of work), and in adiabatic compression d**T** is a positive real number and, d**U** and d**w** are positive real numbers (the surroundings are transferring energy to the surroundings by the process of work).

From Equations (5.10) and (5.11) we can write:

$$-\mathrm{P}\mathbf{dV} = \mathrm{nC_V}\mathbf{dT} \tag{5.12}$$

Rearrangement gives:

$$\mathrm{nC_V}\mathbf{dT} + \mathrm{P}\mathbf{dV} = 0 \tag{5.13}$$

Examination of Equation (5.13) tells us what we already know and that is that **dV** and **dT** have opposite signs. Using the Ideal Gas Equation (remember this assumes the gas is always in equilibrium, i.e., we are considering a reversible process), we can eliminate P from the Equation (5.13):

$$\mathrm{nC_V d}\mathbf{T} + \frac{\mathrm{nRT}}{\mathrm{V}}\mathrm{d}\mathbf{V} = 0 \tag{5.14}$$

Dividing both sides of Equation (5.14) by $\mathrm{nC_VT}$ gives:

$$\frac{\mathrm{d}\mathbf{T}}{\mathrm{T}} + \left(\frac{\mathrm{R}}{\mathrm{C_V}} \times \frac{\mathrm{d}\mathbf{V}}{\mathrm{V}}\right) = 0 \tag{5.15}$$

Integration yields (assuming the indefinite integration of zero yields a constant)[46]:

$$\int \frac{\mathbf{dT}}{\mathrm{T}} + \int \left[\frac{\mathrm{R}}{\mathrm{C_V}} \times \frac{\mathbf{dV}}{\mathrm{V}}\right] = \int \frac{\mathbf{dT}}{\mathrm{T}} + \frac{\mathrm{R}}{\mathrm{C_V}} \int \left[\frac{\mathbf{dV}}{\mathrm{V}}\right] = \ln \mathrm{T} + \frac{\mathrm{R}}{\mathrm{C_V}} \ln \mathrm{V} = \ln \mathrm{TV}^{\frac{\mathrm{R}}{\mathrm{C_V}}} = \int 0 = \text{constant} \tag{5.16}$$

Knowing $\ln \mathrm{TV}^{\frac{\mathrm{R}}{\mathrm{C_V}}} = \text{constant}$ then we know:

$$\mathrm{TV}^{\frac{\mathrm{R}}{\mathrm{C_V}}} = (\text{another})\,\text{constant} \tag{5.17}$$

We know that $\mathrm{C_P} = \mathrm{C_V} + \mathrm{R}$ (see Section 4.3.5, Equation 4.15). Therefore, Equation (5.17) can be rewritten as:

$$\mathrm{TV}^{\frac{\mathrm{C_P}-\mathrm{C_V}}{\mathrm{C_V}}} = \mathrm{TV}^{\frac{\mathrm{C_P}}{\mathrm{C_V}}-1} = (\text{yet another})\,\text{constant} \tag{5.18}$$

In thermodynamics the symbol γ is conventionally used to represent $\frac{C_P}{C_V}$ and Equation (5.18) can be written:

$$TV^{\gamma-1} = \text{constant} \tag{5.19}$$

From the Ideal Gas Equation, we know $T = \frac{PV}{nR}$. Substituting for T into Equation 5.19 we have:

$$\frac{PV}{nR}V^{\gamma-1} = \text{constant} \tag{5.20}$$

Rearranging (remembering $V \times V^{\gamma-1} = V^{\gamma}$):

$$PV^{\gamma} = (\text{again another})\text{constant} \tag{5.21}$$

We know for an ideal gas that $C_P = \frac{5}{2}R$ and $C_V = \frac{3}{2}R$, therefore $\gamma = \frac{5}{3}$, and the equation of the ideal gas adiabat is:

$$PV^{\frac{5}{3}} = \text{constant} \tag{5.22}$$

Integration of Equation (5.11) gives the work done on the system by the surroundings during adiabatic expansion and compression:

$$\mathbf{w} = \int_{T_{initial}}^{T_{final}} nC_V\,\mathbf{dT} = nC_V \int_{T_{initial}}^{T_{final}} \mathbf{dT} \tag{5.23}$$

This is straightforward, the system is closed therefore, n is constant and, for an ideal gas, C_V is constant; in addition, T is a state function,[47] so we have:

$$\mathbf{w} = nC_V(T_{final} - T_{initial}) = nC_V\Delta\mathbf{T} \tag{5.24}$$

Let me clarify using absolute values. For adiabatic expansion $T_{final} < T_{initial}$ and $nC_V\Delta\mathbf{T}$ is negative, therefore we know **w** is negative:

$$-|\mathbf{w}_{\mathbf{expansion}}| = -|nC_V\Delta\mathbf{T}| \tag{5.25}$$

Using the same approach, we can write for adiabatic compression (where $T_{final} > T_{initial}$):

$$+|\mathbf{w}_{\mathbf{compression}}| = +|nC_V\Delta\mathbf{T}| \tag{5.26}$$

Equations (5.25) and (5.26) agree with our sign convention. In adiabatic compression, energy is flowing into the system by the process of work and **w** is a positive real number. Similarly, in adiabatic expansion energy is flowing out of the system by the process of work and **w** is a negative real number.

5.5.4 Discussion

The behaviour of a gas experiencing an irreversible process is very different from that of the same gas experiencing a reversible process. In an irreversible process, the gas deformation is rapid enough that the inertia of the gas microparticles is important and this causes turbulence that results in pressure and temperature to be non-uniform within the gas. As a result, the gas pressure and temperature at the piston face where the work is being done are not equal to the average gas pressure and temperature within the system, and an equation of state like the ideal gas law cannot be used. The reason for this is straightforward, there is no single quantity that can be used to characterise the pressure or temperature within a system undergoing an irreversible process (in other words, any real process). If the change is carried out so that the system is considered to be in a succession of equilibrium states, then the process is called reversible, and the system is always in equilibrium. A reversible process is an idealised concept that allows us to use the ideal gas law that only applies under equilibrium conditions. Reversibility is a concept that allowed thermodynamics to develop into the powerful scientific tool it is now, and it was Carnot who first came up with this concept. This results in a paradox, because for any process to occur the system, by definition, cannot be in equilibrium. However, the classical thermodynamics considered in this book is only concerned with equilibrium states and not dynamics (which must be irreversible). This indicates that the name thermodynamics is a misnomer.

All of this poses an important question – why are we developing a theory of thermodynamics whose objective (apart from energy accountancy) is to understand why processes occur when by definition a process (aka a change) cannot, by definition, be in equilibrium? I am afraid I cannot give you a satisfactory answer, apart from making the following two points:

1. Equilibrium thermodynamics, compared to non-equilibrium thermodynamics (see next point) is relatively simple, and like many gross simplifications in science (ideal gas) it can provide descriptions of a real process with useful accuracy.
2. Non-equilibrium thermodynamics is much more complex, still involves simplifications and at its present stage of development offers little benefit to most working chemists.

This begs the question – why is thermodynamics not called thermostatics? This question was posed on the "StackExchange" website.[48] The answers were not satisfactory, the following one is probably nearer the truth:

> Probably because there is not a need for a sharp distinction between systems without heat flow and those with. The same study covers them both.

5.6 CARNOT'S THOUGHT EXPERIMENT – THE IDEALISED HEAT ENGINE

5.6.1 Introduction

Carnot's model of an idealised heat engine abstracted the essential features of all heat engines, and his insight was brilliant. Consider the following two quotes by RH Thurston who translated and published Carnot's monograph (*Réflexions sur la puissance motrice du feu*) in the late 19th century.[49]

> NICOLAS-LEONARD-SADI CARNOT was, perhaps, the greatest genius, in the department of physical science at least, that this century has produced. By this I mean that he possessed in highest degree that combination of the imaginative faculty with intellectual acuteness, great logical power and capacity for learning, classifying and organizing in their proper relations, all the facts, phenomena, and laws of natural science which distinguishes the real genius from other men and even from the simply talented man. Only now and then, in the centuries, does such a genius come into view. Euclid was such in mathematics; Newton was such in mechanics; Bacon and Compte were such in logic and philosophy; Lavoisier and Davy were such in chemistry; and Fourier, Thomson, Maxwell, and Clausius were such in mathematical physics.
>
> Carnot's "Réflexions sur la puissance motrice du feu," published in 1824, escaped notice at the time, was only now and then slightly referred to later, … and might, perhaps, have still remained unknown to the world except for the fact that Sir William Thomson, that greatest of modem mathematical physicists, fortunately, when still a youth and at the commencement of his own great work, discovered it, revealed its extraordinary merit, and, readjusting Carnot's principles in accordance with the modern views of heat-energy, gave it the place that it is so well entitled to in the list of the era-making books of the age.

Carnot's work attracted little attention during his lifetime and many of the insights he arrived at as a result of his thought experiment have been claimed by others). In his thought experiment he presented a model thermodynamic system that can be analysed to understand and clarify the fundamental principles that apply to all heat engines, independent of their design. His work was recognised in the mid to late-19th century and Carnot is now considered the founder of the science of thermodynamics.

Carnot wrote (remember Carnot, like virtually all scientists of that time, was an adherent of the caloric theory of heat)[50]:

> The phenomenon of the production of motion by heat has not been considered from a sufficiently general point of view. … In order to consider in the most general way the principle of the production of motion by heat, it must be considered independently of any mechanism or any particular agent. It is necessary to establish principles applicable not only to steam-engines, but to all imaginable heat-engines, whatever the working substance and whatever the method by which it is operated.

Carnot's idealised heat engine operates reversibly in a cycle and uses a fluid (usually known as the "working substance") in a frictionless piston/cylinder arrangement. Although in this book I will analyse his idealised heat engine using an ideal gas as the working fluid, because it makes the mathematics easier, it can easily be demonstrated that the resulting conclusions apply to any working substance. Before considering the "ideal heat engine" is worth summarising what we mean by a reversible process:

- Reversible processes occur infinitesimally slowly.
- The system is always at equilibrium, and infinitesimal changes can also reverse the direction of the change.
- Because a reversible process is always at equilibrium, the system is exactly described by the ideal gas law, or, to put it another way, the system is exactly described by the state functions P, V, and T. In an isolated or closed system (i.e., where n is constant) we know $\frac{PV}{T}$ is constant, and we can exactly describe the macroscopic state of the system by two of these three state functions. Conventionally P and V are used.
- A reversible process is unattainable but like an ideal gas, it can be approached. Figures 5.7b and 5.9 show how a reversible process may be approached. The discs (of negligible but finite mass) on the piston exert a negligible but finite force. These discs can be added or removed to change the direction of the process.
- A reversible process is also known as a non-dissipative process (see Section 5.5).

The flow of energy, by the process of heat, "down" a temperature gradient is an irreversible or dissipative process. Therefore, to ensure reversible energy transfer, by the process of heat, the temperature gradient must be zero, i.e., the heat reservoirs and the system must be at the same temperature. Of course, if the system and the reservoir are at the same temperature there would be no overall transfer of energy by the process of heat, but as I have already stated reversible processes are idealised constructs.

The quote below is yet another by Richard Feynman:[51]

> We must also consider the analog of frictionless motion, "frictionless" heat transfer. If we put a hot object at a high temperature against a cold object, so that the heat flows, then it is not possible to make that heat flow in a reverse direction by a very small change in the temperature of either object. But when we have a practically frictionless machine, if we push it with a little force one way, it goes that way, and if we push it with a little force the other way, it goes the other way. We need to find the analog of frictionless motion: heat transfer whose direction we can reverse with only a tiny change. If the difference in temperature is finite, that is impossible, but if one makes sure that heat flows always between two things at essentially the same temperature, with just an infinitesimal difference to make it flow in the desired direction, the flow is said to be reversible (Fig. 44–4). If we heat the object on the left a little, heat will flow to the right; if we cool it a little, heat will flow to the left. So we find that the ideal engine is a so-called reversible engine, in which every process is reversible in the sense that, by minor changes, infinitesimal changes, we can make the engine go in the opposite direction. That means that nowhere in the machine must there be any appreciable friction, and nowhere in the machine must there be any place where the heat of the reservoirs, or the flame of the boiler, is in direct contact with something definitely cooler or warmer.

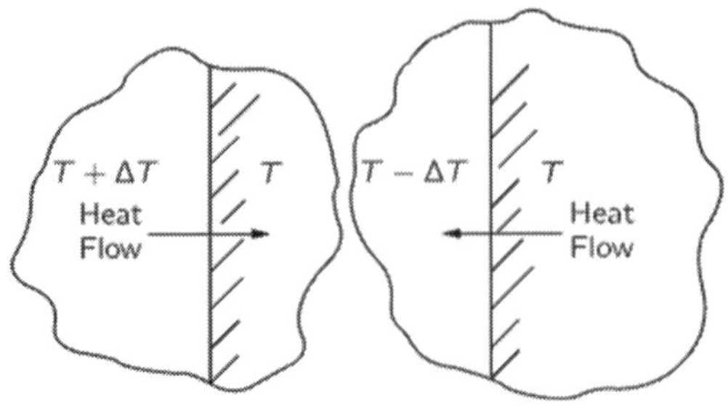

Note Feynman's use of the word "heat" but you should, by now, be able to convert his text into a form that considers heat a process, e.g., "heat flows" or "heat transfer" is "energy transfer by the process of heat" etc.

5.6.2 The Ideal Heat Engine

I will consider the heat engine cycle in detail later in this chapter, but for now, all we need to know is that in one cycle of the heat engine:

1. $+|\mathbf{q_H}|$ of energy flows from the hot reservoir, by the process of heat, to the heat engine (the system)
2. The heat engine apportions a flow of some of this energy ($+|\mathbf{w}|$ where $|\mathbf{w}| < |\mathbf{q_H}|$) to the work reservoir (the surroundings) by the process of work, i.e., the system carries out mechanical (PV) work on the surroundings.
3. To complete the cycle (i.e., to return the system to its original condition), a flow of $+|\mathbf{q_C}|$ of energy (where $|\mathbf{q_C}| = |\mathbf{q_H}| - |\mathbf{w}|$) is transferred from the heat engine to the cold reservoir by the process of heat. This final process returns the engine to its initial state and the cycle repeats itself.

Using our sign convention, in a single cycle, heat engine receives an energy flow of:

1. $+|\mathbf{q_H}|$ from the hot reservoir by the process of heat.
2. $-|\mathbf{w}|$ from the work reservoir by the process of work.
3. $-|\mathbf{q_C}|$ from the hot reservoir by the process of heat.

Knowing the energy change after a single cycle is zero, we can write:

$$|\mathbf{q_H}| - |\mathbf{w}| - |\mathbf{q_C}| = 0 \tag{5.27}$$

A schematic diagram of a heat engine is given in Figure 5.12. The heat engine is the system, and the three reservoirs are the surroundings. The system and the surroundings are the universe which is an isolated system.

In summary, a heat engine converts some of the flow of energy it receives from the surroundings by the process of heat and transfers some of this energy flow back to the surroundings by the process of work. The surplus energy is transferred back to the surroundings by the process of heat. In shorthand, a heat engine converts heat to work. One question that immediately comes to mind

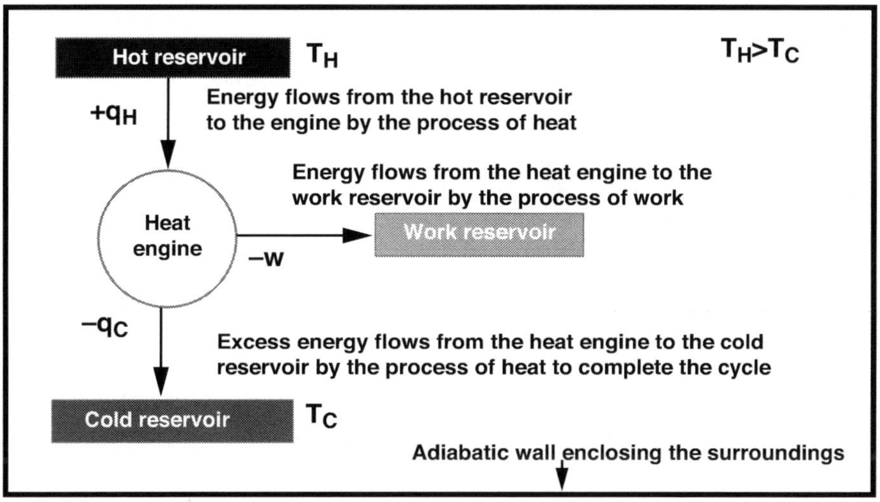

Figure 5.12 Schematic diagram of a heat engine.

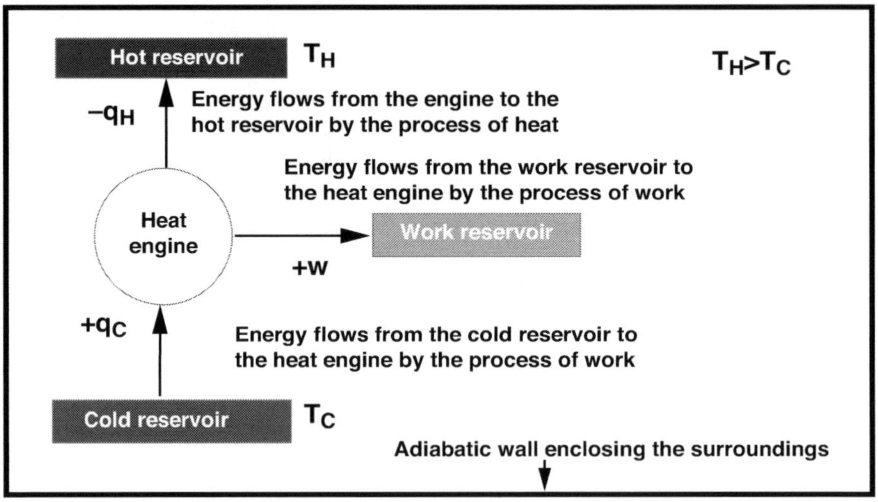

Figure 5.13 Schematic diagram of a heat pump.

is why is there a requirement for an energy flow, by the process of heat, to the surroundings? In other words, why cannot the heat engine convert all of the flow of energy it received from the surroundings by the process of heat, back to the surroundings by the process of work? This question is dealt with later in this chapter.

By definition, a reversible heat engine can be run in reverse (Figure 5.13). In this case, the cycle consists of the following: $+|\mathbf{w}|$ of energy is transferred to the engine from the work reservoir by the process of work, this results in the transfer of $+|\mathbf{q}_C|$ of energy from the cold reservoir to the engine by the process of heat, and the transfer of $+|\mathbf{q}_H|$ of energy from the engine to the hot reservoir by the process of heat. Using our sign convention, the above cycle is written as follows. Energy $\left(-|\mathbf{q}_H|\right)$ is transferred to the system from the hot reservoir by the process of heat, energy $\left(+|\mathbf{w}|\right)$ is transferred to the system from the work reservoir by the process of work and energy $\left(+|\mathbf{q}_C|\right)$ is transferred from the cold reservoir to the system by the process of heat. This type of engine is known as a heat pump or refrigerator because it uses the energy transferred to the system, by the process of work, to transfer energy by the process of heat against the temperature gradient, i.e., from the cold reservoir to the hot reservoir. Since the heat engine is reversible, all the energy arrows in Figure 5.13 are reversed to those in Figure 5.12, and the energy terms are changed in sign but not in magnitude. In shorthand, a heat pump converts work to heat.

Efficiency of a heat engine is defined as the energy transferred to the surroundings by the heat engine, by the process of work $\left(|\mathbf{w}|\right)$, divided by the energy transferred from the hot reservoir to the engine by the process of heat $\left(|\mathbf{q}_H|\right)$, i.e., $\frac{\mathbf{w}}{\mathbf{q}_H}$.[52] Carnot stated that there is no heat engine more efficient than the hypothetical reversible heat engine. To prove that Carnot's reversible heat engine is the most efficient, we need to consider the two ways a super-efficient heat engine can be more efficient. These are:

1. The super-efficient reversible heat engine could transfer more energy, by the process of work, to the surroundings (work reservoir) compared to a normal heat engine for the same quantity of energy transferred to the super-efficient reversible heat engine from the hot reservoir.

2. The super-efficient reversible heat engine could transfer the same energy, by the process of work, to the surroundings compared to a normal heat engine when a lesser quantity of energy is transferred to the super-efficient heat engine from the hot reservoir compared to that transferred to the normal heat engine.

These modes are operation of the super-efficient heat engine are considered in the following two sections.

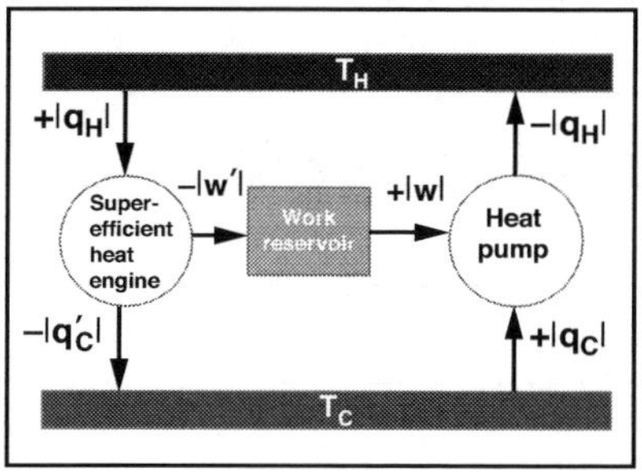

Figure 5.14 Super-efficient reversible heat engine combined with reversible heat pump where $|w'| > |w|$ and $|q'_C| < |q_C|$.

5.6.3 Super-Efficient Heat Engine Operation Mode I

A super-efficient heat engine receiving $+|\mathbf{q}_H|$ of energy from the hot reservoir can transfer more energy $|\mathbf{w}'|$, by the process of work, to the work reservoir than the standard reversible heat engine (which transfers **w** amount of energy by the process of work to the work reservoir) when $+|\mathbf{q}_H|$ of energy is transferred to the heat engine from the hot reservoir. To do this (because of conservation of energy) the super-efficient heat engine would have to transfer less energy $|\mathbf{q}'_C|$ to the cold reservoir by the process of heat than the normal reversible heat engine $(|\mathbf{q}_C|)$, i.e., $\mathbf{q}'_C| < |\mathbf{q}_C|$. Let us now imagine that after the completion of a cycle of this super-efficient heat engine the original "normal" reversible heat engine is run in reverse as a heat pump. Since the original Carnot Cycle is reversible all heat and work terms are changed in sign. This arrangement is shown in Figure 5.14. The heat pump takes in $|\mathbf{q}_C|$ of energy from the cold reservoir, by the process of heat, work $|\mathbf{w}|$ is supplied to the heat pump from the work reservoir, and the heat pump delivers $|\mathbf{q}_H|$ of energy to the hot reservoir by the process of heat. In summary, $|\mathbf{w}'| > |\mathbf{w}|$, $|\mathbf{q}'_C| < |\mathbf{q}_C|$, as required by the Law of Conservation of Energy, and, by definition $|\mathbf{q}_H| = |\mathbf{q}_H|$.

Let us now consider the energy balances of the combined heat engine and heat pump (using our sign convention). For the super-efficient reversible heat engine (remember that after one cycle the system returns to its original condition, i.e., $\Delta\mathbf{U} = 0$:

$$+|\mathbf{q}_H| - |\mathbf{w}'| - |\mathbf{q}'_C| = 0 \tag{5.28}$$

For the "standard" reversible heat pump:

$$-|\mathbf{q}_H| + |\mathbf{w}| + |\mathbf{q}_C| = 0 \tag{5.29}$$

For the overall engine, we add Equations (5.28) and (5.29):

$$-|\mathbf{w}'| - |\mathbf{q}'_C| + |\mathbf{w}| + |\mathbf{q}_C| = 0 \tag{5.30}$$

If we let $-|\mathbf{w}''| = |\mathbf{w}| - |\mathbf{w}'|$ (knowing that, by definition, $|\mathbf{w}'| > |\mathbf{w}|$), and $+|\mathbf{q}''_C| = |\mathbf{q}_C| - |\mathbf{q}'_C|$ (knowing that, by definition, $|\mathbf{q}_C| > |\mathbf{q}'_C|$) then we can write Equation (5.30) as:

$$|\mathbf{q}''_C| - |\mathbf{w}''| = 0 \quad \text{or} \quad |\mathbf{q}''_C| = |\mathbf{w}''| \tag{5.31}$$

The net result of the combined engine is shown in Figure 5.15. Using our sign convention, we say that after one cycle $+|\mathbf{q}''_C|$ of energy has flowed into the combined engine, by the process of heat, from the cold reservoir, and $-|\mathbf{w}''|$ of energy has flowed into the combined engine, by the process of work, from the work reservoir, where $|\mathbf{q}''_C| = |\mathbf{w}''|$. In more everyday language we say the combined engine has extracted energy $(+|\mathbf{q}''_C|)$ from the cold reservoir, by the process of heat, and has transferred all this energy to the work reservoir, by the process of work. In other words, it has

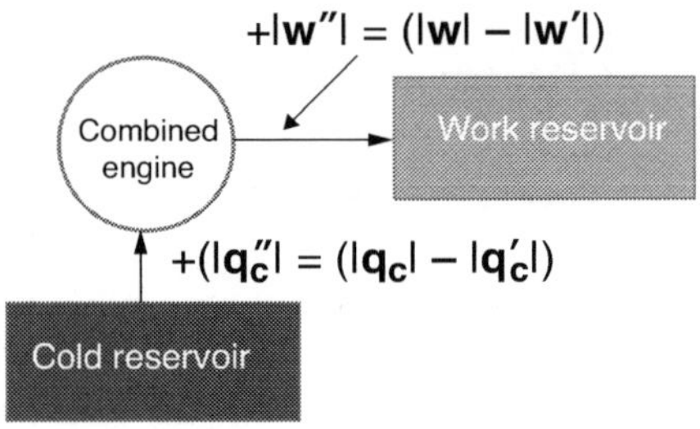

Figure 5.15 The net result of the combined engine shown in Figure 5.12.

extracted energy, by the process of heat, from the surroundings and has transferred all of this energy to the surroundings (wok reservoir) by the process of work. By definition, heat reservoirs are an infinite source of energy, and this means that the engine shown in Figure 5.15 is a perpetual motion machine which Carnot had assumed to be impossible (see Section 5.1).

The following abstract is from Jaynes[53]:

> The excess [work $|\mathbf{w}''|$] can then be used to drive our ships, locomotives, and factories. Once started, this would run forever, delivering an infinite amount of useful work without any further expenditure of fuel. We would have a new kind of perpetual motion machine. We can imagine the economic impact that invention of such a machine would have today. As Max Planck put it, 'we expect to make a most serviceable application' of any physical phenomenon that is found to deviate from Carnot's principle, or from any other principle that can be deduced from it.

Callendar states:

> The extreme improbability of such a result is a sufficient reductio ad absurdum[54] to satisfy any reasonable intelligence. … The condenser [cold heat reservoir] might be the ocean or the whole earth. We should thus obtain the practical equivalent of a perpetual motion, interpreted by Carnot as the continuous creation of motive power.

5.6.4 Super-Efficient Heat Engine Operation Mode II

The other way a super-efficient heat engine can operate is that it can receive less energy from the hot reservoir $|\mathbf{q}'_H|$, by the process of heat, than the standard reversible heat engine and still deliver the same amount of work as the original reversible heat engine (Figure 5.16). We know, by definition, $|\mathbf{q}_H| > |\mathbf{q}'_H|$ and $|\mathbf{w}| = |\mathbf{w}|$; and (Principle of Energy Conservation) $|\mathbf{q}'_C| < |\mathbf{q}_C|$.

Let us now consider the energy balances of the combined heat engine and heat pump. For the super-efficient heat engine:

$$+|\mathbf{q}'_H| - |\mathbf{w}| - |\mathbf{q}'_C| = 0 \tag{5.32}$$

For the heat pump:

$$-|\mathbf{q}_H| + |\mathbf{w}| + |\mathbf{q}_C| = 0 \tag{5.33}$$

For the overall engine, we add Equations (5.32) and (5.33):

$$+|\mathbf{q}'_H| - |\mathbf{q}_H| + |\mathbf{q}_C| - |\mathbf{q}'_C| = 0 \tag{5.34}$$

If we let $-|\mathbf{q}''_H| = |\mathbf{q}'_H| - |\mathbf{q}_H|$ (knowing that $|\mathbf{q}_H| > |\mathbf{q}'_H|$), and $+|\mathbf{q}'''_C| = |\mathbf{q}_C| - |\mathbf{q}'_C|$ (knowing that $|\mathbf{q}'_C| < |\mathbf{q}_C|$) then we can write Equation (5.34) as:

$$|\mathbf{q}'''_C| - |\mathbf{q}''_H| = 0 \quad \text{or} \quad |\mathbf{q}'''_C| = |\mathbf{q}''_H| \tag{5.35}$$

The overall result of the combined engine is the transfer of energy, by the process of heat, from the cold reservoir to the hot reservoir with no other change in the system. This is from Moore[55]:

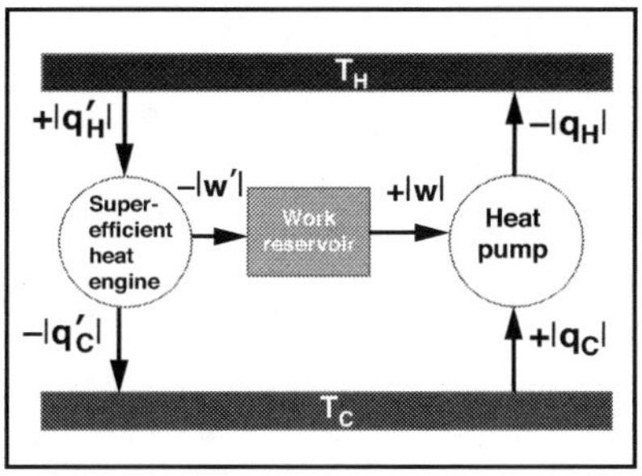

Figure 5.16 Super-efficient reversible heat engine combined with reversible heat pump where $|q'_H| < |q_H|$ and $|q'_C| < |q_C|$.

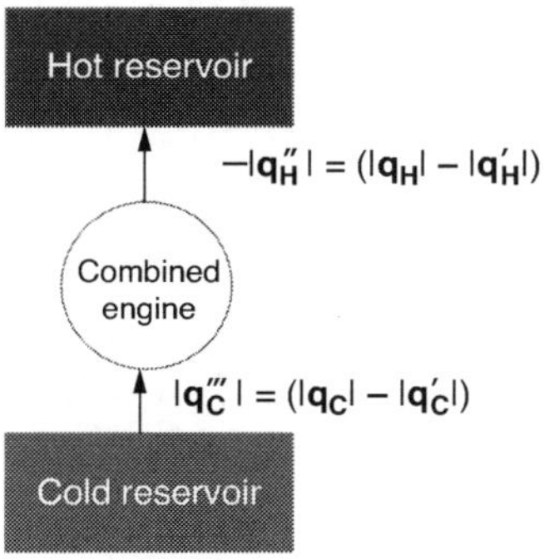

Figure 5.17 The net result of the combined engine shown in Figure 5.16.

> We know that heat always flows from the hotter to the colder region. If we place a hot body and a cold body together, the hot one never becomes hotter, the cold one colder. In fact, considerable work must be expended to refrigerate something, to pump heat out of it. Heat never flows uphill, i.e., against a temperature gradient.

5.6.5 Discussion

The operation of the engines shown in Figures 5.15 and 5.17 does not contravene the First Law of Thermodynamics but both are contrary to experience, i.e., energy does not flow from a hot object to a cold object by the process of heat, and energy cannot be transferred to the surroundings by the process of work by abstracting energy from the surroundings, by the process of heat, without any other changes occurring in the universe. The final phrase is essential, for example in a refrigerator energy flows against the temperature gradient but other changes do occur in the universe (i.e., energy must flow from the work reservoir to the engine). The impossibility of perpetual motion machines and of energy flowing, by the process of heat, from a hot body to a cold body with no outside agency was used by Carnot to prove that a reversible ideal heat engine is the most efficient heat engine, or to put it another way all reversible heat engines, when operating between the same heat reservoirs, run at the same efficiency. In other words, by a *reductio ab absurdum* approach we have proved Carnot's assertion that the reversible heat engine produces the maximum amount of work for a given flow of energy, by the process of heat, i.e., it is the most efficient heat engine.

I thought it worthwhile to set down how Carnot demonstrated that there is no heat engine more efficient than the hypothetical reversible heat engine[56]:

> Now if there existed any means of using heat preferable to those which we have employed, that is, if it were possible by any method whatever to make the caloric produce a quantity of motive power greater than we have made it produce by our first series of operations, - it would suffice to divert a portion of this power in order by the method just indicated to make the caloric of the body B return to the body A from the refrigerator to the furnace, to restore the initial conditions, and thus to be ready to commence again an operation precisely similar

to the former, and so on : this would be not only perpetual motion, but an unlimited creation of motive power without consumption either of caloric or of any other agent whatever. Such a creation is entirely contrary to ideas now accepted, to the laws of mechanics and of sound physics. It is inadmissible. We should then conclude that the maximum of motive power resulting from the employment of steam is also the maximum of motive power realizable by any means whatever.

Not, I would suggest, the clearest of demonstrations. Callendar wrote:

Unfortunately his [Carnot's]original demonstration [see previous extract], ... is not so exactly worded that exception cannot be taken to it. I will, therefore, repeat the proof in a slightly more definite and exact form, as Carnot probably intended it to run.[57]

The rather abstruse double negative in the final sentence is, I guess, Callendar's rather diplomatic way of saying that Carnot's explanation was, at best, ambiguous. Here is Callendar's version[58]:

If it were possible to produce from a given quantity of heat supplied a greater quantity of motive power than that obtained from a reversible engine, it would suffice to divert a portion of this power to return to the source by means of a reversible engine the quantity of heat taken from it. We should thus obtain at each repetition of the cycle a balance of motive power without taking any heat from the source – that is to say, without any consumption of fuel. The extreme improbability of such a result is a sufficient reductio ad absurdum to satisfy any reasonable intelligence. Whether the condenser is heated or cooled in the process is immaterial for practical purposes. The condenser might be the ocean or the whole earth. We should thus obtain the practical equivalent of a perpetual motion, interpreted by Carnot as the continuous creation of motive power.

I think the deficiencies in the above two extracts are resolved when the contents of Sections 7.3 and 7.4 are read and understood.

The same arguments used to prove that a reversible heat engine is the most efficient heat engine can be used to show that the maximum efficiency of the reversible heat engine is independent of the working substance. This leads to the conclusion that all the results obtained by using ideal gas as a working substance apply to all reversible heat engines operating with other working substances (such as steam). This led Carnot to set down his principle: *no heat engine can be more efficient than a reversible one operating between the same temperatures.*

The following is another abstract from Jaynes[59]:

For example, it follows at once that all reversible heat engines have the same efficiency, independent of the working substance or mode of operation; i.e., the reversible efficiency is a universal function [$f(T_H,T_C)$] of the two temperatures. This answered in one stroke all those questions about improvements in steam engines; already Carnot's principle had resolved a mass of ambiguities. Carnot's reasoning is outstandingly beautiful, because it deduces so much from so little – and with such a sweeping generality that rises above all tedious details - but at the same time with such a compelling logical force. In this respect, I think that Carnot's principle ranks with Einstein's principle of relativity.

Before moving on it is worth noting that the most famous example of a machine based on Figure 5.15 is the "Zeromotor" proposed to the US Navy as a way to power its ships. Of course, it did not work, but the proposal did fool the President of the United States and the Chief Engineer of the US Navy. In 1880 Gamgee devised an engine in which energy was transferred from the ocean-boiled liquid ammonia to drive a piston engine. The expansion of the exhaust was then supposed to cause spontaneous condensation of the ammonia in a condenser cooled by sea water. The condensed liquid is returned to the reservoir to complete the cycle. The problem is that ammonia must be cooled to –33°C to make it to condense at atmospheric pressure, and with no part of the engine at that temperature the ammonia simply could not be condensed. It appears Gamgee was proposing that the ammonia was to be condensed using the same heat reservoir (the ocean) used to boil the ammonia, and the process would repeat, forever, with no external fuel source.

The following two quotes are useful (please mentally convert "heat energy" to "energy transferred by the process of heat" in the first quote:

It propels a ship that simply withdraws heat energy from the ocean to power itself. Obviously, this doesn't violate the first law, because the ocean contains plenty of ... energy. The ship

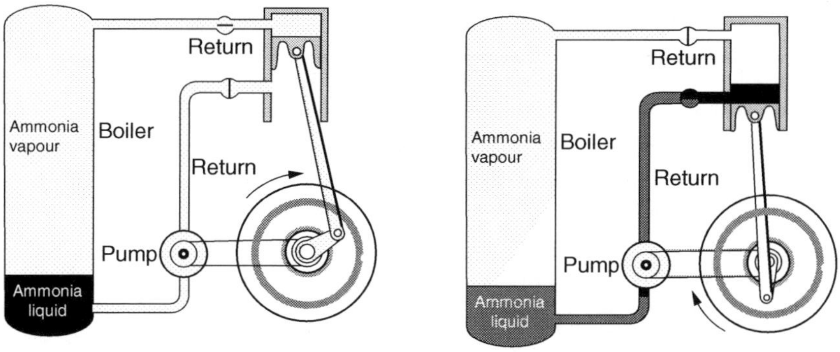

Figure 5.18 The cycle of the Gamgee ammonia engine of the 1880s in which the gas pressure drove a piston. As the gas expanded Gamgee assumed that it would condense and return as liquid to the boiler.[60]

would simply extract some of it, leaving a cold wake behind. Losses from friction in the ship's propellers and shafts would return some of this energy to the ocean immediately. Stopping the ship at its destination port would turn the remainder to heat and return it to the ocean as well. The ship would merely borrow energy for its voyage from the ocean.[61]

Gamgee may have been a crook or he may have been sincere, but if the latter he clearly had not tested the concept in reality. Nonetheless, he persuaded the Chief Engineer of the US Navy, B F Isherwood, of its practicality. Isherwood should have realised that the Zeromotor blatantly violated the Second Law of Thermodynamics,[62] but instead recommended it to several cabinet officers, and even US President Garfield was enthusiastic. A working Zeromotor would have been of enormous benefit to a navy; at this period steamships were fuelled by coal, and long-range naval operations required expensive and far-flung networks of coaling stations scattered over the globe. The range of the ships would only have been limited by provisioning, if they were powered by heat from the very water they floated on.[63]

I do not intend to consider the detailed workings of the machine. We can see (Figure 5.18) that it is a heat engine and that it has only one heat reservoir (the ocean) and is, in essence, the same as the engine shown in Figure 5.15. The Zeromotor requires, for its operation, the condensation of ammonia at temperatures above −33°C and this would require energy spontaneously flowing *up* a temperature gradient by the process of heat. All of human experience (and the Second Law of Thermodynamics) tells us that this never happens. Therefore, even if we do not understand the internal workings of the Zeromotor, a knowledge of thermodynamics immediately tells us it cannot work.

The following abstract completes the story[64]:

> Naval enthusiasm was short-lived; a working model would function only with the condenser disconnected, and then only unsatisfactorily. The problem is that ammonia must be cooled to -33 degrees Centigrade to make it to condense at atmospheric pressure, and with no part of the engine at that temperature the ammonia simply could not be condensed.
>
> It appears Gamgee was proposing that the ammonia was to be recycled through the traditional condenser applied to marine steam engines, but since this is cooled by sea water in normal use, it could hardly have condensed the ammonia since it had previously boiled it.

The US Navy's enthusiasm was understandable. We know the mass of water in the oceans is 1.4×10^{21} kg [65]; the heat capacity of sea water is 3,990 $J.kg^{-1}.K^{-1}$ [66] and the average temperature of ocean water is between 0°C and 4°C [67]. To avoid the complications involved when water freezes, I will calculate the energy that would be released if the Earth's oceans were cooled by 1°C – this is 5.5×10^{20} J. The International Energy Agency estimates that, in 2013, total world energy consumption was 5.67×10^{20} J.[68] Therefore, if we could design an engine (like the Zeromotor) that could harness this energy there would be no global energy crisis (either now or in the future).

I want now to consider Carnot's thought experiment in more detail.

5.7 CARNOT'S THOUGHT EXPERIMENT

Carnot first states that a cyclic process is the underlying principle for an engine and that all parts of the engine, including the working substance or agent, must be, at the end of the process, in precisely the same state as at the beginning, he writes (according to Callendar)[69]:

> When a body has undergone any changes and after a certain number of transformations is brought back identically to its original state, considered relatively to density, temperature and mode of aggregation, it must contain the same quantity of heat[70] as it contained originally.

It will prove helpful if I start, as Carnot did, by utilising an arbitrary temperature scale which I will depict as "θ"[71] (all will become clear). Carnot proceeds to show that work cannot be obtained from the process of heat by alternate expansion and contraction of the working substance, without postulating the existence of two heat reservoirs at different temperatures; and conversely that, wherever a difference of temperature exists, it is possible to utilise it for the production of work. He then sets down, as we have seen in Section 5.7.2, the essential condition for maximum efficiency – reversibility. Carnot realised that even with the elimination of turbulence and friction energy flowing across a non-zero temperature gradient, by the process of heat, will reduce the efficiency of any heat engine.[72] Carnot writes (according to Callendar[73]):

> In order to realise the maximum effect, it is necessary that, in the process employed, there should not be any direct interchange of heat between bodies at sensibly different temperatures.

Direct transference of energy, by the process of heat, between bodies at different temperatures would be equivalent to wasting a temperature difference that might have been utilised for the production of work as well as losses due to friction.

Carnot recognised that for a heat engine operating between two heat reservoirs (θ_H, θ_C) the components of the cycle, for maximum efficiency, are isothermal expansion, adiabatic expansion, isothermal compression, and finally adiabatic compression. This is known as the Carnot Cycle (Figure 5.19). A schematic diagram of a heat engine running through the Carnot cycle is shown in Figure 5.20a and together with a diagram summarising the energy flows in the Carnot Cycle is shown in Figure 5.20b.

One cycle of a heat engine running a Carnot Cycle is briefly described as follows:

1 Step 1: *Reversible isothermal expansion* – the ideal gas at temperature θ_H in the engine (the system) is in thermal contact via a diathermal wall with the hot reservoir at temperature θ_H (state 'a' – Figure 5.19). There is a positive energy flow $\left(+|\mathbf{q_H}|\right)$ into the heat engine, by the process of heat; the gas expands, and the piston moves outwards. There is a negative energy flow $\left(-|\mathbf{w}_1|\right)$ into the engine from the surroundings (the work reservoir), by the process of work. Although this *could* reduce the temperature of the ideal gas, it is maintained at constant temperature, θ_H, because the system extracts $\left(+|\mathbf{q_H}|\right)$ of energy from the hot reservoir, by the process of heat, maintaining the temperature at θ_H. In other words, the heat engine transfers $|\mathbf{w}_1|$ energy to the

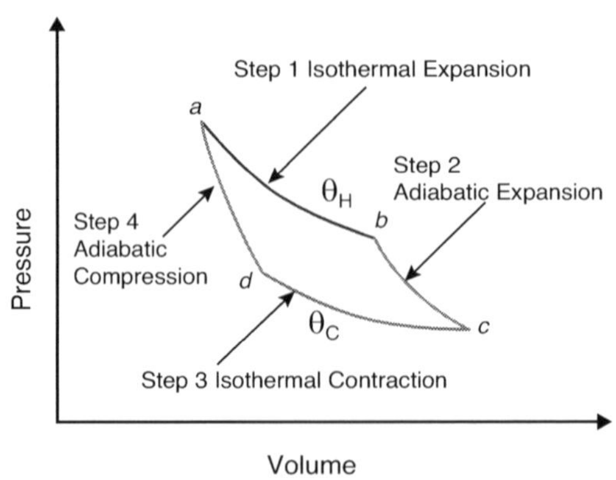

Figure 5.19 The Carnot cycle (a) P-V Carnot cycle, (b) Step 1 isothermal expansion, (c) Step 2 adiabatic expansion, (d) Step 3 isothermal contraction, (e) Step 4 adiabatic contraction.

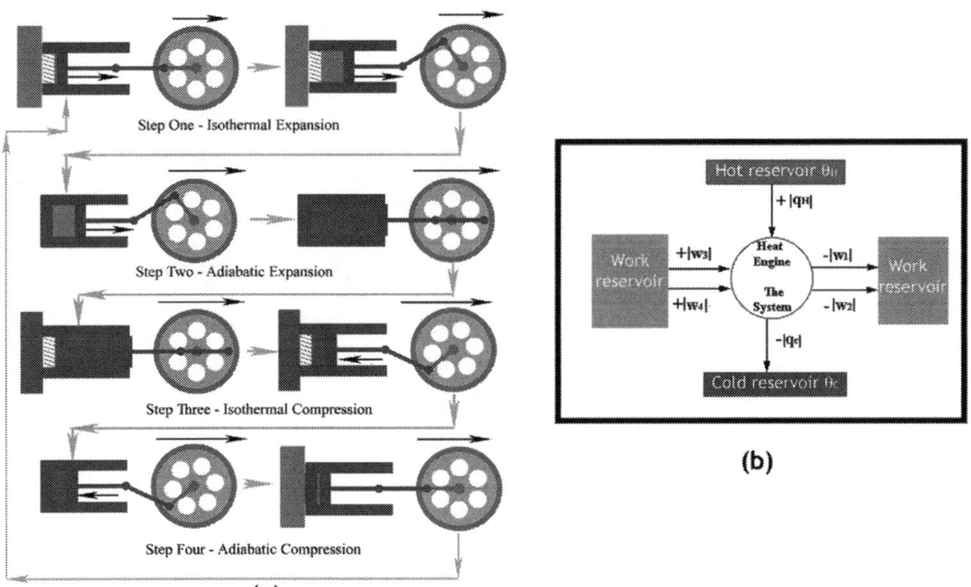

Figure 5.20 (a) Schematic diagram of a heat engine running the Carnot Cycle: (b) Summary of the energy flows in the Carnot Cycle with energy signs meeting our agreed sign convention.

surroundings (the work reservoir), i.e., the heat engine (the system) does work on surroundings. Usually, in a cyclically operating heat engine, the energy transferred to the surroundings, by the process of work, manifests itself by driving a flywheel. The system moves from state a to state b. Note $|\mathbf{q}_H| = |\mathbf{w}_1|$.

2. Step 2: *Reversible adiabatic gas expansion* – at state 'b' the heat engine is thermally insulated from its surroundings. The inertia of the fly wheel continues the expansion of the gas and there is a further energy flow $(-|\mathbf{w}_2|)$ into the heat engine from the surroundings, by the process of work. In other words, the heat engine transfers $(|\mathbf{w}_2|)$ energy to the surroundings (i.e., it does further work on the surroundings). During this adiabatic expansion stage, the ideal gas cools to a lower temperature, θ_C, and the system moves from state b to state c. Note $\mathbf{q} = 0$.

3. Step 3: *Reversible isothermal compression* – at state 'c', the system is placed in thermal contact (via a diathermal wall) with a cold reservoir at temperature θ_C. At this point the clockwise motion of the flywheel (Figure 5.20a) and its inertia results in gas compression. There is now an energy flow of $+|\mathbf{w}_3|$ into the engine, by the process of work, i.e., the surroundings are doing work on the system. Although this *could* raise the temperature of the ideal gas, it is maintained at constant temperature, θ_C, because the system extracts $(-|\mathbf{q}_C|)$ of energy from the cold reservoir, by the process of heat, maintaining the temperature at θ_C. In other words, the heat engine dumps $|\mathbf{q}_C|$ of energy into the cold reservoir by the process of heat to maintain its temperature at θ_C. The system moves from state c to state d. Note $|\mathbf{q}_C| = |\mathbf{w}_3|$.

4. Step 4: *Reversible adiabatic gas compression* – at state 'd', the system is thermally insulated from its surroundings. The inertia of the fly wheel continues to compress the gas and further energy $(+|\mathbf{w}_4|)$ is transferred to the system by the process of work, i.e., the surroundings are doing work on the system, and the temperature of the ideal gas increases to θ_H. At this stage the machine is back in its original state and another Carnot Cycle can commence. Note $\mathbf{q} = 0$.

We know internal energy is a state function.[74] Therefore, after one complete cycle of the heat engine, we know that $\Delta\mathbf{U} = 0$. The First Law of Thermodynamics requires that $\Delta\mathbf{U} = \mathbf{q} + \mathbf{w} = \mathbf{q}_{\text{input}} + \mathbf{w}_{\text{input}}$. For a heat engine and using our sign convention $\mathbf{w}_{\text{input}}$ is a negative real number $(-|\mathbf{w}_{\text{input}}|)$, and that $\mathbf{q}_{\text{input}}$ is a positive real number $(+|\mathbf{q}_{\text{input}}|)$.[75] From the above text and Figure 5.20b we can write:

$$-|\mathbf{w}_{\text{input}}| = (-|\mathbf{w}_1|) + (-|\mathbf{w}_2|) + (+|\mathbf{w}_3|) + (+|\mathbf{w}_4|) \tag{5.36}$$

$$+|\mathbf{q}_{\text{input}}| = (+|\mathbf{q}_{\text{H}}|) + (-|\mathbf{q}_{\text{C}}|) \tag{5.37}$$

Where $\mathbf{w}_{\text{input}}$ is the total flow of energy, by the process of work, into the reversible heat engine in a single cycle, and $\mathbf{q}_{\text{input}}$ is the total flow of energy, by the process of heat, into the reversible heat engine in a single cycle. For one cycle of the heat engine, we know:

$$\Delta\mathbf{U} = (-|\mathbf{w}_{\text{input}}|) + (|\mathbf{q}_{\text{input}}|) = 0 \tag{5.38}$$

The efficiency of a heat engine (ε) is defined as the flow of energy from the heat engine to the work reservoir $(|\mathbf{w}_{\text{output}}|)$ divided by the energy supplied from the hot reservoir to the heat engine by the process of heat $(|\mathbf{q}_{\text{H}}|)$. The work output is the same magnitude as the work input $(|\mathbf{w}_{\text{output}}| = |\mathbf{w}_{\text{input}}|)$, but the sign is changed because we know from the Law of Conservation of Energy that $(+|\mathbf{w}_{\text{output}}|) + (-|\mathbf{w}_{\text{input}}|) = 0$. So, we can write:

$$\varepsilon = \frac{|\mathbf{w}_{\text{output}}|}{|\mathbf{q}_{\text{H}}|} = \frac{|\mathbf{w}_{\text{input}}|}{|\mathbf{q}_{\text{H}}|} \tag{5.39}$$

We know from Equations (5.37) and (5.38) that $|\mathbf{w}_{\text{input}}| = |\mathbf{q}_{\text{input}}| = +|\mathbf{q}_{\text{H}}| - |\mathbf{q}_{\text{c}}|$. This allows Equation (5.39) to be written as:

$$\varepsilon = \frac{|\mathbf{q}_{\text{H}}| - |\mathbf{q}_{\text{C}}|}{|\mathbf{q}_{\text{H}}|} = 1 - \frac{|\mathbf{q}_{\text{C}}|}{|\mathbf{q}_{\text{H}}|} \tag{5.40}$$

We know that this efficiency is a maximum for any heat engine operating between the temperatures θ_{H} and θ_{C}. I now want to revisit the above determination of the equation of efficiency of the heat engine without resorting to absolute values. This is how many thermodynamics texts treat the subject.

Equations (5.36)–(5.38) and (5.40) are written (respectively) as:

$$\mathbf{w}_{\text{input}} = \mathbf{w}_1 + \mathbf{w}_2 + \mathbf{w}_3 + \mathbf{w}_4 \tag{5.41}$$

$$\mathbf{q}_{\text{input}} = \mathbf{q}_{\text{H}} + \mathbf{q}_{\text{C}} \tag{5.42}$$

$$\mathbf{w}_{\text{input}} + \mathbf{q}_{\text{input}} = 0 \tag{5.43}$$

$$\mathbf{w}_{\text{input}} = -\mathbf{w}_{\text{ouput}} \tag{5.44}$$

From Equations (5.43) and (5.44) we can write:

$$\mathbf{w}_{\text{input}} = \mathbf{q}_{\text{input}} \tag{5.45}$$

The efficiency of a heat engine is defined as the energy transferred to the surroundings by the heat engine, by the process of work $(\mathbf{w}_{\text{output}})$, divided by the energy transferred from the hot reservoir to the engine by the process of heat $(\mathbf{q}_{\text{H}})$:

$$\varepsilon = \frac{\mathbf{w}_{\text{output}}}{\mathbf{q}_{\text{H}}} = -\frac{\mathbf{w}_{\text{input}}}{\mathbf{q}_{\text{H}}} = \frac{\mathbf{q}_{\text{H}} + \mathbf{q}_{\text{C}}}{\mathbf{q}_{\text{H}}} = 1 + \frac{\mathbf{q}_{\text{C}}}{\mathbf{q}_{\text{H}}} \tag{5.46}$$

This all works well if the agreed sign convention is rigorously adhered, for example, in Equation (5.46) $\mathbf{q}_{\text{C}}$ is a negative real number and $\mathbf{q}_{\text{H}}$ is a positive real number. Unfortunately, you will often see Equation (5.46) erroneously written as $\varepsilon = 1 - \frac{q_{\text{C}}}{q_{\text{H}}}$, where q_{C} and q_{H} must either be both positive or both negative real numbers, because efficiency must be a positive number that is less than one. This indicates that any agreed sign convention has been ignored, or that q_{C} and q_{H} must represent absolute values and not real numbers.

The following is a good review (by Mach) of Carnot's thought experiment.[76] I say this in spite of his ambiguous use of the word "heat", however, at this stage in the book I have assumed the reader can deal with this and have not changed his wording (but I have changed the symbols for temperature, work, and heat to agree with those used in this book.)

> In the determination of the maximum quantity of work that, generally, a heat-machine, or, to take a special case, a steam-engine, can perform with the expenditure of a given amount of heat of combustion, Carnot is guided by mechanical analogies. A body can do work on being heated, by expanding under pressure. But to do this the body must receive heat from a hotter body. Heat, therefore, to do work, must pass from a hotter body to a colder body, just as water; must fall from a higher level to a lower level to put a [waterwheel][77] in motion. Differences of temperature, accordingly, represent forces able to do work exactly as do differences of height in heavy bodies.[78] Carnot pictures to himself an ideal process in which no heat flows away unused, that is, without doing work. With a given expenditure of heat, accordingly, this process furnishes the maximum of work. An analogue of the process would be a [waterwheel] which scooping its water out of a higher level would slowly carry it to a lower level without the loss of a drop.[79] A peculiar property of the process is, that with the expenditure of the same work the water can be raised again exactly to its original level. This property of reversibility is also shared by the process of Carnot. His process also can be reversed by the expenditure of the same amount of work, and the heat again brought back to its original temperature level.
>
> Suppose, now, we had two different reversible processes A, B, such that in A a quantity of heat, q, flowing off from the temperature T_H to the lower temperature T_C should perform the work w, but in B under the same circumstances it should perform a greater quantity of work $w+w'$; then, we could join B in the sense assigned and A in the reverse sense into a single process. Here A would reverse the transformation of heat produced by B and would leave a surplus of work w′, produced, so to speak, from nothing. The combination would present a perpetual motion.
>
> With the feeling, now, that it makes little difference whether the mechanical laws are broken directly or indirectly (by processes of heat), and convinced of the existence of a universal law-ruled connection of nature, Carnot here excludes for the first time from the province of general physics the possibility of a perpetual motion. But it follows, then, that the quantity of work w, produced by the passage of a quantity of heat q from a temperature T_H to a temperature T_C, is independent of the nature of the substances as also of the character of the process, so far as that is unaccompanied by loss, but is wholly dependent upon the temperature T_H, T_C."
>
> This important principle has been fully confirmed by the special researches of Carnot himself (1824), of Clapeyron (1834), and of Sir William Thomson (1849), now Lord Kelvin. The principle was reached without any assumption whatever concerning the nature of heat, simply by the exclusion of a perpetual motion. … Carnot's principle led to the most remarkable results. W. Thomson (1848) founded upon it the ingenious idea of an 'absolute" scale of temperature'. …

I now want to consider the subject raised in Mach's final sentence "the ingenious idea of an 'absolute' scale of temperature". There was always concern among scientists about the arbitrary nature of temperature scales in use (Celsius, Fahrenheit, ideal gas, etc). As we shall see in the next section, Carnot's thought experiment dealt with this issue once and for all.

5.8 TEMPERATURE SCALES AND THE KELVIN

In Table 1.1 where I listed "thermodynamic temperature" as one of five fundamental quantities in science; I gave little explanation. Thermometers are based on the measurement of a physical parameter of an object that has a simple (hopefully) linear, functional relationship with temperature, e.g., its dimension, electrical resistance, density, etc. The functional relationship between physical property and temperature will never be exactly linear and the different empirical scales may not be compatible with each other, except for small regions of temperature overlap. The Celsius and Fahrenheit temperature scales were originally based on the linear expansion of a narrow mercury column within a limited range of temperature each using different reference points and scale increments. If an alcohol thermometer and a mercury thermometer have the same two fixed points, for example the freezing and boiling point of water, their reading will probably not exactly agree with each other except at the fixed points, as there is no guarantee of a linear relationship of expansion and temperature. Although temperature is a universal attribute of matter, empirical scales only map a narrow range of temperature onto a "linear" relationship. For example, mercury freezes at around −40°C and cannot serve as a temperature scale below this point.

In, Section 3.3.3 I mentioned an "absolute" temperature scale which has zero as its lowest temperature:

> This absolute temperature scale is based solely on the behaviour of an ideal gas. Unlike the arbitrary temperature scales only one fixed point is needed. The fixed point used is … the "triple point" of water, which is the unique pressure and temperature situation where solid, liquid, and vapour all coexist.

The zero point of this absolute temperature scale is based on the temperature when the volume of an ideal gas is zero. I also gave the units of this ideal gas temperature scale as "kelvins" when the temperature difference between the freezing point and boiling point of water was 100 units. To use this scale requires the use of an ideal gas thermometer to establish standards and calibrate other thermometers.

The temperature scales, used by Carnot and by us in the previous section, were arbitrary scales such as Celsius or Fahrenheit. The arbitrariness of temperature scales had long been a cause for concern. Although scientists considered that the ideal gas temperature scale was more fundamental than existing empirical temperature scales, there was no escaping the fact that the ideal gas temperature scale was, in essence, another arbitrary scale.

> The ideal gas scale is in some sense a 'mixed' scale. It relies on the universal properties of gas, a big advance from just a particular substance. But still it is empirical since it puts gas at a special position and thus has limited applicability – at some point no gas can exist.[80]

W. Thomson (Lord Kelvin) realised that Carnot's work could resolve this issue. If the reversible efficiency of a heat engine is a universal function of the temperatures of the hot and cold reservoirs, then it in effect defines a universal temperature scale that is independent of the properties of any particular substance like mercury or ideal gas. Kelvin used Carnot's conclusion that the efficiency of a reversible heat engine is independent of the working substance and is a function only of the temperatures of the two heat reservoirs) to define an absolute or logical temperature scale, which is independent of any thermometric substance. This is known as the thermodynamic temperature scale. I will now set down Kelvin's approach.

We know the efficiency ε of a reversible Carnot heat engine is a function of the temperatures of the hot and cold reservoirs (see Section 7.5).[81] Using Equation (5.40), we can write:

$$\varepsilon = 1 - \frac{|\mathbf{q_C}|}{|\mathbf{q_H}|} = g(\theta_H, \theta_C) \tag{5.47}$$

Since $\{[-g(\theta_H,\theta_C)]-1\}$ is also a function of the temperatures, say $f(\theta_H,\theta_C)$, we can rewrite Equation (5.47) as:

$$\frac{|\mathbf{q_C}|}{|\mathbf{q_H}|} = \{[-g(\theta_H,\theta_C)]-1\} = f(\theta_H,\theta_C) \tag{5.48}$$

Now consider two Carnot cycles sharing an isotherm at θ_M (Figure 5.21). Let the flow of energy, by the process of heat, into the reversible heat engine as the working substance (in our case the ideal gas) as it expands along the isotherms at θ_H, θ_M, and θ_C be $|\mathbf{q_H}|$, $|\mathbf{q_M}|$ and $\mathbf{q_C}|$ respectively. From Equation (5.48) we know:

$$\frac{|\mathbf{q_H}|}{|\mathbf{q_M}|} = f(\theta_H,\theta_M); \qquad \frac{|\mathbf{q_M}|}{|\mathbf{q_C}|} = f(\theta_M,\theta_C); \qquad \frac{|\mathbf{q_H}|}{|\mathbf{q_C}|} = f(\theta_H,\theta_C) \tag{5.49}$$

Knowing $\frac{|\mathbf{q_H}|}{|\mathbf{q_C}|} = \frac{|\mathbf{q_H}|}{|\mathbf{q_M}|} \times \frac{|\mathbf{q_M}|}{|\mathbf{q_C}|}$ we can write:

$$f(\theta_H,\theta_C) = f(\theta_H,\theta_M) \times f(\theta_M,\theta_C) \tag{5.50}$$

Now θ_M is an independent variable (we could have chosen any θ_M provided it meets the requirement $\theta_H > \theta_M > \theta_C$, consequently $f(\theta_H,\theta_C)$ cannot be a function of θ_M, and, consequently, $f(\theta_H,\theta_M) \times f(\theta_M,\theta_C)$ cannot be a function of θ_M. To put it another way, the first function in

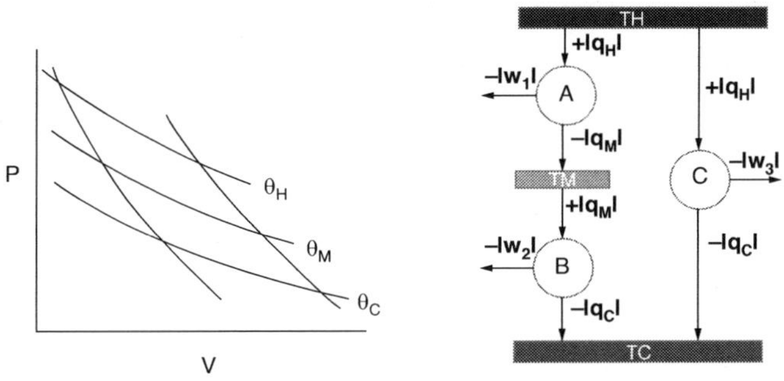

Figure 5.21 Two Carnot cycles sharing a common isotherm θ_M.

Equation (5.50) is not a function of θ_M, therefore the product of the final two functions must result in the removal of θ_M as a variable. The only way to define the function f is as follows:

$$f(\theta_H,\theta_M)=\frac{F(\theta_H)}{F(\theta_M)} \qquad f(\theta_M,\theta_C)=\frac{F(\theta_M)}{F(\theta_C)} \tag{5.51}$$

where F(θ) denotes an arbitrary function of the single variable θ. From Equation (5.49) we know $f(\theta_H,\theta_C)=f(\theta_H,\theta_M)\times f(\theta_M,\theta_C)$, therefore, we can write:

$$f(\theta_H,\theta_C)=\frac{F(\theta_H)}{F(\theta_C)}=\frac{|q_C|}{|q_H|} \tag{5.52}$$

Kelvin decided to use Equation (5.52) as the basis of a non-arbitrary absolute thermodynamic temperature scale and chose the simplest function $F(\theta)=T^{kelvin}$[82]:

$$\frac{|q_C|}{|q_H|}=\frac{T_H^{kelvin}}{T_C^{kelvin}} \tag{5.53}$$

Therefore, a temperature ratio on the thermodynamic scale was defined as equal to the ratio of the energy absorbed from the hot reservoir, by the process of heat, to the energy rejected to the cold reservoir, by the process of heat, in the reversible Carnot cycle. The efficiency of the reversible Carnot heat engine (see Equation 5.47) is:

$$\varepsilon=1-\frac{|q_C|}{|q_H|}=1-\frac{T_C^{kelvin}}{T_H^{kelvin}}=\frac{T_H^{kelvin}-T_C^{kelvin}}{T_H^{kelvin}} \tag{5.54}$$

The zero point of the thermodynamic temperature scale is physically fixed as the temperature of the cold reservoir at which the efficiency becomes equal to unity. From Equation (5.54), in the limit as $T_C^{kelvin}\rightarrow 0$, $\varepsilon\rightarrow 1$. From the efficiency of the reversible heat engine $\left(\varepsilon = 1-\frac{T_C^{kelvin}}{T_H^{kelvin}}\right)$ one can calibrate any thermometer to read T^{kelvin}. Consequently, in the reversible Carnot cycle, the ratio of the energy flows absorbed and rejected, by the process of heat, $\frac{|q_H|}{|q_C|}$, is, by definition, equal to the ratio of the temperatures of the reservoirs as expressed on the thermodynamic scale. In the thermodynamic temperature scale (as with Celsius scale) the triple point of water is chosen to be 273.16 K (see Section 3.3.3) and there are 100 degrees between the triple point of water and the boiling point of water at standard atmospheric pressure. The zero point on the thermodynamic scale is called absolute zero.

To summarise – Equation (5.52) defines an absolute temperature scale, based on the Carnot cycle, that is independent of the behaviour of any particular working substance. Until now the "absolute" temperatures used in this book are based on ideal gas laws with the zero-point defined as the temperature when the volume and pressure of the ideal gas become zero. The thermodynamic absolute temperature scale is independent of substance and is therefore a non-empirical (i.e., a 'true' absolute temperature scale). Also note that Equation (5.53) would be satisfied by $\frac{cT_H^{kelvin}}{cT_C^{kelvin}}$ where c is any real number; Kelvin chose the simplest value for c and that is unity.

In this section, for clarity, I have used absolute values in the development of a thermodynamic temperature scale (see Equation 5.53). We know that in a heat engine $\mathbf{q_H}$ is a positive real number and $\mathbf{q_C}$ is a negative real number. Therefore, we can rewrite Equation (5.53) as:

$$\frac{\mathbf{q_H}}{-\mathbf{q_C}} = \frac{T_H^{kelvin}}{T_C^{kelvin}} \tag{5.55}$$

Rearrangement gives:

$$\frac{\mathbf{q_H}}{T_H^{kelvin}} + \frac{\mathbf{q_C}}{T_C^{kelvin}} = 0 \tag{5.56}$$

5.9 THE RELATIONSHIP BETWEEN THE THERMODYNAMIC AND IDEAL-GAS TEMPERATURE SCALES

Temperature on the Kelvin, or thermodynamic scale is denoted by the symbol T, which is the same as I used for the absolute temperature scale based on the thermal expansion of an ideal gas. It can be shown that these scales are numerically the same by running a Carnot cycle (Figure 5.22a) with ideal gas as the working substance and applying Equations (5.8), (5.9), (5.25), and (5.26) to the four steps of the Carnot cycle. I will use the superscript "ideal gas" to identify the temperature scale as that based on the thermal expansion/compression of the ideal gas. The proof is long winded, but straightforward. I have reduced the use of absolute values to reduce the size of equations and also to reduce the number of step, so remember as we know that $T_H > T_C$, so $T_H - T_C$ is positive and $T_C - T_H$ is negative, and as $V_b > V_a$ and $V_c > V_d$, so $\ln\frac{\mathbf{V_b}}{\mathbf{V_a}}$ is positive and $\ln\frac{\mathbf{V_d}}{\mathbf{V_c}}$ is negative (also note $\ln\frac{\mathbf{V_a}}{\mathbf{V_b}}$ is negative and $\ln\frac{\mathbf{V_c}}{\mathbf{V_d}}$ is positive).

Step 1. Isothermal expansion (Figure 5.22b):

$$-|\mathbf{w}_1| = -nRT_H^{ideal\,gas} \ln\frac{\mathbf{V_b}}{\mathbf{V_a}} \tag{5.57}$$

$$+|\mathbf{q_H}| = nRT_H^{ideal\,gas} \ln\frac{\mathbf{V_b}}{\mathbf{V_a}} \tag{5.58}$$

Step 2. Adiabatic expansion (Figure 5.22c):

$$-|\mathbf{w}_2| = nC_V\left(T_C^{ideal\,gas} - T_H^{ideal\,gas}\right) \tag{5.59}$$

Step 3. Isothermal contraction (Figure 5.22d):

$$+|\mathbf{w}_3| = -nRT_C^{ideal\,gas} \ln\frac{\mathbf{V_d}}{\mathbf{V_c}} \tag{5.60}$$

$$-|\mathbf{q_C}| = nRT_C^{ideal\,gas} \ln\frac{\mathbf{V_d}}{\mathbf{V_c}} \tag{5.61}$$

Step 4. Adiabatic contraction (Figure 5.22e):

$$+|\mathbf{w}_4| = nC_V\left(T_H^{ideal\,gas} - T_C^{ideal\,gas}\right) \tag{5.62}$$

After step 4, the system is back in in its initial state, i.e., one complete cycle has been completed. We can now follow the procedure carried out in Section 5.8 to complete the analysis. Knowing $-|\mathbf{w_{input}}| = -|\mathbf{w}_1| - |\mathbf{w}_2| + |\mathbf{w}_3| + |\mathbf{w}_4|$ (Equation 5.36) we can write:

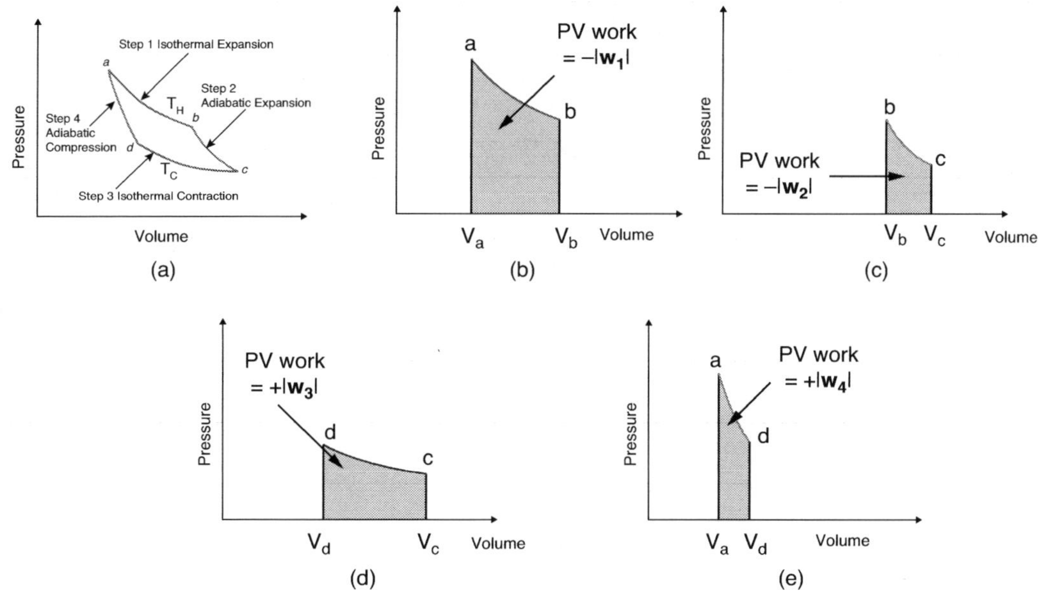

Figure 5.22 The Carnot cycle.

$$-\left|\mathbf{w_{input}}\right| = \left[-nRT_H^{\text{ideal gas}} \ln \frac{\mathbf{V_b}}{\mathbf{V_a}}\right] + \left[nC_V\left(T_C^{\text{ideal gas}} - T_H^{\text{ideal gas}}\right)\right] + \left[-nRT_C^{\text{ideal gas}} \ln \frac{\mathbf{V_d}}{\mathbf{V_c}}\right] + \left[nC_V\left(T_H^{\text{ideal gas}} - T_C^{\text{ideal gas}}\right)\right] \tag{5.63}$$

Knowing $nC_V\left(T_C^{\text{ideal gas}} - T_H^{\text{ideal gas}}\right) = -nC_V\left(T_H^{\text{ideal gas}} - T_C^{\text{ideal gas}}\right)$ we can simplify Equation (5.63):

$$-\left|\mathbf{w_{input}}\right| = \left[-nRT_H^{\text{ideal gas}} \ln \frac{\mathbf{V_b}}{\mathbf{V_a}}\right] + \left[-nRT_C^{\text{ideal gas}} \ln \frac{\mathbf{V_d}}{\mathbf{V_c}}\right] \tag{5.64}$$

If we substitute $\frac{nRT^{\text{ideal gas}}}{V}$ for P in the equation of an ideal gas adiabat - $PV^{\frac{5}{3}} = \text{constant}$ (Equation 5.22) we have:

$$\frac{nRT^{\text{ideal gas}}V^{\frac{5}{3}}}{V} = \text{constant for a fixed mass of ideal gas} \tag{5.65}$$

This can be simplified to give (remember n, R are constants and $\frac{V^{\frac{5}{3}}}{V} = V^{\frac{2}{3}}$):

$$T^{\text{ideal gas}}V^{\frac{2}{3}} = \text{another constant for a fixed mass of gas} \tag{5.66}$$

Therefore, we can write for the adiabatic steps 2 and 4 of the Carnot Cycle:

$$T_H^{\text{ideal gas}}V_b^{\frac{2}{3}} = T_C^{\text{ideal gas}}V_C^{\frac{2}{3}} \quad (\text{Step 2}) \tag{5.67}$$

$$T_H^{\text{ideal gas}}V_a^{\frac{2}{3}} = T_C^{\text{ideal gas}}V_d^{\frac{2}{3}} \quad (\text{Step 4}) \tag{5.68}$$

Dividing Equation (5.67) by (5.68):

$$\frac{V_b^{\frac{2}{3}}}{V_a^{\frac{2}{3}}} = \frac{V_C^{\frac{2}{3}}}{V_d^{\frac{2}{3}}} \tag{5.69}$$

Therefore:

$$\frac{V_b}{V_a} = \frac{V_C}{V_d} \tag{5.70}$$

Equation (5.64) can be simplified using Equation 5.70 (remember $\ln\frac{V_d}{V_c} = -\ln\frac{V_c}{V_d}$)[83]:

$$-\left|\mathbf{w}_{\mathbf{input}}\right| = \left[-nRT_H^{\text{ideal gas}}\ln\frac{V_b}{V_a}\right] + \left[nRT_C^{\text{ideal gas}}\ln\frac{V_b}{V_a}\right] \tag{5.71}$$

We can simplify Equation (5.71):

$$-\left|\mathbf{w}_{\mathbf{input}}\right| = -nR\ln\frac{V_b}{V_a}\left[T_H^{\text{ideal gas}} - T_C^{\text{ideal gas}}\right] \tag{5.72}$$

We know $-\left|\mathbf{w}_{\mathbf{input}}\right| + \left|\mathbf{w}_{\mathbf{output}}\right| = 0$, and we can write Equation (5.72) as:

$$+\left|\mathbf{w}_{\mathbf{output}}\right| = nR\ln\frac{V_b}{V_a}\left[T_H^{\text{ideal gas}} - T_C^{\text{ideal gas}}\right] \tag{5.73}$$

We know the efficiency of a heat engine (ε) is the energy supplied by the heat engine to the work reservoir (work output) divided by the energy supplied from the hot reservoir by the process of heat (Equation 5.39). We can write:

$$\varepsilon = \frac{nR\ln\frac{V_b}{V_a}\left[T_H^{\text{ideal gas}} - T_C^{\text{ideal gas}}\right]}{+\left|\mathbf{q_H}\right|} \tag{5.74}$$

We know from Equation (5.58) that $+\left|\mathbf{q_H}\right| = nRT_H^{\text{ideal gas}}\ln\frac{V_b}{V_a}$, therefore we can write:

$$\varepsilon = \frac{nR\ln\frac{V_b}{V_a}\left[T_H^{\text{ideal gas}} - T_C^{\text{ideal gas}}\right]}{nRT_H^{\text{ideal gas}}\ln\frac{V_b}{V_a}} = \frac{\left[T_H^{\text{ideal gas}} - T_C^{\text{ideal gas}}\right]}{T_H^{\text{ideal gas}}} \tag{5.75}$$

Comparison with Equation (5.54)[84] completes the proof that ideal-gas and thermodynamic temperature scales are the same, i.e., $T^{\text{ideal gas}} = T^{\text{kelvin}} = T$.

5.10 THE CARNOT CYCLE CAN BE EXTENDED TO ANY REVERSIBLE CYCLE

Figure 5.23 shows a general reversible cyclic process represented on a PV diagram, superimposed on the PV plot is a series of connected Carnot cycles. We can see one Carnot cycle consists of a high-temperature isotherm a_1b_1 (at temperature T_{H_1}) and a low-temperature isotherm c_1d_1 (at temperature T_{C_1}) joined by two adiabats a_1c_1 and b_1d_1. We know from Equation (5.53) that:

$$\frac{\left|\mathbf{q_H}\right|}{-\left|\mathbf{q_C}\right|} = \frac{T_H}{T_C} \quad \text{or} \quad \frac{\left|\mathbf{q_H}\right|}{T_H} - \frac{\left|\mathbf{q_C}\right|}{T_C} = 0 \tag{5.76}$$

where $\left|\mathbf{q_H}\right|$ and $-\left|\mathbf{q_C}\right|$ are energy flows to the system, by the process of heat, from the hot and cold heat reservoirs respectively.

In Figure 5.23 the cyclic process is divided into 8 Carnot cycles, i.e., we have 8 hot reservoirs and 8 cold reservoirs at different temperatures. If we consider the first cycle, between a hot reservoir at T_{H_1} and cold reservoir at T_{C_1}, then we have $\frac{\left|\mathbf{q_{H_1}}\right|}{T_{H_1}} - \frac{\left|\mathbf{q_{C_1}}\right|}{T_{C_1}} = 0$, where $+\left|\mathbf{q_{H_1}}\right|$ is the energy flow to

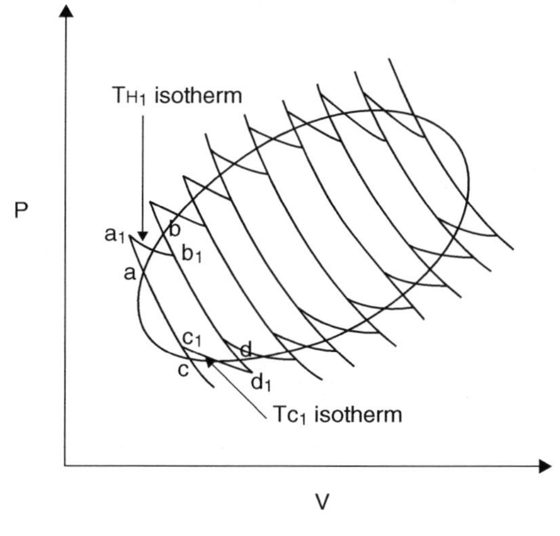

Figure 5.23 A general cyclic process on a PV diagram with superimposed Carnot cycles.

the system, by the process of heat, from hot reservoir 1 at temperature T_{H_1}, and $-|\mathbf{q}_{C_1}|$ is the energy transferred to the system, by the process of heat, from the cold reservoir 1 at temperature T_{C_1}. We can write (using absolute values):

$$\sum_{i=1}^{i=8}\left[\frac{|\mathbf{q}_{H_i}|}{T_{H_i}}-\frac{|\mathbf{q}_{C_i}|}{T_{C_i}}\right]=0 \tag{5.77}$$

Imagine many heat reservoirs at successive temperatures that differ only by infinitesimal amounts and energy is transferred from these reservoirs, by the process of heat, when the working substance is brought successively into contact with them as it traverses the cycle. In other words, on this cycle can be superimposed a system of adiabats and isotherms and they can be drawn as close together as we want. Therefore, as $n \to \infty$ we can write:

$$\int_{i=1}^{i=\infty}\frac{|\mathbf{q}_{H_i}|}{T_{H_i}}-\int_{i=1}^{i=\infty}\frac{d|\mathbf{q}_{C_i}|}{T_{C_i}}=0 \tag{5.78}$$

I have used absolute values in the above three equations to aid clarity, but absolute value equations do have their limits of usefulness. Most texts will write Equations (5.76)–(5.78) as (but not with the "**bold**" notation):

$$\frac{\mathbf{q}_H}{T_H}+\frac{\mathbf{q}_C}{T_C}=0 \tag{5.79}$$

$$\sum_{i=1}^{i=8}\left[\frac{\mathbf{q}_{H_i}}{T_{H_i}}+\frac{\mathbf{q}_{C_i}}{T_{C_i}}\right]=0 \tag{5.80}$$

$$\int_{i=1}^{i=\infty}\frac{d\mathbf{q}_{H_i}}{T_{H_i}}+\int_{i=1}^{i=\infty}\frac{d\mathbf{q}_{C_i}}{T_{C_i}}=0 \tag{5.81}$$

This proves that the Carnot Cycle can be extended to any reversible cycle and Equation (5.79) holds true for any reversible cyclic process whatsoever. In other words, any reversible cycle can be written as a succession of Carnot cycles; therefore, what is true for a Carnot cycle is true for all reversible cycles. Equation (5.79) is usually written as:

$$\oint\frac{dq}{T}=0(\text{reversible}) \tag{5.83}$$

1. The symbol integral symbol in Equation (5.83) is known as the cyclic integral symbol.[85] I have previously mentioned cyclic integral in Section 1.6.9.2, Section 4.6. In Section 4.6, I wrote "The cyclic integral of state function is zero"; this suggests that $\frac{dq}{T}$ is a state function, but I am getting well ahead of myself here.

5.11 HEAT PUMPS

I have included this section on heat pumps as the UK government is considering them as a replacement for gas boilers for space heating. Heat pumps can be used to heat and cool spaces, in this section I will be considering heat pumps used for space heating. As we know energy will not flow spontaneously from lower temperature to higher, by the process of heat, but it can be made to flow in this direction if work is performed on a suitable system, i.e., a heat pump or refrigerator. The energy transferred, by the process of heat, from the cold heat reservoir (the outside of the space to be heated) to the hot heat reservoir (the space to be heated) against the temperature gradient is less than the amount of energy transferred directly, by the process of heat, e.g., by an electrical resistance heater, to the space to be heated (see analysis below). This is the motivation for using heat pumps to heat or cool the interior of buildings. There are air source and ground source heat pumps. Air source heat pumps use the outside air as the cold heat reservoir and ground source heat pumps employ a ground heat exchanger in contact with the ground or groundwater as the cold heat reservoir.[86]

The efficiency of a heat pump is expressed as a coefficient of performance (COP) which is the ratio of the energy flow per cycle, by the process of heat, to the hot reservoir $(\mathbf{q}_H)$ to the energy flow per cycle to the heat pump, by the process of work $(\mathbf{w})$[87]:

$$\mathrm{COP}_{\text{heating}} = \frac{|\mathbf{q}_H|}{|\mathbf{w}|} \tag{5.83}$$

The higher the number, the more efficient a heat pump is and the less energy it consumes. When used for space heating these devices are typically much more energy efficient than simple electrical resistance heaters.

From the First Law of Thermodynamics, we know $-|\mathbf{q}_H| + |\mathbf{q}_C| + |\mathbf{w}| = 0$. Therefore, we can write Equation (5.81) as:

$$\mathrm{COP}_{\text{heating}} = \frac{|\mathbf{q}_H|}{|\mathbf{q}_H| - |\mathbf{q}_C|} \tag{5.84}$$

For a reversible heat pump, we know $\frac{\left|\mathbf{q}_H^{\text{reversible}}\right|}{T_H} = \frac{\left|\mathbf{q}_C^{\text{reversible}}\right|}{T_C}$ and we can write $\left|\mathbf{q}_C^{\text{reversible}}\right| = \frac{\left|\mathbf{q}_H^{\text{reversible}}\right| T_C}{T_H}$. Substituting for $|\mathbf{q}_C|$ in Equation (5.84) and simplifying we have:

$$\mathrm{COP}_{\text{heating}}^{\text{reversible}} = \frac{\left|\mathbf{q}_H^{\text{reversible}}\right|}{\left|\mathbf{q}_H^{\text{reversible}}\right| - \frac{\left|\mathbf{q}_H^{\text{reversible}}\right| T_C}{T_H}} = \frac{1}{1 - \frac{T_C}{T_H}} = \frac{T_H}{T_H - T_C} \tag{5.85}$$

When $T_H > T_C$ the coefficient of performance of a heat pump is greater than unity, the energy, transferred by the process of work, is lesser than the energy transferred by the process of heat, making a heat pump a more efficient form of heating than electrical resistance heating. If we have a COP of say 10, then the heat pump uses 10 times less energy than an electrical resistance space heater.

Table 5.1 gives some $\mathrm{COP}_{\text{heating}}^{\text{reversible}}$ values for cold and hot reservoirs typical of those for heating a home. Note $\mathrm{COP}_{\text{heating}}^{\text{reversible}}$ values are very high when temperature differences between the two reservoirs are small and continually fall as the temperature difference increases. Remember Table 5.1 shows the maximum possible value of the COP under differing operating parameters.

The important question is what does today's heat pump achieve? In Europe, the standard test conditions for ground source heat pump units use 35°C for the hot reservoir and 0°C for the cold reservoir. Maximum theoretical COP is 8.8 but even the best heat pumps only manage 4.5.[88] My reading around the subject (and it is a minefield) indicates that for the above standard conditions,

Table 5.1: Maximum $COP_{heating}$ for Different Cold and Hot Reservoir Temperatures (in °C) Typical of Those Inside and Outside a Building

t°_{C}	t°_{H}	$COP_{heating}$	t°_{C}	t°_{H}	$COP_{heating}$	t°_{C}	t°_{H}	$COP_{heating}$
−10°C	8°C	16	−3°C	8°C	26	5°C	8°C	94
−10°C	16°C	11	−3°C	16°C	15	5°C	16°C	26
−10°C	24°C	9	−3°C	24°C	11	5°C	24°C	16
−10°C	32°C	7	−3°C	32°C	9	5°C	32°C	11

the best ground source heat pump will achieve a COP of about 3. In other words, a heat pump is three times more efficient in terms of energy consumption than using an electric space heater. However, the price for installing an air source heat pump varies from £8,000 to £18,000, while ground source heat pump costs can range anywhere from £20,000 to £35,000. You can buy a cheap electric fan heater for £20-30 and an electric storage radiator (running on off-peak electricity) for £200-300. Also note that heat pumps deliver energy at lower temperatures than gas or oil boilers, this means you need to run them for much longer periods to heat your home to a comfortable temperature and or increase the size of the radiators. There is also more to go wrong with a heat pump.

5.12 DISCUSSION AND CONCLUSIONS

There are two principal ways of approaching the concepts of thermodynamics. The more rigorous and mathematical is to investigate the passage from one thermodynamic equilibrium to another. The second (based on the Carnot cycle) is to study cyclic processes, in which, after a complete cycle, the system is left unchanged and changes in the surroundings are considered. This is less rigorous and requires less mathematics. This latter approach is the principal one taken in this book.

Energy transferred by the process of work can easily be converted into energy transferred by the process of heat. For example, if we rub our hands, the energy transferred by mechanical work increases the temperature of our hands and this energy is transferred by the process of heat to the surroundings (in plain English - work is easily converted into heat). However, the energy transferred by the process of heat cannot be easily converted into energy that can be transferred by work (again in plain English heat cannot so easily be converted into work). In fact, this conversion requires a special piece of equipment known as a heat engine, and this was only developed 300 years ago. A heat engine, which operates in a cycle, receives energy by the process of heat from a high-temperature heat reservoir, converts a part of this energy into energy that can be transferred by the process of work and rejects the remaining energy by the process of heat to a lower temperature sink or reservoir. From now on, in this book, I will use the shorthand notation, e.g., a heat engine converts heat into work and rubbing your hands together converts work into heat.

Carnot considered the following question - how are heat and work connected? The question was triggered by the great economic significance which the British steam engines had already achieved. Carnot was motivated by the practical problem of improving these machines, namely, the delivery of a given amount of work – usually measured by pumping a given quantity of water from the bottom of a shaft (i.e., force x distance) – with the combustion of the least amount of coal. In 1824, Carnot published his theory of heat engines based on the idea that energy transferred by the process of heat from a high temperature to a low temperature can transfer energy by the process of work to the surroundings (i.e., "do useful work").[89]

> Recent studies have confirmed beyond all reasonable doubt that Carnot owed far more to his contact with the world of power engineering than he did to the physics of his day. It is now clear that the precedents for his innovations are to be found not so much in the scientific traditions and textbooks of the early nineteenth-century, as in the practices and beliefs that were current among engineers ...[90]

He turned the question – how are heat and work connected - into a scientific problem: what determines the upper limit for the work delivered by a heat engine? Carnot began by rigorously developing the concept of reversibility. He was the first to distinguish between reversible and irreversible processes in thermodynamics, and to lay down the simple and sufficient rules for the reversibility of a heat engine:

- Such an engine must work without friction (this includes turbulence), which involves direct waste of mechanical work.
- There must be no transfer of heat, from a hotter to a colder body, which involves wasting the energy available from a temperature difference.

A reversible process is one in which the time scale is assumed to be so slow that every intermediate state deviates, only infinitesimally, from equilibrium. Every intermediate state is then considered to be at equilibrium and is exactly described by a set of thermodynamic state functions (e.g., P, V, and T). At any stage a reversible process can be reversed to the same state as before, i.e., reversible processes do not permanently change the universe while irreversible processes permanently change the universe. An important point is that all 'real' processes are irreversible.

Carnot's belief in the caloric theory of heat led him to analyse the heat engine in an analogous manner to a waterwheel. In a waterwheel, the water falls through a gravitational potential difference and that original potential energy is transferred to the surroundings by the process of work by the waterwheel. Carnot assumed that the caloric flow through a heat engine was equivalent to the flow of water through a waterwheel. Consequently, the most efficient heat engines and waterwheels are frictionless and turbulence free, but also, in analogy with the water entering and leaving the wheel gently with no intermediate loss of height, the caloric would enter and leave the heat engine isothermally (remember the temperature is analogous to the gravitational potential energy of the water). By analogy with the waterwheel, where $g\Delta h$ measures the potential energy given up by a unit mass of water, the drop in temperature $\Delta T = T_H - T_C$ in the heat engine measures the potential energy given up by a unit amount of caloric. Carnot's analysis showed that his hypothetical reversible heat engine operated at maximum efficiency. The most efficient heat engine has isothermal heat exchange, this is analogous to the most efficient water wheel where there is zero fall of water before it flows over the waterwheel and zero fall of water after it flows over the waterwheel (see Figure 6.3 in Chapter 6). As we have seen the most efficient heat engine cycle is the reversible Carnot cycle. Carnot's theorem states that no heat engine operating between the same hot and cold heat reservoirs has a higher efficiency than a reversible heat engine. If either of them is operated in reverse, then the reversible waterwheel or heat engine operating in reverse mode can do the same amount of work on the surroundings as the surroundings can do on forward operating reversible waterwheel or heat engine. The following statements of Carnot's Principle[91] are equivalent:

- "Every reversible heat engine operating between the same two temperature reservoirs have [sic] identical efficiency".[92]
- "Every Carnot heat engine between a pair of heat reservoirs is equally efficient, regardless of the working substance employed or the operation details".[93]
- "The efficiency of all reversible engines operating between the same two temperatures is the same, and no irreversible engine operating between these temperatures can have a greater efficiency than this"[94]

Analysis of the Carnot cycle shows that a portion of the energy flowing through a heat engine by the process of heat must always be lost as waste energy (by the process of heat to the cold reservoir) and does not contribute to the overall useful energy output by the process of work. As we shall see this postulate that 100% of the energy transferred by heat can never be converted into work is vital to our understanding of how nature works.

At the end of Section 5.8, there is an extract from Mach. I removed the penultimate sentence from the final paragraph (where the ellipsis is). This sentence stated:

> Carnot, it is true, was an adherent of the theory of Black, according to which the sum-total of the quantity of heat in the world is constant, but so far as his investigations have been hitherto considered the decision on this point is of no consequence.

Mach was making the point that Carnot's conclusions were not based on any particular theory of heat. His comment that Carnot "was an adherent of the theory of Black" is a reference to the caloric theory of heat which considered that heat was a fluid. This theory predated the modern "mechanical" theory of heat (which has been used so far in this book) that considers heat to be an energy transfer process like work. The caloric theory of heat is the subject of the next chapter.

Opinions about Carnot's essay have differed widely during the course of history. At first, for a decade it was completely ignored by the scientific world. Then it was taken up and propagated by Clapeyron (1834), a mining and railroad engineer, without any noticeable response from the scientific community. In the late 1840s, it was finally recognised by Thomson (later Lord Kelvin) in his work on the construction of the absolute temperature scale in which he applied Carnot's line of thinking. The following abstracts are from Kelvin[95]:

> The determination of temperature has long been recognized as a problem of the greatest importance in physical science. It has accordingly been made a subject of most careful attention, and, especially in late years, of very elaborate and refined experimental researches; and we are thus at present in possession of as complete a practical solution of the problem as can be desired, even for the most accurate investigations. The theory of thermometry is however as yet far from being in so satisfactory a state. The principle to be followed in constructing a thermometric scale might at first sight seem to be obvious, as it might appear that a perfect thermometer would indicate equal additions of heat, as corresponding to equal elevations of temperature, estimated by the numbered divisions of its scale. It is however now recognized (from the variations in the specific heats of bodies) as an experimentally demonstrated fact that thermometry under this condition is impossible, and we are left without any principle on which to found an absolute thermometric scale.
>
> ...
>
> The relation between motive power[96] and heat, as established by Carnot, is such that quantities of heat, and intervals of temperature, are involved as the sole elements in the expression for the amount of mechanical effect to be obtained through the agency of heat; and since we have, independently, a definite system for the measurement of quantities of heat, we are thus furnished with a measure for intervals according to which absolute differences of temperature may be estimated.

Many scientists believe that it was the erroneous available experimental data, that prevented Carnot from discovering the First and Second Laws of Thermodynamics.[97] This is from Callendar[98]:

> If Carnot had not been misled by this fundamental error in the work of Delaroche and Bérard,[99] and if his numerical data had been a little more accurate, he could not have failed to notice that the quantity of caloric in a kilocalorie[100] varied inversely as the absolute temperature, or that its equivalent motive power was the same at all temperatures. In other words, that the kilocalories [kilojoules] was a unit of motive power, and not a unit of caloric.

It is now considered that Carnot's essay is one of the great works of science and that it contains the foundations of present-day thermodynamics. Carnot's thought experiment also answered the question of why real processes have directionality and showed why a perpetual motion machine was impossible.

Consider the PV diagram of the Carnot Cycle again (Figure 5.24a). In the expansion step, the work done on the system is the area abche. Using our sign convention this quantity is a negative number because the system has lost energy by the process of work (there is a flow of energy out of the system into the surroundings). In the compression step the work done on the system is the area cdaeh. This quantity is a positive number because the system has gained energy during the compression cycle (there is a flow of energy into the system from the surroundings). If we add these areas together (with their signs) we get the area abcd which is the energy gained by the

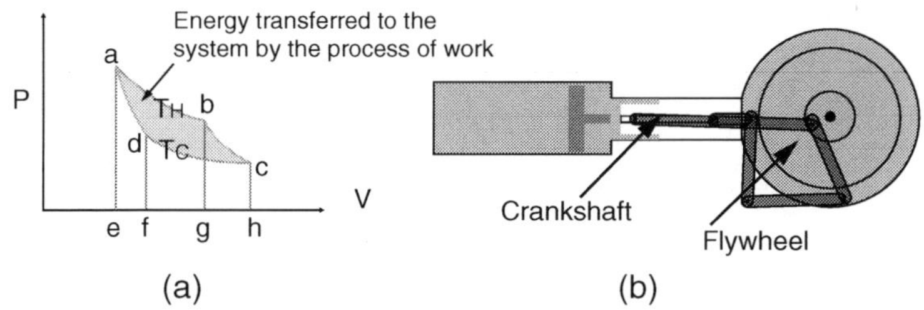

Figure 5.24 (a) The Carnot Cycle and (b) Heat engine with flywheel and connecting crankshaft.

Table 5.2: Efficiencies of Power Stations

Power Station	T_C (°C)	T_H (°C)	Carnot Efficiency	Observed
West Thurrock (UK) coal-fired power station	25	565	0.64	0.36
CANDU (Canada) nuclear power station	25	300	0.48	0.3
Larderello (Italy) geothermal power station	80	250	0.33	0.16

system by the process of work. This quantity must be negative because area abche is greater than area cdaeh – this is the work input. We know $\mathbf{q_H} + \mathbf{q_C} + \mathbf{w_{input}} = 0$, therefore $(\mathbf{q_H} + \mathbf{q_C})$ is positive because over the cycle the system has lost energy by the process work, i.e., $\mathbf{w_{input}}$ is negative. Work output is the energy transferred to the surroundings from the system by the process of work because we are considering reversible systems $\mathbf{w_{output}} = -\mathbf{w_{input}}$. If we consider the area abcd is a measure of the work output of the reversible heat engine, then the area is considered to be a positive quantity. To increase the magnitude of the work output we need to increase T_H and/or decrease T_C. You can also see why a heat engine uses a flywheel, or something similar, using its inertia to store and smooth the intermittent energy from the heat engine and transform the reciprocating motion of the engine into rotary motion (Figure 5.24b).

Let us now consider some real examples: if a heat engine (e.g., a steam engine) is operating between 100°C (373 K) and 20°C (293 K) then the maximum possible efficiency of this engine is $1-\frac{293}{273} = 0.21$ (or 21%), i.e., a maximum of 21% of the internal energy of the steam at 100°C could be converted into work. The efficiency would be improved if we used a colder cold reservoir or a hotter hot reservoir. For example, if the engine operated between 150°C (423 K) and 0°C (273 K) then its maximum efficiency is 35%. If the cold reservoir is at 0 K then the maximum efficiency is 100%. However, in the real world, 100% will never be achieved. Table 5.2 lists the Carnot efficiency (ε) and the observed efficiency of three power stations.[101]

To end this chapter, I want to set down the following extract[102]:

> On reading Carnot's memoir, one is struck by his depth of understanding and the generality of his conclusions. It is all there! As a model of a steam engine, he conceived the generalized heat engine operating between a high-temperature heat source (the furnace) and a low-temperature heat sink (the condensing water). He then deduced that, for a given transfer of caloric, the maximum work output depends not on the working fluid, but solely on the temperatures of the heat reservoirs. He introduced the idea of completing the cycle (because a steam engine is an open circuit device) and further arguments led to the conclusion that the cyclic engine with maximum efficiency is one that can operate reversibly. Taking an ideal gas as the working fluid, he then calculated how the state of the fluid changes as it passes around the cycle. This was followed by a discussion of the principles in the context of real engines and the realization that whenever heat is transferred from one body to another across a finite temperature difference there will be a loss in the possibility for producing mechanical work.

Carnot believed in the caloric theory of heat when he carried out his thought experiment. In the next chapter consider the rise and fall of the caloric theory of heat.

NOTES

1 S. Carnot, *Réflexions sur la puissance motrice du feu*, translated by R.H. Thurston in *Reflections on the Motive Power of Heat*, 2nd Edition, John Wiley and Sons, p. 38, 1897.

2 From http://www.explainthatstuff.com/engines.html.

3 https://mathshistory.st-andrews.ac.uk/Biographies/Carnot_Sadi/.

4 E.T. Jaynes, "the Evolution of Carnot's Principle", The opening talk at the EMBO Workshop on Maximum-Entropy Methods in x-ray crystallography and biological macromolecule structure determination, Orsay, France, April 24–28, 1984. Reprinted in Ericksen & Smith (1988), Vol. 1, pp. 267–282. https://bayes.wustl.edu/etj/articles/ccarnot.pdf.

5 Carnot (père et fils) considered a perpetual motion machine was one that continuously transferred energy to the surroundings by the process of work (see Section 5.7.3).

6 See final paragraph of this section for a discussion on Jaynes' use of the word *heat*.

7 https://mathshistory.st-andrews.ac.uk/Biographies/Carnot_Sadi/.

8 E.T. Jaynes (1922–1998) was one of the foremost physicists/mathematicians of his day. See http://theor.jinr.ru/~kuzemsky/etjaybio.html.

9 See Section 3.6.

10 See, for example, https://en.wikipedia.org/wiki/Engine.

11 Diagrams from http://d1068036.site.myhosting.com/chemGraphics.f/SteamEngines.gif.

12 The efficiency of a heat engine relates how much useful mechanical work is output for a given amount of energy input into the engine by the process of heat.

13 https://en.wikipedia.org/wiki/Industrial_Revolution.

14 A stroke refers to the full travel of the piston along the cylinder, in either direction.

15 Carnot was the first to introduce the concept of what is now known as the "thermodynamic cycle". A thermodynamic cycle is a series of processes where the properties of the system are the same after the cycle as they were prior to the cycle. A major advantage of the use of cycles is they can simplify the analysis of thermodynamic processes (as we shall see).

16 Reflections on the Motive Power of Heat/Chapter 3 - Wikisource, the free online library Endnote 15. This quote does not appear in S. Carnot, *Réflexions sur la puissance motrice du feu* (1824) translated by R.H. Thurston in *Reflections on the Motive Power of Heat*, 2nd Edition, Endnote 15, John Wiley and Sons, 1897. See https://www3.nd.edu/~powers/ame.20231/carnot1897.pdf.

17 Professor H.L. Callendar (1863–1930), Professor of Physics at the Imperial College of Science and Technology is probably best known for his steam tables: published in 1915 these are still widely used by engineers. He also developed the platinum resistance thermometer.

18 H.L. Callendar, "The caloric theory of heat and carnot's principle", *Proc. Phys.* XXIII, 159, 1911.

19 https://www.cora.nwra.com/~werne/eos/text/turbulence.html.

20 Convection is the process by which energy is transferred by movement of volumes of fluid (often called convection currents). Natural convection currents result from the tendency of most fluids to expand when heated, i.e., they become less dense and rise as a result of the increased buoyancy. Circulation caused by this effect accounts for the uniform heating of water in a kettle or air in a heated room.

21 Chaos Theory deals with phenomena that are effectively impossible to predict or control, like turbulence, weather, the stock market, our brain states, and so on. Most people have heard of the "butterfly effect" – a butterfly flapping its wings in, say, London may cause a hurricane in Miami. It may take a very long time, but the connection is real. If the butterfly had not flapped its wings at just the right point in space/time, the hurricane would not have happened where and when it did. A more rigorous way to express this is that small changes in the initial conditions lead to drastic changes in the final conditions.

22 A good example is the use of supercomputers to try and analyse turbulent phenomena is modern day weather forecasting (and explains why they get it wrong so often).

23 https://www.mpg.de/7078493/Einstein-turbulence.

24 A streamline is an imaginary line in a fluid such that the tangent at any point indicates the direction of the velocity of a particle of the fluid at that point. Note the tangent to a straight line is the straight line.

25 See Section 4.6 for an introduction to state and path functions.

26 Many texts use the term "quasi-static" which is defined as a process that happens slowly enough for the system to remain in equilibrium. This term like "reversible" is still impossible to achieve with real processes which begs the question why the term is used.

27 I have already discussed this – see Section 2.9.

28 As we know (Section 2.8), these are elastic collisions – "An elastic collision is that it is an encounter between two bodies in which the total kinetic energy of the two bodies after the encounter is equal to their total kinetic energy before the encounter".

29 A process that is independent of time is known as a stochastic process. Stochastic processes are widely used as mathematical models of systems and phenomena that appear to vary in a random manner.

30 See Sections 6.5.3 and 6.6.2.

31 See Section 7.5.3.

32 https://en.wikipedia.org/wiki/Dissipation.

33 In other words – an engine operating under reversible conditions.

34 H.L. Callendar, "The Caloric Theory of Heat and Carnot's Principle", *Proc. Phys.* XXIII, 153–189, 1911. British physicist probably best known for his steam tables: published in 1915 these are still widely used by engineers. He also developed the platinum resistance thermometer.

35 H.L. Callendar, "The Caloric Theory of Heat and Carnot's Principle", *Proc. Phys.* XXIII, p. 160, 1911. British physicist probably best known for his steam tables: published in 1915 these are still widely used by engineers. He also developed the platinum resistance thermometer. The underlining is mine.

36 Motive power = work.

37 This sentence is probably one of the most important sentences in the whole book, another way of expressing it is to say real (i.e., reversible) processes change the universe while reversible processes do not.

38 E. Mendoza (ed.), Reflections on the motive power of fire by Sadi Carnot, and other papers on the second law of thermodynamics by E Clapeyron and R Clausius (New York, 1960).

39 From second extract in Section 5.1.

40 This is not necessarily the case for other materials, e.g., the internal energy of liquid water is higher than the same amount of ice at 0°C.

41 See Section 4.3.5 Equation 4.14.

42 $\int \frac{dx}{x} = \ln(x) + C$ (a well-known integral found in all textbooks on calculus) therefore

$\int_{x_1}^{x_2} \frac{dx}{x} = [\ln x_2 + C] - [\ln x_1 + C] = [\ln x_2] - [\ln x_1] = \ln \frac{x_2}{x_1}$ – see also Table 1.3 in Section 1.6.6.

43 Remember temperature of a fixed quantity of ideal gas is directly proportional to the mass of the microparticles and the root-mean-square of the speeds of the microparticles –Section 3.5.3.

44 E. Fermi, *Thermodynamics*, Dover Publications, 1956.

45 Although work and heat are inexact differentials (see Section 1.6.9.1 and Section 4.7) in the case of an adiabatic process where q = 0 then dw must be an exact differential.

46 See, for example, https://math.stackexchange.com/questions/287076/what-is-the-integral-of-0. Also note (i) the constant in Equation 5.16 contains the integration constants of $\int \frac{dT}{T}$ and $\int \left[\frac{d\mathbf{V}}{V}\right]$, and (ii) integration has removed all vector-like quantities from Equation 5.16.

47 See Section 4.6.

48 See https://hsm.stackexchange.com/questions/5252/why-isnt-thermodynamics-replaced-with-thermostatics.

49 S. Carnot, *Réflexions sur la puissance motrice du feu* (1824) translated by R.H. Thurston in *Reflections on the Motive Power of Heat*, 2nd Edition, John Wiley and Sons, p. 38, 1897. First quote pages 1 and 2; the second quote page 5.

50 S. Carnot, *Réflexions sur la puissance motrice du feu* (1824) translated by R.H. Thurston in *Reflections on the Motive Power of Heat*, 2nd Edition, John Wiley and Sons, p. 38, 1897.

51 http://www.feynmanlectures.caltech.edu/I_44.html.

52 Note absolute values used here, efficiency can never have a negative number. If the physics sign convention is used there would be no need to use absolute values. I consider the efficiency of heat pumps in Section 5.12.

53 E.T. Jaynes, "the Evolution of Carnot's Principle", The opening talk at the EMBO Workshop on Maximum-Entropy Methods in x-ray crystallography and biological macromolecule structure determination, Orsay, France, April 24–28, 1984. Reprinted in Ericksen & Smith (1988), Vol 1, pp 267–282. https://bayes.wustl.edu/etj/articles/ccarnot.pdf.

54 A method of proving the falsity of a premise by showing that its logical consequence is absurd or contradictory.

55 W.J. Moore, *Physical Chemistry*, 5th Edition, Longman, p. 80, 1972. Please note the use of the word "heat" (as not a process).

56 S. Carnot, Réflexions sur la puissance motrice du feu, translated by R.H. Thurston in Reflections on the Motive Power of Heat, 2nd Edition, John Wiley and Sons, p. 38, 1897.

57 H.L. Callendar, "The caloric theory of heat and carnot's principle", *Proc. Phys.*, XXIII, 61, 1911.

58 H.L. Callendar, "The caloric theory of heat and carnot's principle", *Proc. Phys.*, XXIII, 161–162, 1911.

59 E.T. Jaynes, "the Evolution of Carnot's Principle", The opening talk at the EMBO Workshop on Maximum-Entropy Methods in x-ray crystallography and biological macromolecule structure determination, Orsay, France, April 24-28, 1984. Reprinted in Ericksen & Smith (1988), Vol 1, pp. 267–282. https://bayes.wustl.edu/etj/articles/ccarnot.pdf.

60 A.W.J.G. Ord-Hume, *Perpetual Motion the History of an Obsession*, Adventures Unlimited Press, p. 155, 2006.

61 http://www.kilty.com/pmotion.htm#Section5.

62 I will consider the Second Law of Thermodynamics in Chapter 7.

63 http://www.douglas-self.com/MUSEUM/POWER/ammonia/ammonia.htm.

64 http://www.douglas-self.com/MUSEUM/POWER/ammonia/ammonia.htm.

65 https://hypertextbook.com/facts/1998/AvijeetDut.shtml.

66 http://web.mit.edu/seawater/2017_MIT_Seawater_Property_Tables_r2a.pdf (at salinity 35 $g.kg^{-1}$ at 0°C.

67 https://www.windows2universe.org/earth/Water/temp.html.

68 https://en.wikipedia.org/wiki/World_energy_consumption.

69 H.L. Callendar, "The caloric theory of heat and carnot's principle", *Proc. Phys.* XXIII, 158–159, 1911.

70 i.e., caloric.

71 See Section 3.3.1.

72 Remember if the system and hot reservoir are at the same temperature there would be no overall transfer of energy, but if they are at different temperatures then energy transfer by the process of heat is irreversible.

73 H.L. Callendar, "The caloric theory of heat and carnot's principle", *Proc. Phys.* XXIII, 160, 1911.

74 See Section 4.6.

75 Because a heat engine transforms energy transferred to it from the surroundings, by the process of heat, into, work, transferred from it to the surroundings. Or, using shorthand, it converts heat into work.

76 E. Mach, *Popular Scientific Lectures*, Translated by Thomas J. McCormack, The Open Court Publishing Company, pp. 160–165, 1898.

77 Mach uses "mill-wheel", but in this book I have used the word "waterwheel" – see Section 5.13 and Sections 6.2.2 and 6.4.1.

78 I presume Mach means bodies in a gravitational field.

79 Also, without turbulence.

80 https://en.wikipedia.org/wiki/Scale of temperature.

81 Remember Carnot's Principle - no heat engine can be more efficient than a reversible one operating between the same temperatures.

82 Any function could be used provided it is monotonic ["A monotonic function is a function which is either entirely non-increasing or non-decreasing. A function is monotonic if its first derivative (which need not be continuous) does not change sign". - http://mathworld.wolfram.com/MonotonicFunction.html]. No, I do not understand it either.

83 Note I have eliminated some of the absolute values in this and following equations in the interests of brevity and clarity – see final sentence in the first paragraph of this section.

84 Equation 5.54 is $\varepsilon = \frac{T_H^{kelvin} - T_C^{kelvin}}{T_H^{kelvin}}$.

85 See Section 4.6.

86 In cooling mode, the heat pump uses the inside of the building as the cold heat reservoir and the outside (air/ground/groundwater) as the hot heat reservoir.

87 The COP for heating and cooling are different because the heat reservoir of interest is different. For space cooling ('air conditioning'), the COP is the ratio of the energy flow from the cold reservoir, by the process of heat, to the energy flow to the heat pump by the process of work, i.e., $COP_{cooling}^{reversible} = \frac{|q_C|}{|w|}$.

88 https://en.wikipedia.org/wiki/Coefficient_of_performance.

89 S. Carnot, *Réflexions sur la puissance motrice du feu* (1824) translated by R.H. Thurston in *Reflections on the Motive Power of Heat*, 2nd Edition, John Wiley and Sons, p. 38, 1897.

90 R. Fox, *The Caloric Theory of Gases: From Lavoisier to Regnault*, Oxford University Press, p. 178, 1971.

91 Also known as "Carnot's Theorem", "Carnot's Rule" or "Carnot's Law".

92 http://aether.lbl.gov/www/classes/p10/heat-engine.html.

93 https://en.wikipedia.org/wiki/Carnot's_theorem_(thermodynamics).

94 http://www.physics.louisville.edu/cldavis/phys298/notes/carnot_thcycle.html.

95 "On an Absolute Thermometric Scale founded on Carnot's Theory of the Motive Power of Heat [1], and calculated from Regnault's Observations". Lord Kelvin (William Thomson) Philosophical Magazine October 1848.
https://zapatopi.net/kelvin/papers/on_an_absolute_thermometric_scale.html.

96 "motive power" = work.

97 See Chapter 7.

98 H.L. Callendar, "The caloric theory of heat and carnot's principle", *Proc. Phys.* XXIII, 172, 1911.

99 Delaroche and Berard ("Ann. Chim.," 1813) had found that the specific heat of a gas increased with increase in volume ('specific heats diminished with diminution of volume.'). For an ideal gas there is no change in specific heat with volume or pressure, for real gases any change is insignificant.

100 In the past the calorie was defined as "one hundredth of the heat required to raise the temperature of a gram of water from 0°C to 100°C" or "the heat required to raise the temperature of a gram of water from 14.5°C to 15.5°C". Since 1925 the calorie has been defined in terms of the joule, the present definition is that one calorie is equal to 4.186 J.

101 https://en.wikipedia.org/wiki/Heat_engine.

102 From "Heat, work and subtle fluids: a commentary on Joule (1850) 'On the mechanical equivalent of heat'" John Young Published: 13 April 2015 https://doi.org/10.1098/rsta.2014.0348.

6 Heat & Caloric

> Models should not be true, but it is important that they are applicable, and whether they are applicable for any given purpose must of course be investigated.[1]

> The very word model implies simplification and idealization. The idea that complex physical, biological or sociological systems can be exactly described by a few formulae is patently absurd. The construction of idealized representations that capture important stable aspects of such systems is, however, a vital part of general scientific analysis …[2]

> Essentially, all models are wrong, but some are useful. However, the approximate nature of the model must always be borne in mind….[3]

6.1 INTRODUCTION

In modern science, heat like work is considered a process[4]:

> Heat and work are methods of energy transfer, and when all flow is over, the words heat and work have no longer any usefulness or meaning. As a result of these transfers, the internal energy of the system has either increased, decreased or remained constant, and once the transfers are over, we can speak only of the internal energy of the system. It is impossible to subdivide the internal energy into two parts, one due to a heat transfer and the other to work, just as it is impossible to subdivide the water in a lake into parts resulting from (i) ground seepage, (ii) rain, (iii) flow from a brook, (iv) condensation etc.

Although most modern scientific texts state that heat is a process, the word "heat" (often in these same texts) is used as a synonym for energy. The following quotes demonstrate this.

1. Heat is energy that spontaneously passes between a system and its surroundings in some way other than through work or the transfer of matter. When a suitable physical pathway exists, heat flows spontaneously from a hotter to a colder body.[5]
2. Heat is energy that is transferred from one body to another as the result of a difference in temperature. If two bodies at different temperatures are brought together, energy is transferred – i.e., heat flows from the hotter body to the colder.[6]
3. "Heat is a form of energy that can be transferred from one object to another or even created at the expense of the loss of other forms of energy."[7]
4. "All matter contains heat energy. … Heat energy is the result of the movement of tiny particles called atoms, molecules or ions in solids, liquids and gases. Heat energy can be transferred from one object to another."[8]
5. Heat is internal thermal energy that flows from one body of matter to another—or, more specifically, from a system at a higher temperature to one at a lower temperature.[9]
6. "Heat energy is needed to increase the temperature of an object."[10]

Admittedly, once the student realises that heat, in modern science, is a process, this near-universal use of heat as energy can be usually considered a shorthand way of referring to the concept of heat as a process rather like the use of weight when we mean mass. However, the terms "heat energy" or "thermal energy" are meaningless[11]:

> The main objection to "thermal energy" (aside from the fact that is undefined), is that at one point it means "heat"[12] and two lines later it means "internal energy". It means whatever pleases the author at the time, and the reader is supposed, somehow, to know which of the two meanings is intended. It is a catch-all expression intended to "simplify" by combining 'heat' and 'internal energy'. Thus, to a confirmed user of 'thermal energy', there is no flow of 'thermal energy' into a gas when it is compressed adiabatically, but a few lines later we learn that the 'thermal energy' has increased!

Zemansky (the author of the above extract) makes the point that "thermal energy" means either heat or internal energy (or both). The term "thermal energy" (and the related "heat energy") is meaningless and unnecessary. I will not use either term in this book.

Zemansky discussed using "heat" as a verb:

> It is very difficult to avoid the expression 'to heat a system'. We are accustomed to heating a house, heating a car, heating some water, etc. There are, however, only two possible meanings

DOI: 10.1201/9781003608400-6

when heat is used as verb: sometimes it means 'add heat to' and at other times 'to raise the temperature of' – two expressions that are by no means identical. When we come to the expression used so often by physics and chemistry teachers, namely '... if you heat something, it becomes warmer' you get either a redundant statement ('If you raise the temperature of something, it gets warmer'), or an incorrect statement ('If you add heat to something, it gets warmer.') What about adding heat to a cake of ice at 0°C?[13].

Although in everyday speech we "heat", "warm", or "cool" things, I have not and will not intentionally use the term "heat" as a verb in this book.

Zemansky also recommends not to use heat as noun, usually in a phrase like "heat in a body". He used the following examples:

- "If a quart of water at 50°C is separated into two pints, each pint remains at the original 50°C temperature. But the heat content of a pint of 50°C water is just half the heat content of a quart of 50°C water."
- "Compared to much larger amount of cool water, the small amount of hot water has a higher temperature, but a lower heat content. The hot water contains more heat than an identical volume of cool water."

I also came across the following definition, which many non-scientists (and many scientists) may agree with:

> Temperature is a measure of how much heat is contained in an object.[14]

The following quotes seem instinctively right to me:

1. A cup of hot coffee loses heat when it cools down.
2. Half a brick contains half the heat of a whole brick when both are at the same temperature.
3. Heat flows from one location to another.

This suggests that we have an intuitive feel for heat that does not mesh with the modern science definition of heat as a process; this is illustrated in the following abstract:

> Heat is often described as if it were a fluid (like water) that can be transferred from one object to another. Even scientists talk this way when they say things like '50 J of heat were added to the metal bar.' It is important to realize that this is just a simplification of what is really going on.[15]

Zemansky again:

> It is incredible that writers and teachers are still referring to the 'heat in a body'. When water flows, the hydraulic engineer refers to a current or, more specifically, a 'water current'. Similarly, when electric charge flows, one refers to an electric current. There is, however, a fundamental difference: both water and electricity are matter, which (neglecting relativistic effects) is conserved. Heat, however, is not conserved. When water and electricity enter a system, they don't disappear. When heat enters a system, it has no existence within the system as 'heat'.

In this chapter, I want to explore how our intuitive feel for heat, not only results in the disconnect between the modern definition of heat as a process, and our intuitive feel for heat (as illustrated above), but also how our intuition with regard to "heat in a body" is manifested in caloric theory (even in modern scientific papers heat "flows" or is "transferred" or is "added"). In addition, I want to demonstrate how caloric theory gives us an intuitive understanding of the concept of entropy (considered in Chapter 7). This abstract from Callendar[16] is a good introduction:

> The caloric theory of heat is now so long forgotten that we rarely hear it mentioned, except as an example of primeval ignorance; but it was not really quite so illogical as it is generally represented to be. The science of heat might have been developed along the lines of the caloric theory nearly, if not quite, as well as on the lines of the [modern] theory. Some most important and fundamental ideas, which are obscured in [the modern theory] would, in fact, have been brought into greater prominence and more widely appreciated, and a clearer view of the whole subject might thus have been obtained.

The primary objective of this chapter is to demonstrate how caloric theory allows "a clearer view of the whole subject" to be obtained. Although in everyday speech heat "flows", etc., but work never "flows" nor is "transferred" nor is "added". In thermodynamic texts, work tends to be "done" or "performed" by a system, but heat is never "done" or "performed". These inconsistencies in the use of heat and work are because, in my view, we intuitively consider work (as used in a scientific context) to be a process and heat to be some sort of substance or stuff.

6.2 ANALOGIES FOR HEAT

6.2.1 Introduction

An analogy is a comparison in which an idea, a thing or a process is compared to another that is quite different from it. We know heat is not really a material substance, but neither are gases composed of the dimensionless microparticles used in the theory of ideal gases. Throughout history, analogies and models have played an important role in scientific discoveries, not as proof, but as inspiration. In addition, they allow an intuitive understanding of particularly arcane subjects. Consider the following[17]:

> Analogies and analogical models are popular in science and help scientists understand and communicate the intricacies, beauty and strangeness of the natural world. Consider these examples: First, Stephen Hawking used at least 74 everyday analogies in 'A brief history of time' to explain astrophysics and quantum ideas. To demonstrate that the universe is expanding equally in all directions, he says "the situation is rather like a balloon with a number of spots painted on it being steadily blown up" (p.45). He later muses that we could capture a black hole by "towing a large mass in front of it, rather like a carrot in front of a donkey" (p.115). Second, Robert Oppenheimer of atomic bomb fame claims that most of the significant advances in science used analogy as a thinking tool. He uses the history of science to show that scientific progress is aided by analogical thought and, for example, shows how analogy favoured the discovery of mesons.[18] Third, Bronowski (1973) claims that imagining the workings of a clock helped Johannes Kepler (1571–1630) develop his ideas of planetary motion and fourth, Watson and Crick insisted that they arrived at the double helix structure of DNA by making analogical models that fitted their data. Finally, Peter Atkins' (1995) book 'The periodic kingdom' is one vast analogy.

Richard Feynman used analogies to explain non-observable particle phenomena in his lecture "Atoms In Motion".[19] These included.

- paramecia are like "small football shaped things",
- molecules in water are moving "like a crowd at a football game",
- if an apple is magnified to the size of the Earth, atoms in the apple will be as big as an apple,
- an atom hitting a moving piston is like a ping-pong ball hitting a moving paddle,
- Brownian motion[20] is like a game of pushball.[21]

However, there are pitfalls, and one must be aware, that no analogy is perfect.

In this section, I will use the waterwheel and the electric circuit as analogies for the heat-engine. In addition, like a picture within a picture, use the analogy of water flow in pipes to explain the workings of a simple electric circuit.

6.2.2 The Frictionless and Turbulent Free Waterwheel

A waterwheel converts the fall of a mass of water in the Earth's gravitational field into mechanical work. I want to consider an idealised waterwheel that is 100% efficient, i.e., a machine that converts all the potential energy of the water into work, i.e., the turning of the waterwheel (Figure 6.1). The potential energy of a mass of water, m, at a height h above the ground (taken as our zero-reference level) is mgh. If the waterwheel is at height h′ above the ground (where, of course, $h > h'$), then the fall of the mass of water m from height h to height h′ results in a reduction of potential energy from mgh to mgh′, i.e., $mg(h - h')$. This is (by definition) an idealised frictionless and turbulent-free machine, so all of this reduction in potential energy is converted into work.

There are two ways of increasing the energy the waterwheel can supply to the surroundings. The first way is to increase the mass of water flowing through the waterwheel, and the second

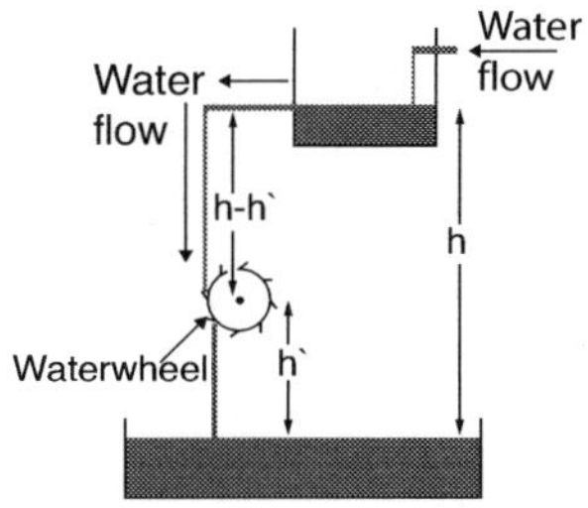

Figure 6.1 Waterwheel.

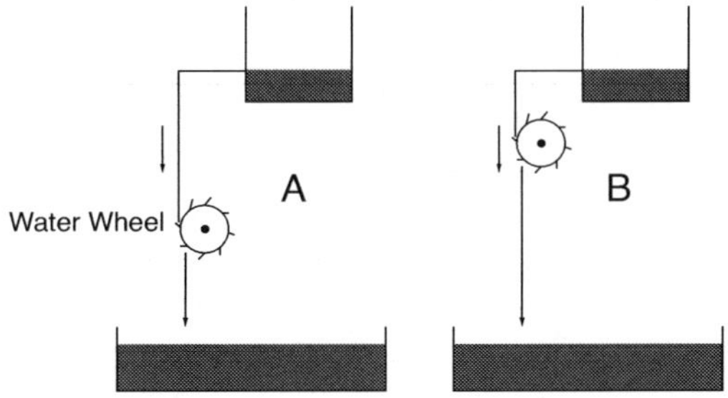

Figure 6.2 Water-wheel A produces more mechanical energy than water-wheel B.

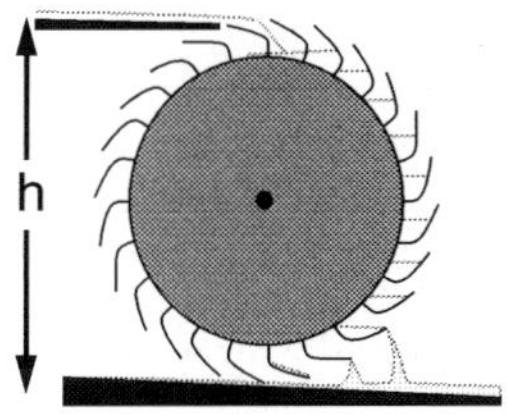

Figure 6.3 A hypothetical 100% efficient waterwheel providing there is no friction, splashing, or turbulence.

way is to increase the force of the water hitting the wheel by increasing h – h′. If the mass of water flowing to the waterwheel is the same, then system A (Figure 6.2) results in the transfer of more energy to the surroundings, by the process of work, than system B. The maximum amount of work that can be obtained from this machine for a constant mass of water is when h′ = 0, i.e., when the waterwheel is at the surface.[22]

Two noteworthy points:

1. The mass of water (or mass per unit time) that flows over the waterwheel is a state function.
2. The potential energy of a mass of water at height h metres is mgh, and the potential energy per metre is $\frac{mgh}{h} = mg$. Note: mgh is energy but mg is not energy (in fact, we know it is force).[23]

The waterwheels shown in Figures 6.1 and 6.2 are obviously not 100% efficient because of splashing and various frictional effects including turbulence. This efficiency could be better approached with an idealised splash-less, turbulence-free, and frictionless waterwheel of a different design (say something like Figure 6.3).

If this hypothetical frictionless and turbulence-free waterwheel is run in reverse by an external power source, it would raise m kg of water by h metres using the same amount of energy that the wheel would generate by the fall of m kg of water by h metres in normal operation, i.e., it would be "reversible". Connecting a "reversible" waterwheel operating in the forward direction with one operating in reverse would produce a perpetual motion machine once started (assuming the

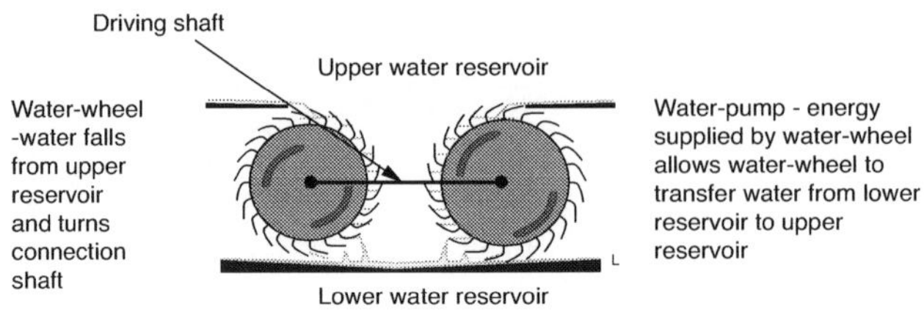

Figure 6.4 A representation of a perpetual motion machine constructed of two reversible waterwheels.

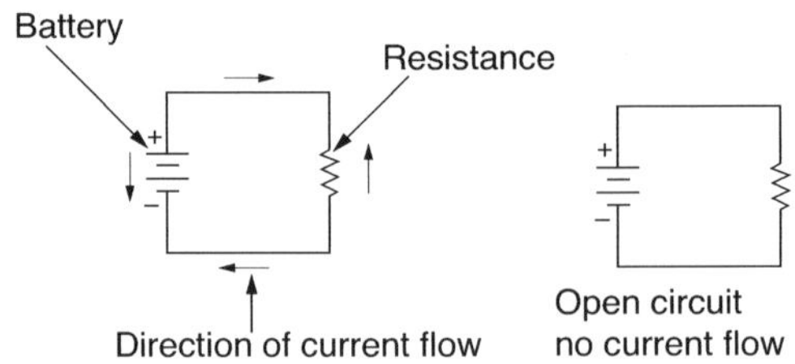

Figure 6.5 A basic electric circuit.

upper water reservoir contains an infinite quantity of water – compare heat reservoirs that are considered infinite sources of energy). Figure 6.4 gives some idea how this may be accomplished.

However, in the real world, there is friction (and turbulence and splashing), and some of the potential energy of the water is converted into energy that is transferred from the system to the surroundings by the process of heat, i.e., converted into heat. The split between the "useful work" (in turning the wheel) and the "lost-work" (work lost to friction and turbulence and ultimately dissipated to the universe by the process of heat) depends on the design and construction of the waterwheel.

6.2.3 Electricity

An electric circuit is the path through which an electric current flows. A very basic electric circuit is shown in Figure 6.5. A battery provides the energy to pump electrons around the electrical circuit, which must be made of electrically conducting material such as metal wire. The circuit must be closed to allow an electrical current flow to flow.

An electric current is a flow of electrons moving through an electrical conductor and is the rate at which a quantity of electricity flows across a surface in a circuit. Quantity of electricity is measured by a physical quantity called "electric charge" (usually shortened to charge). Electric charge is a basic property of matter carried by some elementary particles that govern how those particles are affected by an electric or magnetic field.

> Electric charge is the physical property of matter that causes it to experience a force when placed in an electromagnetic field. Electric charge can be positive or negative (commonly carried by protons and electrons respectively). Like charges repel each other and unlike charges attract each other. An object with an absence of net charge is referred to as neutral.[24]

As with energy and other fundamental quantities, no one can tell you what electric charge is; they can only tell you how charges interact. Charge comes in two types: positive (+) and negative (–) and like charges repel and unlike charges attract. Electrons considered to be negatively charged simply because of Benjamin Franklin[25] made a choice that electrons are negatively charged: he could have quite easily decided that electrons were positively charged. It should be noted, because of Franklin's arbitrary decision, that the direction of current flow is the opposite to which electrons flow. The unit of charge is a coulomb (C), and the magnitude of the charge on an electron is 1.60×10^{-19} coulombs. Electric charge, like mass and energy, is a conserved property; the net charge of an isolated system, the amount of positive charge minus the amount of negative charge, is constant. A battery provides an energy difference (usually termed "voltage" or "potential difference"), and it is this energy difference that drives the charge around the circuit. Electrical resistance is a measure of how a device impedes the flow of charge (current), and it is assumed that the wires connecting various items in an electrical circuit have zero resistance.

Water flowing through pipes can be used as an analogy of how electric circuits work (Figure 6.6). Although it does not explain many aspects of electricity, it gives the student an intuitive feel for current, potential difference and resistance. This is often known as the electronic–hydraulic analogy.[26] In this analogy, charge is represented by the quantity of water, electric current by rate of flow of water, voltage is represented by the water pressure, and current is represented by the water flow rate. Consider a water tank at a certain height above the ground and at the bottom of the tank there is a tube. The pressure difference between the water level in the tank and the bottom of the tube represents the voltage or potential difference. The quantity of water in the tank represents charge: the more water in the tank, the higher the charge, and the greater the potential difference (because the water level is higher): a higher-pressure difference results in a higher flow and *vice versa*.

We can think of this water tank as a battery, a place where we store a certain amount of energy and then release it. If a quantity of water is removed from the tank, the pressure difference decreases, and there will be a decrease in the amount of water flowing through the pipe, less pressure means less water is flowing. We can think of the amount of water flowing through the hose from the tank as current. The higher the pressure, the higher the flow (the current), and *vice-versa*.

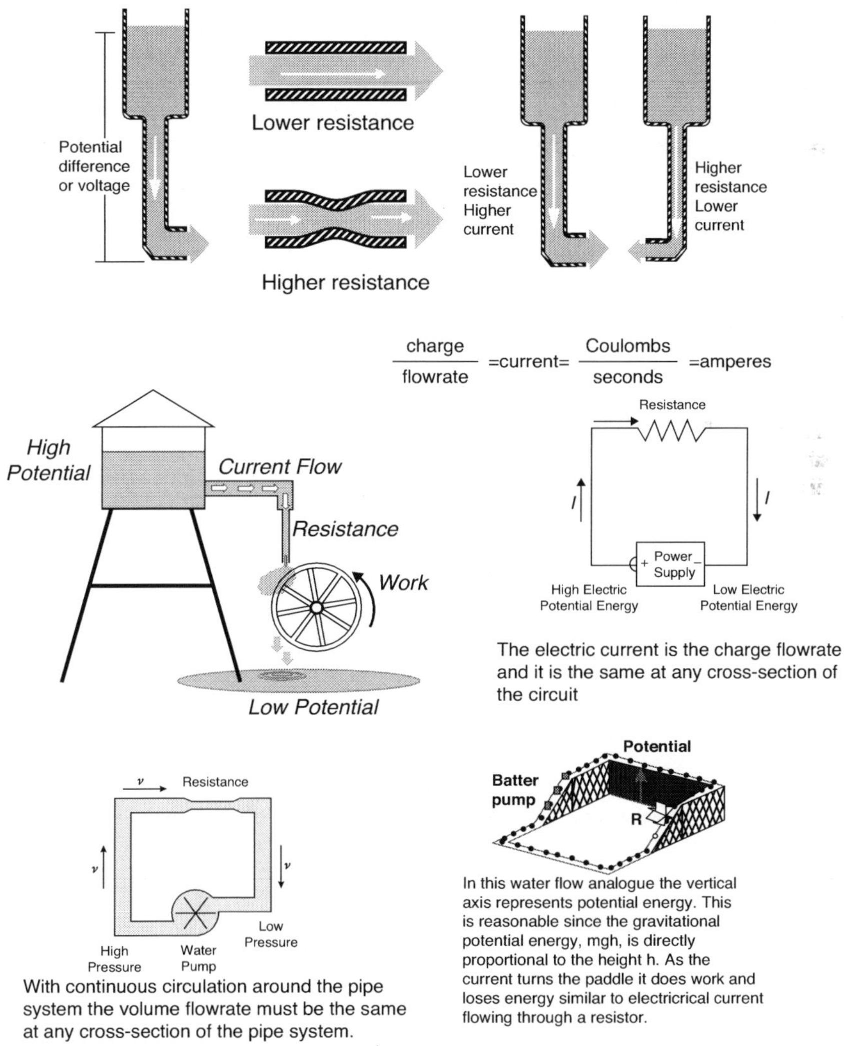

Figure 6.6 Electric-hydraulic analogies.

With water, we would measure the volume of the water flowing through the hose over a certain period of time. With electricity, we measure the amount of charge[27] flowing through the circuit over a period of time. The relationship between voltage or potential difference (V), current (I), and resistance (Ω) is straightforward and is known as Ohm's Law: $I = \frac{E}{R}$. Current is measured in amperes (usually just referred to as "amps"). An ampere is defined as 6.241×10^{18} electrons (1 coulomb of electric charge) per second passing through a surface in an electrical circuit. Voltage or potential difference (V) can be considered as the pressure difference. The greater the voltage, the greater the flow of electrical current through a conducting medium for a given resistance to the flow. Voltage is symbolised by either "E" or "V": one volt will drive one coulomb charge carriers (usually electrons), through a resistance of one ohm in one second. This is as far as I want to go with electrical theory and the electric-hydraulic analogy.

6.2.4 Discussion

Amount of substance (mass or volume) and electrical charge are quantities that scale with the system, i.e., they are extensive properties. If a homogeneous system is divided into two equal halves, each part has half the original quantity. If we want to set up a theory of the flow of water, we need to set up a new quantity which we could call mass current or volume current (units $kg.s^{-1}$ or $m^3.s^{-1}$).[28] In electricity, the analogous quantity is charge current which is usually shortened to current (units coulombs per second, $C.s^{-1}$).

Using volume and volume current, we are able to say something about the amount of water of water stored in the system/reservoir and the rate at which the water is flowing. However, these quantities do not tell us anything about why the water is flowing. Similarly, with electric circuits, charge and current are not sufficient to inform us why there is a charge current.

To explain the flow of water through pipes, we need a pressure difference. If the same rate of flow is to be maintained through two pipes of different diameters, the pressure difference between the inlets and the outlets of the pipes must be different. Similarly, different potential differences (voltages) are required if the same electrical current is to pass through two different resistors. These examples demonstrate the nature of pressure and electrical potential: they are intensive quantities.[29] In contrast to extensive quantities, intensive quantities do not scale with the size of the system, and if a homogeneous system is divided into two parts, the intensive quantities are the same in both.

To understand and quantify a physical phenomenon, we must consider both intensive and extensive physical quantities. Since a difference in pressure or electrical potential is necessary for water or electrical charge to flow, we may call the difference the driving force for the process. In the flow of water on Earth, a difference in altitude, level or height is usually the cause of water flow. This suggests that a driving force may be imagined to be a difference in levels of an intensive quantity such as height, voltage, or (as we shall see) temperature (Figure 6.7).

6.3 A SHORT HISTORY OF HEAT

6.3.1 Introduction

As mentioned in Section 4.1 the Scottish chemist Joseph Black (1728–1799) was the first person to use the word "heat" in a scientific sense. He was also the first to clearly distinguish between the intensive quantity of temperature and the extensive "quantity of heat"[30]:

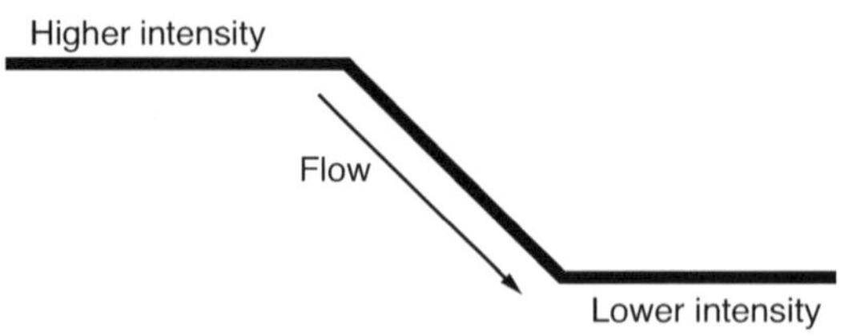

Figure 6.7 Both water in pipes and electrical charge may be imagined to flow by themselves from regions of high intensities of the potential associated with the process to regions of lower intensities, i.e., they flow from regions at high levels to regions at lower level.

> If, for example, we have one pound of water in one vessel, and two pounds of water in another, and these two quantities of water are equally hot, as examined by the thermometer, it is evident, that the two pounds must contain twice the quantity of heat that is contained in one pound. But, undoubtedly, we can suppose that a cubical inch of iron may contain more heat than a cubical inch of wood, heated to the same degree; and we cannot avoid being convinced of this by daily experience.

This is from "The Life and Letters of Joseph Black, MD"[31]:

> Here we have a confusion between heat and temperature; and this was perceived by Black, for he pointed out that a distinction must be drawn between quantity and intensity of heat: the latter being what we now call temperature. He quotes Fahrenheit to show that while equal measures of water at different temperatures acquire a mean temperature when mixed, it requires three measures of quicksilver at a high temperature to convert two measures [volumes] of water at a low temperature to the mean of the two temperatures; and this corresponds to twenty times the weight of the water. Black expressed this by the statement that the capacity for heat of quicksilver is much less than that of water.

Black's statement indicates that his concept of heat is not only the same as our intuitive feel for heat but also that "Black's heat" is a state function ("two pounds must contain twice the quantity of heat that is contained in one pound").[32] It is noteworthy that Black used the term "heat" in the same sense as in "Temperature is a measure of how much heat is contained in an object." – see Section 6.1. Falk in 1985 wrote[33]:

> This explanation is unambiguous and clear. Black's heat is a quantity which is contained in a body and can flow into or out of it, in short, it is an extensive quantity of a substance-like nature.

Mach wrote[34]:

> When a cold body is put in contact with a warm body it is observed that the first body is warmed and that the second body is cooled. We may say that the first body is warmed at the expense of the second body. This suggests the notion of a thing, or heat-substance, which passes from the one body to the other.

In the same book, he writes (and think about a waterwheel when reading this)[35]:

> Water does not disappear when work is done. Why? Because we measure quantity of water with scales, just as we do electricity.[36] But suppose the capacity of water for work were called quantity, and had to be measured, therefore by a mill [waterwheel] instead of by scales; then this quantity also would disappear as it performed the work. …
>
> In the case of heat, now, the historically established measure of "quantity" is accidentally the work-value of the heat. Accordingly, its quantity disappears when work is done. But that heat is not a substance follows from this as little as does the opposite conclusion that it is a substance. …
>
> …. The fact is, heat behaves in some connexions like a substance, in others not.

In relatively elementary science, students are taught about wave-particle duality. This is from Wikipedia[37]:

> Wave–particle duality is the concept in quantum mechanics that every particle … may be partly described in terms not only of particles, but also of waves. It expresses the inability of the classical concepts "particle" or "wave" to fully describe the behaviour of quantum-scale objects.

At first, scientists were reluctant to accept the dual nature of light, as most of us like to have one answer. But Einstein paved the way in 1905 by embracing wave-particle duality, and today, it is an accepted concept by scientists. I can see no reason why we cannot have a dual "substance-energy" concept of heat if it assists in understanding what "heat" is.

6.3.2 Hotness, Heat and Energy

This section is based on pages 51–66 in The Dynamics of Heat by Fuchs.[38] Fuchs uses the word "heat" in the same sense as Black and to distinguish between this "heat" and "modern heat", I will italicise Fuch's/Black's *heat*. Let me start with a question – "How does science express the difference between an accelerating object and an object that is becoming hotter?" The following is what Fuchs wrote (my underlining)[39]:

> As scientists, we have learned that amounts of motion are measured in terms of momentum. Changing the motion of a body requires that we change its momentum. True, we also change the energy of the body at the same time, but knowing that the energy of a system can be different for many reasons, we take momentum as the quantity which tells us why the body is moving rather than becoming hot.
>
> Applying this argument to the hot body, we have to conclude that the [body] is hot not because it contains energy but because it has more heat than if it were colder. A stone gets hot and an ice cube melts because of heat. Even though energy accompanies all processes, heating included, it cannot distinguish between them. The point is this; heat is not energy.[40]

The *heat* content of an accelerating macroscopic object may well be constant (ignoring frictional effects), but the *heat* content of a warming/cooling object is changing. We intuitively know that a hot body contains more *heat* than a cold body or *vice versa*. Even though energy accompanies both changes, we must resort to non-energy concepts to distinguish between them (in this case, momentum and *heat*).

We have an intuitive sense of heat that allows us to place objects in a sequence that we label very cold, cold, cool, warm, hot, and very hot (or something similar). From the sensation of hotness, we abstract a primitive quantity[41] which we call hotness. We can build devices to tell how hot an object is. These devices are thermometers, with which we measure temperature, which can be considered as the characterisation of hotness by a number. There is an important feature of hotness: bodies cannot get colder than "very, very, very cold", i.e., there is a lower limit because we have never found bodies with a hotness below a certain level. To put it another way, hotness is always positive real number.

The process of heating changes hotness, and I will define the process of heating as the transport of *heat*. [42] Consequently, if heating is the transfer of *heat*, then (stating the obvious) some quantity of *heat* is transferred during heating. This suggests that *heat* is something tangible that we can measure as it flows from one location to another when the process of heating occurs, just like we can measure the water that flows from one location to another while it drives a waterwheel. Based on these analogies, one might even be tempted to think of *heat* as something that can be stored in bodies. In this chapter, we consider that *heat* is a quantity that is stored in a body and can (under the right conditions) flow from one body to another body. Consequently, we can say that bodies contain a certain amount of *heat*, and this quantity of *heat* can change because of the flow or transport of *heat*, i.e., by heating (positive flow of heat into a body or system) or cooling (negative flow of heat into a body or system). At this point, we have *heat* and hotness as two primitive concepts. Hotness is an intensive thermal quantity while *heat* is the extensive one. *Heat* flows provided there is a difference in hotness (temperature). This suggests comparisons with other quantities that have been introduced in other theories, most notably electricity and motion (Table 6.1).

Where does *heat* come from? We know *heat* comes from various sources and processes, e.g., the Sun, a fire, rubbing your hands together, and by electricity flowing though conductors. We also know that *heat* flows out of bodies in the process of cooling. There is a distinct difference between the latter process and the others. In cooling, an object loses the *heat* it receives by heating, i.e., cooling is the reverse of heating. In other cases, the body emitting *heat* does not have to change its hotness, e.g., the Sun is not getting cooler, an electric heater does not cool as it heats a room, and the *heat* produced by rubbing your hands together is not put into your hands previously. Where does the *heat* come from in these cases? We know in all of these cases that energy is input into the system by the process of work (from radioactive processes in the Sun; from movement of electrical charge; chemical reaction in a fire; and muscle movement when rubbing your hands together).

Table 6.1: Heating and Similar Processes

	Extensive quantity	Intensive quantity
Waterwheel	Gravitational mass of water (flow of water)	Water level
Electricity	Electrical charge (or current which is flow of electric charge)	Electric potential.
Heat Engine	*Heat* (flow of *heat*)	Hotness (temperature)

If *heat* can be created, does this mean it can be destroyed? Consider this simple example of irreversibility. A moving wooden block comes to rest on a horizontal surface. We know this is because of friction. We also know that friction creates *heat*. Logically, we can state that the creation of *heat* causes the block to come to rest. But this does not explain why the process is irreversible because the motion of the block can be reversed – we can push the body back the way it came, i.e., we have caused the block to accelerate. However, something cannot be changed – *heat* will also be created during the reverse motion. No one has ever seen a block absorbing *heat* from the surroundings, destroy it by converting it into mechanical energy and start to move by itself – this suggests that *heat* cannot be destroyed. This is what we mean when we say a process is irreversible. In other words, an irreversible process irreversibly changes the universe (i.e., the system and surroundings) by the production of indestructible *heat*. Consequently, we can define an irreversible process as one during which *heat* has been created. The corollary is that a reversible process is one in which there is no creation of *heat*. Because the term irreversible can be considered ambiguous, another term is used for processes which produce *heat* – they are called dissipative.[43] It logically follows that the directionality defined by irreversible processes must be matched by a "one-sided" physical quantity, i.e., *heat* that can be created but not destroyed. In other words, we need the concept of *heat* given in this section (together with the concept of energy) to explain the directionality of real thermodynamic processes.[44]

There is another property of *heat* we have not yet considered in this section. *Heat* can be used to do things such as driving engines, i.e., *heat* can do work. To quote Carnot[45]:

> Everyone knows that heat can produce motion. That it possesses vast motive power[46] no one can doubt, in these days when the steam engine is everywhere so well known. ... To heat are also due the vast movements which take place on the Earth. It causes agitations in the atmosphere, the ascension of clouds, the fall of rain and of meteors, the currents of water which channel the surface of the globe, and of which man has thus far employed but a small portion. Even earthquakes and volcanic eruptions are the result of heat.

We now know that *heat* drives a whole range of other processes including electrical and chemical ones.

Carnot suggested that we could compare the ability of *heat* to do work to that of the fall of water driving a waterwheel (see Section 6.2.2) – mentally convert "caloric" into "heat" when reading this extract (I will be considering "caloric" in the next section[47]:

> According to the established principles at the present time, we can compare with sufficient accuracy the motive power of heat to that of a fall of water. ... The motive power of a fall of water depends on its height and on the quantity of the liquid; the motive power of caloric depends also on the quantity of caloric used, and on what may be termed, on what in fact we will call, the height of its fall, that is to say, the difference of temperature of the bodies between which the exchange of caloric is made. In the fall of water the motive power is exactly proportional to the difference of level between the higher and lower reservoirs. In the fall of caloric the motive power undoubtedly increases with the difference in temperature between the warm and the cold bodies, ...

Just as water falls from a high level to drive a water wheel, after which it flows out of the engine at a lower level, *heat* is imagined to fall from a high level to a lower one, thereby driving the heat engine. The principle of operation of steam engines is in accordance with this image: steam takes up *heat* from a hot reservoir, it passes it through the engine where it causes motion in the surroundings, and then it flows out of the steam engine and dumps the *heat* to a cold reservoir. As Carnot stated[48]:

> The steam is here only a means of transporting caloric .... The production of motive power [work] is then due in steam-engines not to an actual consumption of caloric but to its transportation from a warm body to a cold body.

All in all, we form a mental picture of *heat* as a substance or fluid. We picture physical objects as depositories of *heat* and speak of the *heat* content of an object. This substance or fluid flows into and out of bodies. The process of the flow of *heat* into a body is called heating and the opposite is called cooling (or, negative heating). *Heat* is a substance-like quantity which simply means that *heat* can be pictured to be contained in bodies and to flow from one body to another.

By considering *heat* as a substance that flows, we add *heat* to the list made up of electric charge, gravitational mass, and momentum.

I will now introduce and consider caloric theory and will gradually drop the use of the word *heat* (in *italics*) and replace it with the word "caloric".

6.4 THE RISE AND FALL OF CALORIC THEORY

6.4.1 Introduction

By analogy with electricity[49] (all the rage in the 1700s) and based on Black's work, it was proposed by Lavoisier in the 1780s that *heat* was a fluid which he called *calorique*. At its most basic level, caloric represented the invisible cause behind the heating and cooling of a substance. Caloric was characterised as a weightless, and invisible fluid, composed of indestructible self-repelling microparticles. Caloric flows into bodies when they are heated, and out again when they are cooled. Although there are a number of variants of caloric theory, the basic properties of the caloric were considered to be:

1. Caloric is an indestructible fluid that flows into and out of physical bodies.
2. Caloric can be considered to be composed of microparticles of zero mass.
3. The microparticles of the caloric repel each other and, this property explains why caloric spread throughout the body, and why bodies expand with temperature.
4. When any object absorbs caloric without phase change, its temperature increases.
5. Provided there is no phase change, temperature is a measure of the concentration of caloric: the concentration of caloric increases with increasing temperature.
6. There is a net flow of caloric from hot bodies into cold bodies, but not from cold bodies to hot bodies. This was explained by mutual repulsion and larger caloric concentration in physical objects of higher temperature allowing the flow of caloric from a high temperature to a lower temperature.
7. When caloric is added to a solid, it flows between the microparticles of the solid and lessens their attraction to each other until the solid melts. As further caloric is added, the caloric particles continue to accumulate around the microparticles until they are pushed apart into a gas.

The caloric theory was used to deduce nearly all the ideal gas laws and explain many other phenomena. Its greatest success was to theoretically (and correctly) predict the speed of sound through a fluid.[50] If caloric or *heat* is an indestructible fluid, then it should (a) obey a conservation law and (b) be a state function. It is worth noting that Lavoisier in 1789 wrote[51]:

> We are not even obligated to suppose that caloric is a real substance; it is sufficient... that it be any kind of repulsive cause which separates the molecules of matter, allowing us to imagine its effects in an abstract and mathematical way.

Carnot developed his principle of the Carnot cycle from the caloric viewpoint[52]:

> The production of motive power is then due in [heat engines] not to any real consumption of caloric, but to its transport from a warm body to a cold body,

In Section 5.9, I demonstrated that for Carnot's idealised (i.e., reversible) heat engine:

$$\frac{q_H^{rev}}{T_H} = -\frac{q_C^{rev}}{T_C} \qquad \text{or} \qquad \frac{|q_H^{rev}|}{T_H} = \frac{|q_C^{rev}|}{T_C} \tag{6.1}$$

In a reversible process, caloric is transported from a warm body to a cold body but not consumed or created.[53] As a direct result of Carnot's thought experiment, we now have a mathematical definition of caloric (I will use the letter **Q** to represent a flow of caloric):

$$\mathbf{Q} = \frac{q_H^{rev}}{T_H} = -\frac{q_C^{rev}}{T_C} \qquad \text{or} \qquad |\mathbf{Q}| = \frac{|q_H^{rev}|}{T_H} = \frac{|q_C^{rev}|}{T_C} \tag{6.2}$$

Figure 6.8 shows a schematic diagram of Carnot's idealised heat engine showing the flow of caloric together with a diagram of a water wheel. Immediately it can be seen that caloric (**Q**) flowing from the hot reservoir to the cold reservoir is analogous to the weight (mg) of water flowing from the

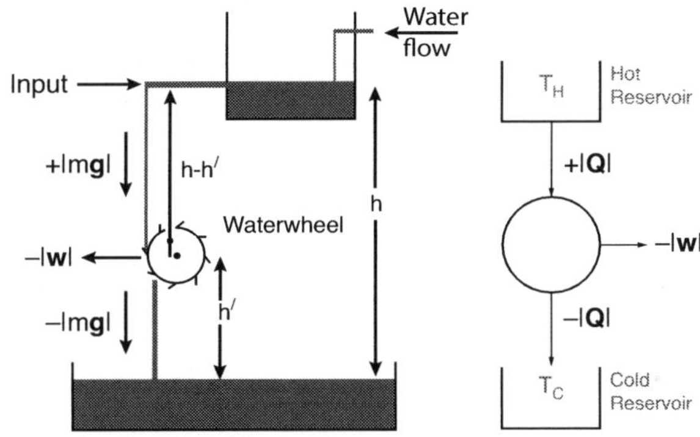

Figure 6.8 Carnot's idealised heat engine and a waterwheel.

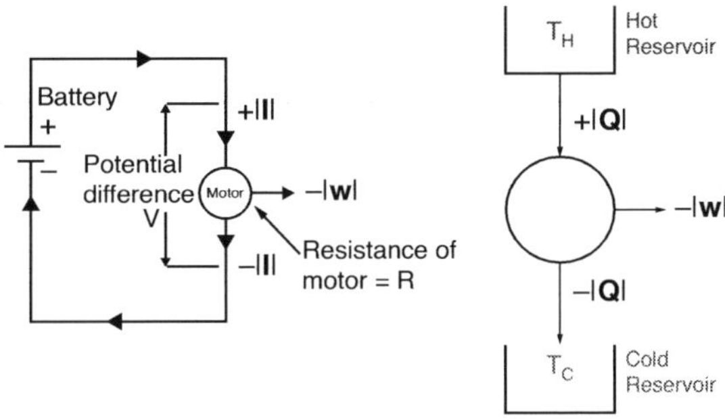

Figure 6.9 Carnot's idealised heat engine and an electric motor.

higher water reservoir to the lower water reservoir. In an idealised waterwheel, the maximum amount of mechanical work per unit volume of water flow $(|\mathbf{w}|)$ that may be extracted is determined just by the difference in "water levels" between the input and the output, i.e., the height difference between the input and the output $(|h - h'|)$. Therefore, the maximum work is:

$$|\mathbf{w}_{max}| = m|\mathbf{g}|(h - h') \tag{6.3}$$

In the idealised Carnot heat engine, the flow of energy to the surroundings by the process of work is determined only by the temperature difference between the two heat reservoirs for a given flow of energy supplied by the hot reservoir to the engine by the process of heat. In other words, no matter how well-designed the engine is, the maximum achievable efficiency does not depend on any details of the engine.

Figure 6.9 shows a simple electric circuit where the resistor is a motor which delivers work to the surroundings. I have labelled the current flow as positive or negative in line with our agreed sign convention (as I have done with '**mg**', '**w**' and '**Q**' in Figure 6.8). Again, we can immediately see the analogy between current and caloric. Therefore, caloric is analogous to weight of water in the waterwheel and current in the electrical circuit. It is worth noting that caloric, weight of water, and current do not have the units of energy (see Section 6.3.2).

6.4.2 The Fall of Caloric Theory

According to nearly all texts, the beginning of the end for caloric theory in mainstream science occurred in 1798 when Benjamin Thompson (a.k.a., Count Rumford) issued a report about the heating of water during cannon manufacture. Cannons were bored by turning an iron bit inside a brass cylinder, the energy was supplied by horses and the bored cannon was immersed in water. The friction of the iron bit on the brass raised the temperature of the brass cylinder/water system. In caloric theory, this increase in temperature was accounted for by the squeezing out of caloric fluid during the boring operation. To quote Rumford[54]:

> At 2 hours and 20 minutes it was 200°; and at 2 hours 30 minutes it ACTUALLY BOILED! It would be difficult to describe the surprise and astonishment expressed in the countenances of the by-standers, on seeing so large a quantity of cold water heated, and actually made to boil without any fire.

Rumford quantitatively analysed the experiment, and he concluded:

> ...we must not forget to consider that most remarkable circumstance, that the source of the Heat generated by Friction, in these Experiments, appeared evidently to be inexhaustible. It is hardly necessary to add, that anything which any insulated body, or system of bodies, can continue to furnish without limitation, cannot possibly be a material substance: ...

A more reasoned view indicates that this is not the case as the following extract demonstrates[55]:

> How could it be so when reputable eighteenth-century writers of the stature of Boerhaave and Lambert, to say nothing of Erasmus Darwin, were able to explain the enormous heating effect associated with friction while maintaining their belief in the materiality of fire or heat without adopting Rumford's view that heat consisted in vibrations of the particles of ordinary ponderable[56] matter?

Rumford, if he was a reasonably proficient scientist/engineer would not have been surprised that the water "ACTUALLY BOILED!". The importance of Rumford's "experiment" was and still is erroneously reported and exaggerated in thermodynamic texts. Compare the following statement – "If caloric was a fluid, it should weigh something, so if it were released, mass should be lost during the experiment. However, the total mass of metal after the experiment equalled the mass of matter before."[57] with my statement in Section 6.4.1 that "Caloric was characterised as a weightless, and invisible fluid, ..." (my subsequent underlining). Also note this comment from Wikipedia[58]:

> Most established scientists, such as William Henry, as well as Thomas Thomson, believed that there was enough uncertainty in the caloric theory to allow its adaptation to account for the new results. It had certainly proved robust and adaptable up to that time. Furthermore, Thomson, Jöns Jakob Berzelius, and Antoine César Becquerel observed that electricity could be indefinitely generated by friction. No educated scientist of the time was willing to hold that electricity was not a fluid.

Rumford's experiments inspired English physicist James Prescott Joule (1818–1889), and it was his work (not Rumford's) in the 1840s that is considered to have demolished the caloric theory. In his famous experiment, he allowed falling weights to turn a paddle wheel that was submerged in a bucket of water (Figure 6.10). The falling weights did work on (i.e., turned) the paddle wheel, which in turn raised the temperature of the water. Joule measured both the amount of work done

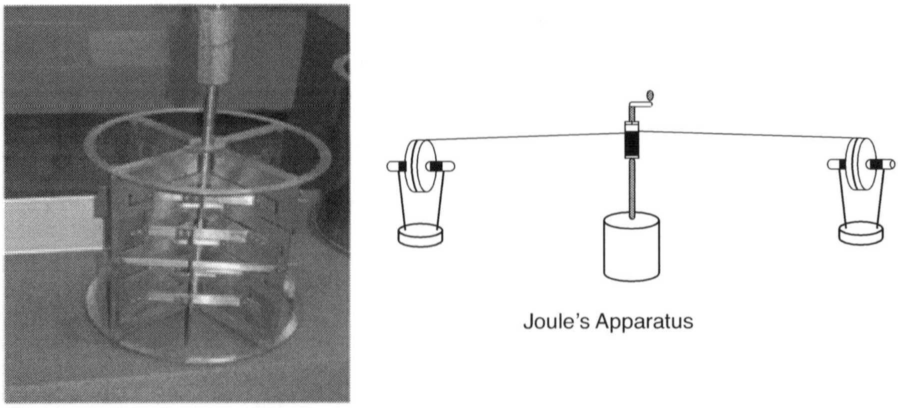

Figure 6.10 The paddlewheel (photograph; based on the one in his original paper) is sealed inside the can in the middle of the diagram. Two 13 kg lead weights were wound up and dropped 1.6 m, this was repeated 20 times and produced a temperature change of about ¼°C. Joule claimed to be able to measure temperatures to an accuracy of around 1/400°C.

on the system (the bucket of water and paddle arrangement) and, by measuring temperature rise, the amount of energy the system gained by the process of work.

Joule did not describe this experiment in detail. What he did report was that he dropped the weights 11m, rewound them and dropped them a further 15 times and he recorded a miniscule temperature increase (around 0.3°C) over this period (in a what was essentially bucket): some believe that technology in 1845 was not advanced enough for this to be possible.[59] Joule also claimed to have measured the difference in water temperature between the top and bottom of a waterfall and accounted for the difference as due to the change in mechanical potential energy of the water falling the height of the waterfall (he was on his honeymoon at the time!). Water falling 50 m under the influence of Earth's gravity is warmed by 0.1°C. Is it reasonable that Joule in the middle of the 19th century could have measured such small temperature differences accurately enough?

Thermodynamic texts also tend to quote the "ice rubbing experiment" by Humphry Davy. The story goes that Davy used a clockwork mechanism to rub two cubes of ice together and thereby melted them. However, there is doubt that he carried out such an experiment. The following abstract is interesting[60]:

> It is commonly stated that [Humphry] Davy melted two pieces of ice by rubbing them together under [a vacuum][61] ..., and thus showed conclusively that heat is not a material substance [i.e., caloric]. In books which I happen to have at hand I find twelve different authors stating that Davy melted two pieces of ice by rubbing them together in a vacuum, and four of them stating in addition that the two pieces of ice were rubbed together by clockwork. In looking to see what Davy himself said about this experiment I have, to my surprise, failed to find any evidence that he ever performed ... this experiment.
>
> Any clockwork which Davy might have placed inside of the receiver would probably not have been sufficiently powerful to melt ice rapidly by rubbing it on ice. I have wondered if some author did not read [published works of Davy] ..., and seeing the words clockwork, [vacuum], ice conclude that two blocks of ice were rubbed together by clockwork under the exhausted[62] receiver. If so, this is an interesting illustration of the ease with which a misstatement may pass from one author to another. If there is evidence that Davy did melt two blocks of ice by causing clockwork to rub them together under the receiver of an air pump I hope someone will [cite] it.

These are two examples of the experiments that were used to disprove the caloric theory of heat.

Knowing the energy transferred to the water by mechanical work and knowing the temperature change, Joule was able to calculate what became known as the "mechanical equivalent of heat". In other words, heat and work are both energy transfer processes: this is the basic tenet of the kinetic or mechanical theory of heat. The mechanical equivalent of heat is now an outdated concept (although I was taught it at school and university in the 1960s and 1970s) because it is just a conversion factor from the units of work (joules) to the then units of heat (calories). Joule's value for the mechanical equivalent of heat of 4.1550 $J.cal^{-1}$ only differs from the modern value of 4.1860 $J.cal^{-1}$ by 3 parts in one thousand.

The work of Joule, Mayer (1814–1878), Helmholtz (1821–1894), Lord Kelvin (1824–1907), and others (known as the "energeticists)" [63] led to the "mechanical" or modern theory of heat (namely heat, like work is an energy transfer process). This work was also considered to have led to the formulation of the First Law of Thermodynamics, i.e., heat, like work, is an energy transfer process, and energy is conserved. The evidence against caloric theory appears damning, but it is built, as we shall see, on the erroneous assumption that *heat*/caloric is energy which is conserved. The energeticists chose the word "heat" for the energy transfer process via a temperature difference. Unfortunately, as we shall see, they had hijacked a word that gave us an intuitive insight into directionality of processes. We can no longer say that a body contains *heat* and that some of this *heat* can flow to another body if they are in thermal contact and the other body is at a lower temperature. Unfortunately, as we shall see, it gets even worse.

Consider the following statement from an 1845 paper by Joule[64] (my comments/amendments are in square brackets and see below about the asterisks, hash, and underlinings):

> The principles I have adopted lead to a theory of the steam-engine very different from the one generally received, but at the same time much more accordant with facts. It is the opinion of many philosophers that the mechanical power of the steam-engine arises simply from the passage of heat* [= caloric – he is commenting on Carnot's thought experiment] from a hot to

> a cold body, no heat* [caloric] being necessarily lost during the transfer. This view has been adopted by Mr. E. Clapeyron.... .... I conceive that this theory, however ingenious, is opposed to the recognised principles of philosophy [science], because it leads to the conclusion that [energy][65] may be destroyed by an improper disposition of the apparatus: thus Mr. Clapeyron draws the inference, that [when] 'the temperature of the fire [hot reservoir] being from 1,000°(C) to 2,000°(C) higher than that of the boiler, there is an enormous loss of [energy] in the passage of the heat* from the furnace into the boiler.' [What Clapeyron is referring to is the caloric generated as energy is transferred across a temperature difference by the process of heat and the reduction in efficiency of the engine – see point '3' below].[66] Believing that the power to destroy belongs to the Creator alone I entirely coincide with Roget and Faraday in the opinion that any theory which, when carried out, demands the annihilation of force [I assume he means energy?], is necessarily erroneous. The principles, however, which I have advanced in this paper are free from this difficulty. From them we may infer that the steam while expanding in the cylinder loses <u>heat</u> in quantity exactly proportional to the mechanical force which it communicates by means of the piston; and that on condensation of the steam, the heat# thus converted into power is not given back. Supposing no loss of <u>heat</u> by radiation etc., the theory here advanced demands that the <u>heat</u> given out in the condenser shall be less than that communicated to the boiler from the furnace, in exact proportion to the equivalent of mechanical power developed.

I find Joule's statement fascinating for a number of reasons:

1. He mixes up the two types of "heat" – his first three mentions of "heat" (I have marked them with an *) would be better translated as "caloric". His four final mentions of the term "heat" are meant to be in the modern sense as a process. Even at the start of the modern use of heat, you can see the difficulties. Joule writes "loses heat", "loss of heat", and "heat given out" where heat is best translated as energy transferred by the process of heat (I have underlined the word heat in these cases). In only one case, which I have marked with a "#", the word heat could be considered a process.

2. The final two sentences from Joule's paper are an early verbal description of the First Law of Thermodynamics. "... that the steam while expanding in the cylinder loses heat in quantity exactly proportional to the mechanical force which it communicates by means of the piston; and that on the condensation of the steam, the heat thus converted into power is not given back. Supposing no loss of heat by radiation etc., the theory here advanced demands that the heat given out in the condenser shall be less than that communicated to the boiler from the furnace, in exact proportion to the equivalent of mechanical power developed."

3. The following statement is revealing – "that [when] "the temperature of the fire [hot reservoir] being from 1,000°(C) to 2,000°(C) higher than that of the boiler, there is an enormous loss of [energy] in the passage of the heat from the furnace into the boiler." Unfortunately, the energeticists jumped to the wrong conclusion. If Joule (and the other energeticists) had recognised that [Black's] heat/[Carnot's] caloric is not energy, then he would have no need to invoke the Creator. This is because energy is not being destroyed in this process, and as we shall see in the next section it is caloric that is being generated. However, in fairness to Joule et al., the early versions of caloric theory did state that caloric (like energy) was conserved (which as we now know is only the case with reversible processes).

However, there is an alternative view because we know that for reversible processes caloric is conserved. Just because one of the tenets of caloric theory was wrong, namely that it does not obey a conservation law for irreversible processes, it does not mean the whole of the caloric theory of heat should be trashed (more on this in Section 6.4.4).

6.4.3 Caloric is a State Function for Reversible Processes

I have defined caloric (**Q**) as $\mathbf{Q} = \frac{q^{rev}}{T}$ (see Equation 6.2). From Carnot's thought experiment (Figure 6.11), we know that the change in the caloric content of the system after one cycle $\left(\Delta S^{one\ cycle}\right)$ is[67]:

$$\Delta S^{one\ cycle} = \frac{q_H^{rev}}{T_H} + \frac{q_C^{rev}}{T_C} = 0 \tag{6.4}$$

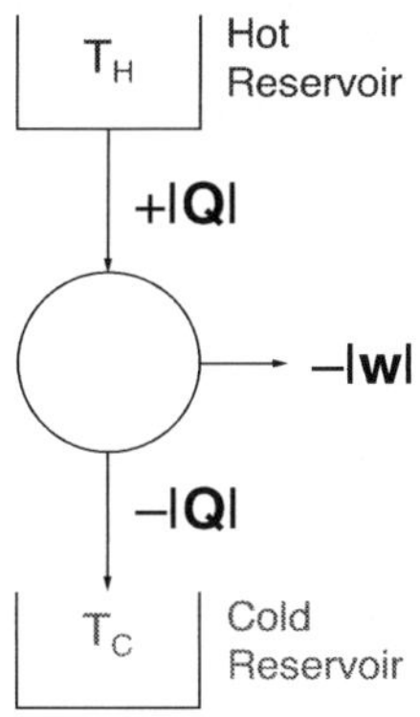

Figure 6.11 Reversible heat engine showing the flow of caloric.

On examination of Figure 6.11, it can be immediately seen that the change in the caloric content of the system (the heat engine) is zero after a single cycle, i.e., $\Delta Q = 0$. In other words, in a reversible heat engine, the quantity of caloric, $|Q|$, supplied at the temperature, T_H, is the same as the quantity rejected at the temperature T_C.

Mathematically $\Delta S^{\text{one cycle}}$ is known as a cyclic integral and is written $\oint \frac{dq}{T} = 0$ (see Section 5.11, Equation 5.79 and Figure 5.16c). In Section 4.6, I stated that one mathematical criterion of a state function is that the cyclic integral of that state function is zero. The corollary of this result is that caloric is not a state function for irreversible processes (considered in the next section). Remember[68]

> A state function is defined as a property of a system (or object) whose value does not depend on the path used to reach a specific state. Another way of defining a state function is that it is a property of a system that depends only on the current state of the system, not on the way in which the system acquired that state.

Consequently, if heat is measured as caloric, then it is correct to state that the production of work in a heat engine is not due to consumption of caloric, but to the transport of caloric from a higher to a lower temperature.

The absolute unit of caloric is that quantity which is capable of doing unit work per degree fall in a Carnot cycle. This unit does not have a name, but there have been proposals to call it the "carnot". Unfortunately, unlike the use of a calorimeter to measure energy (in joules), it is difficult to directly measure caloric in carnots ($J.K^{-1}$).[69] In this book, I will only use the caloric concept as a teaching aid and analogy.

6.4.4 Discussion

Callendar states:

> The original conception of caloric as a subtle imponderable[70] fluid, capable of flowing from one body to another, of expanding bodies, and of producing rise of temperature or change of state, was, in the main, merely a picturesque analogy, invented, like the electric or magnetic fluids, with the object of giving the mind a tangible something to reason about, a measurable entity remaining constant in various transformations or exchanges. The fluid terminology remains a convenient method of expression even if the thing measured has not all the properties of a material fluid.[71]

It is worth remembering that when Carnot was working, there were little hard experimental data on the properties of gases, and what data were available, they were often wrong.

The "energeticists" thought they had demolished the caloric theory by showing that if heat is a fluid and is energy, then it cannot be generated by processes such as friction. We have already seen that the flow of energy (**q**) transferred by the process of heat is not a state function for irreversible processes. In other words, heat, as defined by energeticists, had become a path function or process variable.

> The consequences of this misfortune were catastrophic for thermodynamics. Not only, that the important quantity Black's heat had disappeared from science, but even worse, its name had been stolen, and it was forbidden to say that a system contains Black's heat, just as you cannot state a system contains work.[72]

The old concept of caloric/*heat* as a state function (in reversible processes) had disappeared from science, and as Callendar observed, "caloric possesses the important property, essential to the natural measure of heat quantity, of remaining constant in reversible exchanges (which the energy measure does not)".[73]

After the energeticists had rubbished the caloric concept and hijacked the word heat, "it was inevitable that caloric should make its reappearance sooner or later in the mechanical theory [of heat]."[74] This is because the concept of caloric explains the directionality of processes. As all students of thermodynamics know, this was achieved by the German physicist and mathematician Rudolf Clausius (1822–1888). Callendar states[75]

"Finally, in 1865 ... Clausius ... gave it the name of 'entropy', and defined it as the integral of $\left[\frac{\mathbf{dq}}{\mathrm{T}}\right]$".

By "it" in this quote, Callendar means **Q** or caloric.

No one at that time appears to have realised that entropy was merely caloric under another name. Possibly because Clausius's approach was based on the energy definition of heat rather than Carnot's then discredited caloric approach. Callendar then writes[76]:

> The conception of caloric defined by Carnot's equation would be more readily assimilated, and might be introduced at a much earlier stage. The direct investigation of the properties of caloric thus defined would probably lead to a more general and intimate appreciation of Carnot's principle itself and of many modern developments of thermodynamics which are practically beyond the comprehension of the majority of students under the present system.

I have already mentioned the advantages of analogy, and I now want to compare the equation for maximum work obtainable from the reversible Carnot cycle when using caloric to that obtained from a waterwheel. We know for a reversible heat engine (Section 5.7.3, Equation 5.27) that:

$$|\mathbf{q}_\mathrm{H}| - |\mathbf{w}| - |\mathbf{q}_\mathrm{C}| = 0 \tag{6.5}$$

We know (see Equation 6.2):

$$\mathbf{Q} = \frac{|\mathbf{q}_\mathrm{H}|}{\mathrm{T}_\mathrm{H}} = \frac{|\mathbf{q}_\mathrm{C}|}{\mathrm{T}_\mathrm{C}} \tag{6.6}$$

Therefore:

$$|\mathbf{q}_\mathrm{H}| = |\mathbf{Q}|\,\mathrm{T}_\mathrm{H} \qquad \text{and} \qquad |\mathbf{q}_\mathrm{C}| = |\mathbf{Q}|\,\mathrm{T}_\mathrm{C} \tag{6.7}$$

Substituting for $|\mathbf{q}_\mathrm{H}|$ and $|\mathbf{q}_\mathrm{C}|$ in Equation 6.5 and rearranging, we have:

$$|\mathbf{w}| = |\mathbf{Q}|\mathrm{T}_\mathrm{H} - |\mathbf{Q}|\mathrm{T}_\mathrm{C} = |\mathbf{Q}|(\mathrm{T}_\mathrm{H} - \mathrm{T}_\mathrm{C}) \tag{6.8}$$

Comparing Equation (6.8) with Equation (6.3) is instructive (see Equation 6.9) and reinforces the analogy discussed in Section 6.5.1 where the flow of caloric in a heat engine is compared to the flow of water in a waterwheel.

$$|\mathbf{w}| = |\mathbf{Q}|\mathrm{T}_\mathrm{H} - |\mathbf{Q}|\mathrm{T}_\mathrm{C} \qquad |\mathbf{w}| = \mathrm{mg}(\mathrm{h} - \mathrm{h}') \tag{6.9}$$

Callendar writes[77]:

> Carnot's analogy of the water-wheel is thus exactly justified. Caloric, like water, is not itself [energy], but is capable of performing work under suitable conditions in virtue of its tendency to flow down a temperature gradient. The quantity of [energy transferred to the surroundings by the process of work] is directly proportional to the quantity of caloric and to the fall of temperature. ... Caloric may be said to possess motive power in virtue of its elevation of temperature [i.e., potential energy], just as water possesses motive power in virtue of its available head or pressure. In the case of caloric there is a definite limit set by the absolute zero of temperature to the quantity of motive power obtainable by a cyclical process from a quantity of caloric supplied at any temperature.

With all analogies, there are always limitations, but provided one is aware of the areas where heat, electricity, and water behave very differently, analogies can offer major benefits. Table 6.2 compares the heat engine, waterwheel, and the electrical circuit.

Table 6.2: The Use of Analogy in the Understanding of the Caloric Theory of Heat

	Carnot's Heat Engine	Waterwheel	Electrical Circuit
Flow	Caloric per second	Weight per second	Current (I) = Charge per second
Quantity	Caloric $\left(Q = \frac{q}{T}\right)$	Weight (mg)	Charge (q)[78]
Energy	Q	mgh	qV
Driving Force	Temperature difference $(\Delta \mathbf{T})$. A temperature difference is a driving force for a caloric flow.	Height difference $(\Delta \mathbf{h})$. A gravitational potential difference is a driving force for the water flow.	Potential difference $(\Delta \mathbf{V})$. An electric potential difference is a driving force for an electric current.
Directionality	Caloric flows from a body of higher temperature to a body of lower temperature.	Water flows from a high to low gravitational potential	Electric charge flows from a higher to a lower electric potential
Equilibrium	If the temperature difference between two bodies is zero, there is no caloric flow: known as thermal equilibrium	If the gravitational potential difference between two bodies is zero, there is no water flow: known as mechanical equilibrium.	If the electric potential difference between two bodies is zero, there is no electric current: known as electrical equilibrium.

Carnot ultimately abandoned the caloric theory in favour of the mechanical theory of heat as evidenced by the following:

> Heat is simply motive power, or rather motion which has changed form. It is a movement among the particles of bodies. Wherever there is destruction of motive power there is, at the same time, production of heat in quantity exactly proportional to the quantity of motive power destroyed. Reciprocally, wherever there is destruction of heat, there is production of motive power.
>
> We can then establish the general proposition that motive power is, in quantity, invariable in nature; that it is, correctly speaking, never either produced or destroyed. It is true that it changes form, that is, it produces sometimes one sort of motion, sometimes another, but it is never annihilated.[79]

The first paragraph in the above quote is a statement of the First Law of Thermodynamics by Carnot in 1824, and the following statement by Joule in 1845 (see Section 6.4.2) is also a statement of the First Law:

> ... that the steam while expanding in the cylinder loses heat in quantity exactly proportional to the mechanical force which it communicates by means of the piston; and that on condensation of the steam, the heat thus converted into power is not given back. Supposing no loss of heat by radiation etc., the theory here advanced demands that the heat given out in the condenser shall be less than that communicated to the boiler from the furnace, in exact proportion to the equivalent of mechanical power developed.

However, Wikipedia states:

> The first explicit statement of the first law of thermodynamics, by Rudolf Clausius in 1850, referred to cyclic thermodynamic processes.[80]

As we have just seen, this is not the case – the first statement of the First Law of Thermodynamics was by Carnot in 1824, 20 years before Joule and 25 years before Clausius. Even today Carnot is still being side-lined. We have seen that the concept energy (a state function) is not sufficient in understanding how the world works. I want to spend the rest of this chapter demonstrating that caloric is a concept that allows the directionality of irreversible processes to be predicted. This is in spite of the following abstract (again from Wikipedia)[81]:

> The caloric theory is an obsolete scientific theory that heat consists of a self-repellent fluid called caloric that flows from hotter bodies to colder bodies. Caloric was also thought of as a weightless gas that could pass in and out of pores in solids and liquids. The 'caloric theory' was superseded by the mid-19th century in favor of the mechanical theory of heat,

but nevertheless persisted in some scientific literature – particularly in more popular treatments—until the end of the 19th century.

The original theory of caloric considered that it was a conserved physical quantity. This is the case for reversible processes: when the total quantity of caloric remains constant, i.e., there is no generation (or loss) of caloric in reversible processes. However, caloric is only conserved when all energy transfers are reversible. If the process is irreversible (as are all "real" processes), the quantity of caloric always increases. Callendar states[82]:

> The total quantity of caloric in an isolated system remains constant only if all the transformations are reversible, in which case the motive power developed exactly suffices to restore the initial state. In all other cases there is an increase of caloric. The old principle of the universal conservation of caloric, which is true only for reversible processes, must therefore be modified …

This increase of caloric in real processes is the principal reason that it is not taught in modern science. It was considered absurd to suppose that anything which could be generated without limit could possibly be regarded even by analogy as a fluid. In addition, this property cannot apply to caloric if caloric (or *heat*) is considered to be equivalent to heat in the mechanical theory, i.e., energy. However, caloric is not energy, and if the following modifications are made to caloric theory, then (to requote Callendar) "it was not really quite so illogical as it is generally represented to be." These modifications are:

- Caloric can be created without limit and once generated, it can never be destroyed. This means that:
  - The total quantity of caloric in an open system can only be diminished by the removal of mass or energy.
  - The total quantity of caloric in a closed system can only be diminished by energy removal.
  - The total quantity of caloric in an isolated system can never be diminished.

In other words, as Callendar states:

> It is true that it may be difficult to isolate a particular set of material particles and label them caloric. It is conceivable that caloric may not be material at all, but merely, as Boltzmann puts it, 'the logarithm of the probability of a complexion'.[83] But if this is really the case, it is all the more necessary for our sanity and progress to think and speak of it as a material fluid. Although this conception of caloric may appear at first sight to run counter to some of our most cherished popular illusions with regard to heat, there could be no serious objection to adopting it as a convenient method of expression. The more shadowy the conception to be visualised, the greater the need of a definite material analogy. From this point of view the old picturesque phraseology of the material fluid, implied in Carnot's waterfall, so far from being a valid objection, is one of the chief advantages of the caloric theory.[84]
>
> … since caloric possesses the important property, essential to the natural measure of heat quantity, of remaining constant in reversible exchanges (which the energy measure of heat does not) all our equations and methods of reasoning with regard to questions of equilibrium are greatly simplified and rendered more direct if we adopt caloric as the true measure of heat quantity, and regard it as possessing energy in virtue of its temperature. With all our preconceived notions of heat as a 'form of energy,' it is difficult now to retrace our steps and express everything in terms of caloric. But, if Carnot's solution had not been overlooked 60 years ago, it would have seemed equally absurd to regard a quantity of heat merely as a quantity of energy, and we should probably have been far ahead of our present position.[85]

6.5 CALORIC AND IRREVERSIBILITY

6.5.1 Introduction

We know there is no caloric generation in reversible processes. In this section, I consider caloric generation in irreversible or dissipative processes. I will start by considering the flow of energy, by the process of heat, down a thermal gradient (known as thermal conduction) and the consequent generation of caloric. Caloric generation by frictional effects is discussed in Section 6.5.3. In Section 6.5.4, I will consider the relationship between caloric generation and "lost work".

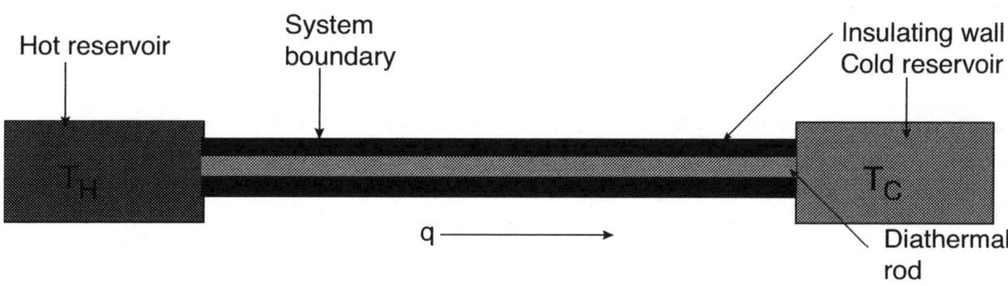

Figure 6.12 Thermal conduction through a finite temperature difference (from hot to cold): where the insulated diathermal rod is the system and thermal reservoirs are the surroundings.

6.5.2 Thermal Conduction[86]

Consider a closed system consisting of a perfectly insulated diathermal (e.g., a copper) rod that is in thermal contact with two heat reservoirs at T_H and T_C (Figure 6.12).[87] At steady state, the net flow of energy along the diathermal rod, by the process of heat, is **q** per unit time and the energy content of the system is, by definition, constant. Note that **q** is an energy flow and is a vector, i.e., it has magnitude and direction. Using our agreed sign convention (from the viewpoint of the system), we can say that in unit time $\mathbf{Q_H} = \frac{\mathbf{q}}{T_H}$ of caloric flows into the system from the hot reservoir and $\mathbf{Q_C} = \frac{\mathbf{q}}{T_C}$ of caloric flows into the system from the cold reservoir. Using absolute values, we can write that $+|\mathbf{Q_H}| = +\left|\frac{\mathbf{q}}{T_H}\right|$ of caloric flows into the system (the diathermal rod) from the hot reservoir and $-|\mathbf{Q_C}| = -\left|\frac{\mathbf{q}}{T_C}\right|$ of caloric flows into the system from the cold reservoir).[88] We can write these equations for caloric flow because the temperatures of the heat reservoirs remain constant. At steady state, not only is the energy content of the closed system unchanged $(\Delta\mathbf{q} = 0)$ so is its caloric content, i.e., $\Delta\mathbf{Q}^{\text{system}} = 0$.

We also know:

i. the energy content of the closed system at steady state is unchanged;

ii. the energy content of both heat reservoirs remains constant because the loss in energy of the hot reservoir is compensated by the gain in energy of the cold reservoir;

iii. the energy content of the universe is unchanged;

iv. caloric cannot be generated in the heat reservoirs because all heat processes in heat reservoirs are carried out at constant temperature

This appears to give rise to a paradox – how can the caloric content of the universe increase when the caloric content of the closed system remains unchanged and no irreversible heat processes can take place in the surroundings (the heat reservoirs)? The solution to the paradox is straightforward, but I have rarely seen it mentioned in thermodynamic texts.

The caloric generation occurs within the diathermal rod, where energy transfer by the process of heat is occurring by way of a temperature gradient.[89] This generated caloric is transferred to the combination of the two reservoirs simultaneously with the flow of energy, by the process of heat, from the hot and cold heat reservoirs.[90] During the process of heat, the hot reservoir loses energy, and the cold reservoir gains an equal amount of energy. In contrast, in caloric transfer, the cold reservoir gains more caloric than the hot reservoir loses. So, even though processes occurring in the heat reservoirs are reversible, the reservoirs receive the generated caloric from the irreversible process taking place in the system, i.e., the diathermal rod.

The more understandable way to proceed from here is by using absolute value equations. The hot reservoir gains $\left(-\left|\frac{\mathbf{q}}{T_H}\right|\right)$ of caloric, the cold reservoir gains $\left(+\left|\frac{\mathbf{q}}{T_C}\right|\right)$ of caloric, and the surroundings (the hot and cold reservoirs) gain $\left(\left|\frac{q}{T_C}\right| - \left|\frac{q}{T_H}\right|\right)$ of caloric. The system's energy and caloric contents remain unchanged as it is at a steady state. Since the energy transfers in the reservoirs took place at constant temperature, they experience a reversible change, and there is no caloric generation within the reservoirs, and we can write:

$$\left|\Delta Q^{universe}\right| = \left|\Delta Q^{system}\right| + \left|\Delta Q^{surroundings}\right| = 0 + \left\{\left(+\left|\frac{q}{T_C}\right|\right) + \left(-\left|\frac{q}{T_H}\right|\right)\right\} = \frac{q(T_H - T_C)}{T_H T_C} \tag{6.10}$$

We know $T_H > T_C$, and we can see the caloric change per unit time of the universe (an isolated system) is positive $\left(+\left|\Delta Q^{universe}\right|\right)$. Therefore, even though the energy content of the universe has remained constant $\left(\left|q_{in}\right| = \left|q_{out}\right|\right)$, the flow of energy, by the process of heat, down the $T_H T_C$ thermal gradient (i.e., thermal conduction) generates caloric.

Consider the following extract from Fuchs (I have substituted "caloric" when Fuchs uses "entropy")[91]:

> [Caloric] is produced when it flows through bodies from hotter to cooler places. In other words, [caloric] is produced in the conductive transport of heat. We are quite familiar with other processes that create [caloric] such as friction ..., but we do not have direct sensory knowledge about what happens in conduction.
>
> Consider a long bar of heat-conducting material. ... Let us wait long enough for steady-state conditions to be established. In other words, we must wait long enough until the currents of [caloric] and energy through then bar become steady. Since energy is conserved, we must conclude that the flux[92] of energy entering the hotter end is equal in magnitude to the flux leaving at [the] lower temperature. This immediately tells us that [caloric] must be produced in the bar: there is a larger current of entropy accompanying the flux of energy at lower temperature. ... This proves our claim: in steady-state conduction, more [caloric] leaves the bar than enters at the hot end. [Caloric] is produced in the body.
>
> If you feel uncomfortable about this result, consider the following arguments. Isn't the flow of [caloric] the prototype of an irreversible phenomenon? Just look at the process that takes place when two bodies of unequal temperatures are brought in thermal contact. ... This is irreversibility in its purest form; [caloric] must be produced as a consequence of the process. Also, you know that we can operate a heat engine between the reservoirs at the hot and cool ends of the bar. We have said that energy can be released for a non-thermal purpose[93] in the fall of [caloric]. If [caloric] flows conductively from one point, it also drops from higher to lower temperatures. The energy released in the fall of [caloric] has been used for the only possible process which is a thermal one: [caloric] is produced continuously. Put in different terms, we might say that we have lost available [work],[94] ... which means that [caloric] must have been created. No matter how we look at the conductive flow of [caloric] it is hard to escape the following conclusion:
>
> Conduction of [caloric] produces [caloric]; i.e., it is a dissipative process. When [caloric] is allowed to fall from points of higher temperature to points of lower temperature without causing motion or some other process, [caloric] is necessarily produced.

Callendar stated[95]:

> Direct transference of [energy by the process of] heat between bodies at different temperatures would be equivalent to wasting a difference of temperature which might have been utilised for the production of motive power.[96]

If the process depicted in Figure 6.12 is reversed, then change in the caloric content of the universe is given by Equation (6.11), and we can immediately see that the caloric change of the universe is negative (because $T_C > T_H$). In other words, the flow of energy, by the process of heat, from a lower to higher temperature is non-spontaneous or forbidden.[97]

$$-\left|\Delta Q^{universe}\right| = -\left|\Delta Q^{closed\,system}\right| = \left\{\left(+\left|\frac{q}{T_C}\right|\right) + \left(-\left|\frac{q}{T_H}\right|\right)\right\} = = \frac{q(T_C - T_H)}{T_H T_C} \tag{6.11}$$

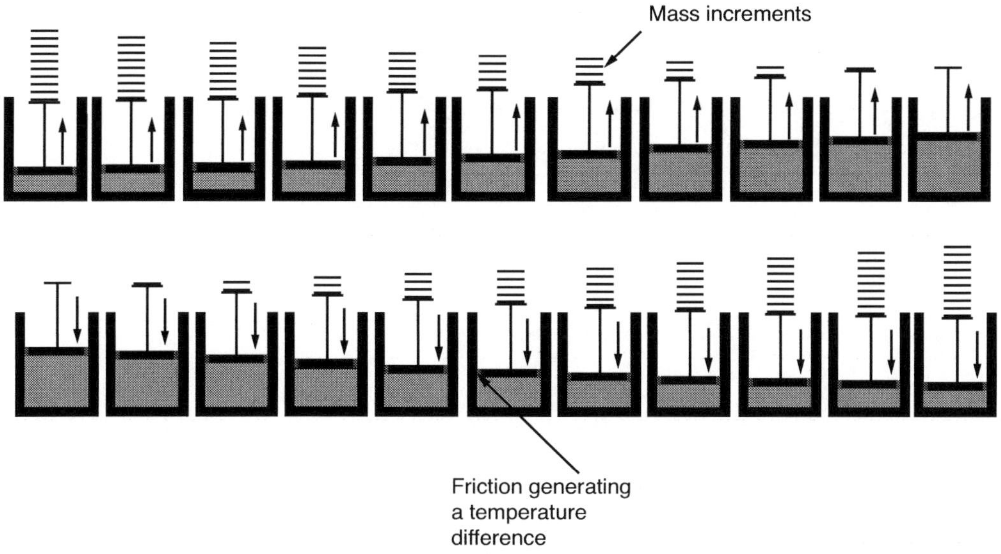

Figure 6.13 Reversible expansion/compression process when friction is present.

6.5.3 Friction and Caloric Generation

Our ubiquitous frictionless piston system is now replaced by a piston system with friction – a "friction piston" as opposed to a "frictionless piston". Consider Figure 6.13 which is based on Figure 5.7 in Chapter 5. When an infinitesimally small mass increment is removed from the piston, it will not move because of the frictional forces affecting the piston. But let us now remove sufficient mass increments to allow the piston to move up a bit, but it does not move to the same height as the frictionless piston when the same number of mass increments have been removed from the piston, i.e., to reach the same position as the frictionless piston more mass increments would need to be removed from the friction piston. Similarly, when the removed mass increments are replaced back on the piston, the piston will not reach its original position.[98] In order to reach its original position, more mass increments would need to be added compared to the frictionless piston system. In other words, with the "friction piston", the magnitude of energy transferred to or from the work reservoir, by the process of work, is less than for the "frictionless piston" when the same force vector (negative and positive) is applied to the piston. What has happened to this "lost" energy? As we shall see, it is all down to frictional effects producing temperature differences.

When the friction piston moves against the cylinder wall, the surface atoms and molecules of the cylinder and piston interact with each other via electromagnetic forces. This results in localised increases in the kinetic energy of various microparticles of the piston materials, this manifests itself as a local increase in temperature and a temperature difference is generated.[99] We know (i) the generation of a temperature difference requires energy (see Section 5.3), and (ii) this energy must be supplied by the process of work (as the piston moves against the opposing force of friction). Once the temperature difference has been created energy flows, by the process of heat, to the lowest temperature available to the system, i.e., the cold reservoir at T_C. Therefore, the "lost" energy has been transferred the lowest accessible temperature in the surroundings (the cold reservoir) by the process of heat. As we now know, this effect must result in an increase in the caloric content of the universe.

Temperature differences vary not only with each individual irreversible engine but also at different points in an individual engine. This results in an extremely complex system that either cannot be analysed or would require the use of the complex mathematics of fluid dynamics and parallel computing power.

6.5.4 Caloric Generation and Lost Work

Consider the reversible heat engine in Figure 6.14a we know $-|\mathbf{w}|+|\mathbf{q_H}|-|\mathbf{q_C}|=0$ and caloric generation is zero. Therefore, we can write:

$$-|\mathbf{w}| = |\mathbf{q_C}| - |\mathbf{q_H}| = |\mathbf{Q}|T_C - |\mathbf{Q}|T_H \tag{6.12}$$

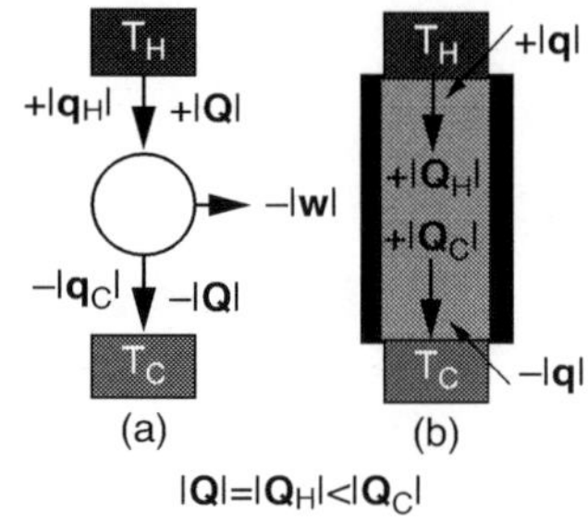

Figure 6.14 The idealised Carnot heat engine and thermal conduction.

where $|\mathbf{Q}| = \frac{|\mathbf{q}_H|}{T_H} = \frac{|\mathbf{q}_C|}{T_C}$. Remember, **w** is the work input of the heat engine and is the energy transferred from the work reservoir (the surroundings) to the heat engine by the process of work (in this case a negative quantity). We also know the work output $(\mathbf{w}^{output})$ of the heat engine (the energy transferred to the surroundings by the process of work) is the negative value of the work input (w), i.e., $\mathbf{w}^{output}$ is a positive quantity.[100] Therefore, we can rewrite Equation (6.12) as:

$$|\mathbf{w}^{output}| = |\mathbf{w}| = |\mathbf{q}_H| - |\mathbf{q}_C| = |\mathbf{Q}|T_H - |\mathbf{Q}|T_C \tag{6.13}$$

We know for the reversible heat engine $\frac{|\mathbf{q}_H|}{T_H} = \frac{|\mathbf{q}_C|}{T_C}$; rearrangement gives $|\mathbf{q}_C| = \frac{T_C|\mathbf{q}_H|}{T_H}$. Therefore, we can rewrite Equation (6.13) as:

$$|\mathbf{w}^{output}| = |\mathbf{q}_H| - \frac{T_C|\mathbf{q}_H|}{T_H} = \frac{|\mathbf{q}_H|(T_H - T_C)}{T_H} \tag{6.14}$$

If the reversible heat engine is replaced by a diathermal conductor (Figure 6.14b) such that at steady state, the energy flow through the conductor is **q** per unit time then work output is zero and caloric generation is (see Equation 6.10):

$$|\Delta\mathbf{Q}^{universe}| = \frac{|\mathbf{q}|}{T_C} - \frac{|\mathbf{q}|}{T_H} = \frac{|\mathbf{q}|(T_H - T_C)}{T_H T_C} \tag{6.15}$$

Note that in the reversible heat engine $\Delta\mathbf{Q}^{universe} = 0$ and $|\mathbf{w}| \neq 0$; and in thermal conduction **q** is constant and $|\mathbf{w}| = 0$. We know that the "lost work" $(|\mathbf{w}^{lost}|)$[101] when $\mathbf{q}_H$ of energy flows directly from the hot to the cold reservoir must be equal to $|\mathbf{w}^{output}| = |\mathbf{w}^{lost}|$. Knowing that, by definition, $|\mathbf{q}| = |\mathbf{q}_H|$ (see Figure6.14), we can write:

$$|\mathbf{w}^{lost}| = \frac{|\mathbf{q}|(T_H - T_C)}{T_H} \tag{6.16}$$

Dividing both sides of Equation (6.16) by T_C, we have:

$$\frac{|\mathbf{w}^{lost}|}{T_C} = \frac{\mathbf{q}(T_H - T_C)}{T_H T_C} \tag{6.17}$$

Comparing Equations (6.15) and (6.17), we can see

$$\frac{|\mathbf{w}^{lost}|}{T_C} = \frac{\mathbf{q}(T_H - T_C)}{T_H T_C} = |\Delta\mathbf{Q}^{universe}| \tag{6.18}$$

From Equation (6.18) we can see that $\frac{|\mathbf{w}^{lost}|}{T_C}$ is equal to the caloric change of the universe in the irreversible process of thermal conduction where $\mathbf{w}^{lost}$ is the energy that could be obtained, by the process of work, from caloric flowing down the temperature difference $(T_H - T_C)$ in a reversible (Carnot) heat engine.

6.5.5 Caloric in Irreversible Processes

I now want to demonstrate that that $\mathbf{Q_{irrev}}$ unlike $\mathbf{Q_{rev}}$ is not a state function. The proof uses the statement, proved in Section 5.7, that an irreversible heat engine is always less efficient than a reversible heat engine. The efficiency of a reversible heat engine is:

$$\varepsilon = \frac{\mathbf{q_{H_{rev}}} + \mathbf{q_{C_{rev}}}}{\mathbf{q_{H_{rev}}}} = 1 + \frac{\mathbf{q_{C_{rev}}}}{\mathbf{q_{H_{rev}}}} = 1 - \frac{T_C}{T_H} \tag{6.19}$$

Because an irreversible heat engine is always less efficient than a reversible heat engine:

$$1 - \frac{T_C}{T_H} > 1 + \frac{\mathbf{q_{C_{irrev}}}}{\mathbf{q_{H_{irrev}}}} \tag{6.20}$$

Subtracting 1 from both sides, multiplying both sides and rearranging yields:

$$\mathbf{Q_H^{irrev}} + \mathbf{Q_C^{irrev}} = \frac{\mathbf{q_{H_{irrev}}}}{T_H} + \frac{\mathbf{q_{C_{irrev}}}}{T_C} < 0 \tag{6.21}$$

Equation (6.21) tells us that for an irreversible process caloric change is not a state function. The reason is straightforward; in a cyclic process, the change in any state function must be zero, and for irreversible processes $\Delta\mathbf{Q_{surroundings}} = \left(+\left|\mathbf{Q_H^{irrev}}\right|\right) + \left(-\left|\mathbf{Q_C^{irrev}}\right|\right) \neq 0$ (while for a reversible process $\Delta\mathbf{Q_{surroundings}} = \left(+\left|\mathbf{Q_H^{rev}}\right|\right) + \left(-\left|\mathbf{Q_C^{rev}}\right|\right) = 0$.

To appreciate what Equation (6.21) is (and is not telling us), we need to go back to basics. We know:

1. After one complete cycle, the caloric content of the system is unchanged $\left(\left|\mathbf{Q_{system}}\right| = 0\right)$.
2. If the flow of energy, by the process of heat, from the hot reservoir into the closed system is $+\left|\mathbf{q_{H_{irrev}}}\right|$, then the flow of energy into the hot reservoir is $-\left|\mathbf{q_{H_{irrev}}}\right|$, i.e., the flow of caloric into the hot reservoir is $-\left|\mathbf{Q_{hot\ reservoir}}\right| = \dfrac{-\left|\mathbf{q_{H_{irrev}}}\right|}{T_C}$.
3. If the flow of energy, by the process of heat, from the cold reservoir into the closed system is $-\left|\mathbf{q_{C_{irrev}}}\right|$, then the flow of energy into the cold reservoir is $+\left|\mathbf{q_{C_{irrev}}}\right|$, i.e., the flow of caloric into the hot reservoir is $+\left|\mathbf{Q_{cold\ reservoir}}\right| = \dfrac{+\left|\mathbf{q_{C_{irrev}}}\right|}{T_C}$.

Let us now consider the caloric change in the universe $(\Delta\mathbf{Q_{universe}})$ during a single cycle of the irreversible heat engine. We know:

$$\Delta\mathbf{Q_{universe}} = \Delta\mathbf{Q_{heat\,engine}} + \Delta\mathbf{Q_{surroundings}} \tag{6.22}$$

We know $\Delta\mathbf{Q_{heat\,engine}} = 0$ and $\Delta\mathbf{Q_{surroundings}} = \mathbf{Q_{hot\,reservoir}} + \mathbf{Q_{cold\,reservoir}}$. Therefore, taking note of points 2 and 3 above, we can write for an irreversible heat engine:

$$\Delta\mathbf{Q_{universe}} = \frac{-\left|\mathbf{q_{H_{irrev}}^{hot\ reservoir}}\right|}{T_H} + \frac{+\left|\mathbf{q_{C_{irrev}}^{cold\ reservoir}}\right|}{T_C} \tag{6.23}$$

We now need to determine whether $\Delta\mathbf{Q_{universe}}$ has increased (i.e., $+\left|\Delta\mathbf{Q_{universe}}\right|$) or decreased (i.e., $-\left|\Delta\mathbf{Q_{universe}}\right|$), i.e., whether $\left(\dfrac{-\left|\mathbf{q_{H_{irrev}}^{hot\ reservoir}}\right|}{T_H} + \dfrac{+\left|\mathbf{q_{C_{irrev}}^{cold\ reservoir}}\right|}{T_C}\right)$ is positive or negative. Equation (6.21) tells us that $\dfrac{\mathbf{q_{H_{irrev}}^{system}}}{T_H} + \dfrac{\mathbf{q_{C_{irrev}}^{system}}}{T_C} < 0$ and we know that $\dfrac{\mathbf{q_{H_{irrev}}^{system}}}{T_H} = -\dfrac{\mathbf{q_{H_{irrev}}^{hot\ resevoir}}}{T_H}$ and $\dfrac{\mathbf{q_{C_{irrev}}^{system}}}{T_C} = -\dfrac{\mathbf{q_{C_{irrev}}^{hot\ resevoir}}}{T_C}$.

Therefore, we know that $\dfrac{\mathbf{q_{H_{irrev}}^{hot\ reservoir}}}{T_H} + \dfrac{\mathbf{q_{C_{irrev}}^{cold\ reservoir}}}{T_C} > 0$. Using absolute values, we can write, for a heat engine, $\dfrac{-\left|\mathbf{q_{H_{irrev}}^{hot\ reservoir}}\right|}{T_H} + \dfrac{+\left|\mathbf{q_{C_{irrev}}^{cold\ reservoir}}\right|}{T_C} > 0$. In other words, after one cycle of the irreversible heat engine, the caloric content of the universe has increased.

Although the caloric content of each of the two reservoirs changes, the caloric generation must occur in the system. However, if the system returns to its original state after one complete cycle, its caloric content does not change. Caloric must be generated in the system either by thermal conduction or by frictional effects (dissipative effects) and is transferred to the two reservoirs concurrently with the flow of heat into or out of the system.

Here, we have another difficulty with understanding thermodynamics. The caloric increase must be generated by the heat engine (i.e., the system), but at the end of a single cycle, the caloric content of the system is unchanged. What is happening is that caloric $\frac{q_H}{T_H}$ flows, along with energy, by the process of heat, from the hot reservoir into the system. Some of this energy is transferred by the process of work, to the surroundings (the work reservoir), there is no caloric flow accompanying this energy flow. The remainder of the energy (q_C) that entered the system from the hot reservoir, at temperature T_H, flows into the cold reservoir, at temperature T_C, concurrent with this energy flow is a caloric flow of $\frac{q_C}{T_C}$. From the above analysis, we know $\frac{|q_C|}{T_C} > \frac{|q_H|}{T_H}$. Although all heat processes occurring in heat reservoirs are reversible (no caloric generation), this does not mean that the caloric content of the heat reservoirs is constant because caloric and energy can flow from the system into the heat reservoirs.

One way of expressing the fundamental difference between reversible and irreversible processes (apart from the fact that one of them does not exist in the real world) is that a reversible process does not change the caloric content of the universe, while an irreversible process increases the caloric content of the universe. As we know that (i) once formed caloric cannot be destroyed, and (ii) reversible processes are always in equilibrium, we can say an irreversible process permanently changes the universe, and a reversible process does not permanently change the universe.

The above argument is essentially the same as that given in Section 6.5.2 when considering the irreversible process of thermal conduction. I will be revisiting these issues again in Chapter 7.

6.5.6 Spontaneous Processes

I have deliberately minimised the use of the word "spontaneous" so far in this book. It is a word that is often used in thermodynamics and chemistry (such as "spontaneous process" or "spontaneous reaction") and, like heat, I would suggest that most people have an intuitive idea of what is meant by the phrase, but we need to be more rigorous. Merriam-Webster Dictionary has six definitions of "spontaneous" only one of which is relevant here – "developing or occurring without apparent external influence, force, cause, or treatment". If you Google "spontaneous process", you either get circular definitions using thermodynamic terms which we have not yet covered[102] or something like the following:

- A spontaneous process is one that occurs naturally under certain conditions.
- A spontaneous process is a process that can actually occur in a finite time period under the existing conditions.
- Any change over time in the state of a system that we observe experimentally is a spontaneous process.
- A spontaneous process is capable of proceeding in a given direction without needing to be driven by an outside source of energy.
- A spontaneous process occurs without the need for a continual input of energy.

The first three definitions are, essentially, meaningless. The second two definitions are less so, but it does not help when you come across quotes such as: "A spontaneous change may be so rapid that it is essentially instantaneous or so slow that it cannot be observed over any practical period of time."[103] How can you tell whether a process is spontaneous or not? If you Google "examples of spontaneous processes", you get something like the following:

- The cooling of an object at a higher temperature than its surroundings.
- The warming of an object at a lower temperature than its surroundings.
- Water flowing downhill.
- Sugar dissolving in a cup of hot coffee.
- A gas diffusing throughout any volume available to it.
- Batteries losing their charge.
- Spilt water evaporating.

- Chemical reactions, such as hydrogen and oxygen gas reacting to form water, or methane spontaneously react with oxygen. However, mixtures of oxygen and hydrogen or oxygen and methane can remain in an unreacted state for extended periods of time under suitable conditions.

A spontaneous process cannot be reversible because, by definition, a reversible system is in dynamic equilibrium, and the process can proceed in either direction. This means that the caloric content of a system in dynamic equilibrium remains constant. Therefore, we can state that all spontaneous processes are irreversible, but that does not help as all real processes are irreversible. However, with the concept of caloric, we have a measure of directionality. This allows a meaningful definition of a spontaneous process as one in which the caloric content of the universe increases, and of a non-spontaneous or forbidden process as one in which the caloric content of the universe decreases. We can also add definitions of reversible process or a process in equilibrium as those in which the caloric content of the universe remains unchanged.

6.6 DISCUSSION, CONCLUSIONS AND SUMMARY

Just as water falls from a high level to drive a turbine, after which it flows out of the engine at a lower level, caloric is imagined falling from a higher temperature to a lower temperature, thereby driving the heat engine. Caloric flows into the working substance of the heat engine from the hot reservoir. It passes through the engine where it transfers energy to the surroundings, by the process of work. The caloric then flows out of the heat engine into the cold reservoir. In other words, the heat engine's working substance is only a means of transporting the caloric. The flow of water down a gravitational gradient, the flow of electric charge down a voltage gradient, and the flow of caloric down a temperature gradient can all be used to do work. If we take caloric flow to be something akin to water flow or to the flow of electric charge, then we have an intuitive feel of what caloric is. Does this make water flow a form of energy? Does it make the flow of electric charge some form of energy? Does it make the flow of caloric some form of energy? In all cases, the answer is no. Callendar wrote in 1911:

> We observe that, if heat is measured as caloric, it is quite correct to state that 'the production of work from heat is due, not to an actual consumption of caloric, but to the transport of caloric from a higher to a lower temperature.' Carnot's analogy of the water-wheel is thus exactly justified. Caloric, like water, is not itself motive power [i.e., energy], but is capable of performing work under suitable conditions in virtue of its tendency to flow down a temperature gradient. The quantity of [work][104] produced is directly proportional to the quantity of caloric and to the fall of temperature.

The final sentence indicates that the temperature at which energy is available is a measure of its quality. Consider a given quantity of energy, if its temperature is say 1,000 K, it will produce more work by the process of heat, then the same amount of energy at 500 K. This same amount of energy at say 300 K is essentially useless for producing useful work in everyday situations on the Earth. [105]

There are only two things that can be done with caloric. It can be moved, and/or it can be created. In a reversible cyclic process, caloric is moved but not created, and in an irreversible cyclic process, caloric is moved and created. The hypothetical Carnot engine moves caloric from a high-temperature heat reservoir to a low-temperature heat reservoir but does not generate caloric. All real engines are irreversible, i.e., they both move and create caloric. Although caloric can be moved or created, there is no known means of destroying it. The quantity of caloric in the universe continually increases and never decreases. As caloric increases, there is less and less energy in the universe available to do work (even though the energy content of the universe remains unchanged). At some point, the temperature of the universe will become uniform, i.e., thermodynamic equilibrium will be attained, and nothing will happen. This is known as the "heat death of the universe".[106] To put this another way, there will be no available energy to operate a heat engine (because there is no temperature difference).[107] Although the Law of Conservation of Energy tells us that the energy content of the universe remains constant, what has changed is the ability of some of this energy to be converted to work by the process of heat (to do work), i.e., the quality of the available energy has changed.

As we have seen, Joule and the energeticists erroneously considered caloric as energy and wanted to demonstrate the equivalence of mechanical energy *and heat*. Joule did this (or at least convinced sufficient numbers of the scientific community) with his famous paddle-wheel experiments, whose conclusions he summarised as:

> The quantity of heat produced by the friction of bodies, whether solid or liquid is always proportional to the quantity of force expended. The quantity of heat capable of increasing the temperature of a pound of water by 1°F requires for its evolution expenditure of a mechanical force represented by the fall of 772 lbs through the space of one foot.

(Notice here the term "force" has evidently meaning of mechanical energy.) Joule considered heat as if it were a mechanical entity, fully equivalent or identical with mechanical energy. His way of thinking was probably due either to influence of Rumford or to the reasoning that in the paddle experiment heat appears just when mechanical work disappears and *ipso facto* these two entities must be identical. Joule rather pompously invokes the "Creator" to support his conclusions (see Section 6.4.2) and the following[108]:

> … being satisfied that the grand agents of nature are, by the Creators's fiat, indestructible; and that wherever mechanical force [presumably he means energy transferred by the process of work] is expended, and exact equivalent of heat is always obtained.

The following abstract is about the two theories of heat[109]:

> Depending on who you ask you will get widely differing answers to the question – what is heat? "For children and non-scientists, heat is something akin to a 'thermal fluid' residing in bodies, and capable of flowing from one body to another. If you had asked physicists and engineers before 1840, they would have given you very similar answers, using for heat the word caloric. Generally their answers would have been colorfully extended by images of 'molecules of heat' and other exotic things. Also, heat was supposed to be a conserved fluid. Since the 1850s it has become clear that this view must be wrong. Heat is a mode of energy transfer (an 'interaction'), and cannot be thought of as being stored in systems. Students have to give up the images they have formed in everyday life. Today, mechanical engineers are the clearest proponents of this view.
>
> Next, ask chemists. For them, heat still resides in bodies. It is something like internal energy or, quite often, enthalpy.[110] Never mind that in this case we do not have a word for thermal interactions, which actually leaves us impotent when it comes to formulating thermodynamics. Finally, ask modern physicists. They are teaching the mechanical interpretation of thermal phenomena. Little particles and their motions relieve us of the task of being clear about macroscopic thermodynamics. Heat may well reside in bodies, while at the same time being a form of energy transfer.

Unfortunately, there were only a few objections to this conclusion of the energeticists, the most prominent was Mach (see Section 6.3.1), who considered the differences between the processes of heat and work.[111] He observed that when energy flows, by the process of heat, it is only partially transformed into "useful" energy (i.e., energy that can be transferred to the surroundings by the process of work). In contrast, it is possible to wholly transfer all the "useful" energy to the cold reservoir (e.g., the surroundings such as the Earth's atmosphere or oceans) by the process of heat. We experience this in our everyday lives with space heating and frictional forces, e.g., a swinging pendulum or a moving billiard ball eventually coming to rest due and their kinetic energy being dissipated to the surroundings by the process of heat. Therefore, there is a trend in the universe where useful energy is transferred into useless energy (heat death of the universe – see above).

Mach considered the differences between energy transferred by the process of heat and that transferred by the process of work. "If a weight sinks, it must create perforce[112] kinetic energy, or heat, or some other form of energy. Also, an electrical charge cannot suffer a fall of potential without loss of energy, …." This is not the case with heat – he writes "Heat can suffer a fall in potential without sustaining a loss of energy". Mach states that it is this property "that gives to heat, besides its property of energy, in many cases the character of a material substance, or quantity." He concluded that

> If we look at the matter in an unprejudiced light, we must ask if there is any scientific sense or purpose in still considering as energy a quantity of heat that can no longer be transformed into mechanical work, (for example, the heat of [an isolated][113] equably warmed material system).[114]

Callendar stated[115]:

> The caloric theory is seen to be perfectly consistent with Carnot's principle and with the mechanical theory for all reversible processes. Caloric is the natural measure of a quantity of

heat in accordance with Carnot's equation, if we adopt the gas-scale of temperature. The only defect in the caloric theory lay in the tacit assumption, so easily rectified, that the ordinary calorimetric units were units of caloric.[116] The quantity measured in an ordinary calorimetric experiment is the motive power or energy of the caloric, and not the caloric itself. If this had been realised in 1850, it would have been quite unnecessary to recast and revolutionise the entire theory of heat. Evolution might have proceeded along safer lines, with the retention of the caloric, and the investigation of its properties, which are such fundamental importance in all questions of equilibrium in physics.

Mach's and Callendar's arguments were ignored and the highly abstract quantity called entropy, invented by Clausius in place of caloric/*heat*, is used in modern science with the result of near-universal confusion. As we will see in Chapter 7, entropy and caloric are not conserved in real (i.e., irreversible) processes, but they are conserved when all changes are reversible. However, in all irreversible processes, the total quantity of entropy or caloric in the universe always increases.

In Chapter 7, I will consider the Second Law of Thermodynamics and entropy, and as we shall see Chapter 7, the final sentence of the previous paragraph is equivalent to the Second Law of Thermodynamics. As Callendar stated (my underlining)[117]:

> Caloric is not conserved in the old sense that its total quantity remains constant in any system of exchanges under all conditions. It remains constant only in the restricted sense, provided that all exchanges are reversible or take place under conditions of equilibrium. But since in any irreversible exchanges the total quantity of caloric is always increased, the principle of the conservation of caloric can still be stated in the modified form that 'the total quantity of caloric in a system cannot be diminished by any internal reactions or by the performance of work by or on the system, provided that no heat is allowed to escape.' Applied in this form, with various modifications to suit special conditions, it has proved to be one of the most fruitful principles of modern thermodynamics; but it might have been reached sooner, and more generally appreciated, if the caloric theory had not been so hastily abandoned.[118]

Finally, consider the following:

> Suppose that we take some firewood and use it to "heat" a sealed room.[119] Because of energy conservation, the energy in the room plus the firewood is the same before and after the firewood has been converted to ash. Which form of the energy is more capable of doing work? You probably realize that the firewood is a more useful form of energy than the 'hot air' and ash that exists after the firewood is burned. Originally the energy was stored in the form of chemical (potential) energy. Afterward the energy is mostly associated with the motion of the molecules in the air. What has changed is not the total energy, but its ability to do work.[120]

Consider now the following question – what do a moving object and a warmer object have in common?[121] We know both have received energy. In each case, the system contains more energy than it would have if the object were not moving or if it were not as warm. We could say the moving body has received an amount of motion and the warmer body an amount of caloric. Scientists would say the momentum of the moving body has changed. It is true the energy of the moving body has changed at the same time. However, we know that the energy of a moving body can be different for many reasons. Therefore, we take momentum as the quantity which tells us why the body is moving faster rather than becoming hot. Applying this argument to the hot body, we have to conclude that the stone is hot not because it contains energy, but because it has more caloric than if it were colder. Even though energy accompanies all processes, heating included, it cannot distinguish between them. Caloric is a quantity that we can imagine being stored in bodies and capable of flowing from body to body.

Carnot developed his theory of heat engines using the caloric theory of heat. However, as discussed in Section 6.4.2, the energeticists considered they had disproved the caloric theory of heat because they erroneously considered Black's heat/caloric to have the units of energy which (unlike caloric[122]) cannot be created. In the mid-19th century, using the mechanical theory of heat and complex (often opaque) mathematics, Clausius gave the world "entropy". Admittedly, Carnot did abandon the caloric theory just before he died but as Callendar has argued this was because of erroneous data[123]:

> If Carnot had not been misled by this fundamental error [in the variation of the specific heat of a gas[124]] ..., and if his numerical data had been a little more accurate, he could not have

failed to notice that the quantity of caloric in a kilocalorie [or joule] varied inversely as the absolute temperature, or that its equivalent motive power [work] was the same at all temperatures. In other words, that the kilocalorie [now kilojoule] was a unit of motive power, and not a unit of caloric.

Clausius' invention of entropy is considered in Chapter 7.

NOTES

1 G. Georg Rasch, *Probabilistic Models for Some Intelligence and Attainment Tests*, Danmarks Paedagogiske Institut, pp. 37–38; republished in 1980 by University of Chicago Press, 1960.

2 D.R. Cox, "Comment on 'Model uncertainty, data mining and statistical inference'", *J. R. Stat. Soc. Ser. A* 158, 455–456, 1995.

3 G.E.P. Box, N.R. Draper, *Empirical Model-Building and Response Surfaces*, John Wiley & Sons, p. 424, 1987.

4 M.W. Zemansky, "The use and misuse of the word "Heat" in physics teaching", *Phys. Teach.*, 295–296, 1970. http://documents.mx/documents/the-use-and-misuse-of-the-word-heat-in-physics-teaching.html.

5 https://en.wikipedia.org/wiki/Heat. I will not rigorously define the term "spontaneous" until Section 6 but most people have an intuitive understanding of what spontaneous means.

6 https://www.britannica.com/science/heat.

7 https://www.physicsclassroom.com/class/thermalP/Lesson-1/What-is-Heat.

8 https://www.sciencelearn.org.nz/resources/750-heat-energy.

9 https://www.encyclopedia.com/science-and-technology/physics/physics/heat.

10 https://www.bbc.co.uk/bitesize/guides/z2gjtv4/revision/5.

11 M.W. Zemansky, "The use and misuse of the word "Heat" in physics teaching", *Phys. Teach.* 295–296, 1970. http://documents.mx/documents/the-use-and-misuse-of-the-word-heat-in-physics-teaching.html.

12 I presume Zemansky means by 'heat' is 'energy transferred by the process of heat'.

13 Zemansky is alluding to latent heat here – which is not covered in this book. You can transfer energy to a mixture of ice and water by the process of heat and its temperature can remain the same at $0^{o}C$ until all the solid ice is converted to liquid water.

14 https://dictionary.cambridge.org/dictionary/english/temperature.

15 https://www.chem.wisc.edu/deptfiles/genchem/netorial/modules/thermodynamics/enthalpy/enthalpy2.htm.

16 H.L. Callendar, "The caloric theory of heat and carnot's principle", *Proc. Phys.* XXIII, 153–154, 1911.

17 A.G. Harrison, D.F. Treagust, Teaching and learning with analogies. In: P.J. Aubusson et al. (eds.), *Metaphor and Analogy in Science Education*, 2006, pp. 11–24. © Springer. Printed in the Netherlands. https://pdfs.semanticscholar.org/8518/367cc1ef090ea263c39e912a4acaf59dd3d5.pdf.

18 Mesons are sub-atomic particles.

19 R. Feynman, *Six Easy Pieces*, Basic Books, 1994.

20 Brownian motion is the random motion of particles suspended in a a liquid or a gas resulting from their collision with the fast-moving atoms or molecules on the gas or liquid.

21 Pushball is a team sport that is played with a very large ball (diameter about six feet) in which the objective is to push the ball through the opponent's goalposts.

22 You could argue that you could increase h – h′ and therefore the amount of work obtained per fixed mass of water by digging a hole (i.e., by making h′ a negative number), but I did say the ground was our zero-reference level. I have also ignored the impracticality of a water-wheel operating half-submerged when h′ = 0.

23 The reason for this statement will become apparent later.

24 https://en.wikipedia.org/wiki/Electric_charge.

25 One of the leading figures of early American history, Benjamin Franklin (1706–1790) was a statesman, author, publisher, scientist, inventor and diplomat. Before the discovery of the electron, scientific thinking in the 17th/18th centuries had described electricity as a flow of charge. Benjamin Franklin made an assumption that charge flowed from electrostatically charged smooth wax to electrostatically charged rough wool in the mid 1700s, his convention became established to the extent that it was too difficult to change by the time actual electron flow was discovered well over a century later.

26 See, for example, https://en.wikipedia.org/wiki/Hydraulic_analogy for further information.

27 In electrical theory the usual abbreviation for charge is, confusingly, q. I will not use this abbreviation apart from in Table 6.2.

28 Flow or flux are equivalent to current in this context.

29 See Section 1.2.2 for definition of extensive and intensive quantities.

30 "Lectures on the Elements of Chemistry Delivered in the University of Edinburgh by the Late Joseph Black, M.D.", John Robison LLD, page 78, Volume 1, 1803.

31 W. Ramsay, *The Life and Letters of Joseph Black, MD*, Constable and Company Ltd, p. 43, 1918.

32 Because in a cyclic process the initial *heat* content must be the same as the final *heat* content (if *heat* is consider substance-like) – see Section 4.6.

33 G. Falk, "Entropy, a resurrection of caloric – a look at the history of thermodynamics", *Eur. J. Phys.* 6, 108–115, 1985.

34 E. Mach, *Popular Scientific Lectures*, Translated by TJ McCormack, 3rd Edition, The Open Court Publishing Company, p. 166, 1898.

35 E. Mach, *Popular Scientific Lectures*, Translated by TJ McCormack, 3rd Edition, The Open Court Publishing Company, p. 170, 1898.

36 The physical quantity which measures an amount of electricity is electrical charge (units coulombs). Coulomb used a torsion balance to measure electric charge – see https://nationalmaglab.org/education/magnet-academy/history-of-electricity-magnetism/museum/torsion-balance-1785.

37 https://en.wikipedia.org/wiki/Wave–particle_duality.

38 H.U. Fuchs, *The Dynamics of Heat*, Springer-Verlag New York Ltd, 1996. Originally published in German – unfortunately the English translation is not a very good one.

39 H.U. Fuchs, *The Dynamics of Heat*, Springer-Verlag New York Ltd, p. 52, 1996. Originally published in German – unfortunately the English translation is not a very good one.

40 If I was writing this my final sentence would be – "The point is this; heat and momentum are not energy".

41 A primitive quantity is one which we do not derive in terms of other more fundamental quantities.

42 All physics and chemistry texts consider the three processes of heating, i.e., conduction, convection, and radiation. In this book we will only be concerned with conduction, where energy is transported by the process of heat between bodies that are in direct thermal contact with each other.

43 See Sections 5.5 and 5.7.1.

44 It may be worthwhile rereading Chapter Three Section 3 on turbulence and generation of localised temperature differences at this point.

45 S. Carnot, *Reflections of the Motive Power of Fire*", Translated by R.H. Thurston, Dover Publications, p. 3, 1960.

46 Remember "motive power" is mechanical work or (being pedantic) the transfer of energy to the surroundings by the process of work.

47 S. Carnot, *Reflections of the Motive Power of Fire*, Translated by R.H. Thurston, Dover Publications, p. 15, 1960.

48 S. Carnot, *Reflections of the Motive Power of Fire*, Translated by R.H. Thurston, Dover Publications, p. 7, 1960.

49 The now obsolete fluid theory of electricity considered electricity as an invisible fluid. Electric current is the continuous flow of electric charge and in fluid theory electric charge was some sort of fluid. If an object had an excess of a normal share of this fluid, it was said to have a positive charge; and if another object had a deficiency of a normal share of this fluid, its charge was considered negative. Based on this theory, it was concluded that electric fluid flowed from positive (excess) to negative (deficiency), i.e. electric current was a flow of fluid. This version of the theory was proposed in 1750 by Benjamin Franklin. The fluid theory was easy to visualise and agreed with all experiments conducted in the 18th and 19th centuries and was universally accepted. The analogy of current as a flowing fluid is still used as a learning aid (see Section 6.2.3). However, the fluid theory cannot explain the formation of a magnetic fields by electric currents.

50 See https://www3.nd.edu/~powers/ame.20231/finn1964.pdf.

51 Quoted in - https://en.wikipedia.org/wiki/Antoine_Lavoisier.

52 S. Carnot, *Reflections of the Motive Power of Fire*, Translated by R.H. Thurston, Dover Publications, p. 46, 1960.

53 Remember – "Consequently, we can define an irreversible process as one during which heat has been created. The corollary is that a reversible process is one in which there is no creation of heat." – Section 6.3.2.

54 From "Lectures on Heat and Thermodynamics" by Michael Fowler of the University of Virginia (30 August 2008): http://galileo.phys.virginia.edu/classes/152.mf1i.spring02/HeatLectures.pdf.

55 R. Fox, *The Caloric Theory of Gases From Lavoisier to Regnault*, Oxford University Press, p. 4, 1971.

56 Ponderable - what can be discerned by the senses.

57 https://en.wikipedia.org/wiki/An_Inquiry_Concerning_the_Source_of_the_Heat_Which_Is_Excited_by_Friction

58 An Experimental Enquiry Concerning the Source of the Heat which is Excited by Friction – Wikipedia.

59 It would appear that the energy increase of the water and apparatus associated with the minute temperature rise would have dispersed into the environment in minutes yet the experiment took at least one hour to perform.

60 https://archive.org/stream/jstor-1645736/1645736_djvu.txt.

61 The author used the phrase "the exhausted receiver of an air pump".

62 "exhausted" = evacuated.

63 Energeticism – the scientific view that energy is the fundamental element in all physical change. Energetics was proposed by Wilhelm Ostwald (1853–1932) and others to replace mechanics as the fundamental theory in physical science.

64 J.P. Joule, "On the changes of temperature produced by the rarefaction and condensation of air", *Phil. Mag.* 26, 369–383, 1845.

65 Joule uses the phrase *vis viva* an obsolete term for what we now call kinetic energy. In this context best translated as "energy".

66 Also see Section 5.7.1, where Feynman uses the phrase '"frictionless" heat transfer'.

67 See Equation 6.1.

68 See Section 4.6.

69 The only direct measurement of caloric I am aware of is made using an ice calorimeter (often called Bunsen's ice calorimeter). It is a device enabling the measurements of total volume of co-existing solid (ice) and liquid phases of water. The exchange of caloric between the system under investigation and Bunsen's calorimeter taking place at the melting temperature of ice (~273 K) is accompanied by a phase change resulting in a volume change of the water–ice system. As the latent caloric of fusion of ice is a constant (C), the amount of caloric corresponding to the observed volume change (ΔV) can be easily computed by a formula $Q = C\Delta V$. See also J.J. Mares, P. Hubık, J. Sestak, V. Spicka, J. Kristofik, J. Stavek, *Thermochim. Acta* 474, 16–24, 2008. Latent heat of fusion = 333.5 $J.g^{-1}$; density of ice at 0°C = 0.92 $g.mL^{-1}$ and water at 0°C is 1.00 $g.mL^{-1}$.

70 'Subtle' – "difficult to understand or perceive" (www.merriam-webster.com); and the Oxford Concise dictionary defines 'imponderable it as "not able to be weighed or evaluated".

71 H.L. Callendar, "The caloric theory of heat and carnot's principle", *Proc. Phys.* XXIII, 153–154, 1911.

72 F. Herrmann, "Entropy from the Beginning", GIREP 2004 Proceedings, Ostrava.

73 H.L. Callendar, "The caloric theory of heat and carnot's principle", *Proc. Phys.* XXIII, 184, 1911. By "energy measure" Callendar means the mechanical or modern definition of heat.

74 H.L. Callendar, "The caloric theory of heat and carnot's principle", *Proc. Phys.* XXIII, 178, 1911. By "energy measure" Callendar means the mechanical or modern definition of heat.

75 H.L. Callendar, "The caloric theory of heat and carnot's principle", *Proc. Phys.* XXIII, 178, 1911. By "energy measure" Callendar means the mechanical or modern definition of heat.

76 H.L. Callendar, "The caloric theory of heat and carnot's principle", *Proc. Phys.* XXIII, 179, 1911. By "energy measure" Callendar means the mechanical or modern definition of heat.

77 H.L. Callendar, "The caloric theory of heat and carnot's principle", *Proc. Phys.* XXIII, 168, 1911.

78 See Endnote 25.

79 S. Carnot, *Reflections of the Motive Power of Fire*, Translated by R.H. Thurston, Dover Publications, pp. 225–226, 1960.

80 https://en.wikipedia.org/wiki/First_law_of_thermodynamics.

81 https://en.wikipedia.org/wiki/Caloric_theory.

82 H.L. Callendar, "The caloric theory of heat and carnot's principle", *Proc. Phys.* XXIII, 188, 1911.

83 Complexion = microstate (see Chapter 8).

84 H.L. Callendar, "The caloric theory of heat and carnot's principle", *Proc. Phys.* XXIII, 185, 1911.

85 H.L. Callendar, "The caloric theory of heat and carnot's principle", *Proc. Phys.* XXIII, 184, 1911.

86 Note – this is the only occasion when I do not consider a cyclic thermodynamic process.

87 Remember a closed system can exchange energy but not matter with its surroundings.

88 We can write these values for caloric because the temperatures of the heat reservoirs remain constant.

89 There are only two ways to change caloric content of a system: one is by the process of heat and the other by the change in the amount of material in an open system. This book is not concerned with open systems, therefore the process of heat is the only caloric change process we need to consider.

90 In other words, the flow of caloric occurs simultaneously with the flow of heat.

91 H.U. Fuchs, *The Dynamics of Heat*, Springer, p. 104, 1996.

92 Flux is a vector quantity, describing the magnitude and direction of the flow of a substance or property.

93 He means work.

94 Fuchs uses the word "power" which is the rate of doing work.

95 H.L. Callendar, "The caloric theory of heat and carnot's principle", *Proc. Phys.* XXIII, 153–189, 1911.

96 Motive power = work.

97 See Section 6.5.6 for a discussion on spontaneous processes.

98 Remember friction has the same effect if the piston is moving up or down by the same incremental distance and we are ignoring any gravitational effects.

99 Every scientific text I have seen states something like *heat is generated from/by friction* when describing this phenomenon.

100 Remember $\mathbf{w} = \mathbf{w}^{\text{input}}$, $\mathbf{w}^{\text{input}} = -\mathbf{w}^{\text{output}}$, and $|\mathbf{w}| = |\mathbf{w}^{\text{input}}| = |\mathbf{w}^{\text{output}}|$.

101 Also called 'degraded energy' or 'energy unavailable to do work'.

102 Examples are: (i) "A spontaneous process is the time-evolution of a system in which it releases free energy and moves to a lower, more thermodynamically stable energy state." (https://en.wikipedia.org/wiki/Spontaneous_process) (ii) "According to the second law of thermodynamics, which states that the entropy of the universe increases for any spontaneous process" (from https://study.com/academy/lesson/spontaneous-process-definition-examples-quiz.html).

103 https://chem.libretexts.org/Textbook_Maps/General_Chemistry/Map%3A_Chemistry_-_The_Central_Science_(Brown_et_al.)/19%3A_Chemical_Thermodynamics/19.1%3A_Spontaneous_Processes.

104 Callendar again used the phrase "motive power"

105 Average temperature of the Earth's atmosphere and oceans is about 290 K.

106 First proposed by W Thomson (Lord Kelvin) in 1852 – although French astronomer Jean-Sylvain Bailly had conjectured something similar in 1777 – see https://en.wikipedia.org/wiki/Heat_death_of_the_universe.

107 Heat death of universe is also discussed in Section 7.9.

108 Joule, "Joint Scientific Papers of James Prescott Joule", Vol 2, The Physical Society of London, 1887, "Mechanical Value of Heat", page 158. https://classes.oc.edu/PhysicsLab/JOULE.htm,Joint.

109 H.U. Fuchs, *The Dynamics of Heat*, Springer, p. 290, 1996.

110 Enthalpy a thermodynamic quantity equal to the internal energy of the system plus the product of pressure and volume. It incorporates the PV work done by a system at constant pressure. Enthalpy is introduced and discussed in Section 8.8.

111 E. Mach, *Popular Scientific Lectures*, Cambridge University Press, 175–177, 1898.

112 "perforce" = unavoidably or necessarily.

113 Mach wrote "a closed system" but he meant an isolated system – see Section 1.13.2 – "Unfortunately, a closed system in classical mechanics is considered an isolated system in thermodynamics."

114 E. Mach, *Popular Scientific Lectures*, Cambridge University Press, 109, 1898. This is the last time I will "translate" the word "heat". The final sentence would be better read "If we look at the matter in an unprejudiced light, we must ask if there is any scientific sense or purpose in still considering as energy [that quantity of energy] that can no longer be transformed into mechanical work by the process of heat, (for example, the energy of [an isolated] equably

warmed material system)." But Mach's writing makes sense and my previous sentence does not. You can begin to see why the word heat is used in a wrong manner even by good scientists. This is because there is no consensual definition of heat. Essentially, heat can be distinguished as a "transfer" of energy and "energy" and may be used in both ways occasionally in the same sentence.

115 H.L. Callendar, "The caloric theory of heat and carnot's principle", *Proc. Phys.* XXIII, 188–189, 1911.

116 Callendar means that caloric was erroneously considered to have the units of energy.

117 H.L. Callendar, "The caloric theory of heat and carnot's principle", *Proc. Phys.* XXIII, 182, 1911.

118 H.L. Callendar, "The caloric theory of heat and carnot's principle", *Proc. Phys.* XXIII, 153–189, 1911.

119 Read "isolated system" and the fire is internally heating the room.

120 https://pdfs.semanticscholar.org/2e04/20ecae32a8a4f354f9baf62c43e35ec1b064.pdf.

121 I have already considered this question in Section 6.3.2.

122 In its more modern guise.

123 H.L. Callendar, "The caloric theory of heat and carnot's principle", *Proc. Phys.* XXIII, 172, 1911.

124 Delaroche and Berard ("Ann. Chim.,"1813) had found that the specific heat of a gas increased with increase in volume ('specific heats diminished with diminution of volume.'). For an ideal gas there is no change in specific heat with volume or pressure, for real gases any change is insignificant.

7 Entropy

Perhaps more nonsense has been written in attempting to explain the second law to students than anything else I have ever come across in the literature.[1]

7.1 INTRODUCTION

In Chapter 6, we have seen that caloric is a state function, but the energeticists, in their enthusiasm for the mechanical theory of heat, had rubbished the concept of caloric which, they erroneously considered, had the units of energy. So now, the energeticists needed a state function that allowed directionality of a process to be predicted. Clausius (1822–1888) provided this state function in the 1860s: he called it entropy, from the Greek – *τρέπειν* – *to give direction*. Unfortunately, unlike caloric, Clausius' entropy is an arcane mathematical quantity that lacks a clear physical meaning. The following two extracts are of interest[2]:

> The entropy introduced into physics by Clausius was, contrary to general belief, not a new physical quantity but the reconstruction of the "quantity of heat" conceived about one hundred years earlier by the Scottish chemist Black. The same quantity was also used under the name "calorique" by Carnot in his work, which laid the foundations of thermodynamics. That entropy and Black's "quantity of heat" are only two names for the same physical quantity is not only of historical interest but is of significance to the teaching of thermodynamics as well. It asserts that entropy can be visualised as a kind of substance which obeys "half a conservation theorem": It can be created but not destroyed.
>
> It is most remarkable that it was not recognised, and is not so even today, that the entropy constructed by Clausius with great effort and admirable scientific tact is nothing but Black's [heat] of old.
>
> Our considerations have made it plausible that the notoriety of entropy as being unvisualisable [sic] and incomprehensible is due to historical accident, namely that the Black-Carnot theory was held to be wrong by the discoverers of the energy principle. The idea of classifying energy into energy forms and calling one of them … heat, in a sense entirely different from that used formerly, seemed very plausible at the time but turned out to be unfortunate. First the concept of heat lost its simple visualisability [sic] and became an "incomplete differential" that is a "process quantity"[3] (as opposed to a 'state quantity').[4] However, as was already recognised by Black and Carnot, since a single quantity does not suffice for the description of heat phenomena it became necessary to introduce the old [heat] of Black somehow. Clausius succeeded in this with his complicated and opaque construction of entropy. Basically, one arrived where one had been before after a difficult and steep detour.

Callendar stated[5]:

> … it was inevitable that Carnot's caloric should make its reappearance sooner or later in the mechanical theory. It first reappears, disguised as a triple integral, in Kelvin's solution ("Phil. Mag.," 4, p. 305, 1852) of the problem of finding the available work in an unequally heated body. The solution (as corrected later[6]) is equivalent to the statement that the total quantity of caloric remains constant when the equalisation of temperature is effected reversibly. Caloric reappeared next as the "thermodynamic function" of Rankine, and the "equivalence-value of a transformation" (Clausius "Pogg. Ann.," 93, p. 497, 1854). Finally, in 1865, when its importance was more fully recognised, Clausius ("Pogg. Ann.," 125, p. 390) gave it the name of "entropy," and defined it as the integral of $\left[\frac{dq}{T}\right]$. Such a definition appeals to the mathematician only. In justice to Carnot, it should be called caloric, and defined directly by his equation $\left[|\mathbf{w}| = |\mathbf{Q}|(T_H - T_C)\right]$,[7] which any schoolboy could understand. Even the mathematician would gain by thinking of caloric as a fluid, like electricity, capable of being generated by friction or other irreversible processes. Conduction of caloric is closely associated with the electrons, and the science of heat would gain, like the science of electricity, by attaching a more material conception to the true measure of a quantity of heat, as distinguished from a quantity of …[8] energy.

The concept of entropy is part of the reasoning used by Rudolf Clausius to tackle the problem of finding a mathematical expression to describe all transformations of a body through heat exchange between that body and another one or the surroundings. In 1865, he published an

DOI: 10.1201/9781003608400-7

integrated volume with several previous and new works (I will reference the English translation published in 1867).[9] Since 1850 Clausius had been considering ways in which to represent in words and mathematically the idea that heat can never pass from a colder to a warmer body without some other change, connected therewith, occurring at the same time.[10] In the Fourth Memoir of 1854, he introduces the equation[11]:

$$N = \int \frac{dq}{T} \tag{7.1}$$

where T denotes absolute temperature. In his 1865 paper, he also introduces the word entropy, deriving it from the Greek τροπη (a turning, or a change), for which Clausius uses the German *Verwandlung* and for which Clausius's translators generally use the English transformation. Also, in 1865, Clausius set down the relationship between N of the 1854 paper and his newly coined entropy[12]:

$$N = \int \frac{dq}{T} = S - S_0 \tag{7.2}$$

where S is entropy of the final condition of a body, and S_0 denotes entropy of the initial condition of the body. Many commentators have highlighted the influence of Thomson's 1852 note entitled "On a Universal Tendency in Nature to the Dissipation of Mechanical Energy"[13] upon Clausius's notion of entropy. In 1865, he was able to set down concisely his now famous: *The entropy of the universe tends to a maximum.*

It appears that it took Clausius 11 years (1854–1865) to symbolise and name an integral, 15 years (1850–1865) to encapsulate into a single word the physical meaning of the second law and possibly 25 years (1850–1875) to do the same with the first law.

Consider the following[14]:

> So, even though criticisms and rivalries and "a long and bitter priority controversy" greeted Clausius's work; even though Clausius's math engendered confusion; even though Max Planck's work was required to explain some of Clausius's statements; and even though some of Clausius's cleverly coined terms have receded into obscurity; still, more than 150 years after Clausius first published on "The Moving Force of Heat and the Laws Which May Be Deduced Therefrom", every general chemistry student is introduced to Clausius's symbols, terms, and concepts. U is for Energy; S is for Entropy. Why? Because – after years of considering the physical significance of his own mathematical constructs of U and S - Clausius said so.

Unlike caloric, one cannot appreciate that Clausius's entropy can flow, can be stored, and cannot be destroyed. Unsurprisingly, because of the obscure mathematical definition of entropy nobody at the time was aware that Clausius's entropy was not new at all but only a resurrection of Black's heat and Carnot's caloric. Even today over 150 years since Clausius introduced the concept of entropy, you can find quotes such as "physicists cannot tell you what it was that Clausius discovered."[15]

However, we are where we are and, universally, students are taught the mechanical theory of heat with its associated incomprehensible concept of entropy as developed by Clausius, who formulated his theory of thermodynamics in his papers of 1850, 1854, 1862, and 1865. While he accepted Carnot's principle, he rejected the caloric theory and accepted Joule's mechanical theory of heat.[16] In the next few paragraphs, I hope to give some insight into how Clausius invented entropy.

Clausius defined entropy (S) as:

$$S = \int \frac{dq_{rev}}{T_{reservoir}} \tag{7.3}$$

Why did Clausius spend years considering this issue when the answer seems to be staring him in the face? Of course, I do not know, and I do not consider I have the knowledge or expertise to set down any meaningful comments. However, I do consider ending this introduction to the invention of entropy by setting down again part of Falk's extract used earlier in this section[17]:

> Clausius succeeded in this with his complicated and opaque construction of entropy. Basically, one arrived where one had been before after a difficult and steep detour.

In the next section, I will consider the "Clausius Inequality," which applies to any real (i.e., irreversible) engine cycle. As we shall see, it allows a definition of entropy and is one way to state the Second Law of Thermodynamics.

7.2 CLAUSIUS' INEQUALITY AND THE SECOND LAW OF THERMODYNAMICS

7.2.1 Introduction

The Clausius' Inequality Theorem (also known as the "Clausius Theorem") applies to a system exchanging energy, by the process of heat, with external reservoirs and undergoing a cyclic process (i.e., one that ultimately returns a system to its original state). It was first proposed in 1855 (and amended in 1862) by Clausius. For reversible cycle, we can write:

$$\oint \frac{\mathbf{dq_{rev}}}{T} = 0, \tag{7.4}$$

and for a cycle into which irreversibility enters at any stage:

$$\oint \frac{\mathbf{dq_{irrev}}}{T} < 0, \tag{7.5}$$

where the cyclic integral symbol $\oint$ shows the integration should be done over the entire cycle where $\mathbf{dq}$ is the infinitesimal amount of energy absorbed by the system, by the process of heat, from a heat reservoir and T is the temperature of the reservoir at a particular instant in time. The sign of $\mathbf{q}$ is determined with respect to the system (positive, if the system's energy increases and, negative, if the system's energy decreases). Combining Equations (7.4) and (7.5) gives the usual form of the inequality:

$$\oint \frac{\mathbf{dq}}{T} \leq 0 \tag{7.6}$$

I have already introduced the concept of the cyclic integral (see Section 4.6) and before considering Clausius' Inequality in detail, I want to review and consider further the concept of the cyclic integral.

7.2.2 The Cyclic Integral

The value of any state function of a system depends only on the present state of the system. In a cyclic process, the initial and the final states of a system are the same, and so are the values of a state function at these two states, i.e., the change in a state function will be zero in a cyclic process. Because the change in any state function depends only on the initial and the final states of the system, it does not depend on the path followed to carry out the change. Therefore, for both a reversible and an irreversible cyclic process, the change in any state function will be zero.

If a state function depends only on the present state of the system, then we can plot a continuous graph of the state function against the state of the system (e.g., pressure, temperature, etc.). Consequently, another way to think of state functions is integrals. Integrals depend on only three things: the function, the lower limit, and the upper limit. Similarly, state functions depend on three things: the property, the initial value, and the final value. In other words, integrals illustrate how state functions depend only on the final and initial value and not on the system's history or the path taken to get from the initial to the final value.

Consider Figure 7.1 which represents a cyclic process in (a) the system moves from state A to state B, and in (b) the system moves back to state A along the same path. The integral $\int_A^B f(x)$ represents the shaded area in (a), and the integral $\int_B^A f(x)$ represents the shaded area in (b). Therefore, the cyclic integral is:

$$\oint f(x) = \int_A^B f(x) + \int_B^A f(x) = 0 \tag{7.7}$$

The area in Figure 7.1a is negative and in Figure 7.1b is positive (see Sections 1.6.7 and 4.6). Using absolute values Equation (7.7) can be written:

$$\left|\oint f(x)\right| = \left(-\left|\int_A^B f(x)\right|\right) + \left(+\left|\int_B^A f(x)\right|\right) = 0 \tag{7.8}$$

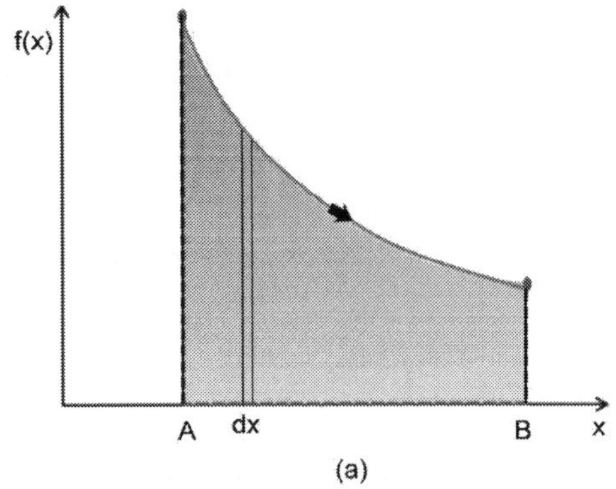

Figure 7.1 (a and b) Graphical representation of a simple cyclic process.

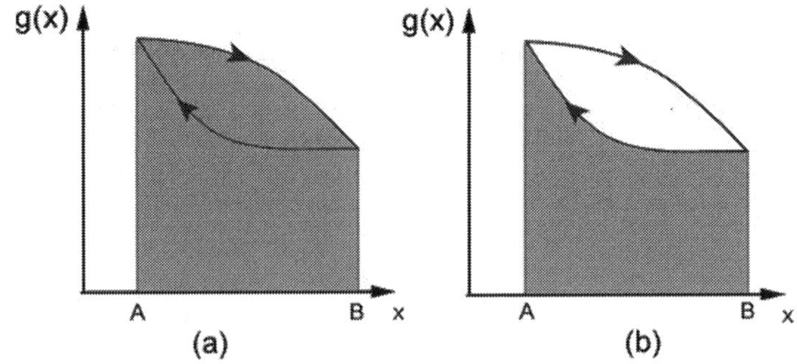

Figure 7.2 A graphical description of a cyclic process.

Therefore, $f(x)$ is a state function. Now consider Figure 7.2 which again represents a cyclic process but, in this case, the forward and reversible paths are different. It can immediately be seen that

$$\oint g(x) = \int_A^B g(x) - \int_A^B g(x) \neq 0 \tag{7.9}$$

The only way to calculate the area enclosed between the two curves is by knowing the paths taken in the cycle. Therefore, in contrast to a state function $\oint g(x) \neq 0$ and $g(x)$ is a path function.

7.2.3 Proof of the Clausius' Inequality

The proof of the Clausius' Inequality for a cyclic process is relatively straightforward. The efficiency (ε) of the reversible Carnot heat engine is the amount of energy obtained by the process of work divided by the amount of energy supplied by the hot reservoir by the process of heat. We know ε can be written (see Equation 6.22):

$$\varepsilon = \frac{\left(q_{H_{rev}} + q_{C_{rev}}\right)}{q_{H_{rev}}} = \frac{\left(T_H - T_C\right)}{T_H} \tag{7.10}$$

Therefore, we can write:

$$\frac{\left(q_{H_{rev}} + q_{C_{rev}}\right)}{q_{H_{rev}}} = \frac{\left(T_H - T_C\right)}{T_H} \tag{7.11}$$

Rearranging yields:

$$\frac{q_{H_{rev}}}{T_H} + \frac{q_{C_{rev}}}{T_C} = 0 \tag{7.12}$$

Since every step in the cycle is carried out reversibly, we know ε is a maximum for the two temperatures. In other words, we know that an irreversible engine that operates between two temperatures T_H and T_C is less efficient than the idealised Carnot engine operating between the same two temperatures. Therefore, for the irreversible case, we can conclude:

$$\frac{(q_{H_{irrev}} + q_{C_{irrev}})}{q_{H_{irrev}}} < \frac{(T_H - T_C)}{T_H} \tag{7.13}$$

Rearranging we have for the irreversible case:

$$\frac{q_{H_{irrev}}}{T_H} + \frac{q_{C_{irrev}}}{T_C} < 0 \tag{7.14}$$

This relation is extended to any number of compound reversible cycles by the argument given in Section 5.11 but using $\frac{q_H}{T_H} + \frac{q_C}{T_C} \leq 0$ instead of $\frac{q_H}{T_H} + \frac{q_C}{T_C} = 0$. Therefore, for any number of compound reversible cycles:

$$\sum \frac{q_{rev}}{T} = 0 \tag{7.15}$$

Supposing the temperature to change continuously, we have:

$$\oint \frac{dq_{rev}}{T} = 0 \tag{7.16}$$

If the process is not reversible, then Equations (7.15) and (7.16) can be written as:

$$\sum \frac{q_{irrev}}{T} < 0 \tag{7.17}$$

$$\oint \frac{dq_{irrev}}{T} < 0 \tag{7.18}$$

or more generally:

$$\oint \frac{dq}{T} \leq 0 \tag{7.19}$$

This is a "proof" of the Clausius Inequality and is essentially the same as set out in Sections 6.4.4 and 6.5.5 where I demonstrated that $Q_H^{rev} + Q_C^{rev} = 0$ and $Q_H^{irrev} + Q_C^{irrev} < 0$. On the face of it, knowing the entropy of an isolated system must always increase and can never decrease the inequality appears to be the wrong way round. Unfortunately, to add to the confusion, you may see the Clausius Inequality equation written the other way round, *viz* $\int \frac{dq}{T} \geq 0$. Consider the following extract (I have changed the upper-case Q's to lower case q's to be consistent with this book, used the symbol $q^{\text{heat reservoir}}$ to emphasise the author is considering energy flows from the heat reservoir's point of view rather than the system, and the underlining is mine)[18]:

> Quantitatively, Clausius states the mathematical expression for this theorem is as follows. Let dq be an element of the heat[19] <u>given up by the body to any reservoir of heat</u> during its own changes, heat which it may absorb from a reservoir being here reckoned as negative, and T the absolute temperature of the body at the moment of giving up this heat, then the equation:
>
> $$\int \frac{dq^{\text{heat reservoir}}}{T} = 0$$
>
> must be true for every reversible cyclical process, and the relation:
>
> $$\int \frac{dq^{\text{heat reservoir}}}{T} \geq 0$$
>
> must hold good for every cyclical process which is in any way possible.

You can understand why the author has changed the sign convention from the system's point of view to the reservoirs' point of view (because the inequality "looks" the right way round), but the end result is probably even more confusion for students when they come across, what on the face, of it and it appears to be two contradictory versions of the inequality.

7.2.4 Entropy Determination

Entropy is defined as $S = \int \frac{dq_{rev}}{T_{reservoir}}$ (see Equation7.3). Therefore, we can write the Clausius Inequality (Equation 7.19) in terms of entropy:

$$\oint dS = 0 \tag{7.20}$$

Consider a general irreversible process in an isolated system, leading from state A to state B represented by a dashed line in Figure 7.3. Then consider that the system returns to its original state A by a reversible path represented by the solid line from B to A. During this reversible process, the system need not be isolated and can exchange energy with its environment by the processes of heat and work (i.e., it can be a closed system). Writing the cycle in terms of its two sections, we obtain:

$$\oint \frac{dq}{T} = \int_A^B \frac{dq_{irrev}}{T} + \int_B^A \frac{dq_{rev}}{T} = \int_A^B \frac{dq_{irrev}}{T} + \int_B^A dS \tag{7.21}$$

Since the entire cycle is in part irreversible, the Clausius Inequality $\oint \frac{dq}{T} = dS < 0$ applies:

$$\int_A^B \frac{dq_{irrev}}{T} + \int_B^A dS < 0 \tag{7.22}$$

Because entropy is a state function, we can write:

$$\int_B^A dS = S_A - S_B \tag{7.23}$$

Therefore, we can rewrite Equation (7.22) as:

$$\int_A^B \frac{dq_{irrev}}{T} + (S_A - S_B) < 0 \tag{7.24}$$

Rearrangement of Equation 7.24 gives:

$$\int_A^B dS^{irrev} < -(S_A - S_B) < (S_B - S_A) \tag{7.25}$$

If we define the entropy change (ΔS) on going from A to B as $S_B - S_A$, then we can write Equation (7.25) as:

$$\Delta S > \int_A^B \frac{dq_{irrev}}{T_{reservoir}} \tag{7.26}$$

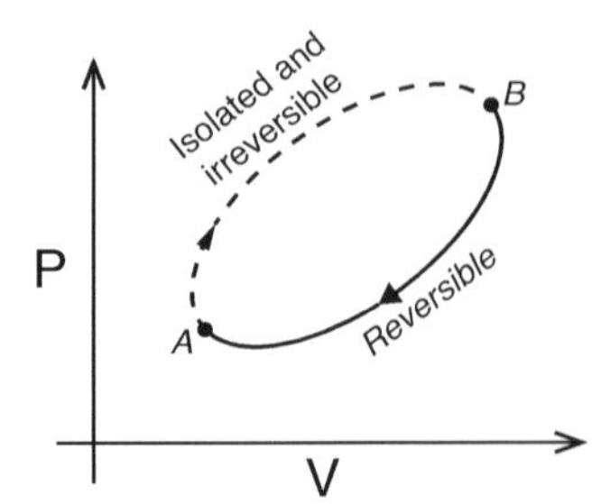

Figure 7.3 A cyclic process in which the isolated system passes from A to B by an irreversible process and returns to its initial state by a reversible process during which it is not isolated.

The integral in Equation (7.26) is equal to zero, since during the process A →B the system is, by definition, isolated and no transfer of energy by the process of heat is possible $\left(q_{irrev} = 0\right)$, and Equation (7.26) can be written:

$$\Delta S > 0 \tag{7.27}$$

Therefore, the entropy change of any process occurring in an isolated system (also known as an adiabatic process[20]) is always greater than or equal to zero, where the equality applies to reversible processes. Although not all processes are adiabatic, we can always enclose a system in a surrounding system which is isolated, e.g., the universe is isolated. Therefore, we have proved that if A passes to B by an irreversible process in an isolated system, then the entropy of the final state B is always greater than that of the initial state A, i.e., for any process the total change in entropy of an isolated system (universe) is always greater than or equal to zero.

The change in entropy going from equilibrium state A to equilibrium state B is always the same irrespective of the path between A and B; this is because entropy is a state function. It makes no difference whether the path is reversible or irreversible. However, only if the path is reversible, the entropy change is given by:

$$\Delta S = S_B - S_A = \int_A^B \frac{dq_{rev}}{T} \tag{7.28}$$

Consequently, to evaluate the entropy change for an irreversible process, we must devise a reversible method going from the same initial to the same final state and then calculate the entropy change. Since the system is in the same equilibrium state at the start as at the end of the cycle, the entropy change of the system is zero, as entropy is a state function (remember for a state function, its change only depends on the initial and final conditions not on the path).

To find the change in S of the universe (heat reservoirs plus system), we need to know the change in S of the surroundings.[21] This can be carried out because all heat processes in heat reservoirs are, by definition, reversible, and any entropy change in the surroundings is a result of entropy flows (positive or negative) from the system concurrent with associated energy flows by the process of heat. The change in the entropy of the universe for an irreversible system, after one complete cycle, is given by:

$$\Delta S_{irrev}^{universe} = \Delta S_{irrev}^{heat\ reservoirs} + \Delta S_{irrev}^{system} = \oint \frac{dq_{irrev}^{heat\ reservoirs}}{T_{reservoirs}} \geq 0 \tag{7.29}$$

For a reversible process, we know the entropy decrease of the hot reservoir is counterbalanced by the same increase in the cold reservoir. In other words, the entropy changes in the surroundings (hot and cold reservoirs), like the entropy change of the system, is zero. Thermodynamic texts often state this by saying a reversible process does not change/alter the universe. In contrast, for an irreversible process, the entropy decrease of the hot reservoir is lesser than the entropy increase in the cold reservoir, and we can write:

$$\frac{+\left|q_{C\ irrev}^{cold\ reservoir}\right|}{T_C} + \frac{-\left|q_{H\ irrev}^{hot\ reservoir}\right|}{T_H} > 0 \tag{7.30}$$

Knowing that $\frac{+\left|q_{H\ irrev}^{system}\right|}{T_H} = \frac{-\left|q_{H\ irrev}^{hot\ reservoir}\right|}{T_H}$ and $\frac{+\left|q_{C\ irrev}^{system}\right|}{T_C} = \frac{-\left|q_{C\ irrev}^{cold\ reservoir}\right|}{T_C}$, therefore Equation (7.28) when written from the system's point of view is:

$$\frac{-\left|q_{C\ irrev}^{system}\right|}{T_C} + \frac{+\left|q_{H\ irrev}^{system}\right|}{T_H} < 0 \tag{7.31}$$

Equation (7.31) is the Clausius Inequality written using absolute values.

7.2.5 The Irreversible Carnot Cycle and the Clausius Inequality

Essentially, there are two sources of irreversibility in the irreversible Carnot cycle:

1. Irreversibility due to lack of thermodynamic equilibrium between the system and the surroundings, i.e., there is a temperature gradient. This is known as "external irreversibility".

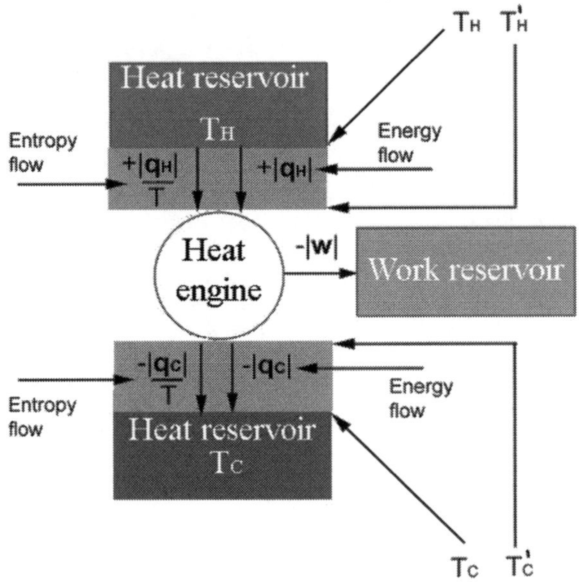

Figure 7.4 Carnot engine with external irreversibility.

2. Irreversibility due to dissipative effects such as friction and turbulence within the interior of the system or between the system and the surroundings. This type of irreversibility is known as "internal irreversibility".

I will first consider an internally reversible heat engine but with external irreversibility due to finite temperature differences between the heat reservoirs and the working substance of the heat engine. The schematic diagram of such a heat engine is depicted in Figure 7.4. We know from the Clausius Inequality that:

$$\oint \frac{dq_{sys}}{T_{res}} = \frac{\left|q_H^{system}\right|}{T_H} - \frac{\left|q_C^{system}\right|}{T_C} < 0 \tag{7.32}$$

$$\oint \frac{dq_{sys}}{T_{sys}} = \frac{\left|q_H^{system}\right|}{T'_H} - \frac{\left|q_C^{system}\right|}{T'_C} = 0 \tag{7.33}$$

where T_H and T_C are the temperatures of the hot and cold heat reservoirs and T'_H and T'_C are the upper and lower temperatures of the heat engine/working substance (I have used T_{sys} instead of T_{res} in Equation (7.33) to account for the temperature gradients between the heat reservoirs and the heat engine).

If the entropy flow during each cycle from the hot reservoir to the heat engine is $\frac{-\left|q_H^{reservoir}\right|}{T_H}$, then we know the entropy flow from the heat engine to the hot reservoir is $\frac{+\left|q_H^{system}\right|}{T_H}$ where $\frac{\left|q_H^{reservoir}\right|}{T_H} = \frac{\left|q_H^{system}\right|}{T_H}$. Similarly, if the entropy flow during each cycle from the cold reservoir to the heat engine is $\frac{-\left|q_C^{reservoir}\right|}{T_C}$, then we know the entropy flow from the heat engine to the cold reservoir is $\frac{+\left|q_C^{system}\right|}{T_C}$ where $\frac{\left|q_C^{reservoir}\right|}{T_C} = \frac{\left|q_C^{system}\right|}{T_C}$. We know from Equation (7.32) that $\frac{\left|q_H^{system}\right|}{T_H} - \frac{\left|q_C^{system}\right|}{T_C} < 0$. Therefore, we can write (from the reservoirs' point of view)[22]:

$$\frac{-\left|q_H^{reservoir}\right|}{T_H} - \frac{-\left|q_C^{reservoir}\right|}{T_C} = \frac{-\left|q_H^{reservoir}\right|}{T_H} + \frac{+\left|q_C^{reservoir}\right|}{T_C} > 0 \tag{7.34}$$

Rearrangement gives:

$$\frac{\left|q_C^{\text{reservoir}}\right|}{T_C} > \frac{\left|q_H^{\text{reservoir}}\right|}{T_H} \tag{7.35}$$

This equation shows that as a result of external irreversibilities, the entropy flow from the hot reservoir to the system is less than in the entropy flow from the system to the cold reservoir. We know:

1. Entropy cannot be generated in the heat reservoirs because all heat processes that occur in the heat reservoirs are reversible.
2. Entropy generation does not occur during the operation of the heat engine because it is internally reversible (Equation 7.36).

$$\frac{\left|q_C^{\text{reservoir}}\right|}{T'_C} = \frac{\left|q_H^{\text{reservoir}}\right|}{T'_H} \tag{7.36}$$

Therefore, entropy generation must occur during the flow of energy, by the process of heat, down a temperature gradient. In our example (Figure 7.4) this occurs as energy $\mathbf{q_H}$ flows down the $T_H T'_H$ temperature gradient, and as energy $\mathbf{q_C}$ flows down the $T'_C T_C$ temperature gradient.

Let me set down something that I have adapted from a Physics Stack Exchange website in answer to the following question – "If both reservoirs are internally reversible, since their state don't change and the "partition" is also internally reversible for the same reason. What is causing the entropy generation (irreversibilities)?"[23]:

> The entropy of each of the reservoirs changes, because one loses [energy by the process of] heat and the other gains [energy by the process of] heat. So their states change, but, since they are internally reversible, there is no entropy generation within the reservoirs. So, where does the entropy generation occur? It occurs within the [diathermal] partition, where finite heat conduction is occurring via a temperature gradient. ...[24] The entropy generated within the partition is transferred to the combination of the two reservoirs. So, even though the reservoirs are internal reversible, they receive the generated entropy from the partition.

This indicates that whenever energy flows from one location to another by the process of heat there is always an accompanying entropy flow. In the next section, I will consider the entropy change during thermal conduction in a bit more detail, but before then, I want to briefly consider "internal irreversibility".

Friction/turbulence results in temperature differences in the heat engine (remember heat reservoirs are, by definition, reversible, i.e., there are no temperature differences and there is no turbulence). We know that temperature differences must reduce the work produced by a heat engine operating between the same hot and cold temperatures. We also know that these temperature differences are eliminated as energy flows, by the process of heat, down the temperature gradient to the lowest temperature available (T_C) with the consequent production of entropy.

7.2.6 Entropy Change during Thermal Conduction

We cannot place the two reservoirs in direct contact since the temperature at a point can have only one value, i.e., it cannot be T_H on one side of the point of contact and T_C on the other side. Mathematically speaking, the temperature function cannot have a jump discontinuity, so the two reservoirs must be separated by a diathermal partition (the "body") in which the temperature drops from T_H on one side to T_C on the other side (Figure 7.5). This process is known as

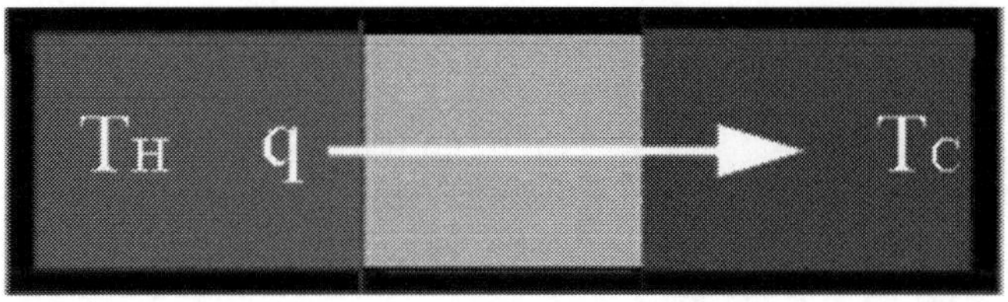

Figure 7.5 Transfer of q quantity of energy by the process of heat from a hot reservoir to a cold reservoir by the process of heat.

thermal conduction and is the energy flow within and through the body itself and is the transfer of internal energy within an object, by the process of heat, via the collisions of microparticles within that body.[25]

We know:

1. Conductive heat transfer via a temperature difference is an irreversible process and is, therefore, accompanied by entropy production.
2. At steady state, the energy and entropy content of the diathermal partition or conductor is constant, i.e., $\Delta S^{conductor} = 0$.
3. There is no entropy generation within the reservoirs as they are internally reversible.
4. Entropy, like energy, spontaneously flows from warmer to cooler places.
5. Entropy generation occurs in the diathermal partition (the 'conductor'), and entropy must flow into or out of the heat reservoirs. In other words, the diathermal partition is an entropy generating system that does not store anything that it creates. The conductor takes $\left|\frac{q}{T_H}\right|$ entropy out of the hot reservoir and transfers $\left|\frac{q}{T_C}\right|$ into the cold reservoir.

Therefore:

$$\left|\Delta S^{surroundings}\right| = \left(-\left|\frac{q^{reservoir}}{T_H}\right|\right) + \left(+\left|\frac{q^{reservoir}}{T_C}\right|\right) = \left|\frac{q^{reservoir}}{T_C}\right| - \left|\frac{q^{reservoir}}{T_H}\right| \tag{7.37}$$

Rearrangement of Equation (7.37) gives:

$$\Delta S^{surroundings} = \frac{\left|q^{reservoir}\right|(T_H - T_C)}{T_H T_C} \tag{7.38}$$

Remember the sign convention in Equation (7.37) is from the point of view of the surroundings. The hot reservoir has lost energy and entropy by the process of heat, and the cold reservoir has gained energy and entropy by the process of heat. We know that if $T_H > T_C$ then $\Delta S^{surroundings} > 0$. We can write:

$$\left|\Delta S^{universe}\right| = \left|\Delta S^{surroundings}\right| + \left|\Delta S^{conductor}\right| = \frac{\left|q^{reservoir}\right|(T_H - T_C)}{T_H T_C} \tag{7.39}$$

Equation (7.39) is the same as Equation 6.10 in Section 6.5.2, apart from the substitution of "Q" (caloric) for "S" entropy.

7.2.7 Numerical Example

To clarify, I want to consider a numerical example. I will start with the by-now-familiar reversible Carnot heat engine (Figure 7.6a). Let us assume a hot reservoir at 400 K, a cold reservoir at 300 K and an energy flow of +4,000 joules per cycle from the hot reservoir to the heat engine. An energy flow of +4,000 joules per cycle at a constant 400 K results in an entropy flow into the system of $\frac{4000}{400} = +10\ \text{J.K}^{-1}$ per cycle. We know the entropy flow into the system from the cold reservoir is $-10\ \text{J.K}^{-1}$ per cycle (remember we are using our sign convention from the system's point of view). Therefore, the energy flow into the system from the cold reservoir at 300 K is –3,000 J per cycle. By the First Law of Thermodynamics, we know the energy flow from the work reservoir into the system is –1,000 J per cycle.

Now let us consider a heat engine with external irreversibility (Figure 7.6b). We will assume that heat reservoirs at 420 K and 280 K will, after thermal conduction, result in the heat engine operating reversibly as per Figure 7.6a, but with the additional entropy sources associated with thermal conduction. We must also assume that the thermal conduction processes have attained steady-state conditions so that there is no change in the energy and entropy content of the diathermal materials. If 4,000 J of energy "falls" down a temperature gradient of 420–400 K, we know from Equation (7.36) that hot reservoir loses $\frac{400 \times (420 - 400)}{420 \times 400} = 0.476\ \text{J.K}^{-1}$ per cycle. Similarly, if

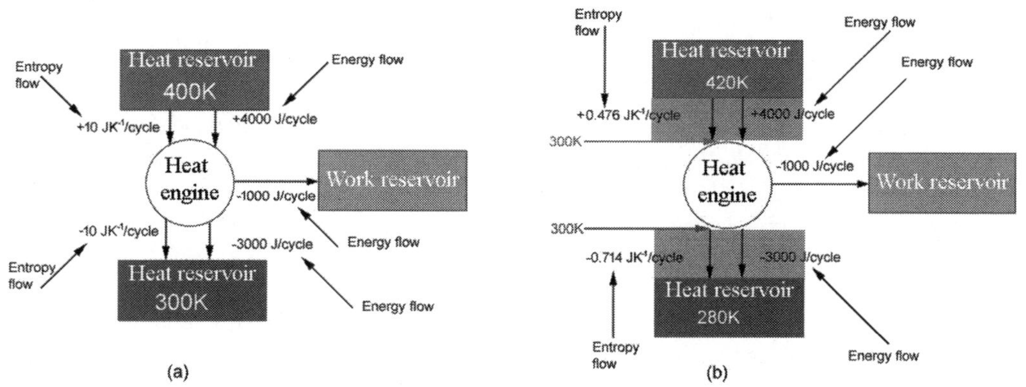

Figure 7.6 (a and b) Entropy generation: a numerical example.

3000J of energy "falls" down a temperature gradient of 300–380 K we know from that cold reservoir gains $\frac{3000\times(300-280)}{300\times 280} = 0.714\ \mathrm{J.K^{-1}}$ per cycle. Therefore, in contrast to the reversible heat engine, the entropy content of the universe increases by 0.238 $\mathrm{J.K^{-1}}$ per cycle.

7.3 THE SECOND LAW OF THERMODYNAMICS

In the previous section, I demonstrated that any naturally occurring process in an isolated system results in an increase in the entropy content of the system. The corollary of this is that any process that would result in a decrease in the entropy content of an isolated system is an "impossible" process and cannot occur. The universe is an isolated system, and the above conclusions led Clausius to his concise statement of the Second Law of Thermodynamics[26]:

> Die Entropie der Welt strebt einem Maximum zu.
> The entropy of the world[27] tends always toward a maximum.

The Second Law of Thermodynamics is, as far as I am aware, unique in science because of the large number of different ways the law can be stated[28]:

> While there is no agreement in the scientific community about how best to state the second law, there is general agreement that the current melange of statements taken en masse, pretty well covers it. This, of course, gives fits to mathematicians, who insist on precision and parsimony.[29]
>
> The Second Law made its appearance around 1850, and almost a century later, the physicist and philosopher Bridgman still complained that "there are almost as many formulations of the Second Law as there have been discussions of it." Even today, the Second Law remains so obscure, due to the lack of its comprehension, that it continues to attract new efforts at clarification, …

These different ways of stating the Second Law can be divided into two categories. One category uses the concept of entropy, and the other category does not. Listed below is a selection of statements of the Second Law that do not include the word entropy:

- *No process is possible whose sole result is cooling a colder body and heating a hotter body.*[30]
- *No device, operating in a cycle, can produce the sole effect of extraction of a quantity of energy from a heat reservoir and the performance of an equal quantity of work.*[31]
- *All Carnot engines operating between the same two temperatures have the same efficiency.*
- *All heat engines have efficiencies satisfying* $0 < \varepsilon < 1$.
- *Perfectly efficient heat engines* $(\varepsilon = 1)$ *are impossible.*
- *All natural processes are irreversible.*
- *Energy transfer by the process of heat only spontaneously occurs from hot to cold.*

However, in my view, definitions that include the word "entropy" are the most straightforward ways of stating the Second Law. The three definitions below are some of the better ones:

1. *In any real process, the entropy content of the universe or that of an isolated system increases.*
2. *For an isolated system that undergoes a change from one equilibrium state to another, if the thermodynamic process is reversible, then the entropy change is zero; if the process is irreversible, the entropy change is positive.*
3. *There exists a function of state known as the entropy S that can never decrease in an isolated system.*

It is worth noting that replacing "entropy" with "caloric" does not change the meaning of the above statements.

Although statements of the Second Law in terms of the entropy are applicable only to isolated systems, most systems of interest can exchange energy and mass with their surroundings, i.e., they are either closed or open systems. However, we can devise a composite system that is isolated; for example, we can take the surroundings of a cup of hot water to be the air in the room, and the composite system to be the cup of hot water and the air in the room and treat this composite system to be isolated. For this composite system, we have for any process, $\Delta S_{composite} = \Delta S_{system} + \Delta S_{surroundings} = \Delta S_{isolated\ system} \geq 0$.

In Section 2.6, I showed that the position of equilibrium in a purely mechanical situation is the position of minimum potential energy. In a thermodynamic system, the position of equilibrium is when the entropy of the system is at a maximum. We can say this because (i) the entropy of an isolated system can never decrease, and (ii) any naturally occurring process is irreversible and proceeds with an increase in entropy. Consequently, naturally occurring processes continue in an isolated system until the system's entropy reaches its maximum value. At equilibrium, the entropy of the system cannot increase (because it is already at a maximum) and it cannot decrease (because entropy cannot be destroyed). Therefore, no further change can occur. This equilibrium can be disturbed if the system can interact with its surroundings, i.e., when the system is no longer isolated. The entropy of the system may then decrease, provided the entropy of the surroundings increases by at least as much, ensuring that there is no decrease in the entropy of the universe, i.e., the only changes allowed are those in which the entropy content of the universe increases.[32]

The Second Law of Thermodynamics allows us to understand why real processes occur only in a certain direction and that direction is the one in which the entropy of the universe increases – there are no exceptions. Irreversibility is an essential behaviour of the physical world, and changes proceed in the direction that increases the entropy of the universe.

You should now be able to appreciate that the concept of entropy developed by Clausius is, in essence, the same as the "discredited" caloric. Entropy like caloric:

- is produced in every real process;
- cannot be destroyed;
- is not produced in reversible processes;
- is at a maximum when the system is in equilibrium.

7.4 CALCULATION OF ENTROPY CHANGE IN SOME IDEAL GAS PROCESSES

7.4.1 Introduction

This section is devoted to calculating and discussing the entropy changes that occur in the basic ideal gas processes previously covered in this book. The calculation of entropy changes in an ideal gas is relatively simple because the internal energy of an ideal gas is independent of volume or pressure changes, and energy changes due to cohesive forces do not need to be considered. Most of this is standard fare and versions can be found in most thermodynamic texts.

7.4.2 Entropy Changes in an Ideal Gas[33]

For a reversible process in an ideal gas, the First Law of Thermodynamics requires that

$$\mathbf{dq} = \mathrm{d}\mathbf{U} + \mathrm{P}\mathbf{dV} = \mathrm{C_V}\mathbf{dT} + \frac{\mathrm{nRT}\mathbf{dV}}{\mathrm{V}} \tag{7.40}$$

Therefore:

$$\mathbf{dS} = \frac{\mathbf{dq}}{T} = C_V \frac{\mathbf{dT}}{T} + \frac{nR\mathbf{dV}}{T} \tag{7.41}$$

On integration (for an ideal gas, C_V is independent of temperature):

$$\Delta\mathbf{S} = S_{final} - S_{initial} = C_V \int_{T_{initial}}^{T_{final}} \frac{\mathbf{dT}}{T} + nR \int_{V_{initial}}^{V_{final}} \frac{\mathbf{dV}}{V} \tag{7.42}$$

$$\Delta\mathbf{S} = S_{final} - S_{initial} = C_V \mathbf{ln} \frac{\mathbf{T_{final}}}{\mathbf{T_{initial}}} + nR\mathbf{ln} \frac{\mathbf{V_{final}}}{\mathbf{V_{initial}}} \tag{7.43}$$

For the special case of a temperature change at constant volume, the change in entropy with change in temperature is:

$$\Delta\mathbf{S} = C_V \mathbf{ln} \frac{\mathbf{T_{final}}}{\mathbf{T_{initial}}} \tag{7.44}$$

If the temperature of 1 mole of ideal gas with $C_V = 12.5\ J.K^{-1}.mol^{-1}$ doubles, the entropy increases by $12.5 \times \ln 2 = 8.63\ J.K^{-1}$.

For the case of an isothermal expansion/compression, the entropy change is:

$$\Delta\mathbf{S} = nR \ln \frac{\mathbf{V_{final}}}{\mathbf{V_{initial}}} = nR \ln \frac{\mathbf{P_{initial}}}{\mathbf{P_{final}}} \tag{7.45}$$

If 1 mole of ideal gas expands to twice its original volume (or half of its original pressure), its entropy increases by $R \times \ln 2 = 5.74\ J.K^{-1}$.

7.4.3 Free Expansion of an Ideal Gas

Free expansion is an irreversible process in which a gas expands, under adiabatic conditions, into an insulated evacuated chamber and is typically visualised by opening a stopcock that allows a gas to expand into a vacuum (Figure 7.7). During the free expansion process, the gas goes through states that are not in thermodynamic equilibrium before reaching its final state, e.g., the pressure changes locally from point to point, and the volume occupied by the gas is not a defined quantity. For an ideal gas where there is no inter-particle interaction, the temperature does not change, and the conditions before and after adiabatic free expansion satisfy $P_{initial} \times V_{initial} = P_{final} \times V_{final}$. If the

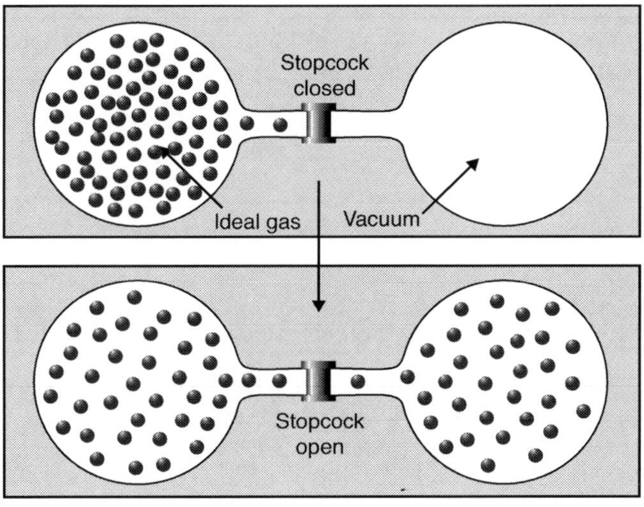

Figure 7.7 Free expansion of an ideal gas.

temperature remains constant, then for an ideal gas, the internal energy (ΔU) is constant ($\Delta U = 0$),[34] and because there is no energy exchanged with the surroundings by the processes of work and heat then $\mathbf{q} = \mathbf{w} = 0$.

To determine the change of entropy in the free expansion of n moles of ideal gas, we need to use a reversible process to restore the system to its original state, i.e., to go back to the original volume ($V_{final} \rightarrow V_{initial}$) at constant temperature T. To do this, we use reversible isothermal compression where the surroundings transfer energy to the system by the process of work. Please note that for the reversible compression process the volume of the initial state of the system is V_{final} and the volume of the final state of the system is $V_{initial}$.

Using Equation (7.43), we can write:

$$\Delta \mathbf{S}^{compression} = nR\ln\frac{\mathbf{V}_{initial}}{\mathbf{V}_{final}} \tag{7.46}$$

Entropy is a state function,[35] i.e., $\Delta \mathbf{S}^{compression} = -\Delta \mathbf{S}^{free\,expansion}$. Therefore:

$$-\Delta \mathbf{S}^{free\,expansion} = nR\ln\frac{\mathbf{V}_{initial}}{\mathbf{V}_{final}} \tag{7.47}$$

Equation (7.45) can be rewritten as:

$$\Delta \mathbf{S}^{free\,expansion} = nR\ln\frac{\mathbf{V}_{final}}{\mathbf{V}_{initial}} \tag{7.48}$$

We know for expansion that $V_{final} > V_{initial}$ we know $\Delta \mathbf{S}^{free\,expansion}$ is a positive number.

Using absolute values, we can write:

$$-\left|\Delta \mathbf{S}^{compression}\right| = nR\left(-\left|\ln\frac{\mathbf{V}_{initial}}{\mathbf{V}_{final}}\right|\right) \tag{7.49}$$

Entropy is a state function, i.e., $\left|\Delta \mathbf{S}^{free\,expansion}\right| = -\left|\Delta \mathbf{S}^{compression}\right|$. Therefore:

$$+\left|\Delta \mathbf{S}^{free\,expansion}\right| = nR\left(-\left|\ln\frac{\mathbf{V}_{initial}}{\mathbf{V}_{final}}\right|\right) = nR\left(+\left|\ln\frac{\mathbf{V}_{final}}{\mathbf{V}_{initial}}\right|\right) \tag{7.50}$$

Let us now consider the process when n moles of ideal gas expand in an isothermal and reversible fashion between the same initial and final states as those in the irreversible free expansion, *and in doing so a flow* $+\left|\mathbf{q}_{rev}\right|$ *of heat is converted into a flow of* $-\left|\mathbf{w}_{rev}\right|$ *of work where* $\left|\mathbf{q}_{rev}\right| = \left|\mathbf{w}_{rev}\right|$ *(remember the sign convention is from the system's point of view).*[36] From Equation 5.7 the total amount of work performed (transferred) to the system from the mechanical reservoir is[37]:

$$\mathbf{w}_{rev} = -nRT\ln\frac{\mathbf{V}_{final}}{\mathbf{V}_{initial}} = -\mathbf{q}_{rev} \tag{7.51}$$

The entropy change is:

$$\frac{\mathbf{q}_{rev}}{T} = \Delta \mathbf{S}^{rev\,expansion} = nR\ln\frac{\mathbf{V}_{final}}{\mathbf{V}_{initial}} = \Delta \mathbf{S}^{free\,expansion} = -\frac{\mathbf{w}_{rev}}{T} \tag{7.52}$$

We can immediately see that the quantity $T\Delta \mathbf{S}$ has a physical meaning as "lost work" in the sense of work that the universe lost because of irreversible free expansion rather than reversible isothermal expansion.

7.4.4 Mixing of Two Ideal Gases

Consider an isolated system separated into two chambers by a removable partition (Figure 7.8). One chamber (volume V_A) contains n_A moles of ideal gas A and the other chamber (volume V_B) contains n_B moles of ideal gas B. Both gases are at the same temperature (T) and pressure (P). When the partition is removed, the gases will irreversibly mix. Gibbs stated that for ideal gas mixtures "...every gas is a vacuum to every other gas...".[38] From this perspective, volume V_B is nothing but empty space for gas A, and *vice versa*. Under these conditions, the mixing process becomes

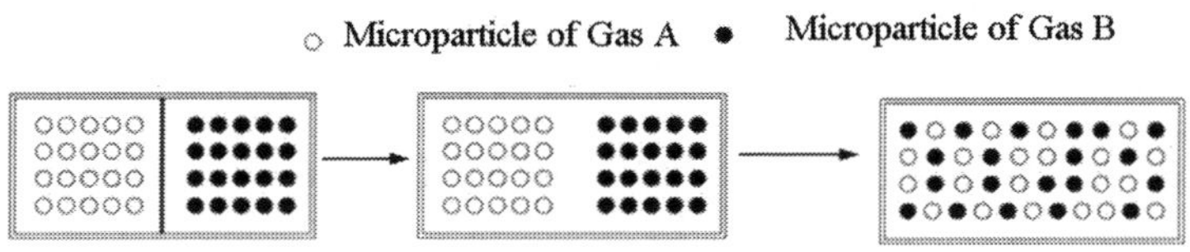

Figure 7.8 Irreversible mixing of two ideal gases.

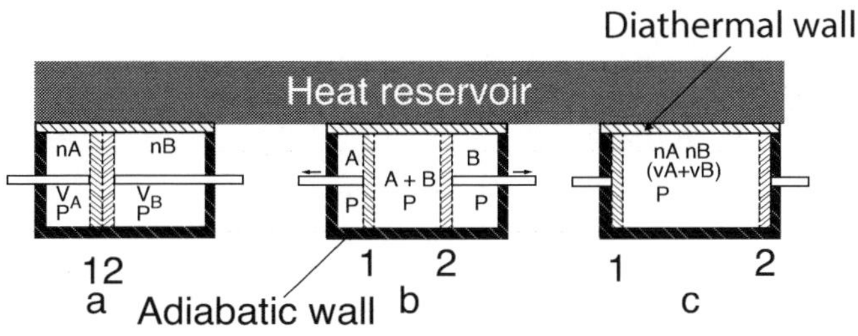

Figure 7.9 (a–c) Reversible mixing thought experiment.

nothing more than the simultaneous free expansions of gases A and B from their respective initial volumes to a common, shared volume equal to $V_A + V_B$. These isothermal free-expansion processes occur without any interaction with its surroundings (by definition the system is isolated). The entropy change for each of these expansions can then be written in accord with that expressed by Equation (7.48). When this is done, and the results combined to produce the total or universe entropy change for this irreversible mixing, we obtain:

$$\Delta \mathbf{S}_{\mathbf{mixing}} = \left[n_A R \ln\left(\frac{\mathbf{V_A} + \mathbf{V_B}}{\mathbf{V_A}}\right)\right] + \left[n_B R \ln\left(\frac{\mathbf{V_A} + \mathbf{V_B}}{\mathbf{V_B}}\right)\right] \tag{7.53}$$

We know $\left(\frac{V_A + V_B}{V_A}\right) > 1$ and $\left(\frac{V_A + V_B}{V_B}\right) > 1$. Therefore, $\ln\left(\frac{\mathbf{V_A} + \mathbf{V_B}}{\mathbf{V_A}}\right)$ and $\ln\left(\frac{\mathbf{V_A} + \mathbf{V_B}}{\mathbf{V_B}}\right)$ are positive real numbers, therefore $\Delta \mathbf{S}_{\mathbf{mixing}} > 0$

Let us now consider a reversible mixing process and carry out the following thought experiment. Consider a cylinder in thermal equilibrium with a heat reservoir at temperature T so that the experiment is isothermal (Figure 7.9). Initially, both gas A and gas B are present in separate compartments at the same pressure P. The two gases in the cylinder are separated by two pistons fitted with semi-permeable membranes of negligible volume.[39] The one on the left ("piston 1") is permeable only to A, and the one on the right ("piston 2") is permeable only to B. Although the gases exert no pressure on their respective pistons, they do exert pressure P on the opposite ones, i.e., gas A exerts pressure on piston 2 and gas B exerts pressure on piston 1.

The initial state shown in Figure 7.9a consists of n_A moles of ideal gas A at pressure P and volume V_A, and n_B moles of ideal gas B at pressure P and volume V_B. Let the hypothetical semipermeable pistons move apart reversibly and isothermally to create, via Figure 7.9b, an ideal gas mixture, as shown in Figure 7.9c. To carry out the mixing process in a reversible manner, the external pressure on the right piston is kept infinitesimally less than the pressure (P) of B in the mixture, and the external pressure on the left piston is kept infinitesimally less than the pressure of A (P) in the mixture. The gases mix (without turbulence) upon reaching the space between the pistons. The continued diffusion of the gases through these membranes marks the transport of gases A and B from their respective smaller initial volumes V_A and V_B to the larger and common volume $V_A + V_B$. The movement of the left piston has expanded gas B with the same reversible work as if gas A were absent, and likewise, the reversible work to expand gas A with the right piston is the same as if B were absent. Therefore, we can write (see Equation 7.51):

$$\mathbf{w}_{\mathbf{B,rev}} = -\left[n_B RT \ln\left(\frac{\mathbf{V_A} + \mathbf{V_B}}{\mathbf{V_B}}\right)\right] \tag{7.54}$$

$$\mathbf{w}_{A,rev} = -\left[n_A RT \ln\left(\frac{\mathbf{V}_A + \mathbf{V}_B}{\mathbf{V}_A} \right) \right] \tag{7.55}$$

The reversible mixing process is equivalent to two isothermal expansions and the total energy transferred by the process of work in the reversible mixing process $(\mathbf{w}_{rev})$is:

$$\mathbf{w}_{rev} = \mathbf{w}_{A,rev} + \mathbf{w}_{B,rev} = -\left\{ \left[n_A RT \ln\left(\frac{\mathbf{V}_A + \mathbf{V}_B}{\mathbf{V}_A} \right) \right] + \left[n_B RT \ln\left(\frac{\mathbf{V}_A + \mathbf{V}_B}{\mathbf{V}_B} \right) \right] \right\} \tag{7.56}$$

Multiplication of Equation (7.53) by T gives:

$$T\Delta\mathbf{S}_{mixing} = \left[n_A RT \ln\left(\frac{\mathbf{V}_A + \mathbf{V}_B}{\mathbf{V}_A} \right) \right] + \left[n_B RT \ln\left(\frac{\mathbf{V}_A + \mathbf{V}_B}{\mathbf{V}_B} \right) \right] \tag{7.57}$$

Comparison of Equation (7.58) with (7.57) allows us to write:

$$T\Delta\mathbf{S}_{mixing} = -\mathbf{w}_{rev} \tag{7.58}$$

Again, we can see that the quantity TΔS has a physical meaning as "lost work" in the sense of work which the universe lost because of free expansion. In other words, the reduction in the system's ability to do useful work is reflected by an increase in the entropy of the universe.

7.4.5 Spontaneous Flow of Energy by Process of Heat

One way of stating the Second Law of Thermodynamics is – "Energy transfer by the process of heat only spontaneously occurs from hot to cold." (see Section 7.4). The following example illustrates this aspect of the Second Law.

Consider a closed system consisting of a vessel of negligible mass containing n moles of ideal gas at T_{system}K in surroundings at $T_{surroundings}$K; where $T_{system} \neq T_{surroundings}$. We assume the vessel is constructed of impermeable diathermal walls of negligible mass. By definition (as $T_{system} \neq T_{surroundings}$) the system is not in equilibrium and there will be a flow of energy $(\mathbf{q})$, by the process of heat, until it reaches the same temperature as the surroundings. The entropy of the system (the ideal gas) and the surroundings will change, so we have to calculate them separately. For each, we need to find the initial and final states A and B and calculate the entropy change for a reversible process that takes us between the same states. I will start with the system.

For a reversible process, we know the energy must be transferred from the surroundings to the system at an infinitesimal rate. Therefore, we can write[40]:

$$\mathbf{dq}_{system} = nC_V\mathbf{dT} \tag{7.59}$$

Then the total entropy change will be

$$\Delta\mathbf{S}_{system} = \int_{T_{system}}^{T_{surroundings}} \frac{\mathbf{dq}_{system}}{T} = nC_V \int_{T_{system}}^{T_{surroundings}} \frac{\mathbf{dT}}{T} = nC_V \ln \frac{\mathbf{T}_{surroundings}}{\mathbf{T}_{system}} \tag{7.60}$$

Remember this reversible process uses an infinite number of heat reservoirs at temperatures varying infinitesimally from each other $(T + \mathbf{dT})$etc. from T_{system} to $T_{surroundings}$. But **ΔS** is a state function and is therefore independent of the path taken.

I will now consider the surroundings for the irreversible process we are considering. Energy **q** flows from the system into the surroundings (aka the heat reservoir at $T_{surroundings}$), by the process of heat. However, by definition, the surroundings/heat reservoir remains at $T_{surroundings}$. We can calculate the energy flow into the surroundings, by the process of heat, by working out how much energy the ideal gas gains $(\mathbf{q}_{system})$, by the process of heat, knowing that $(\mathbf{q}_{system}) = -(\mathbf{q}_{surroundings})$:

$$\mathbf{q}_{system} = nC_V\left(T_{surroundings} - T_{system}\right) \tag{7.61}$$

Knowing $(\mathbf{q}_{system}) = -(\mathbf{q}_{surroundings})$ we can write:

$$\mathbf{q}_{surroundings} = nC_V\left(T_{system} - T_{surroundings}\right) \tag{7.62}$$

We know all energy transfer processes that occur in heat reservoirs are reversible. Therefore, the entropy change of the surroundings, at constant temperature $T_{surroundings}$ is

$$\Delta S_{surroundings} = \frac{q_{surroundings}}{T_{surroundings}} = \frac{nC_V\left(T_{system} - T_{surroundings}\right)}{T_{surroundings}} \quad (7.63)$$

By definition $nC_V\left(T_{surroundings} - T_{system}\right) = \Delta U_{system}$. Therefore, we can write:

$$\Delta S_{surroundings} = \frac{-\Delta U_{system}}{T_{surroundings}} \quad (7.64)$$

Consequently, the entropy change of the universe can be written as

$$\Delta S_{universe} = \Delta S_{system} + \Delta S_{surroundings} = \Delta S_{system} - \frac{\Delta U_{system}}{T_{surroundings}} \quad (7.65)$$

Equation (7.65) expresses the entropy change of the universe at constant temperature T in terms of the thermodynamic properties of the system exclusively. In other words, we are able to write an equation for the change of entropy of the universe in terms of the parameters of the system only, i.e., the entropy change of the universe can be calculated only by knowing the system's thermodynamic parameters. Why are we able to do this? The answer is straightforward but is rarely explained: the reason is that, by definition, all energy transfer processes that occur in heat reservoirs are reversible.

There are two scenarios to consider $T_{system} > T_{surroundings}$ (Scenario I) and $T_{system} < T_{surroundings}$ (Scenario II). In Scenario I both ΔU_{system} and ΔS_{system} are negative.[41] For a spontaneous reaction ($\Delta S_{universe}$ is positive) we know $\Delta S_{system} > \frac{\Delta U_{system}}{T_{surroundings}}$. Using absolute numbers, we know $\left|\Delta S_{system}\right| < \left|\frac{\Delta U_{system}}{T_{surroundings}}\right|$. Although by now we should be used to this, I will illustrate with a numerical example.

Consider a closed system containing 1 mole of ideal gas at 100°C (373 K) in surroundings at 20°C (293 K). We know that C_V (molar heat capacity) of an ideal gas is 12.48 J.kg^{-1}.mol^{-1} (see Section 4.3.2).

We know $\Delta S_{system} = (1)(12.48)\ln\frac{293}{373} = -3.0\ \text{J.K}^{-1}$, i.e., the entropy of the system decreases, as expected, since energy is flowing out by the process of heat. We know $\Delta S_{surroundings} = \frac{12.48(373-293)}{293} = +3.4\ \text{J.K}^{-1}$, which is positive, and larger than the entropy decrease of the system. The entropy change of the universe is $\Delta S_{universe} = \Delta S_{systen} + \Delta S_{surroundings} = (-3.0)+(+3.4) = +0.4\ \text{J.K}^{-1}$.

Let us now consider Scenario II. In Scenario II both ΔU_{system} and ΔS_{system} are positive. Therefore, for a spontaneous reaction $\Delta S_{system} > \frac{\Delta U_{system}}{T_{surroundings}}$ (the same as Scenario I). However, as in this case both ΔU_{system} and ΔS_{system} are positive this means that, in contrast to Scenario I, $\left|\Delta S_{system}\right| > \left|\frac{\Delta U_{system}}{T_{surroundings}}\right|$. Again, I will illustrate with a numerical example. Consider a closed system containing 1 mole of ideal gas at 20°C (293 K) in surroundings at 100°C (373 K). We know $\Delta S_{system} = (1)(12.48)\ln\frac{373}{293} = +3.0\ \text{J.K}^{-1}$, i.e., the entropy of the system increases, as expected, since energy is flowing into the system by the process of heat. We know $\Delta S_{surroundings} = \frac{12.48(293-373)}{373} = -2.6\ \text{J.K}^{-1}$, which is negative, and less than the entropy increase of the system. The entropy change of the universe is $\Delta S_{universe} = \Delta S_{system} + \Delta S_{surroundings} = (+3.0)+(-2.6) = +0.4\ \text{J.K}^{-1}$.

All of the above tells that for the entropy of the universe to increase, energy must flow, by the process of heat, down a temperature gradient. In other words, spontaneous energy transfer, by the process of heat, only flows down a temperature gradient.

7.5 THE KELVIN-PLANCK STATEMENT

I want now to consider the assembly shown in Figure 7.10 operating in a reversible cyclic process.[42] To begin with I will ignore the practical details of this hypothetical engine. In one cycle $+\left|q_{system}\right|$ of energy flows into the heat engine (the "frictionless piston system"), by the process of heat, and $+\left|w_{system}\right|$ of energy flows by the process of work[43] to the work reservoir. During the cycle the

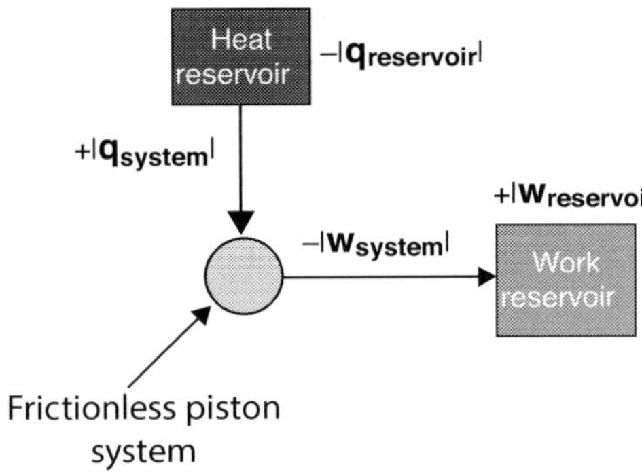

Figure 7.10 100% conversion of energy, transferred by the process of heat, into energy, transferred by the process of work – a forbidden process.

system rejects no energy by the process of heat, therefore, we know $|\mathbf{q}_{\mathbf{system}}| = |\mathbf{w}_{\mathbf{system}}|$. The entropy change of the system after one cycle is zero, i.e., $\Delta S_{system} = 0$, (remember entropy is a state function). Consequently $|\Delta \mathbf{S}_{\mathbf{universe}}| = |\Delta \mathbf{S}_{\mathbf{heat\,reservoir}}|$. We know for isothermal expansion that $\Delta \mathbf{S}_{\mathbf{heat\,reservoir}}$ is a negative real number $\left(-|\Delta \mathbf{S}_{\mathbf{heat\,reservoir}}|\right)$ therefore we know $\Delta S_{universe}$ is negative $\left(-|\Delta \mathbf{S}_{\mathbf{universe}}|\right)$. This process reduces the entropy of the universe which is not allowed.

By this argument, we have demonstrated the Kelvin-Planck statement which states that it is *impossible to devise a cyclically operating heat engine, the sole effect of which is to absorb energy by the process of heat from a single heat reservoir and to deliver an equivalent amount of work.*[44] The Kelvin-Planck statement is another way of stating the Second Law of Thermodynamics, i.e., the Second Law tells us that we cannot get work from a single reservoir only. The "sole effect" is important; it means without any other changes occurring. Using the Second Law, it is guaranteed, that somewhere else in the surroundings a positive entropy change will occur that is as large or larger than the decrease of the entropy of the reservoir. We know (for a heat engine) that this is a cold reservoir acting as an energy sink. If the practical details of such a heat engine operating with only a single heat reservoir are considered it is immediately apparent that a cold reservoir is required to allow the volume of the system to return to its initial value and for this, we need either a cold reservoir or energy transferred by the process of work from another machine (in other words, no "sole effect").

If the cyclic process shown in Figure 7.10 using a single heat reservoir and a single work reservoir was not a forbidden process, we would have a perpetual motion machine converting the infinite quantity of energy in the heat reservoir into an infinite amount of energy in the work reservoir.[45] A perpetual motion machine that contravenes the Second Law of Thermodynamics is known as a perpetual motion machine of the second type.

Finally, let us consider the reverse process (isothermal compression) where energy from the work reservoir is transported to provide a continuous source of energy by the process of heat to the heat reservoir (see Figure 7.11). Please note that in contrast to a heat reservoir, a work reservoir is not the source of infinite amounts of energy.[46] In one cycle the engine receives $+|\mathbf{w}_{\mathbf{system}}|$ of energy by the process of work and gains $-\mathbf{q}_{\mathbf{system}}$ of energy, by the process of heat, from the heat reservoir (or rejects$++|\mathbf{q}_{\mathbf{system}}|$ of energy, by the process of heat, to the heat reservoir). We know for isothermal compression $\Delta \mathbf{S}_{\mathbf{heat\,reservoir}}$ is positive $\left(+|\Delta \mathbf{S}_{\mathbf{heat\,reservoir}}|\right)$. Therefore, we know $\Delta \mathbf{S}_{\mathbf{universe}}$ is positive $\left(+|\Delta S_{universe}|\right)$ for a cyclic process. The transfer of all the energy from the work reservoir to the heat reservoir is an allowed process because it increases the entropy of the universe.

When writing this I was reminded of the following piece of kit that I used at school immediately came to mind (see Figure 7.12)[47]: The following extract explains how it all works:

> Using this apparatus ..., a measurable amount of work is performed by turning the crank, which turns the aluminum[48] cylinder. A nylon rope is wrapped several times around the cylinder so that, as the crank [handle] is turned, the friction between the rope and the cylinder is just enough to support a mass hanging from the other end of the rope. This insures [sic – should be ensures] that the torque acting on the cylinder is constant and measurable. A counter keeps track of the number of turns of the crank.

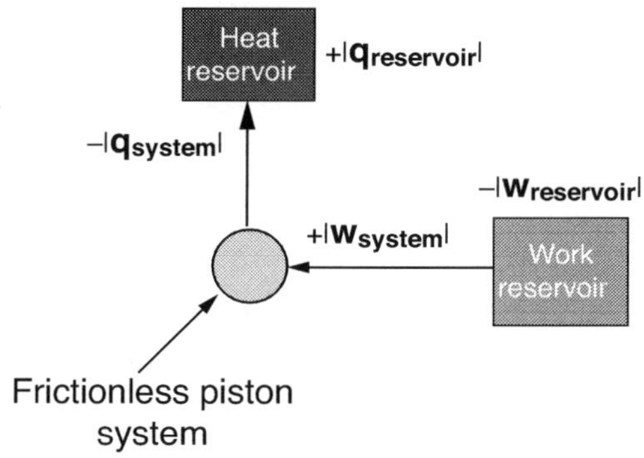

Figure 7.11 100% conversion of energy, transferred by the process of work, to energy transferred by the process of work – an allowed process.

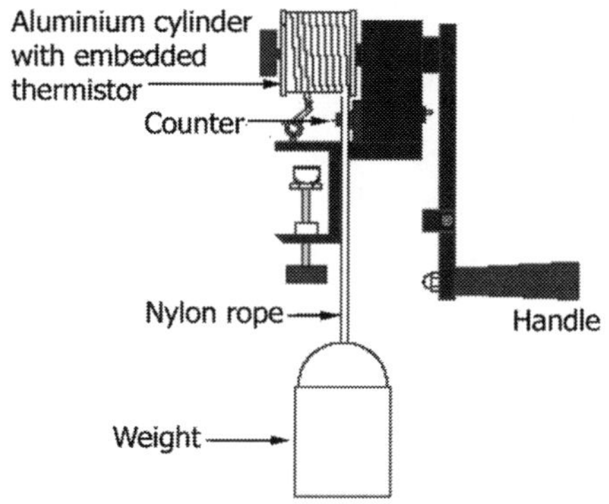

Figure 7.12 A device that converts work into heat.

> As the cylinder turns, the friction between the cylinder and the rope converts the work into [heat],[49] which raises the temperature of the aluminum cylinder. A thermistor is embedded in the aluminum so that, by measuring the resistance of the thermistor, the temperature of the cylinder can be determined. By monitoring the temperature change of the cylinder, the …[50] energy transferred into the cylinder can be calculated. Finally, the ratio between the work performed and the … energy transferred into the cylinder determines J, the mechanical equivalent of heat.[51]

If the handle is turned for long enough the temperature of the aluminium cylinder reaches a steady state and energy is transferred from the cylinder to the cold reservoir by the process of heat. At this point, all the energy transferred to the system (cylinder plus rope), by the process of work, during one cycle of the handle is transferred to the cold reservoir (the room) by the process of heat.

The above analysis demonstrates that energy transferred by the process of heat cannot be completely converted into energy transferred by the process of work, but energy transferred by the process of work can be completely converted into energy transferred by the process of heat. This is often written - *work can be completely converted into heat, but heat cannot be completely converted into work.*[52] This is yet another way of defining the Second Law of Thermodynamics. Note:

1. Joule carried out experiments that completely converted work into heat when he measured the mechanical equivalent of heat (Section 6.4.2).
2. Heat transfers entropy as well as energy, work only transfers energy.
3. The complete conversion of heat to work would require the destruction of entropy, and the complete conversion of work to heat creates entropy.
4. We know from the Second Law that for all processes, entropy either stays the same (reversible processes) or increases (real/dissipative processes).

7.6 SOME THOUGHTS ON ENTROPY

I want to start this section on entropy by considering the concept of energy. How well do you understand energy? Can you visualise energy? Can you touch it? Can you see it? In other words, do you know what energy is? In Section 2.3 I used the following quotes by Richard Feynman:

> There is a fact, or if you wish, a law, governing all natural phenomena that are known to date. There is no known exception to this law - it is exact, so far as we know. The law is called the conservation of energy. It states that there is a certain quantity, which we call energy, that does not change in manifold changes which nature undergoes. That is a most abstract idea, because it is a mathematical principle; it says that there is a numerical quantity, which does not change when something happens. It is not a description of a mechanism, or anything concrete; it is just a strange fact that we can calculate some number, and when we finish watching nature go through her tricks and calculate the number again, it is the same.
>
> It is important to realize that in physics today, we have no knowledge of what energy is. We do not have a picture that energy comes in little blobs of a definite amount.

As we moved through this book we differentiated between various types of energy: kinetic energy; potential energy; the incoherent microscopic energy of microparticles (internal energy) and the coherent energy of moving macroscopic physical objects. We have also considered the two fundamental energy transfer processes – heat and work. However, the bottom line is that "we have no knowledge of what energy is" apart from the fact that is conserved, and the universe has a finite fixed quantity of this "energy". Wikipedia states[53]:

> In physics, energy is the quantitative property that must be transferred to an object in order to perform work on, or to heat, the object. Energy is a conserved quantity; the law of conservation of energy states that energy can be converted in form, but not created or destroyed.

Physics textbooks tell you that "energy is the capacity to do work" but "This is a limited idea incorrectly used as a definition in many middle school science courses and text books".[54] For example, we know that the complete conversion of heat to work is impossible. This is because every time energy is transferred by the process of heat, some of its ability to do work is irretrievably lost. The measure of this loss of work is caloric/entropy generation. No such limitation applies to the transfer of work to heat where all of the organised or coherent energy can be converted into the random energy of microparticles. In other words, the amount of work that can be obtained from energy depends on the degree of organisation of the energy. This is from a paper by Lehrman entitled "Energy is not the ability to do work"[55]:

> …, it is clear that it is misleading to leave heat out of any definition of energy. The concept that energy is the ability to do work dates back to the seventeenth and was put into question when energy was defined quantitatively as a conserved quantity in the 1840s. Within ten years, the enunciation of the second law of thermodynamics had shown this definition to be false. It is time that it was abandoned. Energy is a quantity having the dimensions of work which is conserved in all interactions. It must be defined in terms of a set of algebraic expressions, written in such a way that their sum does not change when a system is isolated. Energy transferred from one system may be called work, heat, radiation, or a variety of other names, depending on the mode of transfer. And at each conversion, the amount of work that can be done diminishes.
>
> A modern definition of energy, then, must be based on both the first and second laws of thermodynamics. Anything less falsifies the picture. If it is not possible to write a satisfactory definition in a few words, we will have to learn to get along without any such neat package.

Lehrman's conclusion is that is probably impossible to define energy because it is an abstract notion, and the concept of energy is really just a kind of shorthand, a tool to help balance the books. Energy is always conserved[56] and the Law of Conservation of Energy is extremely useful in working out the results of any kind of physical or chemical process.

Despite all of this I would suggest the average scientist thinks he/she has a good handle on energy. Now let us consider entropy. Entropy is considered one of the most difficult physical quantities to understand, and I could have listed dozens of interesting quotes on the subject. I have limited myself to four, the first three because of the eminence of the respective authors, and the fourth because it echoed my feelings as a student:

- ...the notion of entropy, ...may repel beginners as obscure and difficult of comprehension. (Gibbs, 1873).[57]
- ...entropy...it is to be feared that we shall have to be taught thermodynamics for several generations before we can expect beginners to receive as axiomatic the theory of entropy. (Maxwell, 1878).[58]
- You should call it entropy, for two reasons. In the first place your uncertainty function has been used in statistical mechanics under that name, so it already has a name. In the second place, and more important, no one really knows what entropy really is, so in a debate you will always have the advantage. [John von Neumann, 1939].[59]
- As a young man I tried to read thermodynamics, but I always came up against entropy as a brick wall that stopped my further progress. I found the ordinary mathematical explanation, of course, but no sort of physical idea underlying it. No author seemed even to try to give any physical idea. Having in those days great respect for textbooks, I concluded that the physical meaning must be so obvious that it needs no explanation, and that I was especially stupid on the particular subject. [James Swinburne, 1904][60]

I feel empathy with Swinburne's quote for two reasons:

1. As a student I felt my lack of understanding of thermodynamics was down to my limitations as a scientist.
2. His need or wish to have a physical meaning of the concept of entropy.

I cannot resist adding this offering entitled "Lamenting a list of confused interpretations of the meaning of entropy, being hotly debated in journals at the time" by Sydney Herbert Evershed[61]:

> In despair, I offer your readers their choice of the following definitions of entropy. My authorities are such books and journals as I have by me at the moment.
>
> a. Entropy is that portion of the intrinsic energy of a system which cannot be converted into work by even a perfect heat engine. - Clausius.
>
> b. Entropy is that portion of the intrinsic energy which can be converted into work by a perfect engine. - Maxwell.
>
> c. Entropy is that portion of the intrinsic energy which is not converted into work by our imperfect engines. - Swinburne.
>
> d. Entropy (in a volume of gas) is that which remains constant when heat neither enters nor leaves the gas. - W. Robinson.
>
> e. Entropy may be called the 'thermal weight', temperature being called the 'thermal height.' - W. Robinson.

Why is entropy still "considered one of the most difficult physical quantities to understand"? There are, in my view, a number of reasons:

1. Thermodynamics was born in the 19th century as scientists were first discovering how to build and operate steam engines and is not a mathematically rigorous area of science (despite the reams of arcane mathematical equations used in many texts).
2. The language used in many texts is careless and lazy as I have indicated in this book. In all probability, this is related to my first point.
3. How entropy is introduced and taught at university (usually by lecturers who do not understand, or care about, the subject).
4. The history of thermodynamics and poor decisions made, e.g., the total ditching of the caloric concept, the banging on that entropy is a measure of disorder but its units are $J.K^{-1}$ (see below).
5. The abstruse mathematical approach taken by Clausius forms the foundation of the modern-day teaching of entropy.

Consider the following quotes:

- "Setting aside for the moment the abstractness of Clausius' procedure the most difficult hurdle for the beginner is that the entropy which is constructed by expressly using reversible processes is not restricted to reversible processes but is a 'state function' of the physical system under consideration and therefore a valid concept for any process of the system, reversible or not."[62]

- "The mathematical definition of entropy, as the integral of $\frac{dq}{T}$ under certain restrictions, is unintelligible to the average student, for whom the conception of entropy possesses an artificial atmosphere of unreality. The conception of caloric defined by Carnot's equation would be more readily assimilated, and might be introduced at a much earlier stage. The direct investigation of the properties of caloric thus defined would probably lead to a more general and intimate appreciation of Carnot's principle itself and of many modern developments of thermodynamics which are practically beyond the comprehension of the majority of students under the present system."[63]
- "In fact, no other discipline in mathematical science is riddled with so many logical and mathematical inconsistencies, differences in definitions, and ill-defined notation as classical thermodynamics. With a notable few exceptions, more than a century of mathematicians have turned away in disquietude from classical thermodynamics, often overlooking its grandiose unsubstantiated claims and allowing it to slip into an abyss of ambiguity."[64]

These difficulties/misunderstandings have fuelled further misconceptions. Let me consider what is arguably the most important misconception - "Entropy is disorder". If you want to get totally confused and bewildered, Google "entropy is disorder" or "entropy is not disorder" and spend time perusing the identified websites. All I will do here is to set down two, hopefully, useful and enlightening quotes.

> ..., if entropy is associated with disorder and if the entropy of the universe is headed towards maximal entropy, then many are often puzzled as to the nature of the "ordering" process and operation of evolution in relation to Clausius' most famous version of the second law, which states that the universe is headed towards maximal "disorder". In the recent 2003 book SYNC – the Emerging Science of Spontaneous Order by Steven Strogatz, for example, we find "Scientists have often been baffled by the existence of spontaneous order in the universe. The laws of thermodynamics seem to dictate the opposite, that nature should inexorably degenerate toward a state of greater disorder, greater entropy. Yet all around us we see magnificent structures—galaxies, cells, ecosystems, human beings—that have all somehow managed to assemble themselves.[65]
>
> Let us dispense with at least one popular myth: 'Entropy is disorder' is a common enough assertion, but commonality does not make it right. Entropy is not 'disorder', although the two can be related to one another. For a good lesson on the traps and pitfalls of trying to assert what entropy is, see Insight into entropy by Daniel F. Styer, American Journal of Physics 68(12): 1090–1096 (December 2000). Styer uses liquid crystals to illustrate examples of increased entropy accompanying increased 'order', quite impossible in the entropy is disorder worldview. And also keep in mind that 'order' is a subjective term, and as such it is subject to the whims of interpretation.[66]

Often a pack of playing cards is used to support an author's contention that entropy is synonymous with disorder but it is artificial to say, for example, that A♠, 2♠, ..., Q♠, K♠, A♦, 2♦, ..., A♣, 2♣, ..., Q♥, K♥ is any more ordered (or disordered) than any of the other 8×10^{67} permutations.[67]

Consider the following[68]:

> It was the physicist Ludwig Boltzmann who in 1905 proposed that the nature of entropy is related to the probability of the state of the atoms or molecules in a system. That is, if the order or arrangement of molecules is unique – has a low probability – then the entropy is low. In contrast, there are many possible combinations of arrangements to obtain a disorderly system – with a high probability – then the entropy is high. For example, if a deck of cards were arranged with the four suits separated and ordered ♠–♦–♣–♥ and all the cards in each suit in ascending sequence (A-K), the arrangement is unique, the probability of accidentally achieving such an arrangement is very low, and the entropy is very low. However, for disorderly arrangements of the cards (suits not in order, cards not in sequence), there are an increasing number of ways to achieve increasingly random arrangements and the entropy increases correspondingly.

The fact, that the author of the above quote appears not to be aware of, is that any arrangement of cards has the same (very low probability) and, by definition, that arrangement is unique. It is a subjective view of the author whether one arrangement of cards is more disordered than another, and science is objective, not subjective. Although the deck of cards analogy is often used, you will come across others. A popular example is the tidy room (low entropy) and the untidy room (high entropy) analogy. Please do not be fooled by the "entropy is disorder" trope.[69] I will return briefly to this subject when I have introduced the concept of the microstate (see Chapter 8 where I will consider statistical thermodynamics, a concept that demonstrates spontaneous processes occur because of probability).

To return to the question asked earlier in this section – "Why is entropy still considered one of the most difficult physical quantities to understand?" I would suggest a major factor is that there is more rubbish written about entropy than anything else in science.

7.7 THE DETERMINATION OR MEASUREMENT ENTROPY

The Third Law of Thermodynamics states that the entropy of a system approaches a constant value (typically zero) as its temperature approaches absolute zero. As we shall see in Chapter 8, entropy is related to the number of accessible microstates.[70] To keep things simple, I will assume that at 0 K there is only one accessible microstate, and the entropy of the system is zero.[71] For there to be only one accessible microstate at 0 K, we have to imagine a perfect crystal (no holes, all the atoms in their exact place in the crystal lattice). This means that there is an absolute entropy, rather than a value relative to some arbitrary standard (there is no comparable zero for internal energy and therefore, no absolute value of internal energy). The ideal gas is unique; in that, we can calculate the absolute entropy using the Sackur-Tetrode equation:

$$S_{\text{ideal gas}} = Nk_B\left[\ln\left\{\frac{V}{N}\left(\frac{4\pi mU}{3Nh^2}\right)^{\frac{3}{2}} + \frac{5}{2}\right\}\right] \tag{7.66}$$

where

- N = number of microparticles
- k_B = Boltzmann's constant
- V = volume
- m = mass of the microparticle
- h = Planck's constant

The Sackur–Tetrode equation is named after H. M. Tetrode (1895–1931) and Otto Sackur (1880–1914), who developed it independently in 1912 as a solution of Boltzmann's statistical thermodynamics and entropy equations (see Chapter 8). I do not intend to set down the derivation or proof of this equation.

The absolute entropy of a real substance at any temperature above 0 K must be determined by the addition of the increments of energy $\mathbf{q}_{\mathbf{rev}}$ required to bring the substance, reversibly, from 0 K to the temperature of interest, and then summing the ratios $\frac{\mathbf{q}_{\mathbf{rev}}}{T}$ because we know:

$$S = \int_0^{T_{\text{final}}} \frac{\mathbf{dq}_{\mathbf{rev}}}{T} \tag{7.67}$$

Even though Equation (7.67) applies when the heating process is reversible, it is approximately correct when the temperature change is small, i.e., by using a quasi-reversible or quasi-static process. The entropy of the substance can then be obtained by measuring the energy required to raise the temperature of the substance by a given amount. Melting of the solid phase and vaporisation of a liquid correspond to increases in entropy, and this corresponds to the heats of fusion and vaporisation which I have not and will not cover in this book because we are primarily interested in the ideal gas. A representation of the change of entropy with temperature for a real substance is shown in Figure 7.13. Although possible the experimental measurement of either absolute entropy or of entropy change between one equilibrium state and another is fraught with practical difficulties and in most cases, it is much easier to calculate the entropy change (based on identifiable reversible paths).

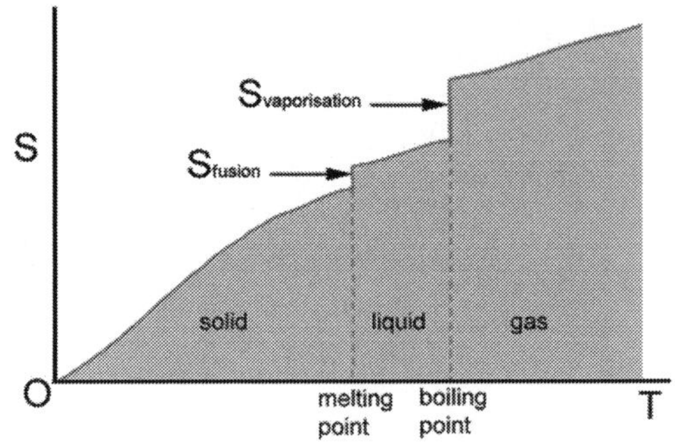

Figure 7.13 Entropy vs temperature plot for a real substance.

7.8 HEAT DEATH OF THE UNIVERSE, THE ARROW OF TIME, AND CAUSE AND EFFECT

7.8.1 Introduction

I now want to mention other related consequences of the Second Law of Thermodynamics:

- The eventual heat death of the universe[72]
- The arrow of time
- Cause and effect.

7.8.2 Heat Death of the Universe

The universe is an isolated system, and the Second Law of Thermodynamics tells us that it is inexorably approaching a state of maximum entropy. At this point, the whole universe will be in a state of equilibrium, i.e., the temperature is uniform, and there is no prospect of extracting useful work (and generating further entropy). In a phrase made popular in the 1930s by the Cambridge cosmologist Sir Arthur Eddington,[73] the universe is said to be approaching a final state of "heat death". As you, the reader, now know, in all real processes, some of the energy transferred by the process of heat always becomes unavailable to do useful work. Consequently, it is inevitable that, as more and more energy is converted to forms unavailable to do work, the ability to do work anywhere in the universe will gradually vanish. Consider the analogy of a clockwork spring. If you wind up a spring, then all of the mechanical energy you put into the spring is now potential energy concentrated in the spring. Let it unwind and do some work – maybe make a clockwork toy move. When the toy stops running, the formerly concentrated energy is not gone, but has been converted through friction and the process of heat into low-grade energy in the air, the toy, and the ground. But it is all spread out now. The amount of energy is the same, but we cannot use it again. We can re-wind the spring, but that takes a new source of concentrated energy. For example, the energy in the food we ate, which we converted to muscle mechanical energy, and then converted to the stored concentrated potential energy of the spring in the clockwork toy which is then finally converted to low-grade energy again. Just like that spring the universe is winding down and unless some divine being 'rewinds' it,[74] there will be a time when the universe is at equilibrium and the ability to do work anywhere in the universe has vanished, i.e., no change can ever occur.

7.8.3 The Arrow of Time

Human experience perceives time flow as unidirectional; the present is forever flowing towards the future and away from a forever fixed past. However, all laws of nature, with one exception, are symmetric in time. For example, reversing the time in the dynamical equations of motion (see Chapter 1) simply describes everything going backwards. The Second Law of Thermodynamics is different; in that, it appears to have a preferred direction for time: the so-called 'arrow of time'.[75] It is in this important aspect that thermodynamics is different from classical mechanics. Time remains one of science's most perplexing mysteries and, raises the following question – why is the concept of time hardly ever found in the most fundamental physical laws of nature and the universe? The following abstract is from Mach, which I will not (or more realistically cannot) comment on[76]:

> If we could really determine the entropy of the world it would represent a true, absolute measure of time. In this way is best seen the utter tautology of a statement that the entropy of the world increases with time. Time, and the fact that certain changes take place only in a definite sense are one and the same thing.

Most physicists have assumed the universe must have begun in a low entropy state and it has been 'running down' (entropy increasing) ever since. This view has the unfortunate implication that all the information in the current universe was present at the beginning, which supports a theological proposal like pre-destination, but not human free will. I will finish this subject with the following quote[77]:

> So where does the arrow of time that correlates with our perception come from? We don't know. What we do know, however, is that the thermodynamic arrow of time isn't it. Our measurements of entropy in the Universe know of only one possible tremendous decrease in all of cosmic history: the end of cosmic inflation[78] and its transition to the hot Big Bang. We know our Universe is headed to a cold, empty fate after all the stars burn out, after all the black holes decay, after dark energy drives the unbound galaxies apart from one another and gravitational interactions kick out the last remaining bound planetary and stellar remnants. This thermodynamic state of maximal entropy is known as the "heat death" of the Universe. Oddly enough, the state from which our Universe arose – the state of cosmic inflation – has exactly the same properties, only with a much larger expansion rate during the inflationary epoch than our current, dark energy-dominated epoch will lead to.[79]
>
> We do understand the arrow of time from a thermodynamic perspective, and that's an incredibly valuable and interesting piece of knowledge. But if you want to know why yesterday is in the immutable past, tomorrow will arrive in a day and the present is what you're living right now, thermodynamics won't give you the answer. Nobody, in fact, understands what will.

7.8.4 Cause and Effect

This is an extremely complicated area, and all I want to do here is to show cause and effect only occurs in thermodynamic systems, i.e., systems in which the "arrow of time" occurs. A cause is the reason something happened, and the effect is what happened (the "event"). Consider[80]:

> Things influence other things. That's a basic statement of any dynamic world where things change, and things would be very dull if it weren't the case – not that we'd exist to know about it, without a cause. Causality is the study of how things influence one other, how causes lead to effects. In the classical world we live in, it comes with a few basic assumptions.
>
> The first big rule of classical causality is that things have causes. They don't just happen of their own accord. If a ball moves, the likelihood is someone kicked it; if an apple falls from a tree, it's because its weight became too great for the branch it was hanging from.
>
> Second, effects follow causes in a predictable, linear manner. You swing your leg, make contact with the ball, and off it moves, in that order and no other.
>
> Third, big effects grow up from little causes. A piston, for example, starts to move when a lot of individual hot atoms hit against it and push it a certain way. The laws of thermodynamics, which govern the way atoms move, then provide certain rules about what causes can precipitate what effects, and so an overall direction for causality – a flow of time.

For the principle of cause and effect to be valid, we must have a concept of time. This can be demonstrated by the following example.

- Pollination allows flowers and plants to (eventually) produce seeds. The cause is the bees moving pollen and the effect is the fertilisation, or ability of flowers to produce seeds.

If the world is time-symmetric it would be just as valid to say that fertilisation causes pollination, or the production of seeds causes pollination. This suggests, that if the laws of nature are time-symmetric, then "initial" and "final" conditions, may be a circular concept, since "initial", and "final" states are all defined with respect to time, i.e., there is no cause and effect. Classical mechanics is time-symmetric and, based on the above argument cause and effect cannot apply in classical mechanics. The purpose of a scientific experiment is to ask a question about what is causing something to happen. You can think of an experiment as a test to see if something you have control over or can change is the cause of something else to happen. The above argument suggests you can only carry out scientific experiments on time asymmetric systems. The differences between time-symmetric and time-symmetric processes are essentially the same as those between reversible and irreversible processes. We know that in time asymmetric systems the entropy of the universe either remains constant (the system is at equilibrium) or the entropy of the universe increases (spontaneous change is occurring). We also know that the concept of entropy cannot be applied to a single microparticle. Therefore, based on this simple observation, we must conclude that the behaviours of macroscopic bodies and single microparticles are different. At the subatomic scale, where the odd rules of quantum mechanics hold sway, physicists describe the state of single particle through a mathematical construct called a wave function. This function is an expression of all the possible states the particle could be in – including all the possible locations it could be in – and the probability of the system being in any of those states at any given time. One of the weirder aspects of quantum mechanics could be explained by an equally weird idea – that causation can run backwards in time as well as forwards. I do not intend to take this any further – if interested Google "quantum entanglement". The following extract[81] is interesting and neatly leads us on to the of statistical thermodynamics - a subject covered in the next chapter.

> …, causality has a definite direction in time: causes have to happen before their effects. Yet the basic laws of physics (as distinct from such higher-level statistical generalisations as the laws of thermodynamics) appear to be time-symmetric: if a certain process is allowed under the basic laws of physics, a video of the same process played backwards will also depict a process that is allowed by the laws.
>
> …
>
> Imagine a stone thrown into a still pond: after the stone breaks the surface, waves spread concentrically from the point of impact. A common enough sight.
>
> Now, imagine a video clip of the spreading waves played backwards. What we would see are concentrically converging waves. For some reason this second process, which is the time-reverse of the first, does not seem to occur in nature. The process of waves spreading from a source looks irreversible. And yet the underlying physical law describing the behaviour of waves – the wave equation – is as time-symmetric as any law in physics. It allows for both diverging and converging waves. So, given that the physical laws equally allow phenomena of both types, why do we frequently observe organised waves diverging from a source but never coherently converging waves?
>
> Physicists and philosophers disagree on the correct answer to this question – which might be fine if it applied only to stones in ponds. But the problem also crops up with electromagnetic waves and the emission of light or radio waves: anywhere, in fact, that we find radiating waves. What to say about it?
>
> On the one hand, many physicists (and some philosophers) invoke a causal principle to explain the asymmetry. Consider an antenna transmitting a radio signal. Since the source causes the signal, and since causes precede their effects, the radio waves diverge from the antenna after it is switched on simply because they are the repercussions of an initial disturbance, namely the switching on of the antenna. Imagine the time-reverse process: a radio wave steadily collapses into an antenna before the latter has been turned on. On the face of it, this conflicts with the idea of causality, because the wave would be present before its cause (the antenna) had done anything.
>
> …
>
> On the other hand, some physicists (and many philosophers) reject appeals to causal notions and maintain that the asymmetry ought to be explained statistically. The reason why we find coherently diverging waves but never coherently converging ones, they maintain, is not that wave sources cause waves, but that a converging wave would require the co-ordinated behaviour of 'wavelets' coming in from multiple different directions of space – delicately co-ordinated behaviour so improbable that it would strike us as nearly miraculous.

Processes that increase the entropy content of the universe are not time symmetrical, i.e., they do not look right when shown in reverse. If we were to throw a ball, classical mechanics predicts that the ball will follow a parabolic trajectory - trajectories (a) and (b) in Figure 7.14. The trajectory is symmetric about its peak so that a video shot of the ball during its flight will look the same whether run forward or backward (apart from the change in the direction of the trajectory). This symmetry is the result of the underlying equations governing the motion of the ball that make no distinction between time moving forward and time moving backward. This symmetry is lost once

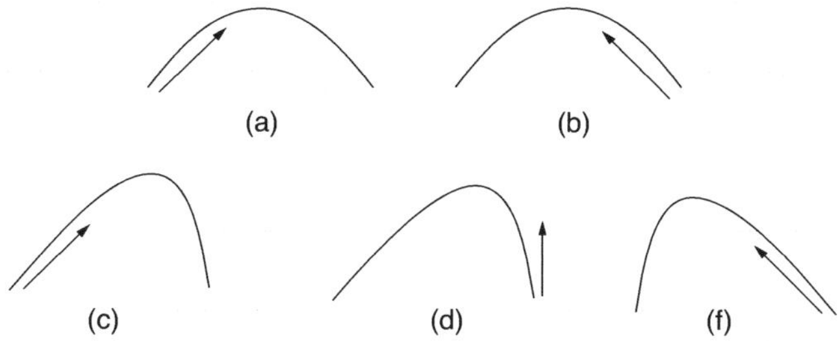

Figure 7.14 Trajectories a and b depict a ball thrown in the absence of air resistance with air resistance, the trajectory would look like c (exaggerated for effect). However, we would never see a thrown ball follow the trajectory in d (it would follow trajectory e).

air resistance becomes appreciable, i.e., in a thermodynamic system with dissipation of energy from a coherent form to an incoherent form (i.e., friction followed by the process of heat). We have now moved from a mechanical system to a thermodynamical system. The drag force on the ball is dependent upon its velocity. Most importantly, it is always directed opposite to the ball's velocity and tends to slow the ball down no matter what its direction of travel. When the ball is thrown with air resistance (trajectory (c) in Figure 7.14), the friction between the ball and the air molecules has increased the entropy/caloric content of the universe, i.e., air resistance makes this an irreversible process. Mathematically the result is that the equations lose their time reversal symmetry. The Second Law of Thermodynamics tells us that we cannot undo the effect throwing the ball with air resistance has on the universe, and that, fundamentally, is why we will not see the ball follow the reversed trajectory (d) depicted in Figure 7.14.[82] The Second Law of Thermodynamics breaks time symmetry in the sense that it uniquely defines what we mean by forward in time and backward in time. Forward in time is the direction associated with increases in entropy of the universe, i.e., the direction a spontaneous process proceeds: and, as we know from experience, we can only travel in one direction on the time axis. From an observational perspective, this lack of symmetry would show up in a video that does not look right when played in reverse.

Another perspective on this asymmetry is the following: we can increase the entropy content of a closed system either by carrying out an irreversible process or by allowing energy to flow into it by the process of heat. On the other hand, we can only decrease the entropy/caloric content of a closed system in one way, and that is by allowing energy to flow out of it by the process of heat. Since the universe is an isolated system, it is not possible for energy to flow out of it by the process of heat. Consequently, the universe as a whole does not have access to the one process that could decrease its entropy – loss of energy by the process of heat. Its entropy can only change through irreversible processes, and these always result in entropy increasing over time.[83]

Newton's Second Law for a single particle, $\mathbf{F} = \frac{d\mathbf{p}}{dt}$ (see Section 1.12, Equation 1.66), is time reversal invariant; in other words, Newton's second law looks the same if the time t is replaced by −t and the momentum p by −p. This indicates there is no direction of time at the microscopic level. However, we know that if a ball is dropped onto a floor, it will bounce and eventually come to rest. Nobody has observed a ball at rest spontaneously begin to bounce, and then bounce higher and higher. Conservation of energy does not explain why the latter process, where the ball rises higher and higher with each bounce, does not occur because such a process also could conserve the total energy of the system and surroundings.[84] By thinking about the constituent microparticles, we gain some insight into the nature of entropy. Initially, the energy of the bouncing ball is associated with the motion of its centre of mass (macroscopic coherent motion). Gradually this energy becomes more and more associated with the individual microparticles near the surface of the ball and the floor (microscopic incoherent motion), and less and less with the macroscopic motion of the ball. Therefore, we can conclude that the entropy is a measure of how energy is distributed.

7.9 DISCUSSION

In the "old-fashioned" and discarded caloric theory heat is an "imponderable subtle fluid"[85] which can be stored in bodies, can flow, cannot be destroyed but can be created. In the modern mechanical or kinetic theory of heat, it is an energy transfer process analogous to work in mechanical energy transfer processes: What I have attempted to do in this chapter and the previous chapter is to demonstrate that the old caloric theory is equivalent to modern mechanical theory of heat, i.e., caloric and entropy are different approaches to the same concept.

Unlike the First Law of Thermodynamics whose formulation began with efforts by scientists to invent a formula to unify a variety of phenomena under a single designation, the Second Law of Thermodynamics is a concept which is a part of basic human experience, i.e., an intuitive appreciation of the uniqueness of the direction of a change. For example, no one has to be told that when the Earth's gravitational potential is the driving force it will cause water to flow from a tank on top of the hill down a pipe and into another tank at the bottom of the hill, but gravity will never cause the reverse to occur. This direction of water flow is always the same unless we supply energy to the system, e.g., by using a pump. We identify the Earth's gravitational attraction at a given water level as a driving force because (i) when the water levels are the same in each tank there is no further transfer of water and, (ii) the rate of transfer increases with an increase in the difference in elevation. An analogous example occurs when energy flows from one system to another by a difference in their temperatures. We know that when this temperature difference is large, the rates of change of the systems' temperatures are greater than when the temperature difference is small, and when the two have the same temperature we observe no further changes. Consequently, we identify temperature as a driving force that causes energy to be transferred by a process we now, in modern science, call heat. No theoretical knowledge of any kind is required for us to know that if we bring two objects into close contact and exclude any interaction between them and their surroundings (an isolated system), the cold one will always get hotter and the hot one cooler but never the opposite. This direction is always the same unless we do some work on the system, as with a refrigerator. This is not a concept in any way contained within the First Law of Thermodynamics, but one involving a completely new requirement. A flow of energy by the process of heat from a cooler to a hotter object would not necessarily violate the Principle of Energy Conservation (aka the First Law of Thermodynamics). It requires the Second Law of Thermodynamics and its associated concept of caloric or entropy to allow this uniqueness of direction to be determined.

The following abstract is a section entitled "WHAT IS ENTROPY?" from Schrödinger's book entitled "What is Life?" (remember to mentally convert – where appropriate – "heat" to "energy supplied by the process of heat" and noting that "cal" (calorie) is a unit of energy – 4.18 cal is equivalent to one joule)[86]:

> Let me first emphasize that it [entropy] is not a hazy concept or idea, but a measurable physical quantity just like of the length of a rod, the temperature at any point of a body, the heat of fusion of a given crystal or the specific heat of any given substance. At the absolute zero point of temperature (roughly –273°C) the entropy of any substance is zero. When you bring the substance into any other state by slow, reversible little steps (even if thereby the substance changes its physical or chemical nature or splits up into two or more parts be of different physical or chemical nature) the entropy increases by an amount which is computed by dividing every little portion of heat you had to supply in that procedure by the absolute temperature at which it was supplied -and by summing up all these small contributions. To give an example, when you melt a solid, its entropy increases by the amount of the heat of fusion divided by the temperature at the more melting-point. You see from this, that the unit in which entropy is measured is cal./C (just as the calorie is the unit of heat or the centimetre the unit of length).

The Second Law of Thermodynamics is able to characterise an extraordinary range of phenomena under one law. Remarkably, whether these phenomena are gases filling their available volumes, iron bars in contact coming to the same temperature, vinegar and oil separating, milk mixing in your coffee, electrical processes, or chemical reactions they all have an observable property in common: their entropy increases.

On working through this book, we soon discovered that knowing the total quantity of energy in a given system is not sufficient to fully characterise that system, we also need to know how the energy is distributed among the different parts of the systems. Consider an isolated system consisting of two identical blocks of metal in thermal contact. If the blocks have the same internal energy they will have the same temperature, and the system will be in equilibrium. However, if the same amount of energy is distributed asymmetrically, so one block is hotter than the other, the system is not in equilibrium. In this case, there is a net flow of energy from the hot to the cold block, by the process of heat, and the condition of the system changes with time.

Eventually, equilibrium is attained once enough energy has been transferred to equalise the temperatures. Consequently, in order to know how a system will change or could change we need to know not only the total amount of energy in the system but also how that energy is distributed. This example also illustrates that the distribution of energy in an isolated system tends to become more uniform as time passes, never less uniform. In essence, this is the Second Law of Thermodynamics. We intuitively "know" that if the system is in thermal equilibrium that energy transfer such that one block heats up and the other cools down does not and will not occur.

As I have demonstrated in Chapter 6 and this chapter the state function – entropy – is the same as the caloric of old. Carnot's and Clausius' analysis shows the complete conversion of heat to work (or, more rigorously) of energy input to the heat engine, by the process of heat, to energy output to the surroundings, by the work process of work), in a cyclic process is impossible, but that no such limitation applies to the conversion of work to heat. In other words, every time energy is converted or transported, some of its ability to do work is irretrievably lost. This equivalence of the state functions (entropy and caloric) arrived at using the two theories of heat suggests that the two heat theories are equivalent, and that modern science was too quick to abandon the caloric theory (at the very least as a teaching aid).

The following is another extract from Callendar's address[87]:

> Finally, in 1865, when its [caloric's] importance was more fully recognised, Clausius [...] gave it the name of "entropy," and defined it as the integral of [dq/T].[88] Such a definition appeals to the mathematician only. In justice to Carnot, it should be called caloric, and defined directly by his equation [...], which any schoolboy could understand. Even the mathematician would gain by thinking of caloric as a fluid, like electricity, capable of being generated by friction or other irreversible processes. Conduction of caloric is closely associated with the electrons, and the science of heat would gain, like the science of electricity, by attaching a more material conception to the true measure of a quantity of heat, as distinguished from a quantity of [energy transferred by the process of heat].[89]

Callendar showed that entropy is basically what had already been introduced by Carnot and had been called caloric, and that the properties of entropy coincide almost perfectly with the layman's concept of *heat*. As we shall see if you take any sentence that includes the word entropy and change it to caloric, in most cases you will get a scientifically correct statement. This is what led Callendar to make the statement 'which any schoolboy could understand.' Unfortunately, Callendar's ideas remained almost unnoticed by the scientific community, and further generations of scientists have continually been confused by an inappropriate heat concept and an incomprehensible concept of entropy. Sixty years later, in 1972, Georg Job published a textbook about thermodynamics, in German language, with the title: "Neudarstellung der Wärmelehre: die Entropie als Wärme" [Restatement of Thermodynamics: Entropy as Heat]. The author was not, at that time, aware of Callendar's work. In 1985 Falk[90] proved Callendar's assertion once again and concluded:

> The entropy introduced into physics by Clausius was, contrary to general belief, not a new physical quantity but the reconstruction of the 'quantity of heat' conceived about one hundred years earlier by the Scottish chemist Black. The same quantity was also used under the name 'calorique' by Carnot in his work, which laid the foundations of thermodynamics. That entropy and Black's 'quantity of heat' are only two names for the same physical quantity is not only of historical interest but is of significance to the teaching of thermodynamics as well. It asserts that entropy can be visualised as a kind of substance which obeys 'half a conservation theorem': It can be created but not destroyed.

Mach was one of the few scientists who questioned the so-called invalidation of the caloric theory of heat[91]:

> The generally accepted notion of a caloric, or heat-stuff, was strongly shaken by the work of Mayer and Joule. If the quantity of heat can be increased and diminished, people said, heat cannot be a substance, but must be a motion. The subordinate part of this statement has become much more popular than all the rest of the doctrine of energy. But we may convince ourselves that the motional conception of heat is now as unessential as was formerly its conception as a substance. Both ideas were favoured or impeded solely by accidental historical circumstances. It does not follow that heat is not a substance from the fact that a mechanical equivalent exists for quantity of heat. We will make this clear by the following question which

> bright students have sometimes put to me. Is there a mechanical equivalent of electricity as there is a mechanical equivalent of heat? Yes, and no. There is no mechanical equivalent of quantity of electricity as there is an equivalent of quantity of heat, because the same quantity of electricity has a very different capacity for work, according to the circumstances in which it is placed; but there is a mechanical equivalent of electrical energy.[92]
>
> Let us ask another question. Is there a mechanical equivalent of water? No, there is no mechanical equivalent of quantity of water, but there is a mechanical equivalent of weight of water multiplied by its distance of descent.

Mach then gives an account of the history of the development of the theory of electricity and demonstrates that it is an accident of history that we measure electricity and water with scales.[93] Because we measure water and electric charge with scales they do not disappear when work is done. Mach then goes on to say[94]:

> But suppose the capacity of water for work were called quantity, and had to be measured, therefore, by a mill instead of by scales; then this quantity also would disappear as it performed the work. It may, now, be easily conceived that many substances are not so easily got at as water. In that case we should be unable to carry out the one kind of measurement with the scales whilst many other modes of measurement would still be left us.
>
> In the case of heat, now, the historically established measure of "quantity" is accidentally the work-value of the heat.[95] Accordingly, its quantity disappears when work is done. But that heat is not a substance follows from this as little as does the opposite conclusion that it is a substance.

Mach concludes[96]:

> If anyone to-day should still wish to think of heat as a substance, we might allow that person this liberty with little ado. He would only have to assume that that which we call quantity of heat was the energy of a substance whose quantity remained unaltered, but whose energy changed.[97] In point of fact we might much better say, in analogy with the other terms of physics, energy of heat, instead of quantity of heat.
>
> When we wonder, therefore, at the discovery that heat is motion, we wonder at something that was never discovered. It is perfectly indifferent and possesses not the slightest scientific value, whether we think of heat as a substance or not. The fact is, heat behaves in some connections like a substance, in others not.

Since entropy is a thermodynamic state property, its value depends only on the state of a system, not on the history of the system. Therefore, to determine the change in entropy of a system from one state to another, it is sufficient to evaluate the change for a reversible process between those two states; the change in entropy for any other process connecting the same two states will be the same as the change for the reversible process.

Because entropy content cannot be destroyed, we can now redefine reversible and irreversible processes:

1. If a process is carried out which does not increase the entropy content of the universe, then in principle it would be possible to return to the state the universe had at the start of the process, since doing so would not require a decrease in entropy content. A process that does not increase the entropy content of the universe is a reversible process.

2. If a process is carried out in such a way that the entropy content of the universe increases, we know it can never be restored to its earlier (smaller) value and the universe can never return to its earlier state. This type of process is therefore called an irreversible process.

The above can be condensed to read - *an irreversible process permanently changes the universe, and a reversible process does not.*

Since entropy is a thermodynamic state property, its value depends only on the state of a system, not on the history of the system. Therefore, to determine the change in entropy of a system from one state to another, it is sufficient to evaluate the change for a reversible process between those two states; the change in entropy for any other process connecting the same two states will be the same as the change for the reversible process.

Valentine: Like a film.
Hannah: What did she see?

Valentine: *That you can't run the film backwards. Heat was the first thing which didn't work that way. Not like Newton. A film of a pendulum, or a ball falling through the air - backwards, it looks the same.*
Hannah: *The ball would be going the wrong way.*
Valentine: *You'd have to know that. But with heat - friction - a ball breaking a window –*
Hannah: *Yes. Valentine: It won't work backwards.*
Hannah: *Who thought it did?*
Valentine: *She saw why. You can put back the bits of glass but you can't collect up the heat of the smash. It's gone.*
Septimus: *So the Improved Newtonian Universe must cease and grow cold. Dear me.*
Valentine: *The heat goes into the mix.*
(He gestures to indicate the air in the room, in the universe.)
Thomasina: *Yes, we must hurry if we are going to dance.*
Valentine: *And everything is mixing the same way, all the time, irreversibly. . .*
Septimus: *Oh, we have time, I think.*
Valentine: *. . . till there's no time left. That's what time means.*
Septimus: *When we have found all the mysteries and lost all the meaning, we will be alone, on an empty shore.*

In the second paragraph of this section, I wrote: "It requires the Second Law of Thermodynamics and its associated concept of caloric or entropy to allow this uniqueness of direction to be determined." I nearly wrote – "It requires the Second Law of Thermodynamics and its associated concept of caloric or entropy to allow this uniqueness of direction to be determined and understood." I then realised that although the Second Law of Thermodynamics predicts the directionality of a process it does not tell you why a process proceeds in a certain direction. To understand why we need to consider statistical thermodynamics. This is the subject of the next and final chapter of this book.

NOTES

1 R. Battino, "'Mysteries' of the first and second laws of thermodynamics", *Chem. Educ. Today,* 84(5), 2007. https://www.academia.edu/27040844/Mysteries_of_the_First_and_Second_Laws_of_Thermodynamics.

2 G. Falk, "Entropy, a resurrection of caloric – a look at the history of thermodynamics", *Eur. J. Phys.* 6, 108–115, 1985.

3 Process quantity ≡ path function; state quantity ≡ state function.

4 Only, of course, in reversible processes.

5 H.L. Callendar, "The caloric theory of heat and carnot's principle", *Proc. Phys.* XXIII, 178, 1911.

6 The corrected paper is "XVII. On the Restoration of Mechanical Energy from an unequally heated space", By Prof. W Thomson, *Phil. Mag. S. 4,* 5(30), 102–105, 1853.

7 For clarity I have altered Callendar's equation to Equation 6.6 from Chapter 6.

8 I have removed the word "thermal".

9 R. Clausius, *The Mechanical Theory of Heat, with its Applications to the Steam-Engine and to the Physical Properties of Bodies,* John van Voorst, 1867.

10 This is known as the Clausius Statemen and is the original version of the Second Law of Thermodynamics. See Fourth Memoir (originally published in 1854).

11 R. Clausius, Fourth Memoir, 1854; in R. Clausius, *The Mechanical Theory of Heat,* W. Browne, Translator, Macmillan, 1879. 1867; Chapter V.

12 R. Clausius, Ninth Memoir, 1865; in R. Clausius, *The Mechanical Theory of Heat,* W. Browne, Translator, London, 1879. 1867; Chapter X.

13 W. Thomson, *Philos. Mag.* 4, 511, 1852.

14 I.K. Howard, "S Is for Entropy. U Is for Energy. What Was Clausius Thinking?", Department of Chemistry, Houghton College, Houghton, NY 14744. http://www.physics.smu.edu/scalise/P3374fa20/notes/Sentropy.pdf.

15 https://www.lindahall.org/about/news/scientist-of-the-day/rudolf-clausius/.

16 As did Carnot himself in a posthumous manuscript.

17 G. Falk, "Entropy, a resurrection of caloric – a look at the history of thermodynamics", *Eur. J. Phys.* 6, 108–115, 1985.

18 https://www.chemeurope.com/en/encyclopedia/Disgregation.html.

19 Remember "heat" in this context should be read as "... energy given up by the body, by the process of heat, to any reservoir ...".

20 An adiabatic process is defined as a process in which no heat transfer takes place, in other words, it is a process that takes place in an isolated system.

21 Remember it is the total entropy change in the heat reservoirs (the surroundings) that is not equal to zero in an irreversible process (but is equal to zero in a reversible process).

22 Remember if $\left[(+|3|)-(+|4|)=-|1|<0\right]$ then $\left[(-|3|)-(-|4|)=+|1|>0\right]$, i.e., we must reverse the inequality if we replace positive absolute values with negative absolute values.

23 https://physics.stackexchange.com/questions/358142/entropy-generation-during-heat-transfer-processes.

24 Because I disagree with it, I have removed the following sentence *"But, if the partition returns to its original temperature in the end, its entropy doesn't change."* I would suggest that at steady state the entropy content of the conductor does not change (and this applies even if entropy is generated in the partition).

25 See also Section 6.5.2

26 This extract is from https://en.wikipedia.org/wiki/Entropy_(order_and_disorder). "For the present I will confine myself to the statement of one result. If for the entire universe we conceive the same magnitude to be determined, consistently and with due regard to all circumstances, which for a single body I have called entropy, and if at the same time we introduce the other and simpler conception of energy, we may express in the following manner the fundamental laws of the universe which correspond to the two fundamental theorems of the mechanical theory of heat:

1. The energy of the universe is constant.

2. The entropy of the universe tends to a maximum."

If nothing else, it gives an indication of the opacity of his writing and/or of the English translation.

27 The "world" here refers to an isolated system, i.e., the universe.

28 https://www.ncbi.nlm.nih.gov/pmc/articles/PMC7517180/.

29 "Challenges To The Second Law Of Thermodynamics Theory And Experiment" by Vladislav Capek and Daniel P Sheehan (2005 Springer). This book gives 25 statements of the Second Law of Thermodynamics and they did not include one I was taught at university – "The entropy of the universe is continually increasing."

30 See Section 5.7.4 Figure 5.17. Known as the "The Clausius Statement".

31 See Section5.7.3 Figure 5.15. Known as "The Kelvin-Planck Statement".

32 Only for hypothetical reversible processes does the entropy content of the universe remain unchanged.

33 This section is based on Section 2.8 (pp 86–87) in W.J. Moore, *Physical Chemistry*, 5th Edition, Longman, 1972.

34 For an ideal gas the internal energy is a function of temperature only.

35 "The change in entropy in going from equilibrium state A to equilibrium state B is always the same irrespective of the path between A and B, since entropy is a state function. It makes no difference whether the path is reversible or irreversible. However, only if the path is reversible is the entropy change is given by: $\Delta S = S_B - S_A = \int_A^B \frac{[\delta q_{rev}]}{T}$" From Section 7.3.3.

36 I have used our agreed shorthand here, previously I would have used the following words for the italicised script – "and in doing there is a flow of $+|\mathbf{q}_{rev}|$ energy, by the process of heat, into the system from the heat reservoir and a flow of an equivalent amount of energy $-|\mathbf{w}_{rev}|$ (where $|\mathbf{q}_{rev}| = |\mathbf{w}_{rev}|$) by the process of work into the system from the work reservoir (remember the sign convention is from the system's point of view)."

37 Remember (i) $V_{initial} < V_{final}$ are initial and final values for irreversible free expansion, (ii) in Equations (7.49) and (7.50) **w** is a negative real number and **q** a positive real number, $\ln \frac{\mathbf{V}_{final}}{\mathbf{V}_{initial}}$ is a positive real number and $\Delta \mathbf{S}^{free\ expansion}$ is a positive real number, (iii) for reversible isothermal processes $\mathbf{q} + \mathbf{w} = 0$.

38 J.W. Gibbs, *The Scientific Papers of J. Willard Gibbs, Volume One, Thermodynamics*, Longmans, Green, and Co., p. 157, 1906. http://books.google.com/books/about/Scientific_Papers_of_J_Willard_Gibbs_The.html?id=-neYVEbAm4oC.

39 A semi-permeable membrane allows certain microparticles to pass through it by diffusion and acts as a barrier to certain other microparticles.

40 Why C_V in Equation (7.31)? The short answer is the definition of the internal energy of n moles of ideal gas at temperature T is $U = nC_VT$. This from Wikipedia (my underlining): "The internal energy of a thermodynamic system is the energy contained within it. It is the energy necessary to create or prepare the system in any given internal state. It does not include the kinetic energy of motion of the system as a whole, nor the potential energy of the system as a whole due to external force fields, including the energy of displacement of the surroundings of the system." "The energy of displacement of the surroundings of the system See next section" https://en.wikipedia.org/wiki/Internal_energy.

41 Remember energy is only transferred by the process of heat from a higher temperature to a lower temperature.

42 I have already considered this type of machine in Section 5.7.3 Figure 5.15 where the heat reservoir is labelled a cold reservoir.

43 Or using our sign convention "in each cycle there is a flow $++|\mathbf{q}_{system}|$ energy into the system by the process of heat and a flow of$-|\mathbf{w}_{system}|$ of energy into the system by the process of work.

44 The original statement by Kelvin is "It is impossible by means of inanimate material agency to derive mechanical effect from any portion of matter by cooling it below the temperature of the coldest of the surrounding objects." Planck wrote "It is impossible to construct an

engine, which working in a complete cycle, will produce no effect other than the raising of a weight and the cooling of a heat reservoir." M.W. Zemansky and R.H. Dittman, *Heat and Thermodynamics*, McGraw-Hill, 1997.

45 See Sections 5.1, 5.3, 5.7.3, 5.7.5, and 5.8.

46 See Section 4.2

47 We used it to determine the mechanical equivalent of heat (see Section 6.5.2). Diagram adapted from https://physics.nyu.edu/~physlab/GenPhysII_PhysIII/Intermediate_Physics_I_writeups/Mechanical%20equivalent%20of%20heat-Physics-Majors%20REVISED.pdf.

48 US spelling of aluminium.

49 The extract uses the term thermal energy which is a meaningless term. The shorthand "converted work into heat" is more rigorous – see Endnote 75 in Chapter 5and Endnote 50 in this chapter.

50 I have deleted the word "thermal".

51 See Section 6.4.2.

52 Rigorously – energy transferred by the process of work can be completely converted into energy transferred by the process of heat, but energy transferred by the process of heat cannot be completely converted into energy transferred by the process of work.

53 https://en.wikipedia.org/wiki/Energy.

54 https://www.quora.com/%E2%80%9CEnergy-is-the-capacity-to-do-work-%E2%80%9D-Is-this-statement-totally-true-Explain.

55 R.L. Lehrman, "Energy is not the ability to do work", *Phys. Teach.* 11, 15, 1973, https://doi.org/10.1119/1.2349846.

56 Or converted into or from mass.

57 J.W. Gibbs, "Graphical methods in the thermodynamics of fluids", *Trans. Conn. Acad.* I, 309–342, 1873.

58 J.C. Maxwell, "Tait's 'Thermodynamics' (I)", Nature, 257–59, 1878; "Tait's 'Thermodynamics' (II)", (pp. 278–81). Nature, Feb. 07. It should also be noted that Maxwell was wrong – several generations (and more) on and the concept of entropy is still bewildering not only students but also experienced scientists.

59 J. von Neumann suggesting to Claude Shannon a name for his new uncertainty function, as quoted in Scientific American Vol. 225 No. 3, (1971), p. 180]. Neuman (1903–1957) was a brilliant mathematician and all-round genius. He ranks with Richard Feynmann (another all-round genius) as one of the great scientists of the 20th century.

60 James Swinburne, "Entropy: or Thermodynamics from an Engineer's Standpoint and the Reversibility of Thermodynamics", Westminster: Archibald Constable & Co (1904) pp. 3–4. [Swinburne (1858–1958) was a British electrical engineer and manufacturer. He worked on the first electric lightbulb and was one of the founders of the plastics industry.]

61 In "The Electrician" (9 Jan 1903) - I am afraid I have no definite information about Sydney Herbert Evershed. There was Sir Sydney Herbert Evershed (13 January 1861–7 March 1937) an English brewer and cricketer who played first-class cricket for Derbyshire from 1880 to 1901.

62 G. Falk, "Entropy, a resurrection of caloric – a look at the history of thermodynamics", *Eur. J. Phys.* 6, 108–115, 1985.

63 H.L. Callendar, "The caloric theory of heat and carnot's principle", *Proc. Phys.* XXIII, 153–189, 1911.

64 W. M. Haddad, "Thermodynamics: The unique universal science", *Entropy*, 19(11), 621, 2017. https://doi.org/10.3390/e19110621.

65 https://en.wikipedia.org/wiki/Entropy_(order and disorder).

66 http://www.tim-thompson.com/entropy1.html

67 This an approximation of an enormous number. The number of permutations is 52 factorial (usually written 52!), which equals $52 \times 51 \times 50 \ldots \times 3 \times 2 \times 1$.

68 http://civil.colorado.edu/~silverst/aren2110/CLAUSIUS%20INEQUALITY%20AND%20 ENTROPY.pdf.

69 Trope - a common or overused theme or device (Merriam-Webster Dictionary - https://www.merriam-webster.com/dictionary/trope).

70 Classical thermodynamics describes macroscopic systems in terms of a few variables (functions of state): temperature, pressure, volume, etc. But such a system is really made of microparticles (atoms and /or molecules), and another description must be possible in principle: we could specify the position and energy of all the microparticles – this is microstate. Of course, as the microparticles interact, this state changes very rapidly – perhaps 10^{35} times a second. But the observed macrostate does not change. Many different microstates all correspond to the same macrostate. This subject is considered in Chapter 8.

71 This is from https://en.wikipedia.org/wiki/Third_law_of_thermodynamics: "Entropy is related to the number of accessible microstates, and there is typically one unique state (called the ground state) with minimum energy. In such a case, the entropy at absolute zero will be exactly zero. If the system does not have a well-defined order (if its order is glassy, for example), then there may remain some finite entropy as the system is brought to very low temperatures, either because the system becomes locked into a configuration with non-minimal energy or because the minimum energy state is non-unique. The constant value is called the residual entropy of the system."

72 First mentioned in Section 6.6.

73 Sir Arthur Stanley Eddington, (1882–1944) was an English astronomer, physicist, and mathematician of the early 20th century who is famous for his work regarding the theory of relativity. He conducted an expedition to observe the Solar eclipse of May 29, 1919 that provided one of the earliest confirmations of relativity, and he became known for his popular expositions and interpretations of the theory.

74 As scientists we must reject the intervention of a divine being and consider it 'magic'.

75 The deep philosophical problem here (which I am not going to address is) – *how can macroscopic irreversibility result from microscopic processes that are fundamentally reversible?*

76 E. Mach, *Popular Scientific Lectures*, (translated by Thomas J McCormack) Cambridge University Press, p. 178, 2014.

77 https://www.forbes.com/sites/startswithabang/2016/10/12/where-does-our-arrow-of-time-come-from/#6982f18277ff.

78 This is from the New Scientist website. "Cosmic inflation is a faster-than-light expansion of the universe that spawned many others".

"Inflation was invented to explain a couple of features of the universe that are really hard to explain without it. The first is that Einstein's general theory of relativity famously makes mass bend space and time – so you'd expect a universe like ours, containing mass, to be overall curved in some way, either in on itself like a ball ("positive" curvature) or out on itself like a saddle ("negative" curvature)".

"In fact, it's darn nearly flat. Not only that, but even bits of it far off in different directions as observed from Earth have roughly the same temperature, although in an expanding universe there wouldn't have been time for heat to pass between them to even things out. That seems a naked assault on the laws of thermodynamics".

"Cosmic inflation solves these problems at a stroke. In its earliest instants, the universe expanded faster than light (light's speed limit only applies to things within the universe). That ironed out wrinkles in its early chaotic self and meant that even now far-flung parts were once in close contact, so could swap heat."

79 Dark energy is hypothesised to be responsible for the accelerated expansion of the Universe – an observed effect cannot be fully explained by gravity.

80 https://www.newscientist.com/definition/causality/.

81 https://aeon.co/essays/could-we-explain-the-world-without-cause-and-effect.

82 Trajectories exaggerated to show the effect.

83 I have ignored open systems here which can decrease their entropy by loss energy by process of heat and/or loss of mass (note that the universe is not an open system).

84 You might be tempted to say the cause is "friction", but friction is just a name for a phenomenological force. Just like the label dyslexic does not tell you why the labelled person is a poor reader.

85 Callendar's words – see Section 6.4.4.

86 "What is Life?" by E. Schrödinger (First published 1944 and based on lectures delivered under the auspices of the Dublin. Institute for Advanced Studies at Trinity College, Dublin, in February 1943). http://www.whatislife.ie/downloads/What-is-Life.pdf.

87 Both Callendar and the German group at Karlsruhe have suggested that the unit of heat or caloric should be the 'carnot' where one carnot is that quantity of heat which is capable of producing one joule of work in a Carnot cycle per 1 K fall. My personal view is that the more general use of the concept of heat or caloric arising from Carnot's work should be the goal and the use of the carnot will probably not help and may, in fact hinder, the attainment of this goal because the direct measurement of the carnot is not so straightforward as calorimetry.

88 Callendar wrote dQ/T.

89 Callendar used the unnecessary/meaningless term "thermal energy".

90 G. Falk, "Entropy, a resurrection of caloric – a look at the history of thermodynamics", *Eur. J. Phys.* 6, 108–115, 1985. It is interesting that Falk was not aware of Callendar's Presidential Address until a referee of his paper informed him.

91 E. Mach, *Popular Scientific Lectures*, translated by Thomas J McCormack, Chicago, The Open Court Publishing Company, pp. 160–161, 1898.

92 Remember quantity of electricity (or quantity of electric charge) is measured in coulombs (C). Electrical work w = IVt, where the units of I are coulombs per second $\left(C.s^{-1}\right)$. Therefore, w = CV where V is the potential difference. In other words, there is no mechanical equivalent of electric charge.

93 Electric charge is measured with a torsion balance developed by Coulomb.

94 E. Mach, *Popular Scientific Lectures*, translated by Thomas J McCormack, Chicago, The Open Court Publishing Company, pp. 170–171, 1898.

95 i.e., energy.

96 E. Mach, *Popular Scientific Lectures*, translated by Thomas J McCormack, Chicago, The Open Court Publishing Company, pp. 170–171, 1898.

97 This sentence is best understood by considering quantity of water in a watermill or quantity of electricity in an electrical system. The energy of these quantities is decided by their potential – height in the watermill and voltage in the electric circuit.

8 Statistical Thermodynamics

> The general struggle for existence of animate beings is not a struggle for raw materials, these for organisms are air water & soil, all abundantly available, nor for energy which exists in plenty in the sun and any hot body in the form of heat, but rather a struggle for entropy, which becomes available through the transition of energy from the hot sun to the cold earth.[1]

8.1 INTRODUCTION

In this final chapter, I want to consider another approach to thermodynamics using statistics. Statistical thermodynamics is a theory that uses the properties of microparticles to predict the behaviour of macroscopic quantities of compounds. Although the origins of statistical thermodynamics predate the development of quantum mechanics, today's statistical thermodynamics assumes the quantisation of energy and position (see Section 8.2.3). The following extract is from Wikipedia[2]:

> In physics, statistical mechanics is a mathematical framework that applies statistical methods and probability theory to large assemblies of microscopic entities. It does not assume or postulate any natural laws, but explains the macroscopic behavior of nature from the behavior of such ensembles.
>
> Statistical mechanics arose out of the development of classical thermodynamics, a field for which it was successful in explaining macroscopic physical properties – such as temperature, pressure, and heat capacity – in terms of microscopic parameters that fluctuate about average values and are characterized by probability distributions. This established the fields of statistical thermodynamics and statistical physics.
>
> The founding of the field of statistical mechanics is generally credited to three physicists:
>
> - Ludwig Boltzmann, who developed the fundamental interpretation of entropy in terms of a collection of microstates.
> - James Clerk Maxwell, who developed models of the probability distribution of such states.
> - Josiah Willard Gibbs, who coined the name of the field in 1884.

8.2 STATISTICAL MECHANICS

8.2.1 Introduction

In classical thermodynamics, we specify the state of a system of a gas using the temperature (T), the pressure (P), the volume (V), internal energy (U), and the number of microparticles (N).[3] A state specified in this manner is known as a macrostate. A macrostate does not change over time if the system is in equilibrium, i.e., its observable properties (T, P, V, U, and N) do not change with time. Because we know such systems are made up of microparticles, there is another way of describing the system: we could specify the position and energy (or momentum), at one instant in time, for all the microparticles in the system – this is known as a microstate. However, in a macroscopic multi-particle system, because the microparticles are moving and continually colliding with each other and with the walls of the container, a particular microstate exists only for an instant – perhaps 10^{-35} of a second.[4] It is then replaced by another state, and there is no experimental way to measure the set of positions and motions which comprise one state among the vast number of them which occur sequentially.

Helium (diameter 2.3×10^{-11} m)[5] atoms at atmospheric pressure and 298 K have a most probable velocity of about 1.8×10^{3} m.s^{-1} with a range of velocities from zero to around 5×10^{3} m.s^{-1}.[6] Typically, a helium atom travels about 200 times its diameter before colliding with another atom and changing kinetic energy; in other words, it typically collides with another helium atom in $\frac{200 \times 2.3 \times 10^{-11}}{1.8 \times 10^{3}} = 2.6 \times 10^{-12}$ seconds or it undergoes about 4×10^{11} collisions every second. However, despite this continuous change of position and kinetic energy, the internal energy of an isolated macroscopic system of helium gas at equilibrium does not change with time. Remember the internal energy of an ideal gas is the sum of the kinetic energies of all constituent microparticles where kinetic energy equals $\frac{1}{2}mv^2$, i.e., $U = \sum_{i=1}^{i=N} \frac{1}{2}mv_i^2$ for a system containing N microparticles. This analysis tells us that for each macrostate of a macrosystem, there are many, many

DOI: 10.1201/9781003608400-8

trillions of microstates, e.g., if you interchange energy of two gas microparticles, the macrostate does not change, but you have a different microstate. I will clarify these points later. But now, I want to introduce the subject of statistical thermodynamics by considering coin tossing thought experiments, and then I will consider the free expansion of an ideal gas from a statistical point of view using the concepts developed from the coin tossing thought experiments.

8.2.2 Macrostates and Microstates Illustrated by Coin Tossing

We know that the macrostate of a system determines its thermodynamic properties. It must also be the case (by definition) that the microstate of a system determines its thermodynamic properties. We know that one way to describe the microstate of an ideal gas system is to set down the positions and kinetic energy of every microparticle in the system, and, in addition, we know that this microstate changes within an instant (about 10^{-35} seconds). On the other hand, when the system is at thermodynamic equilibrium, the macrostate of the system can be uniquely defined using only a few thermodynamic variables. Consequently, one macrostate must correspond to many different available microstates, i.e., a macrostate defined by a given pressure and temperature is consistent with a huge number of microstates. This fundamental difference between microstate and macrostate can be illustrated by considering a system that contains N coins; each of them is either heads up or tails up. To define a microstate of the system, we must know the state (heads or tails) of all the coins in the system: this implies that all coins must be identifiable (or distinguishable). In contrast, the macrostates of the system can be defined by specifying the total number of the coins that are heads up and tails up: this implies that individual coins are not identifiable. This is shown in Figure 8.1 for n = 4.

If each coin is identifiable, then there are 16 "available microstates" (known as permutations[7] in mathematics). The fundamental assumption of statistical thermodynamics is that "all microstates are equally probable".[8] If the coins are not identifiable, then there are 5 "macrostates" (known as combinations[9] in mathematics). In contrast to microstates, each macrostate is not equally probable. If all four coins are tossed, then you would expect the odds of getting four heads or four tails to be 1/16, three heads and one tail, or three tails and one head to be 4/16, and two heads and two tails 6/16 (see Table 8.1).

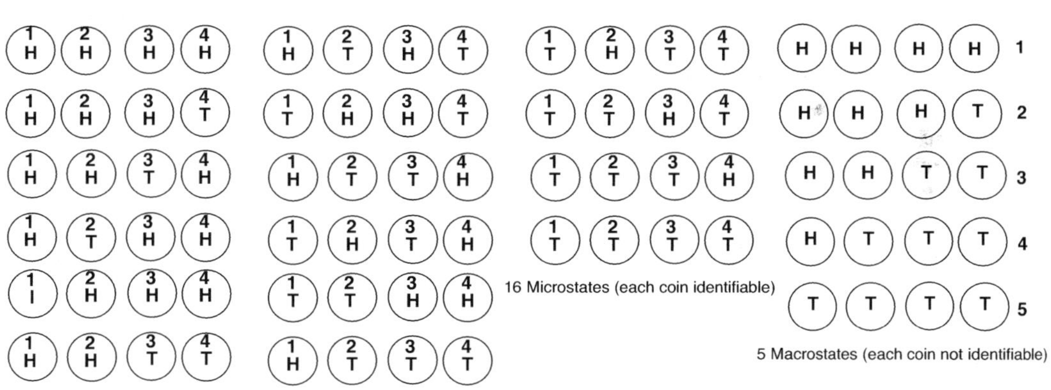

Figure 8.1 Illustration, using coins, of the difference between macrostates and microstates.

Table 8.1: Number of Microstates Corresponding to Each Macrostate When Four Coins Are Tossed

Macrostate	Number of Microstates Corresponding to Macrostate
4H	1
3H/1T	4
2H/2T	6
1H/3T	4
4T	1

Table 8.2: The Permutations of T and H for Six Coins

1	HHHHHH	2	THHHHH	3	HTHHHH	4	HHTHHH	5	HHHTHH	6	HHHHTH
7	HHHHHT	8	TTHHHH	9	THTHHH	10	THHTHH	11	THHHTH	12	THHHHT
13	HTTHHH	14	HTHTHH	15	HTHHTH	16	HTHHHT	17	HHTTHH	18	HHTHTH
19	HHTHHT	20	HHHTTH	21	HHHTHT	22	HHHHTT	23	TTTHHH	24	TTHTHH
25	TTHHTH	26	TTHHHT	27	THTTHH	28	THTHTH	29	THTHHT	30	THHTTH
31	THHTHT	32	THHHTT	33	HTTTHH	34	HTTHTH	35	HTTHHT	36	HTHTTH
37	HTHTTH	38	HTHHTT	39	HHTHTT	40	HHTTTH	41	HHTTHT	42	HHHTTT
43	HHTTTT	44	HTHTTT	45	HTTHTT	46	HTTTHT	47	HTTTTH	48	THHTTT
49	THTHTT	50	THTTHT	51	THTTTH	52	TTHHTT	53	TTHTHT	54	TTHTTH
55	TTTHHT	56	TTTHTH	57	TTTTHH	58	THHHHH	59	HTHHHH	60	HHTHHH
61	HHHTHH	62	HHHHTH	63	HHHHHT	64	TTTTTT				

Table 8.3: The Combinations of T and H for Six Coins

Combination				
Number of Heads	Number of Tails	Number of Permutations per Combination	Odds	Probability
6	0	1	1/64	0.015625
5	1	6	6/64	0.09375
4	2	15	15/64	0.234375
3	3	20	20/64	0.3125
2	4	15	15/64	0.234375
1	5	6	6/64	0.09375
0	6	1	1/64	0.015625

Now consider tossing six distinguishable (by order) coins. There are 64 different states or permutations (microstates), these are listed in Table 8.2.

Tossing six indistinguishable coins results in seven different combinations (macrostates); these are listed in Table 8.3 with their associated probabilities.

If you throw six coins onto the floor, there are 20 chances out of 64 that there will be 3 heads and 3 tails: this is the most probable combination. But this combination is made up of 20 permutations each of which has a 1 chance in 64 of occurring.[10] In other words, the macrostate of 3 heads and 3 tails is made up of 20 microstates. In Table 8.4, I have set down data for systems containing 2–16 coins in increments of 2 (so we can have an odd number of microstates and therefore, a unique most probable "macrostate").

The following statement is one of the most fundamental assertions in statistical thermodynamics:

> When a macroscopic sample of gas, i.e., large enough to be chemically significant, is considered only a single macrostate needs be considered; even the next-most probable macrostate can be neglected and the probability of a deviation in an observed (i.e., measured) macrostate as a result a change in microstates is so small as to be ignored.

Unfortunately, it is not straightforward to demonstrate the above statement with relatively small numbers. Let me start by considering an incredibly large number of coin tosses. In this example, I have used 10^{23} coin tosses as this is a number of the same order as the number of microparticles in a mole of material.[11] Assuming each toss is fair (i.e., there is an evens chance that it will be either a head or a tail), then 10^{23} tosses will result in exactly 50% heads and 50% tails. But this does not mean that after 10^{23} tosses of a coin you will have exactly counted exactly 5×10^{22} heads and 5×10^{22} tails. Consider the following, if after 10^{23} tosses of a coin we had, say, 10^{13} more heads than tails (odds of this happening are astronomical) then, using a normal hand-held calculator we calculate $\frac{0.5\times10^{23}+10^{13}}{10^{23}} = 0.500000000$, i.e., 50.0000000% heads. If we had 10^{14} more heads than tails, then

Table 8.4: Macrostate and Microstate Data When up to 16 Coins Are Tossed

Number of Coins	Number of Available Microstates	Number of Macrostates	Most Probable Macrostate	Number of Microstates Corresponding to the Most Probable Macrostate
2	4	3	1H &1T	2
4	16	5	2H & 2T	6
6	64	7	3H & 3T	20
8	256	9	4H & 4T	70
10	1024	11	5H & 5T	252
12	4096	13	6H & 6T	924
14	16384	15	7H & 7T	3432
16	65536	17	8H & 8T	12870

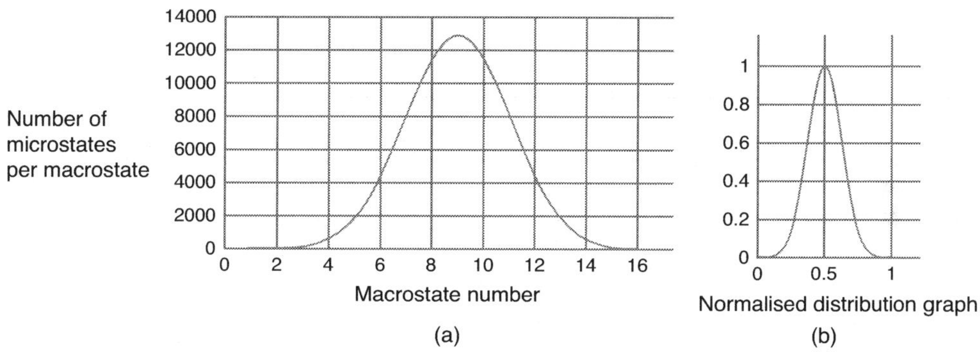

Figure 8.2 Distribution graphs: (a) Plot of number of microstates in each (numbered) macrostate and (b) a normalisation of (a).

we would 50.0000001% heads and 49.9999999% tails. Such is the property of the enormous numbers that we deal with in statistical thermodynamics.

The best way I have come up with to demonstrate the above fundamental assertion ("only a single macrostate needs to be considered") is to expand the example given earlier using coins. From Table 8.4 we know that 16 identifiable coins can be arranged in 2^{10}=65536 ways or microstates and we know we have 17 ways (macrostates) of arranging 16 non-identifiable coins. Let us number the macrostates in the same manner as I have done on the right-hand side of Figure 8.1, e.g., macrostate 1 is 16 heads, macrostate 2 is 15 heads and 1 tail, ..., macrostate 17 is 16 tails. Then we can plot a graph of the number of microstates in each macrostate against each macrostate number for 16 coins (Figure 8.2a). We can only compare these "distribution" graphs if we "normalise" the axes.[12] In this case, I have (i) normalised the x-axis by dividing each numbered macrostate by the total number of macrostates, and (ii) have normalised the y-axis by dividing the number of microstates per macrostate by the total number of available microstates. This procedure converts the graph shown in Figure 8.2a into that shown in Figure 8.2b. I will call Figure 8.2b a "normalised distribution graph".

Using an Excel spread sheet I have carried out a similar exercise for 40 and 100 identifiable coins. Figure 8.3 shows the normalised distribution graphs from these calculations together with the one with 16 coins: all graphs are shown on the same scale. Note that the maximum number of coins is only 100 in Figure 8.3 and macroscopic quantities of material contain greater than 10^{20} microparticles. This suggests that as the number of coins tends to infinity the distribution curve tends to a vertical straight line which I have also shown in Figure 8.3 (but note that the final straight line graph is illustrative and not based on calculation).

If you think about it, you already knew this because a macrostate is the thermodynamic state of any system (in equilibrium) that is exactly characterised by the measurement of the system's macroscopic properties such as P, V, T, and the number of microparticles present. We would be rather taken aback if its pressure, volume, and/or temperature spontaneously changed because

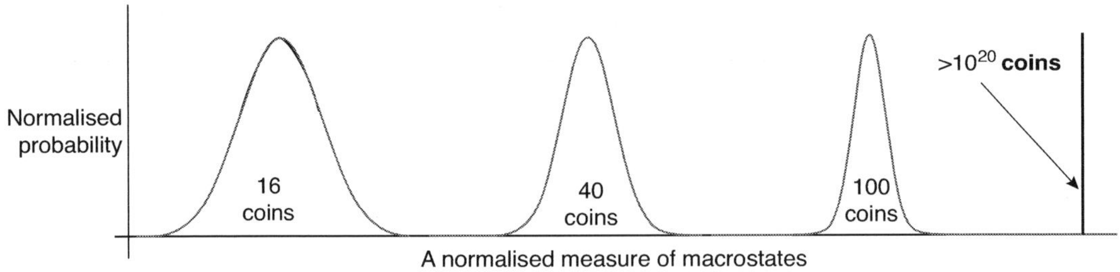

Figure 8.3 Normalised distribution graphs.

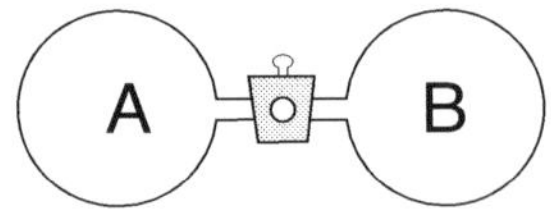

Figure 8.4 Two vessels of equal volume connected by a valve.

its macrostate had changed while the system was in equilibrium. It is exactly analogous to why we do not expect ever to observe significantly different numbers of microparticles of a chemically significant (i.e., macroscopic) amount of an ideal gas in two connected vessels under equilibrium conditions (see next section).

The above analysis answers the following frequently asked question: *If the microstate determines the thermodynamic properties of the system and we know that each microstate lasts only for an* "instant", *then why do not the thermodynamic properties of a macroscopic quantity of ideal gas at equilibrium (such as pressure, temperature or volume) change in an instant?* We now know the answer is to do with the incredibly large number of microparticles in a macroscopic quantity of gas.

8.2.3 Free Expansion Considered from a Statistical Point of View

Consider two vessels A and B of equal volume connected by a closed valve (Figure 8.4). Numbered non-interacting gas microparticles are introduced into volume A and volume B is under vacuum. The valve is opened and after a suitable time to ensure equilibrium is reached the number of microparticles in A and B are counted. At equilibrium each microparticle has 50/50 chance of being in either vessel; this probability is independent of the position of the other microparticles in the system. As far as we are concerned here, a microstate of the system consists of the position of each of the numbered microparticles (vessel A or vessel B) and a macrostate of the system consists of how many microparticles are in each of the two vessels Also note that the microparticle(s) are constantly moving from A to B or B to A.

If we place one gas microparticle into vessel A and open the valve. Once equilibrium is reached there are two equally likely microstates and there is a 50% chance the microparticle is in A and a 50% chance the microparticle is in B, i.e., there are two macrostates (Figure 8.5).

If we add two gas microparticles to A and open the valve, then once equilibrium is reached there are four equally likely microstates and three macrostates (Figure 8.6).

Although there are four equally probable microstates,[13] the three macrostates are not equally likely because there is a 25% (one in four) probability the two microparticles are in A, 25% (one in four) probability the two are in B and 50% (two in four) one is in A, and one is in B.

If we start with three microparticles, then there are eight equally likely microstates and four macrostates (Figure 8.7). The probability of finding all three microparticles in A is12.5% (one in eight), two gas microparticles in A and one in B is 37.5% (three in eight), one gas microparticle in A and two in B is 37.5% (three in eight), and three gas microparticles in B is 12.5% (one in eight).

Finally consider four microparticles. There are 16 equally likely microstates and five macrostates (Figure 8.8). The probability of finding all four microparticles in A is 6.25% (one in sixteen), three microparticles in A and one in B is 25% (four in sixteen), two microparticles in A and two in B is 37.5% (six in sixteen), one microparticle in A and three microparticles in B is 25% (four in sixteen), and all four microparticles in B is 6.25% (one in sixteen), i.e., there are five macrostates.

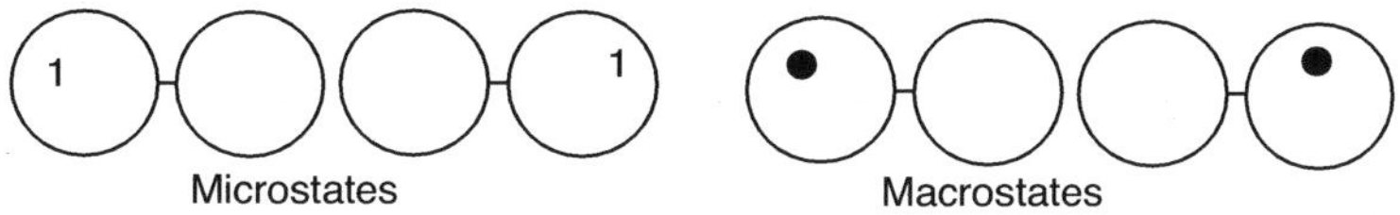

Figure 8.5 Microstate and macrostate representations when the system contains one numbered microparticle.

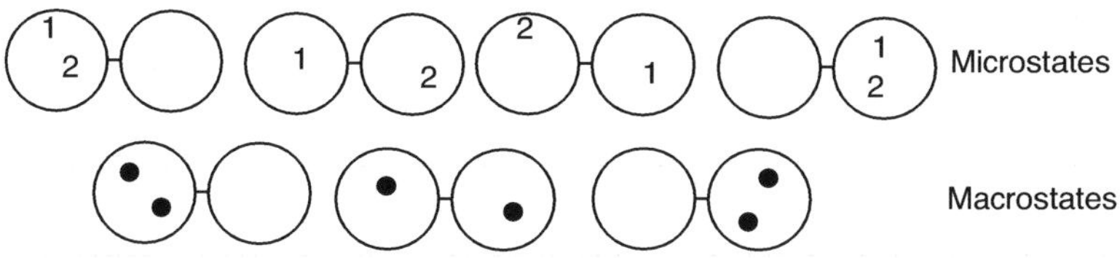

Figure 8.6 Microstate and macrostate representations when the system contains two numbered microparticles.

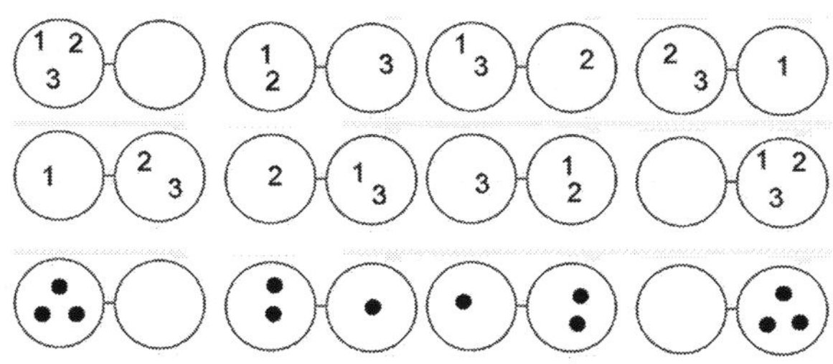

Figure 8.7 Microstate and macrostate representations when the system contains three numbered microparticles.

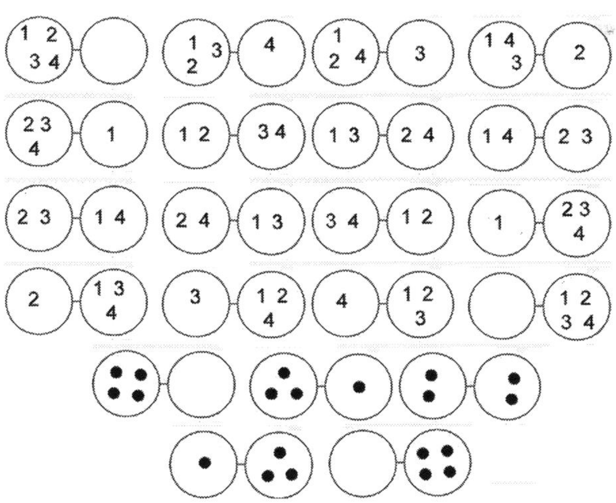

Figure 8.8 Microstate and macrostate representations when the system contains four numbered microparticles.

The easiest way to calculate the number of macrostates, microstates, and their probabilities is by using the binomial theorem (or binomial expansion). This describes the algebraic expansion of a polynomial such as $(x+y)^n$ – known as a binomial.[14] The binomial theorem states it is possible to expand the polynomial $(x+y)^n$ into a sum involving terms of the form ax^by^c, where the exponents

```
                              1   n=0
                           1     1   n=1
                        1     2     1   n=2
                     1     3     3     1   n=3
                  1     4     6     4     1   n=4
               1     5    10    10     5     1   n=5
            1     6    15    20    15     6     1   n=6
         1     7    21    35    35    21     7     1   n=7
      1     8    28    56    70    56    28     8     1   n=8
   1     9    36    84   126   126    84    36     9     1   n=9
1    10    45   120   210   252   210   120    45    10     1   n=10
```

Figure 8.9 Pascal's Triangle: each entry is the sum of the two above it.

b and c are non-negative integers with b+c=n, and the binomial coefficient, a, of each term is a positive integer depending on n and b. For example,

$$(x+y)^5 = x^5 + 5x^4y + 10x^3y^2 + 10x^2y^3 + 5xy^4 + y^5 \tag{8.1}$$

I will not be covering the mathematical aspects of the binomial theorem apart from saying that Pascal's triangle (Figure 8.9) is a triangular array of binomial coefficients. The rows of Pascal's triangle are enumerated starting with row n = 0 at the top (the 0th row). The entries in each row are usually staggered relative to the numbers in the adjacent rows. The triangle may be constructed in the following manner: in row 0, there is a unique non-zero entry 1. Each subsequent row is constructed by adding the number above and to the left with the number above and to the right, treating blank entries as 0. For example, the initial number in the first (or any other) row is 1 (the sum of 0 and 1), whereas the numbers 1 and 3 in the third row are added to produce the number 4 in the fourth row. When n=5, we can see Pascal's triangle gives us the binomial coefficients 1, 5, 10, 10, 5, and 1 comparing with Equation (8.1) we can see these are the binomial coefficients of $(x+y)^n$.

From Pascal's triangle, we can see that the expansion of $(x+y)^8$ is:

$$(x+y)^8 = x^8 + 8x^7y + 28x^6y^2 + 56x^5y^3 + 70x^4y^4 + 56x^3y^5 + 28x^2y^6 + 8xy^7 + y^8 \tag{8.2}$$

From Equation (8.2) and using the same procedure as earlier in this section we can see that if we have 8 microparticles (n = 8) we have 9 macrostates and 256 microstates $(1+8+28+56+70+56+28+8+1=256)$. The probability of finding four microparticles in each vessel is now 70/256 which is around 27%, and the probability of finding just one of the microparticles in one vessel (and the rest in the other vessel) is 1/256 which is around 0.4%. Using a mathematical formula[15] to determine the binomial coefficients for the expansion of $(x+y)^{99}$ we can see that if the number of microparticles is 99 then the number of macrostates is 100 and the total number of available microstates is 6.33825×10^{29}.

If we plot the probability of finding a certain percentage of microparticles in vessel A against the percentage of microparticles in vessel A for n particles, then we get the graphs shown in Figure 8.10. These plots are those of a normal or Gaussian distribution.[16] A better way to display the results is to plot the probability divided by the maximum probability against the percentage of microparticles in vessel A. If we have four microparticles we have seen that the maximum probability is 0.375 (37.5%) for finding 2 microparticles in vessel A and 2 microparticles in vessel B therefore, the $\frac{(\text{probability})}{(\text{maximum probability})} = \frac{0.375}{0.375} = 1$. Finding 0 or 4 microparticles in vessel A, gives $\frac{(\text{probability})}{(\text{maximum probability})} = \frac{0.0625}{0.375} = 0.167$, and similarly finding 1 or 3 (75%) of microparticles in vessel A gives $\frac{(\text{probability})}{(\text{maximum probability})} = \frac{0.25}{0.375} = 0.667$. Using this approach, we get the "distribution curves" shown in Figure 8.11.

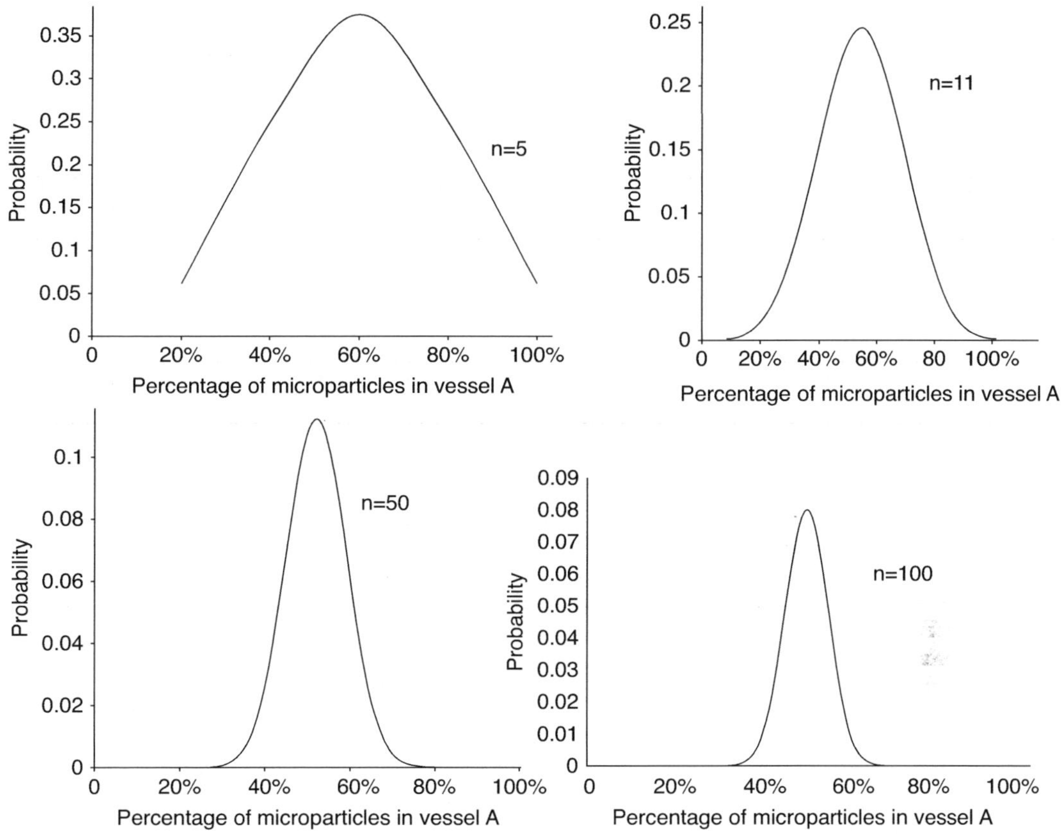

Figure 8.10 Probability plots.

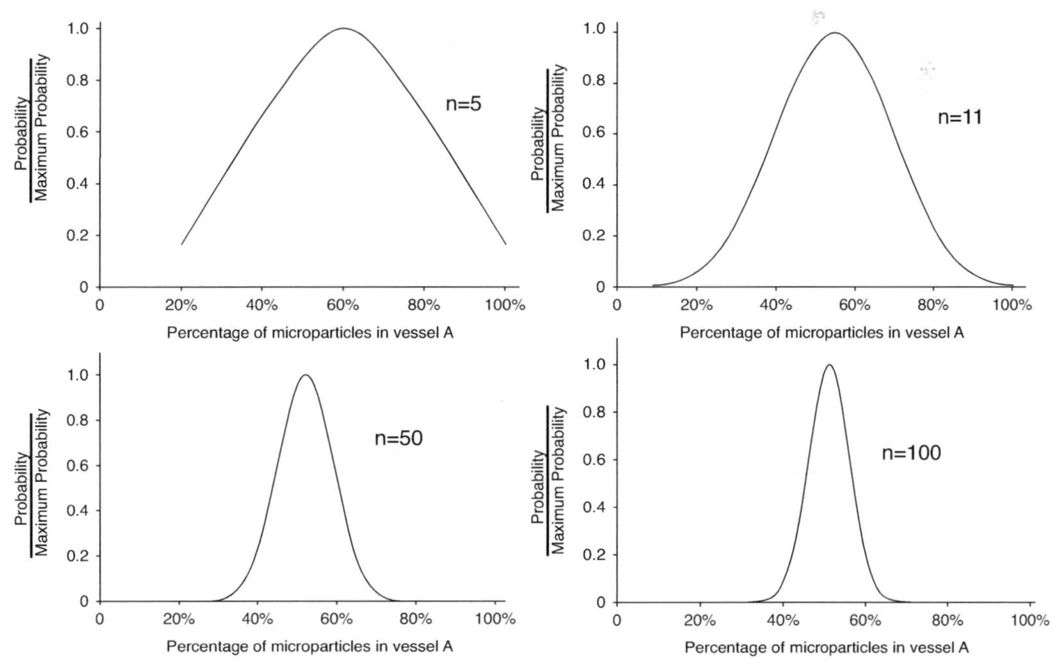

Figure 8.11 Distribution curves for various numbers of microparticles.

You can see from the distribution curves that the probability of the macrostate of 50% of the microparticles in A and 50% in B is always the maximum and the distribution curves inclines towards a narrower distribution curve as the number of microparticles increases. As previously mentioned, one unavoidable fact of statistical mechanics is the immensity of the numbers involved, and, in fact, your pocket calculator can quickly reach its limit of 10^{100}.[17] Therefore, I must use systems containing very small numbers of microparticles as illustrative examples and attempt to extend the results to the immensely large numbers found in the macro-world. When the number of microparticles in the system is of the order 10^{20} (i.e., of macroscopic proportions) then the distribution "curve" is a vertical straight line. This means there is only one macrostate that needs to be considered and that is the one where 50% of the microparticles are in each vessel. This situation is exactly what we experience experimentally. When a macroscopic quantity of gas is present in vessel A and the valve is opened then, once the system is at equilibrium, a difference in pressure (a measure of the number of microparticles) between A and B has never been observed. In other words, if we assume all possible microstates of a given system are equally probable (a fundamental assumption of statistical thermodynamics), then the system will and unless disturbed will remain, in the statistically most probable macrostate.

I now need to introduce the concept of quantisation. If we consider a single microparticle at a single instant in time, then it is in a certain position and has a certain kinetic energy, i.e., it is in a particular microstate.[18] At another instant in time, it will (almost certainly) be in a different position and/or have a different energy. Classical physics (mechanics and thermodynamics) explains matter and energy on a scale familiar to human experience, and in our experience position and kinetic energy are continuous functions. If the position and kinetic energy are continuous variables, then the particle could be in an infinite number of different available microstates. However, towards the end of the 19th century, scientists discovered phenomena in both the large (macro) and the small (micro) worlds that classical physics could not explain. This led to the theory of relativity (the science of the very large) and the development of quantum mechanics (the science of the very small). All we need to know about quantum mechanics, at this stage, is that all physical quantities are "quantised". This means they can have only certain "allowed" values and not the continuum of values permitted in classical physics. Quantisation is like having only certain speeds at which a car can travel, e.g., your car can only travel at 0, 5, 10, … mph (and let us forget about the acceleration as it switches from one allowed speed to another). Quantisation allows us to state a microparticle can have a finite number of kinetic energies available to it because its velocity and therefore, its energy is quantised. Likewise, because the position is also quantised, the microparticle has a finite number of positions available to it. Quantisation is illustrated by Figure 8.12 which indicates that the microparticle "jumps" from one "allowed" energy or position level to another. A useful analogy is walking up steps – only certain levels (the steps) are available to you.

8.2.4 Discussion

The principal objective of this section is to show that when a macroscopic sample of gas, i.e., large enough to be chemically significant (i.e., > 10^{18} microparticles) then only a single macrostate need be considered; even the next-most probable macrostate can be neglected and the probability of a deviation in an observed macrostate as a result of a change in microstates is so extremely small as to be ignored. The Second Law of Thermodynamics can now be restated in terms of microstates:

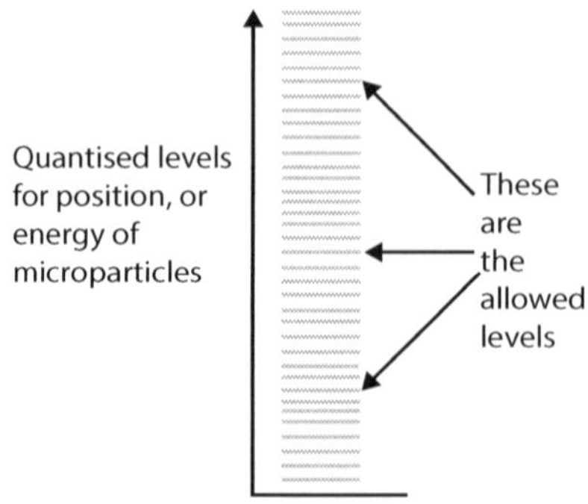

Figure 8.12 Quantised energy or position values.

> An isolated system tends toward an equilibrium macrostate with maximum entropy because then the number of available microstates is the largest and this state is statistically the most probable.

In other words, the Second Law of Thermodynamics simply states that we do not observe microscopically unlikely events. Unlike most other laws of physics, the Second Law of Thermodynamics is statistical in nature, and its reliability arises from the huge number of microparticles present in macroscopic systems. You would not be surprised to toss 2 heads in a row, but you would be very suspicious if someone tossed 100 heads in a row. To put things in perspective, if you did 10 tosses every second, it would take roughly 10^{12} times the current age of the universe before you had a reasonable chance to get a run that long.[19] When systems become so large it becomes not merely "unlikely", but "nigh on impossible" that these types of events will occur. For example, the entropy of the universe will never decrease because the microparticles in a mole of gas will never, never, never, etc., move to one half of a container leaving a vacuum in the other half of the container (or even imperceptibly approach such a situation). Although it is not impossible, in principle, for all 6×10^{23} microparticles in a mole of a gas to spontaneously migrate to one half of a container; it is so unlikely that no macroscopic violation of the Second Law has ever been or ever will be observed. In theory, the Second Law could be violated, just as, in theory, the same person could win every draw of every lottery he/she purchased a ticket for, but it is not going to happen. We use exactly the same arguments when considering a closed system whose temperature is higher than its surroundings. We all know what happens; over time the system spontaneously cools down to the same temperature as the room, i.e., the system loses energy to the surroundings by the process of heat. Now consider a closed system whose temperature is the same as its surroundings. Would we expect the temperature of the system to spontaneously increase or, to put it another way, would we expect energy to flow from the surroundings to the system by the process of heat? No, it has never been observed to happen even though it would not contravene the First Law of Thermodynamics.

Consider the free expansion of an ideal gas. Immediately after the barrier between the two containers is removed, all of the microparticles are on one side of the container and we have a system that is not in thermodynamic equilibrium. By analogy with the coin tossing examples of Section 8.2.2, there are many more microscopic states that correspond to an even distribution of microparticles in the container than there are microscopic states in which all the microparticles are in one half of the container. This process is irreversible in the sense that we can wait however long we wish, but we are not going to see those microparticles spontaneously re-congregate on the left side of the container. This is no more likely than getting all heads if we toss the same number of pennies as there are microparticles.

The number of microstates that correspond to a particular macrostate is often represented by 'W'. Some texts refer to W as thermodynamic probability. We can use W as a general criterion for deciding whether a process is spontaneous or not. A spontaneous process corresponds to movement from a state in which W is smaller to a state where W is larger. If we can find some way of calculating or measuring the initial and final values of W, the problem of deciding in advance whether a process is spontaneous or not is solved. Although there is nothing wrong in principle with this approach to spontaneous processes, in practice, it turns out to be very cumbersome. For macroscopic amounts of matter, the values of W are on the order of $10^{10^{24}}$. The logarithm of W, however, is "only" of the order of 10^{24}. I now want to consider the relationship between the available microstates of a thermodynamic system and its entropy. This relationship is usually known as the Boltzmann equation, but it is also known as the Boltzmann formula, or the Boltzmann-Planck equation.

8.3 THE BOLTZMANN EQUATION[20]

8.3.1 Introduction

The Boltzmann equation $S = k_B \ln W$, relating the entropy of an ideal gas (S) to the number of microstates (W) corresponding to the gas's macrostate, is one of the important ones in thermodynamics because it provides a fundamental link between microscopic and macroscopic thermodynamic properties.[21] Despite the importance of the Boltzmann equation

> The actual full rigorous step-by-step derivation of the logarithmic formulation or interpretation of entropy $\left(S = k_{[B]} \ln W\right)$ is a bit difficult to track down for a number of reasons, the first of which being that it was originally done in German and readily available English translations are wanting. [22]

The same site also states – "Some consider $S = k_{[B]} \ln W$ to be easily the second most important formula of physics, next to $E = mc^2$ or at par with it". It then continues – "A crude approximate derivation [was] given in 1946 by Belgian-born English thermodynamicist Alfred Ubbelohde".[23] This self-same website also stated that the proof of the Boltzmann's equation is a "a bit murky".

The following is from "Statistical Mechanics A survival guide"[24]:

> It must be stressed that it [the Boltzmann equation] is a postulate, indeed the most important postulate of statistical mechanics (it is even engraved on Boltzmann's tombstone); nevertheless it remains a postulate.[25] As such, it is no more provable than the laws of thermodynamics, …. It is a guess – a very intelligent guess – which can be justified ultimately because it works. That said, we should note that there are good reasons for believing that the entropy of a system should be of this form. For instance (i) we know from thermodynamics that as systems go towards equilibrium the entropy is maximised, and we have already stated that the equilibrium macrostate is the one with the most microstates, and (ii) we also know from thermodynamics that S is additive, but the laws of probability say that [W] must be multiplicative – the [Boltzmann] equation meets this criterion as well because of the logarithm.

Boltzmann never used the term microstate or anything similar, and after reading around the subject, I have grave reservations that in Boltzmann's mind his 'W' (from the German 'Wahrscheinlichkeit' - meaning probability) bore any resemblance to the microstates as defined above. Consider the following[26]:

> The statement that Boltzmann defined entropy as function of the number of microstates rather than of a probability, according to some arguments, is a misrepresentation, supposedly owing to the fact that Boltzmann's writing was fairly difficult and few people understood it at the time; and that most of Boltzmann's original papers have never been translated from German, so most information about them is second or third hand interpretations. According to American physicist Robert Swendsen, Boltzmann, supposedly, interpreted the second law to mean that a system tends toward more probable macrostates. Boltzmann, supposedly, originally – incorrectly (some have argued) – identified the entropy as the probability (wahrscheinlichkeit) of a system's macrostate. Later, in the same paper, he considers an ideal gas as a special case and calculates the logarithm of the macrostate's probability as proportional to the volume in phase space.[27] In later papers he re-identifies the entropy as the logarithm of a macrostate's probability, but as he is usually considering the special case where all microstates are equally likely, he calculates it as proportional to the volume in phase space. Readers likely misidentified this special case as the definition of entropy since he seemed to have rarely mentioned what his definition was.

I came across this quote from another source[28]:

> All existing attempts to give a general proof of this postulate [$S=k_B \ln W$] must be considered as an aggregate of logical and mathematical errors superimposed on a general confusion in the definition of the basic quantities.

The same paper went on:

> The lack of a crystal-clear proof of Boltzmann's principle puts physics students, teachers, and all physicists in the uncomfortable position of being forced to accept the relation as a postulate which is necessary to link thermodynamic entropy to microscopic dynamics.

8.3.2 Proof

I have set down a 'proof' of Boltzmann's Equation, so you may appreciate why the Hmolpedia website used the phrase "a bit murky". The following 'proof' of the Boltzmann Formula is based on that of Moore[29] who appears to have based it on Ubbelohde's approach (although Moore does not reference or acknowledge him). The approach is reasonably straightforward.

Consider one mole of an ideal gas system divided into two parts, if the entropy contents of the two systems are S_1 and S_2, then the entropy content of a larger system formed by a combination of the two must be $S_1 + S_2 = S$. The number of microstates, however, does not combine in this manner. If W_1 and W_2 represent the total number of all possible microstates in each system before their combination, then the number for the larger combined system is $W_1 \times W_2$. This is because each of the W_1 microstates of one system can be arranged with W_2 microstates of the other system and the total number in the combined system is then obtained when all W_1 microstates of the first system have each been multiplied by W_2 microstates of the other so that this total number is then the product of the W values.

Because of this different behaviour for S and W, the only way that an increase in the entropy content of the system can be related to an increase in microstates is for the entropy content to be proportional to the logarithm of the number of microstates. In this way, the increases in both S and W will be described correctly when these increases are produced by altering only the size of the system. The most general form of the relationship is:

$$S = a \times \ln(W) + b \tag{8.3}$$

The value of the constant 'a' may be derived by analysing, from the viewpoint of probability, a simple change for which ΔS is known. Consider the free expansion of 1 mole of ideal gas, originally at pressure P_1 in a container of volume V_1, into an evacuated container of volume V_2. The final pressure is P_2 and the final volume $V_1 + V_2$. We know (Section 7.5.3, Equation 7.48):

$$\Delta S = S_2 - S_1 = R \times \ln\left[\frac{V_1 + V_2}{V_1}\right] \tag{8.4}$$

Using Boltzmann's constant k_B,[30] we can rewrite Equation 8.4 as (where A is equal to the total number of microparticles in one mole, i.e., Avogadro's number):

$$\Delta S = A \times k_B \times \ln\left[\frac{V_1 + V_2}{V_1}\right] \tag{8.5}$$

Equation 8.5 can be written as[31]:

$$\Delta S = k_B \times \ln\left[\frac{V_1 + V_2}{V_1}\right]^A \tag{8.6}$$

After the two containers are connected and equilibrium is achieved, the probability of finding one given microparticle in the first container is simply the ratio of V_1 to the total volume $V_1 + V_2$. Since probabilities are multiplicative, the probability (Probability_{V_1}) of finding all A microparticles in the first container is:

$$\text{Probability}_{V_1} = \left[\frac{V_1}{V_1 + V_2}\right]^A \tag{8.7}$$

Note Probability_{V_1} is the probability of the system remaining in its original condition before free expansion, i.e., in its original state. Let $\text{Probability}_{V_1+V_2}$ equal the probability of finding all microparticles in the volume $V_1 + V_2$. We know this must be one (unity) as $V_1 + V_2$ is the only volume available to the microparticles.

Ubbelohde then made the following assumption:

$$\frac{\text{Probability}_{V_1+V_2}}{\text{Probability}_{V_1}} = \frac{W_2}{W_1} \tag{8.8}$$

where W_1 and W_2 are the available microstates at volumes V_1 and $V_1 + V_2$, respectively. In other words, the ratio of the probability $(\text{Probability}_{V_1})$ that the microparticles will all remain in the volume V_1, to the probability $(\text{Probability}_{V_1+V_2})$ of the microparticles occupying, uniformly, the larger volume $V_1 + V_2$ is equal to the ratios of the number of microstates available to the respective systems. This is where the derivation becomes 'a bit murky' according to Hmolpedia. Ubbelohde said it (his "proof") was 'A crude approximate derivation'. Using this assumption, we can write:

$$\frac{\text{Probability}_{V_1+V_2}}{\text{Probability}_{V_1}} = \frac{W_2}{W_1} = \frac{1}{\left[\frac{V_1}{V_1 + V_2}\right]^A} = \left[\frac{V_1}{V_1 + V_2}\right]^{-A} = \left[\frac{V_1 + V_2}{V_1}\right]^A \tag{8.9}$$

As $S_1 = a \times \ln(W_1) + b$ and $S_2 = a \times \ln(W_2) + b$ (see Equation 8.3) we have:

$$\Delta S = S_2 - S_1 = [a \times \ln(W_2) + b\ ] - [a \times \ln(W_1) + b\ ] = a[\ln(W_2) - \ln(W_1)] = a \times \ln\left(\frac{W_2}{W_1}\right) \quad (8.10)$$

Comparing Equation 8.10 with Equation 8.6, we can write:

$$\Delta S = S_2 - S_1 = a \times \ln\left(\frac{W_2}{W_1}\right) = k_B \times \ln\left[\frac{V_1 + V_2}{V_1}\right]^A \quad (8.11)$$

Therefore 'a' in Equation 8.3 is equal to 'k_B' the Boltzmann constant, i.e., for a change from state 1 to state 2:

$$\Delta S = k_B \times \ln\left(\frac{W_2}{W_1}\right) \quad (8.12)$$

and,

$$S = k_B \times \ln(W) + \text{constant} \quad (8.13)$$

Hmolpedia's website has the following statement[32]:

> In 1901, German physicist Max Planck, …, introduced the $S = k \log W + c$ formula, in semi deus ex machina[33] fashion, via citation of Austrian physicist Ludwig Boltzmann ….

Note the constant which is almost always neglected (for reasons I am unclear about)[34] and the Boltzmann formula is nearly always given as $S = k_B \ln W$.[35] This abstract is also from the Hmolpedia website:

In 1904, and thereafter, German physicist Albert Einstein began to repeatedly criticize the formula $S = k \log W$; in 1910, Einstein had the following to say on the matter:

> 'The equation $S = k \log W + \text{cons}$ appears without an elementary theory – or however one wants to say it – devoid of any meaning from a phenomenological point of view.'

I am not qualified to comment on the Moore/Ubbelohde "crude approximate derivation", but, as you will now be aware, this is not the first time in thermodynamics things are "bit murky". It does appear we are missing a few end steps in the proof of the Boltzmann equation as well as the details of how Planck arrived at the exact formulation of $S = k \times \log W$.[36]

I would like to finish off this section with two more quotes from Hmolpedia, the first one is:

> In the decades to follow Planck's probabilistic 'disorder-dependent' model of entropy seems to have been adopted with abandon to be the universal measure of entropy, … Ubbelohde even seems to corroborate on this, in his concluding statement that 'the probability increase [in number of microstates] accompanying any process is quite general, and holds for all the diversity of spontaneous happenings for which the measurement of entropy has any meaning.'

The second is:

> The discerning theorist, however, should hold reserve in using this formulation as an absolute universal measure of entropy, for a number of obvious reasons, the foremost of which being that the equation contains the [ideal gas] constant R hidden in its body:
>
> $$S = \frac{R}{n}(\log W)$$
>
> When we thus see entropy calculations using this equation in non-ideal gas situations, … one begins to question the validity of such a calculation.

Despite the reservations regarding the rigorousness of $S = k \times \log W$, the Boltzmann equation establishes a link between the behaviour of microparticles and the macroscopic entropy/caloric content, i.e., entropy/caloric content is directly proportional to the number of accessible microstates (at least for the ideal gas). A formulation of the Second Law of Thermodynamics can now be made in terms of the change in the number of microstates that produce a change in a thermodynamic state of an isolated system. In these terms, the Second Law can be stated as:

> A change in the thermodynamic state of an isolated system can occur only if there is an increase in the number of equally probable microstates in the system.

Finally, I want to briefly return to the "entropy is disorder" discussion (see Section 7.7). In this section, you can appreciate how entropy has become associated with disorder. One of the Hmolpedia website extracts used above states "In the decades to follow Planck's probabilistic 'disorder-dependent' model of entropy seems to have been adopted with abandon to be the universal measure of entropy". Ask yourself this question – will my knowledge about a system increase or decrease when the system cools down? When it is hot, there is high uncertainty about the position and energy of the microparticles composing the system. The lower the temperature, the more precise your knowledge about position and energy will be. At 0K, the macroscopic description gives you exactly the same information than the microscopic one, therefore the entropy is zero.[37] This leads to the following consideration: entropy is always a property of a set of microstates – it is proportional to the logarithm of the total number of accessible microstates. In other words, a particular microstate itself does not have a defined entropy. This in contrast to internal energy where a particular microstate does have an internal energy associated with it.

If the higher uncertainty about the position and energy of the microparticles in a higher entropy system compared to that in a lower entropy system is considered to be greater disorder, then so be it. However, many texts use analogies such as a pile of bricks or compare a tidy room and to the (more probable) untidy room. A typical quote is "disorder is more probable than order", but there is often no mention of why (assuming we can agree definitions of "disorder and order"). Consider the problem that this website has got itself into (Figure 8.13).[38] It is inarguable that work is required to move from the "disordered" state on the left to the "ordered" state on the right. However, it is also inarguable that work is required to move from the "ordered" pile of bricks on the right to the "disordered" stack on the left.[39] Do not conflate entropy with macroscopic disorder.

8.4 THERMAL CONDUCTIVITY[40]

8.4.1 Introduction

If we accept that energy is quantised and can only take on discrete values – one quantum of energy, two quanta of energy, three quanta of energy, etc., then I want to show that energy flowing from a high temperature to a lower temperature occurs spontaneously because it is statistically the most probable outcome. In these examples, I will assume that the difference in energy between these quanta is the same. I will also show how this result leads to a rigorous definition of temperature. As I have already mentioned, the unavoidable fact of statistical mechanics is the immensity of the numbers involved, and that with immense numbers involved statistical maximum probability becomes a near certainty. The enormity of these numbers means that examples must use small numbers and extrapolate.

8.4.2 Example of a Thermal Interaction

Consider an isolated and fixed volume system of N = 4 microparticles that is divided into two subsystems. The subsystems of this 'composite system' are also initially isolated and of fixed volume (see Figure 8.14a) so that there can be no change in the energy, number of microparticles and the

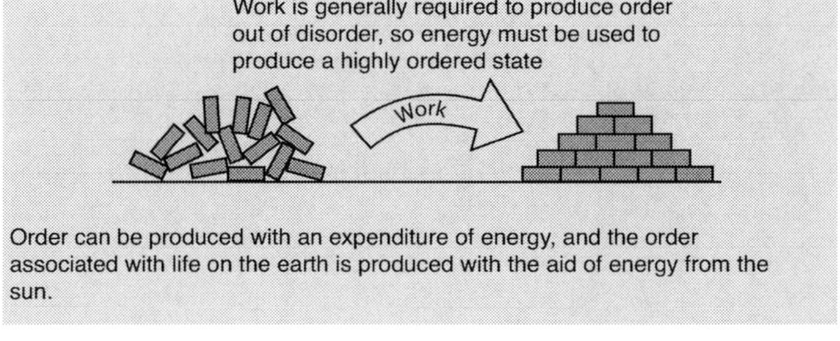

Figure 8.13 You need to do work on the system of bricks in whichever direction you proceed.

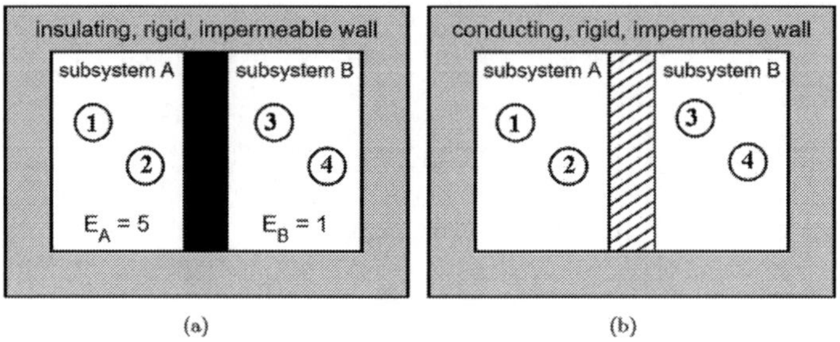

Figure 8.14 An isolated composite system consisting of two subsystems, each with two distinguishable microparticles. In (a), the wall between the two subsystems is insulating, rigid, and impermeable. In (b), the wall is conducting, rigid, and impermeable. The total energy of the composite system is six energy units.

Table 8.5: The 12 Equally Probable Microstates of the Isolated Composite System Composed of Subsystems A and B with $N_A=2$, $E_A=5$, $N_B=2$, and $E_B=1$

E_A	Accessible Microstates of Subsystem A		E_B	Accessible Microstates of Subsystem B	
	5,0	0,5			
5	4,1	1,4	1	1,0	0,1
	3,2	2,3			

The 12 microstates consist of all combinations of the microstates of A and B.

volume of each subsystem. Subsystem A consists of two microparticles, '1' and '2' sharing 5 quanta of energy units (E_A); subsystem B consists of two microparticles, '3' and '4', with 1 quantum of energy (E_B). The total energy E_{tot} of the isolated composite system consisting of subsystem A plus subsystem B is 6 quanta of energy units.

The accessible microstates for the isolated composite system are shown in Table 8.5. We see that subsystem A has $W_A=6$ accessible microstates and subsystem B has $W_B=2$ accessible microstates. The total number of microstates W_{tot} accessible to the composite system is 12.

Now suppose that the insulating, rigid, impermeable partition separating subsystems A and B is changed to a diathermal, rigid, impermeable partition (see Figure 8.14b). The rigid partition maintains the volumes V_A and V_B and, because the partition is impermeable, the microparticles still cannot penetrate the partition and go from one subsystem to the other. However, energy can be transferred from one subsystem to the other by the process of heat, subject only to the constraint that the total energy of the isolated composite system is a constant 6 quanta. The microstates of subsystems A and B are listed in Table 8.6 for all the possible values of E_A and E_B.

The total number of microstates W_{TOT} accessible to the isolated composite system whose subsystems have energy E_A and $\mathbf{q}_{ice}$ is

$$W_{TOT} = W_{A(E_A)} \times W_{B(E_B)} \tag{8.14}$$

where:

- W_{TOT} is the number of accessible microstates of the composite system when the quanta of energy present in the system is $E_A + E_B$.
- $W_{A(E_A)}$ is the number of accessible microstates of subsystem A when the quanta of energy present in the system is E_A.
- $W_{B(E_B)}$ is the number of accessible microstates of subsystem B when the quanta of energy present in the system is E_B.

For example. if $E_A = 4$ and $E_B = 2$, then subsystem A can be in any one of five microstates $\left(W_{A(4)} = 5\right)$ and subsystem B can be in any of three microstates $\left(W_{B(2)} = 3\right)$. These two sets of microstates of subsystems A and B can be combined to give $5 \times 3 = 15$ microstates of the composite system (W_{TOT}).

The total number of microstates $\left(\sum W_{TOT}\right)$ accessible to the composite system can be found by summing the product $W_{A(E_A)} \times W_{B(E_B)}$ over the possible values of E_A and E_B consistent with the condition that $E_A + E_B = 6$, i.e., $E_A = 0/E_B = 6$; $E_A = 1/E_B = 5$; $E_A = 2/E_B = 4$; $E_A = 3/E_B = 3$; $E_A = 4/E_B = 2$; $E_A = 5/E_B = 1$; and $E_A = 6/E_B = 0$. Hence, $\left(\sum W_{TOT}\right)$ can be expressed as:

$$\sum_{E_A+E_B=6} W_{TOT} = \sum_{E_A+E_B=6} \left(W_{A(E_A)} \times W_{B(E_B)}\right) \tag{8.15}$$

From Table 8.6, we see that $\sum_{E_A+E_B=6} W_{W_{TOT}} = 84.$

The isolated composite system is equally likely to be in any one of its 84 accessible microstates, i.e., the total number of microstates accessible to the isolated composite system increases from 12 to 84 when the insulating internal constraint is removed. Let $P(E_A)$ be the probability that subsystem A has energy E_A (this is also the probability that subsystem B has energy $E_{tot} - E_A = E_B$. $P(E_A)$ is given by

$$P(E_A) = \frac{\left(W_{A(E_A)} \times W_{B(E_B)}\right)}{\sum_{E_A+E_B=6} W_{TOT}} \tag{8.16}$$

These data are shown in Table 8.7. Although our simple model has only four microparticles, we can ask questions that are relevant to much larger systems. For example, what is the probability that energy is transferred from one subsystem to the other? Given that initially $E_A = 5$ and $E_B = 1$, we see from Table 8.7 that the probability of subsystem A gaining energy when the internal constraint is removed is 7/84. The probability that its energy remains unchanged is 12/84. In the remaining 65/84 cases, subsystem A loses energy and subsystem B gains energy. We can also see that the most probable macrostate is when both subsystems have 3 energy quanta each.

Let us now consider a slightly larger isolated composite system consisting of two subsystems with subsystem A having three microparticles and subsystem B having four microparticles, and the composite system has ten quanta of energy ($E_{tot} = 10$) (Figure 8.15).

In this example, I will only consider the accessible microstates when the system's partition is diathermal. The number of microstates for the two subsystems for the various possible values of E_A and E_B are summarised in Table 8.8. There are 8008 equally likely microstates. The most

Table 8.6: The 84 Equally Probable Microstates Accessible to the Isolated Composite System Composed of Subsystems A and B After the Removal of the Internal Insulating Constraint

E_A	Accessible Microstates	$W_{A(E_A)}$	E_B	Accessible Microstates	$W_{B(E_B)}$	W_{TOT}
6	6,0 0,6 5 ,1 1,5 4,2 2,4 3,3	7	0	0,0	1	7
5	5,0 0,5 4,1 1,4 3,2 2,3	6	1	1,0 0,1	2	12
4	4,0 0,4 3,1 1,3 2,2	5	2	2,0 0,2 1,1	3	15
3	3,0 0,3 2,1 1,2	4	3	3,0 0,3 2,1 1,2	4	16
2	2,0 0,2 1,1	3	4	4,0 0,4 3,1 1,3 2,2	5	15
1	1,0 0,1	2	5	5,0 0,5 4,1 1,4 3,2 2,3	6	12
0	0,0	1	6	6,0 0,6 5,1 1,5 4,2 2,4 3,3	7	7

Table 8.7: Reduced Table 6 Plus the Probability $P_A(E_A)$ that Subsystem A Has Energy E_A

E_A	$W_{A(E_A)}$	E_B	$W_{B(E_B)}$	W_{TOT}	$P(E_A) = \frac{(W_{A(E_A)} \times W_{B(E_B)})}{\sum_{E_A+E_B=6} W_{TOT}}$
6	7	0	1	7	$\frac{7}{84} = 0.083$
5	6	1	2	12	$\frac{12}{84} = 0.143$
4	5	2	3	15	$\frac{15}{84} = 0.179$
3	4	3	4	16	$\frac{16}{84} = 0.190$
2	3	4	5	15	$\frac{15}{84} = 0.179$
1	2	5	6	12	$\frac{12}{84} = 0.143$
0	1	6	7	7	$\frac{7}{84} = 0.083$
		$\sum_{E_A+E_B=6} W_{TOT}$		84	

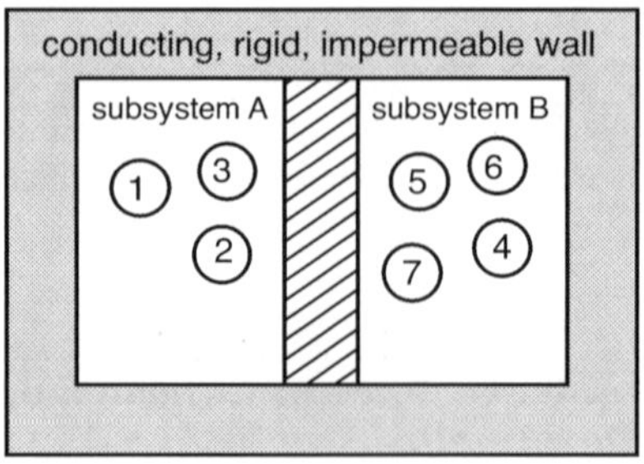

Figure 8.15 An isolated composite system consisting of two subsystems, one with three distinguishable microparticles and one with four distinguishable microparticles. In (a), the wall between the two subsystems is conducting, rigid, and impermeable. The total energy of the composite system is ten units.

probable macrostate corresponds to $E_A = 4$ and $E_B = 6$. In the previous example (when the number of microparticles in each subsystem was equal) the most probable macrostate was when the number of energy units in each subsystem was equal. In this case, the most probable microstate is when 4 energy quanta are in subsystem A and 6 energy quanta are in subsystem B. In both examples, the most probable microstate is when the energy quanta in the subsystems are in the same proportion as the proportion of microparticles in the subsystems. Because of our association of the most probable microstate with thermal equilibrium, we know that the quantity that is the same for the two subsystems is the temperature. This leads nicely to the next section.

Table 8.8: The 8008 Equally Probable Microstates Accessible to the Isolated Composite System Composed of Subsystems A and B After the Removal of the Internal Insulating Constraint

E_A	$W_{A(E_A)}$	E_B	$W_{B(E_B)}$	W_{TOT}	$P(E_A) = \frac{(W_{A(E_A)} \times W_{B(E_B)})}{\sum_{E_A+E_B=6} W_{TOT(E_A+E_B)}}$
10	66	0	1	66	$\frac{66}{8008} = 0.008$
9	55	1	4	220	$\frac{220}{8008} = 0.027$
8	45	2	10	450	$\frac{450}{8008} = 0.056$
7	36	3	20	720	$\frac{720}{8008} = 0.090$
6	28	4	35	980	$\frac{980}{8008} = 0.122$
5	21	5	56	1176	$\frac{1176}{8008} = 0.147$
4	15	6	84	1260	$\frac{1260}{8008} = 0.157$
3	10	7	120	1200	$\frac{1200}{8008} = 0.150$
2	6	8	165	990	$\frac{990}{8008} = 0.124$
1	3	9	220	660	$\frac{660}{8008} = 0.082$
0	1	10	286	286	$\frac{286}{8008} = 0.036$
	$\sum_{E_A+E_B=6} W_{TOT}$			8008	

The total energy is $E_{tot}=10$ with 3 microparticles in subsystem A and 4 microparticles in subsystem B. Also shown are the number of accessible microstates in each subsystem and the probability $P(E_A)$ that subsystem A has energy E_A.

8.5 TEMPERATURE REVISITED

8.5.1 Thermodynamic Temperature

So far in this book, we have defined the temperature of a physical object in terms of the average kinetic energy of its constituent microparticles ("kinetic temperature"). This approach works fine when considering an ideal gas. However, there are problems with this approach when considering polyatomic molecules with their internal degrees of freedom such as rotation and vibration, and with the internal interactions in solid materials. To overcome these problems, temperature is defined in terms of two other state functions, entropy (S) and internal energy (U) – this is known as "thermodynamic temperature". I do not want to go into too much detail here because the mathematics involves inexact and partial derivatives (see Section 1.6.9). |||Let us start by considering the differential form of the internal energy[41]:

$$d\mathbf{U} = Td\mathbf{S} - Pd\mathbf{V} \tag{8.17}$$

If we consider U as a function S and V, we can write Equation 8.17 as a partial differential equation:

$$\mathbf{dU} = \left(\frac{\partial \mathbf{U}}{\partial \mathbf{S}}\right)_V \mathbf{dS} + \left(\frac{\partial \mathbf{U}}{\partial \mathbf{V}}\right)_S \mathbf{dV} = \mathbf{dU} = \mathbf{T}\mathbf{dS} - \mathbf{P}\mathbf{dV} \tag{8.18}$$

Equating the coefficients yields:

$$\left(\frac{\partial \mathbf{U}}{\partial \mathbf{S}}\right)_V = T \qquad \left(\frac{\partial \mathbf{U}}{\partial \mathbf{V}}\right)_S = -P \tag{8.19}$$

The relationship $\left(\frac{\partial \mathbf{U}}{\partial \mathbf{S}}\right)_V = T$ defines thermodynamic temperature. I do not plan on taking this much further, but the following extract entitled "What is the physics behind temperature?"[42] is interesting.

Temperature is a statistical property of systems of large numbers of particles that describes the way that heat will flow when such a system exchanges energy with another system. Only in very unusual cases is temperature equal to the average kinetic energy of the particles. In some cases, temperature is proportional to the average kinetic energy of the particles. In general though, temperature is related to average energy in a more complex way.

If you don't believe me, just consider a glass of ice water. Both the ice and the water are at the same temperature, but the particles of the water are moving about like crazy in the liquid, while the particles of the ice are frozen in place, just vibrating a bit. The average kinetic energy of the particles in the water and the ice are not at all the same, even though the two are at the same temperature. The key to this puzzle is entropy.

A good start for defining temperature is: **the quantity that is the same when two objects are in thermal equilibrium**. That is, if you have two objects and you put them together so they can exchange energy, after a while they will have the same temperature.

This leads to the question, how is the total energy divided up between the objects at equilibrium? It turns out that with extremely high certainty, the energy will be divided up such that the constituents of the objects can be arranged in the largest number of ways consistent with that division of energy. Another way of putting this is that the total energy will be divided between the two objects in the way that maximizes the *total entropy*.

Putting the preceding statement into mathematics, we have a condition for thermal equilibrium:

$$\frac{\partial S_1}{\partial U_1} = \frac{\partial S_2}{\partial U_2}$$

where S and U are the entropy and energy of the two objects. So the two sides of this equation are quantities that are equal when in thermal equilibrium, and thus must be related to temperature in some way.

Now we might be tempted to define temperature as $\frac{\partial S}{\partial U}$. That would be a highly reasonable thing to do. Unfortunately, someone a long time ago decided that cold things have low temperature and hot things have high temperature. In order to agree with this, we have to define temperature as the inverse of that partial derivative:

$$\frac{1}{T} = \frac{\partial S}{\partial U}$$

This is the standard definition of temperature. You can interpret this graphically: if you have a plot of entropy versus energy, the temperature is just the inverse of the slope of that curve.

...

So how does the formal definition of temperature above agree with the usual intuitive definition: 'a measure of the average energy'? In many systems, it turns out that as you add more energy the system, there are more possible ways the constituents of the system can be arranged, and therefore, there is higher entropy. In fact, in the vast majority of familiar systems, the graph of entropy vs. energy looks something like this:

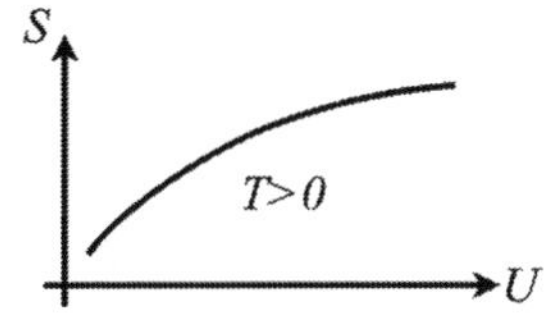

First, note that the slope is always positive, so temperature is always positive. Also, the slope becomes smaller as energy increases, meaning that temperature becomes bigger. So this works out then that temperature is directly related to the energy. Sometimes you even get lucky, and entropy is related to energy logarithmically. In that case, since the derivative of log x is 1/x, we get temperature proportional to average energy. This is the case in the familiar ideal gas, where the average energy of a particle is $U = \frac{3}{2}k_BT$, where k_B is a constant (namely, the Boltzmann constant). One example of where temperature is actually equal to the average kinetic energy would be in a classical ideal gas confined somehow to move only in two dimensions - in that case the average energy of a particle would be $U = k_BT$, where you can consider k_B to just be a conversion factor between units of energy and temperature.

8.5.2 Definition of Temperature

I want now to use the data from Table 8.8 to illustrate the concept of thermodynamic temperature. Clausius defined entropy as (where the function S is the entropy of the system)[43]:

$$S = \int \frac{\mathbf{dq_{rev}}}{T_{reservoir}} \tag{8.20}$$

Differentiation yields:

$$\mathbf{dS} = \frac{\mathbf{dq_{rev}}}{T} \tag{8.21}$$

Rearrangement gives:

$$\frac{1}{T} = \frac{\mathbf{dS}}{\mathbf{dq_{rev}}} \tag{8.22}$$

We know if the system gains energy by the process of heat $(+|\mathbf{dq_{rev}}|)$ then its entropy increases $+|\mathbf{dS}|$). In other words, T is always a positive real number (also see previous section). The simple systems used in Section 8.4.2 are constant volume systems, and we know at constant volume all of the energy entering a system by the process of heat increases the internal energy, whereas at constant pressure some of this energy is used to do $P\mathbf{dV}$ work on the surroundings.[44] Therefore, we know at constant volume $\mathbf{dq_{rev}} \equiv \mathbf{dE}$[45]:

$$\frac{1}{T} = \frac{\mathbf{dS}}{\mathbf{dE}} \tag{8.23}$$

Differentiation of the Boltzmann equation $(S = k_B \ln W)$ yields:

$$\mathbf{dS} = k_B\mathbf{d}(\mathbf{lnW}) \tag{8.24}$$

We can write:

$$\frac{1}{T} = \frac{k_B\mathbf{d}(\mathbf{lnW})}{\mathbf{dE}} \tag{8.25}$$

Therefore, $\frac{1}{T}$ is the slope of a plot of k_B**d (lnW)** against **dE**.

Table 8.9: This is Table 8.8 with Reciprocal Temperature Values Calculated Using Equation 8.26

$_{E}A$	(ΔH^{o}_{soln})	ΔH^{o}_{f}	$\frac{1}{T_{A(E_A)}}$	E_B	$W_{B(E_B)}$	$\ln W_{B(E_B)}$	$\frac{1}{T_{B(E_B)}}$	W_{TOT}	$\ln[W_{TOT}]$
10	66	4.19	–	0	1	0	–	66	4.19
9	55	4.01	0.19	1	4	1.39	1.15	220	5.39
8	45	3.81	0.215	2	10	2.30	0.805	450	6.11
7	36	3.58	0.24	3	20	3.00	0.63	720	6.57
6	28	3.33	0.27	4	35	3.56	0.515	980	6.89
5	21	3.04	0.31	5	56	4.03	0.435	1176	7.07
4	15	2.71	0.37	6	84	4.43	0.38	1260	7.14
3	10	2.30	0.46	7	120	4.79	0.34	1200	7.09
2	6	1.79	0.60	8	165	5.11	0.305	990	6.90
1	3	1.10	0.895	9	220	5.39	0.275	660	6.49
0	1	0	–	10	286	5.66	–	286	5.66

As these simple numerical examples are only illustrative and as k_B is only a scale factor, I will use $k_B = 1$ in the following equations. In column 4 in Table 8.9 I have calculated $\frac{1}{T_{A(E_A)}}$ values for subsystem A in the system illustrated in Figure 8.15, by using the central difference approximation[46] for the slope at E_A:

$$\frac{1}{T_{A(E_A)}} \approx \frac{\left[\ln\left(W(E_A+1)\right)-\ln\left(W(E_A-1)\right)\right]}{2\Delta E_A} \tag{8.26}$$

where $\frac{1}{T_{A(E_A)}}$ is the slope of the curve of the plot of $\ln W(E_A)$ versus E_A at a specific value of E_A. If we consider $E_A = 8$ in the system represented by Figure 8.15 then

$$\frac{1}{T_{A(E_A=8)}} = \frac{4.01-3.58}{2} = 0.215 \tag{8.27}$$

I have used the same procedure for calculating $\frac{1}{T_{B(E_B)}}$ values:

$$\frac{1}{T_{B(E_B)}} \approx \frac{\left[\ln\left(W(E_B+1)\right)-\ln\left(W(E_B-1)\right)\right]}{2\Delta E_B} \tag{8.28}$$

The results are tabulated in Table 8.9. Figure 8.16 is a plot of $\frac{1}{T_{A(E_A)}}$ and $\frac{1}{T_{B(E_B)}}$ (calculated using Equations 8.26 and 8.28) against E_A or E_B. We can see $\frac{1}{T_{A(E_A)}}$ and $\frac{1}{T_{B(E_B)}}$ values are approximately equal at $E_A = 4$, and this corresponds with the maximum number of microstates accessible to the composite subsystem with a rigid diathermal barrier between subsystems A and B (see Table 8.8 and Figure 8.17).

Consider a microstate away from equilibrium, say $E_A = 2$ and $E_B = 8$. If energy is transferred from B to A, the total number of accessible microstates of the composite system increases until two energy quanta have been transferred when the number of accessible microstates has reached a maximum. Similarly, if we considered another microstate away from equilibrium, say E_A=8.

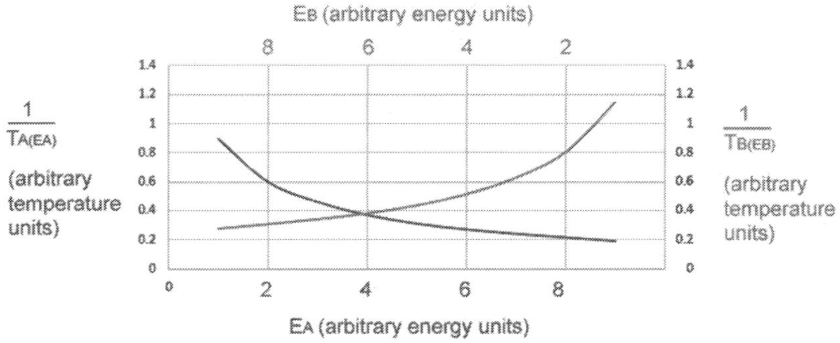

Figure 8.16 A plot of $\frac{1}{T_{A(E_A)}}$ and $\frac{1}{T_{B(E_B)}}$ against E_A or E_B.

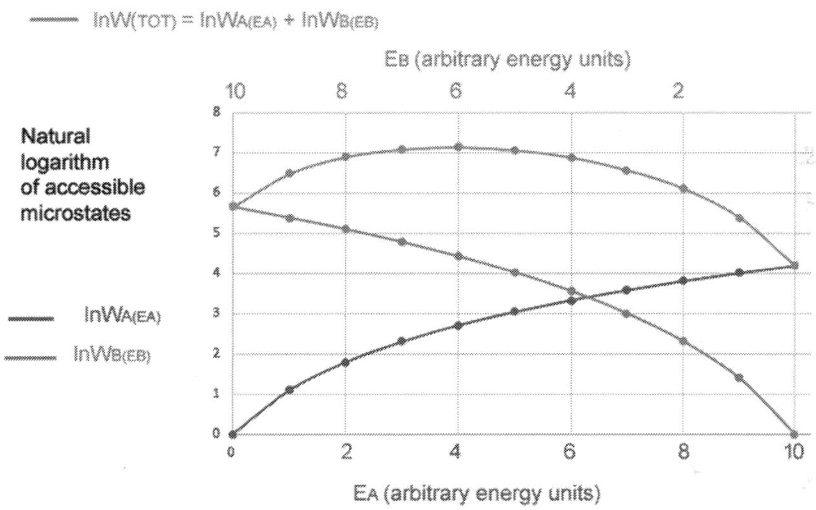

Figure 8.17 Plot of the natural logarithm of accessible microstates of the two subsystems (A and B) and of the composite subsystem probable microstates accessible to the isolated composite system composed of subsystems A and B after the removal of the internal insulating constraint and replacement by a rigid diathermal partition.

The total number of accessible microstates of the composite system will only increase if energy quanta are transferred to B.

I do not want to take this any further, but hopefully, you can understand (or at least appreciate) the following extract[47]:

> … when the body is described by stating its internal energy U, an extensive variable, as a function of its entropy S, also an extensive variable, and other state variables V, N, with U=U (S, V, N), then the temperature is equal to the partial derivative of the internal energy with respect to the entropy:
>
> $$T = \left(\frac{\partial U}{\partial S}\right)_{V,N} \qquad [2]$$
>
> Likewise, when the body is described by stating its entropy S as a function of its internal energy U, and other state variables V, N, with $S = S(U, V, N)$, then the reciprocal of the temperature is equal to the partial derivative of the entropy with respect to the internal energy:

$$\frac{1}{T} = \left(\frac{\partial S}{\partial U}\right)_{V,N} \quad [3]$$

... the equations (2) and (3) above are actually alternative definitions of temperature.

8.6 THE GIBBS' PARADOX

I want to finish this section with a consideration of, what is commonly known, as the Gibbs Paradox. Although I have mentioned Gibbs only a couple of times in Chapter 7 and, so far, only once in this chapter, he is one of the most important (along with Carnot, Clausius, and Kelvin), if not the most important figure in the development of thermodynamics. Accordingly, a short biographical note is in order[48]:

> Josiah Willard Gibbs; February 11, 1839–April 28, 1903) was an American scientist who made significant theoretical contributions to physics, chemistry, and mathematics. His work on the applications of thermodynamics was instrumental in transforming physical chemistry into a rigorous inductive science. Together with James Clerk Maxwell and Ludwig Boltzmann, he created statistical mechanics (a term that he coined), explaining the laws of thermodynamics as consequences of the statistical properties of ensembles of the possible states of a physical system composed of many particles. Gibbs also worked on the application of Maxwell's equations to problems in physical optics. As a mathematician, he invented modern vector calculus (independently of the British scientist Oliver Heaviside, who carried out similar work during the same period).

In this section I want only to consider one aspect of the paradox known as the 'mixing paradox'. Consider the following thought experiment. There are n_1 moles of an ideal gas of type 1, n_2 of another non-interacting type 2, confined in two volumes V_1, V_2 and separated by a diaphragm; both volumes are at the same temperature and pressure. The diaphragm is then removed, allowing the gases to diffuse through each other. Eventually, we reach a new equilibrium state with $n = n_1 + n_2$ moles of a gas mixture, occupying the total volume $V = V_1 + V_2$ with uniform composition, the temperature, pressure, and total energy remaining unchanged. If the gases are different, the entropy change due to the diffusion is given by (see Chapter 7 Equation 7.52):

$$\Delta S_{mixing} = \left[n_1 R\ln\left(\frac{V_1 + V_2}{V_1}\right)\right] + \left[n_2 R\ln\left(\frac{V_1 + V_2}{V_2}\right)\right] = \left[n_1 R\ln\left(\frac{V}{V_1}\right)\right] + \left[n_2 R\ln\left(\frac{V}{V_2}\right)\right] \quad (8.29)$$

Gibbs considers the case where $V_1 = V_2 = \frac{1}{2}V$ and because both volumes are at the same temperature and pressure, we know $n_1 = n_2 = \frac{1}{2}n$. Therefore, we can write Equation (8.29) as

$$\Delta S_{mixing} = \left[\frac{n}{2} R\ln\left(\frac{V}{\frac{V}{2}}\right)\right] + \left[\frac{n}{2} R\ln\left(\frac{V}{\frac{V}{2}}\right)\right] = nR\ln\left(\frac{V}{\frac{V}{2}}\right) = nR\ln 2 \quad (8.30)$$

If the two gases are identical there would be no entropy change when the partition is removed according to classical thermodynamics (because entropy is an extensive state function):

$$\Delta S = S_{final} - S_{initial} = 2S - (S + S) = 0 \quad (8.31)$$

This begs the question of how similar do two samples of gas have to be for this conclusion to follow? This is the paradox[49]:

> The entropy of mixing has the same value ... however alike are the two substances, but suddenly collapses to zero when they are the same. It is the absence of any 'warning' of the impending catastrophe, as the substances are made more and more similar, which is the truly paradoxical feature.

E.T. Jaynes argued that there is no paradox.[50] I will now try to set down Jaynes' reasoning with extensive use of extracts from his paper.

> When unlike gases mix under the conditions described above, there is an entropy increase $\Delta S = nR \log 2$ independent of the nature of the gases. When the gases are identical they still mix, but there is no entropy increase.

He then goes on to say that when two unlike gases mix, they could be separated and brought back to their original state, but this would require energy and an irreversible change in the entropy of the universe. He then considers what is meant by original state:

> But by the 'original state' we do not mean that every molecule has been returned to its original position, but only a state which is indistinguishable from the original one in the macroscopic properties that we are observing. For this we require only that a molecule originally in V_1 returns to V_1. In other words, we mean that we can recover the original thermodynamic state, defined for example by specifying only the chemical composition, total energy, volume, and number of moles of a gas; and nothing else. It is to the states of a system thus incompletely defined that the propositions of thermodynamics relate.
>
> But when we say that two identical gases mix without change of entropy, we do not mean that the molecules originally in V_1 can be returned to V_1 without external change. The assertion of thermodynamics is that when the net entropy change is zero, then the original thermodynamic state can be recovered without external change. Indeed, we have only to reinsert the diaphragm; since all the observed macroscopic properties of the mixed and unmixed gases are identical, there has been no change in the thermodynamic state. It follows that there can be no change in the entropy or in any other thermodynamic function.

He then argues that considering the discontinuity as the unlike gas particles become more and more alike and then identical "misses the point".

> The principles of thermodynamics refer not to any properties of the hypothesised microstates, but to the observed properties of macrostates; there is no thought of restoring the original microstate. We might put it thus: when the gases become exactly the same, the discontinuity is in what you and I mean by the words "restore" and "reversible".

Jaynes then considers what is meant by "identical" and that the entropy change $(\Delta S = nR \ln 2)$. is independent of the nature of the particles. In the future we may find differences in gas particles that we now consider identical:

> [We may] imagine other gases than those presently known, and there does not appear to be any limit to the resemblance which there might be between two of them; but ΔS would be independent of it.
>
> We may even imagine two gases which are absolutely identical in all properties which come into play while they exist as gases in the diffusion cell, but which differ in their behaviour in some other environment. In their mixing an increase of entropy would take place, although the process, dynamically considered, might be absolutely identical in its minutest details (even the precise path of each atom) with another process which might take place without any increase of entropy. In such respects, entropy stands strongly contrasted with energy.

He then considers the mixing of identical gases:

> From the fact that they mix without change of entropy, we must not conclude that they can be separated again without external change. On the contrary, the "separation" of identical gases is entirely impossible with or without external change. If "identical" means anything, it means that there is no way that an "unmixing" apparatus could determine whether a molecule came originally from V_1 or V_2, short of having followed its entire trajectory.

I have included the "Gibbs' Paradox" in this book as I consider it gives further insight into statistical mechanics and entropy (and it shows its conflicts and contradictions). It also indicates how far a simple model of matter (the ideal gas) can take us. Finally, I have included the following extract from Jaynes' paper on the Gibbs' paradox for no other reason than I found it interesting:

> Presumably, nobody doubts today that the measurable properties of argon … are described correctly by conventional thermodynamics which ascribes zero entropy change to the mixing of two samples of argon at the same temperature and pressure. But suppose that, unknown

to us today, there are two different kinds of argon, A1 and A2, identical in all respects except that A2 is soluble in Whifnium, while A1 is not (Whifnium is one of the rare superkalic[51] elements; in fact, it is so rare that it has not yet been discovered).

Until the discovery of Whifnium in the next Century, we shall have no way of preparing argon with controllably different proportions of A1 and A2. And even if, by rare chance, we should happen to get pure A1 in volume V_1, and pure A2 in V_2, we should have no way of knowing, or of detecting any difference in the resulting diffusion process. Thus all the thermodynamic measurements we can actually make today are accounted for correctly by ascribing zero entropy of mixing to argon.

Now the scene shifts to the next Century, when Whifnium is readily available to experimenters. What could happen before only by rare chance, we can now bring about by design. We may, at will, prepare bottles of pure A1 and pure A2. Starting our mixing experiments with $n_1 = fn$ moles of A1 in the volume $V_1 = fV$, and $n_2 = (1-f)n$ moles of A2 in $V_2 = (1-f)V$, the resulting actual diffusion may be identical in every detail, down to the precise path of each atom, with one that could have happened by rare chance before the discovery of Whifnium; but because of our greater knowledge we shall now ascribe to that diffusion an entropy increase ΔS given by Eq (1),[52] which we write as

$$\Delta S = \Delta S_1 + \Delta S_2 \tag{3}$$

where

$$\Delta S_1 = -nRf \log f \tag{4a}$$

$$\Delta S_2 = (1-f)\log(1-f) \tag{4b}$$

But if this entropy increase is more than just a figment of our imagination, it ought to have observable consequences, such as a change in useful work we can extract from the process.

There is a school of thought which militantly rejects all attempts to point out the close relation between entropy and information, claiming that such considerations have nothing to do with energy; or even that they would make energy "subjective" and it could therefore have nothing to do with experimental facts at all. We would observe, however, that the number of fish you can catch is an "objective experimental fact"; yet it depends on how much "subjective" information you have about the behaviour of fish.

But if this is not obvious, it is easily demonstrated in our present scenario. The diffusion will still take place without any change of temperature, pressure, or internal energy; but because of our greater information we shall now associate it with a free energy decrease $\Delta F = -T\Delta S$. Then according to the principles of thermodynamics, if instead of allowing the uncontrolled irreversible mixing we could carry out the same change of state reversibly and isothermally, we should be able to obtain from it the work

$$W = -\Delta F = T\Delta S \tag{5}$$

Let us place beside the diaphragm a movable piston of Whifnium. When the diaphragm is removed, the A2 will then diffuse through this piston until its partial pressure is the same on both sides, after which we move the piston slowly (to maintain equal A2 pressure and to allow heat to flow in to maintain constant temperature), in the direction of increasing V_1. From this expansion of A_1 we shall obtain the work

$$W_1 = \int_{V_1}^{V} P_1\, dV = n_1 RT \log(V/V_1)$$

Or from (4a)

$$W_1 = T\Delta S_1 \tag{6}$$

The term ΔS_1 in the entropy of mixing therefore indicates the work obtainable from reversible isothermal expansion of component A1 into the full volume $V = V_1 + V_2$. But the initial diffusion of A2 still represents an irreversible entropy increase ΔS_2 from which we obtain no work.

Spurred this partial success, the other superkalic element Whafnium is discovered, which has the opposite property that is permeable to A1 but not to A2. Then we can make an apparatus with two superkalic pistons; the Whifnium moves to the right, giving the work $W_1 = T\Delta S_1$, while the Whafnium moves to the left, yielding $W_2 = T\Delta S_2$. We have succeeded in extracting just the work $W = T\Delta S$ predicted by thermodynamics. The entropy of mixing does indeed represent human information; just the information needed to predict the work available from the mixing.

At this stage, I consider that I have gone as far as I can with the thermodynamics of ideal gases without the use of what is at least to me advanced mathematics (principally calculus). The next chapter introduces the concept of thermodynamic potentials, and at this stage, I must make the following important point that thermodynamic potentials are often used for systems where chemistry or other internal energy changes occur that are independent of temperature: to put this another way they are used for processes where work involves other energy transfer processes other than mechanical PV work. In this book, the only way the internal energy change of a system can occur is when its temperature changes. The reasons I am considering thermodynamic potentials are two-fold, firstly to reinforce the thermodynamic concepts already addressed in this book and, secondly to act as an introduction to the broader aspect of thermodynamics.

8.7 SUMMARY

I introduced the concepts of macrostate and microstate by coin tossing thought experiments. The number of microstates is the number of permutations of distinguishable (i.e., labelled) coins. The number of macrostates is the number of combinations of indistinguishable (i.e., unlabelled) coins. This exercise demonstrates that all microstates are equally probable, and if we toss an extremely large number of coins only a single macrostate needs to be considered, i.e., if I tossed a coin once I may get heads or tails, but as the number of tosses increases to an extremely large number the outcome of 50% heads and 50% tails (macrostate) becomes more and more precise. I considered the free expansion of an ideal gas, and in order to count the number of microstates, we need to introduce the concept of quantisation. I then demonstrated that only one macrostate needs to be considered (pressure of the gas in both volumes is equal) even though the number of accessible microstates is unbelievably large (even for just 100 microparticles expanding into an equal volume the accessible number of microstates approaches 10^{30}). All of this led us to the three fundamental postulates of statistical thermodynamics:

1. All possible microstates of a given system, which satisfy the given boundary conditions such as temperature, volume, and number of particles, are equally likely to occur.
2. For a system containing macroscopic quantities of microparticles (typically between 10^{22} and 10^{24}) the system will be in the statistically most probable microstate.
3. Kinetic energy and position are quantised, i.e., there are only discrete energies and positions that the microparticle can have and we say that the microparticle occupies a certain energy level.

In Section 3 I introduced the Boltzmann equation ($S = k_B \ln W$) which provides a link between microscopic and macroscopic thermodynamic properties.

I then considered, in Sections 4 and 5, thermal conductivity and temperature from a statistical point of view. Using two subsystems divided by a diathermal wall, with each subsystem containing a small number of microparticles sharing a small number of energy quanta, I demonstrated the most probable macrostate is when the energy quanta in the subsystems are in the same proportion as the proportion of microparticles in the subsystems. We know that at thermal equilibrium a quantity that is the same for the two subsystems is the temperature. This allowed an understanding of how a rigorous definition of temperature ("thermodynamic temperature") could be developed (albeit using relatively advance calculus). In other words, I used this thought experiment to demonstrate that energy flowing down a temperature gradient until thermal equilibrium was obtained is the most probable macrostate.

In Section 8.6 I discussed Gibbs' paradox mainly because it is an interesting story that helps clarify some aspects of statistical mechanics and entropy. It also indicates how far a simple model of matter (the ideal gas) can take us. Sections 8.7 and 8.8 introduced the concepts of thermodynamic potentials. This subject took us outside the remit of the book but does, I hope, act as a bridge between this book and the more traditional/advanced thermodynamic texts.

NOTES

1 Ludwig Boltzmann 1875.

2 https://en.wikipedia.org/wiki/Statistical_mechanics

3 For an ideal gas you only need a sub-set of these parameters, e.g., P, V and T, or P, V and N, U and T etc. Also remember a measurement of N is also given by the number of moles (n) where $n = \frac{N}{A}$ where A is Avogadro's number.

4 This figure is given in http://theory.physics.manchester.ac.uk/~judith/stat_therm/node55.html

5 https://www.chegg.com/homework-help/questions-and-answers/25-molecular-diameter-helium-atom-0225-nm-estimate-0-c-101325-kpa-helium-self-diffusion-co-q26164775

6 https://socratic.org/questions/which-of-the-following-graphs-represents-the-velocity-distribution-of-nitrogen-m

7 Permutation – all possible arrangements of a collection of things, where the order is important. A permutation focuses on the arrangement of objects with regard to the order in which they are arranged, i.e., each object is distinguishable either by labelling or order.

8 This is known as the "Ergodic Hypothesis" – "all accessible microstates are equally probable over a long period of time" or "all states of a system with a given energy are equally likely to occur over long periods of time".

9 Combination – all possible arrangements of a collection of things, where the order does not matter. A combination focuses on the selection of objects without regard to the order in which they are selected, i.e., each object is indistinguishable (unlabelled).

10 Returning to the "entropy is disorder" trope - it is only a subjective view that three heads and three tails is more disordered than six heads or six tails.

11 Assuming one toss every second, 10^{23} tosses would take about 3×10^{15} years. The universe is considered to be about 10^{10} years old.

12 The word "normalisation" is used informally in statistics, and so the term normalised data can have multiple meanings. In most cases, when you normalise data you eliminate the units of measurement for data, enabling you to compare distribution data from different systems.

13 Remember this is one of the fundamental assumptions of statistical thermodynamics.

14 A binomial is a polynomial that is the sum of two terms, each of which is a monomial (which are polynomials with only one term such as ax^by^c. A binomial is the simplest kind of polynomial after the monomial.

15 For example, see https://en.wikipedia.org/wiki/Binomial_theorem

16 All normal distributions are symmetric and have bell-shaped curves with a single peak.

17 To give some idea of the enormity of these numbers the number of atoms in the "known observable" universe is estimated to be between 10^{78} and 10^{82}.

18 The unwritten assumption here is that we are dealing with classical physics. In other words, we are ignoring Heisenberg's Uncertainty Principle which tells us there is a fundamental limit or fuzziness to what we can know about the behaviour of microparticles aka "quantum particles" i.e., their position and energy. Consideration of the implications of this soon leads to complexity and difficulty.

The following abstract from https://www.oxfordreference.com/view/10.1093/oi/authority.20110810105929664 is helpful to those with some background in the subject:

"In classical statistical mechanics, each particle is regarded as occupying a point in phase space, i.e. to have an exact position and momentum at any particular instant. The probability that this point will occupy any small volume of the phase space is taken to be proportional to the volume. The Maxwell-Boltzmann law gives the most probable distribution of the particles in phase space."

"With the advent of quantum theory, the exactness of these premises was disturbed (by the Heisenberg uncertainty principle). In the quantum statistics that evolved as a result, the phase space is divided into cells, each having a volume h^f, where h is the Planck constant and f is the number of degrees of freedom of the particles. This new concept led to Bose-Einstein statistics, and for particles obeying the Pauli exclusion principle, to Fermi-Dirac statistics."

Also see Endnote 27.

19 Odds of getting 100 heads in a row is about 8×10^{-31}. The universe is 13.8×10^{9} years. Therefore, the number of coin tosses over the time the universe has been in existence is about 4×10^{18}. Assuming, a reasonable chance of getting 100 heads in row was if there were say 10^{30} coin tosses, then it would require a time equivalent to about 2×10^{12} times the age of the universe.

20 "Ludwig Boltzmann (1844–1906) is generally acknowledged as one of the most important physicists of the nineteenth century. Particularly famous is his statistical explanation of the second law of thermodynamics. The celebrated formula $S = k\log W$, expressing a relation between entropy S and probability W has been engraved on his tombstone (even though he never actually wrote this formula down). Boltzmann's views on statistical physics continue to play an important role in contemporary debates on the foundations of that theory." Stanford Encyclopedia of Philosophy. https://plato.stanford.edu/

21 "Even though Boltzmann died in 1906, there is no evidence that he ever saw, and thus certainly never calculated entropy values via the equation Planck published in a 1900 article, $S = \frac{R}{N} \ln W$. It was first printed in 1906 in a book by Planck as $S = K_B \ln W$ and subsequently carved on Boltzmann's tombstone. Planck's nobility in allowing $\frac{R}{N}$ to be called 'Boltzmann's constant', k_B, was uncharacteristic of most scientists of that day, as well as now." Abstract from https://www.claymath.org/wp-content/uploads/2023/09/Lebowitz.pdf.

22 The website of Hmolpedia or Encyclopedia of Human Thermodynamics, Human Chemistry, and Human Physics at https://www.eoht.info/page/S%20=%20k%20ln%20W

23 It was Ubbelohde himself who called it a "crude derivation".

24 A.M. Glazer, J.S. Wark, *Statistical Mechanics A Survival Guide*, Oxford University Press, p. 18, 2001. https://archive.org/details/statisticalmecha0000glaz

25 In other words – a hypothesis.

26 From the Hmolpedia website, https://www.eoht.info/page/S%20=%20k%20ln%20W

27 "In classical mechanics, the phase space is the space of all possible states of a physical system; by "state" we do not simply mean the positions q of all the objects in the system (which would occupy physical space or configuration space), but also their velocities or momenta p (which would occupy momentum space)."

From https://www.math.ucla.edu/~tao/preprints/phase_space.pdf. The origins of the phrase "phase space" are obscure – see https://hsm.stackexchange.com/questions/11521/what-is-the-etymology-of-phase-space-of-a-dynamical-system

Each momentum and position coordinate counts as one dimension of phase space. So, for a single particle, there are three degrees of freedom (x, y, and z); and the state of the particle is defined by those three coordinates, and also by the three coordinates of the momentum. Consequently, one would have a 6-dimensional phase space.

28 M.Campisia, D.H. Kobeb, "Derivation of the Boltzmann principle", *Am. J. Phys.* 78, 6, 2010.

29 W.J. Moore, *Physical Chemistry*, 5th Edition, pp. 170–171, 1972.

30 Remember $k_B = \frac{R}{A} = 1.4 \times 10^{-23} J.K^{-1}$, where A is Avogadro's number.

31 Remember $x \times \ln A = (\ln A)x$ – see Section 1.6.9 Table 1.5.

32 https://www.eoht.info/page/S%20=%20k%20ln%20W

33 *Deus ex machina* ('god from the machine') - the term has evolved to indicate whereby a seemingly unsolvable problem is suddenly and abruptly resolved by the contrived and unexpected intervention of some new event, character, ability, or object.

34 I googled the phrase "Boltzmann entropy formula" and 19 of the first 20 set down the equation without the constant. The only exception was the Hmolpedia site previously mentioned.

35 The equation $S = k\log W$ (base e must be assumed) that is famously displayed on the Boltzmann tombstone, which was erected in the 1930s, at the Central Cemetery (Zentralfriedhof), Vienna, Austria. Hmolpedia states "The modern formulation, $S = k \ln W$ (or $S = k \log W$, base e assumed), as it is now known as a staple of science, was proposed or rather stated as a matter-of-fact by German physicist Max Planck as being the standard formula for the measure of the entropy …."

36 If the constant is an 'arbitrary constant' (an arbitrary constant is a constant to which various values may be assigned but which remains unaffected by the changes in the values of the variables of the equation) then it cancels out in the calculation of ΔS. However, if it is not an arbitrary constant (and I am not able to tell from the 'proof' given by Ubbelohde/Moore) then it will not cancel out and I have no idea what the implications of that are. The constant could of course be zero as Planck must have decided, but if it is not and if we do not know it, then we cannot calculate absolute values of entropy content. See ET Jayne "The Gibbs Paradox" pages 15–16 https://www.damtp.cam.ac.uk/user/tong/statphys/jaynes.pdf

37 Adapted from https://www.quora.com/If-heat-is-released-from-a-system-will-entropy-increase. Also, I am only considering the formation of a perfect crystal at 0 K (see Section 7.8).

38 http://hyperphysics.phy-astr.gsu.edu/hbase/thermo/bioentropy.html

39 There are good or useful analogies and there are bad and confusing analogies – this one falls into the latter class.

40 Based on the paper found at http://stp.clarku.edu/notes/chap4.pdf; also see http://personal.rhul.ac.uk/UHAP/027/PH2610/PH2610_files/SECT2.pdf

41 Remember $\mathbf{dU} = \mathbf{dq} + \mathbf{dw}$ or, more rigorously, $\mathbf{dU} = \delta\mathbf{q} + \delta\mathbf{w}$ (where δ indicates an inexact differential – see Section 1.6.9.1). For gases, $\mathbf{dw} = -\mathbf{PdV}$ and $\mathbf{dq} = \mathbf{TdS}$. Note that there are no inexact differentials in $\mathbf{dU} = \mathbf{TdS} - \mathbf{PdV}$.

42 https://www.quora.com/What-is-the-physics-behind-temperature

43 See, Section 7.1, Equation 7.3

44 See Section 4.3.5.

45 $\mathbf{dq_{rev}} \equiv \mathbf{dE}$ means $\mathbf{dq_{rev}}$ is identical to $\mathbf{dE}$.

46 The most common numerical approximation to the derivative $f'(x)$ of a function $f(x)$ is to take the gradient of the chord joining the point and another point where x has been increased

by a small amount h. The central difference approximation uses the chord joining the two points whose x values are a small amount h from the value x, giving $f'(x)=\frac{f(x+h)-f(x-h)}{2h}$.

47 https://en.wikipedia.org/wiki/Temperature

48 https://en.wikipedia.org/wiki/Josiah_Willard_Gibbs

49 K. Denbigh, M. Redhead, "Gibbs' paradox and non-uniform convergence", *Synthese*, 81, 283–312, 1989.

50 E.T. Jaynes, The Gibbs paradox. In: C.R. Smith, G.J. Erickson, & P.O. Neudorfer (Editors.) *Maximum Entropy and Bayesian Methods*, Kluwer Academic Publishers, Dordrecht, Holland, 1992; pp. 1–21. Found at https://www.damtp.cam.ac.uk/user/tong/statphys/jaynes.pdf

51 As far as I am aware "superkalic" is a made-up word, as are the more obvious Whifnium and Whafnium.

52 Present author's note – Eq (1) is $\Delta S=-nR\left[f\log f+(1-f)\log(1-f)\right]$.

9 Free Energy

9.1 INTRODUCTION

The principal objective of this chapter is to provide a link between this book and the traditional thermodynamic texts, aimed at experimental chemists, which concentrate on subjects such as enthalpy and free energy. From the previous chapters, we know that a spontaneous process results in the entropy of the universe increasing. We also know that the change in the entropy of the universe is the summation of the change in the entropies of the system and its surroundings. In this section and the next, I want to consider the concept of "thermodynamic potential" which gives rise to four terms all or some of which will be known by anyone with a basic knowledge of thermodynamics. We have already considered one of these – internal energy (U) – the other three are enthalpy (H), Helmholtz free energy or function (F) and Gibbs free energy or function (G). Thermodynamic potentials are analogous to mechanical potential energy and are defined as the capacity to do work.

To deal rigorously with thermodynamic potentials requires advanced (at least, to me) mathematics. Consider the following extract[1]:

> A thermodynamic potential (or more accurately, a thermodynamic potential energy) is a scalar quantity used to represent the thermodynamic state of a system. The concept of thermodynamic potentials was introduced by Pierre Duhem in 1886. Josiah Willard Gibbs in his papers used the term fundamental functions. One main thermodynamic potential that has a physical interpretation is the internal energy U. It is the energy of configuration of a given system of conservative forces (that is why it is called potential) and only has meaning with respect to a defined set of references (or data). Expressions for all other thermodynamic energy potentials are derivable via Legendre transforms from an expression for U. In thermodynamics, external forces, such as gravity, are typically disregarded when formulating expressions for potentials.

The objective of this chapter is to introduce the concept of thermodynamic potentials without the arcane mathematics, or, for the mathematically literate without "using elementary multivariate calculus",[2] and do not worry I will not be carrying out any Legendre transformations.[3]

Those readers who have no previous experience in thermodynamics are probably none the wiser at this point. However, as thermodynamics potentials play such an important role in practical thermodynamic calculations I felt that I should end this book with this rather flawed "foundation course" on thermodynamic potentials.

To state the obvious, any process that provides work must (by definition) transfer energy from the system to the surroundings by the process of work. This means that for a system to provide useful work (e.g., a falling weight in the Earth's gravitational field which could, e.g., drive a waterwheel) its thermodynamic potential must decrease. This is exactly analogous to potential energy in classical mechanics where an object falling under the influence of gravity can provide useful work and its potential energy decreases.

As far as we are concerned:

- Internal energy is the maximum energy that is available to perform thermodynamic work at constant volume.
- Enthalpy is the maximum energy that is available to perform thermodynamic work at constant pressure.
- Free energy is the maximum energy that is available to perform thermodynamic work at constant temperature. Two free energy functions may be formulated – Helmholtz free energy which is the maximum energy that is available to perform thermodynamic work at constant temperature and constant volume, and Gibbs free energy which is the maximum energy that is available to perform thermodynamic work at constant temperature and constant pressure.
- In all cases for a spontaneous process, there must be a decrease in the thermodynamic potential.

9.2 INTERNAL ENERGY

Consider a closed system consisting of a vessel of negligible mass containing n moles of ideal gas at T_{system}K in surroundings at $T_{surroundings}$K; where $T_{system} \neq T_{surroundings}$. We assume the vessel is constructed of rigid impermeable diathermal walls of negligible mass, i.e., the volume

DOI: 10.1201/9781003608400-9

is constant, and energy can flow in and out of the system by the process of heat. By definition (as $T_{system} \neq T_{surroundings}$), the system is not in equilibrium, and there will be a flow of energy $(\mathbf{q_V})$, by the process of heat, until it reaches the same temperature as the surroundings, i.e., it is an irreversible process. The entropy of the system (the ideal gas) and the surroundings will change, so we have to calculate them separately. For each, we need to find the initial and final states A and B and calculate the entropy change for a reversible process that takes us between the same states. I will start with the system.

For a reversible process, we know the energy must be transferred from the surroundings to the system at an infinitesimal rate. Therefore, we can write (where $\mathbf{q_V}$ is the flow of energy to the system by the process of heat at constant pressure)[4]:

$$\mathbf{dq_V} = nC_V\mathbf{dT} \tag{9.1}$$

Then the total entropy change will be

$$\Delta\mathbf{S_{system}} = \int_{T_{system}}^{T_{surroundings}} \frac{\mathbf{dq_V}}{T} = nC_V \int_{T_{system}}^{T_{surroundings}} \frac{\mathbf{dT}}{T} = nC_V \ln \frac{\mathbf{T_{surroundings}}}{\mathbf{T_{system}}} \tag{9.2}$$

Remember this reversible process uses an infinite number of heat reservoirs at temperatures varying infinitesimally from each other $(T + \mathbf{dT})$, etc. from T_{system} to $T_{surroundings}$. But $\Delta\mathbf{S}$ is a state function and is independent of the path taken.

I will now consider the surroundings for the irreversible process we are considering. Energy $\mathbf{q}$ flows from the system into the surroundings (also known as the heat reservoir at $T_{surroundings}$) by the process of heat. However, by definition, the surroundings/heat reservoir remains at $T_{surroundings}$. We can calculate the energy flow into the surroundings, by the process of heat, by working out how much energy the ideal gas gains $(\mathbf{q_{system}})$, by the process of heat, knowing that $(\mathbf{q_{system}}) = -(\mathbf{q_{surroundings}})$:

$$\mathbf{q_V} = nC_V\left(T_{surroundings} - T_{system}\right) \tag{9.3}$$

Knowing $(\mathbf{q_V}) = -(\mathbf{q_{surroundings}})$ we can write:

$$\mathbf{q_{surroundings}} = nC_V\left(T_{system} - T_{surroundings}\right) \tag{9.4}$$

We know all energy transfer processes that occur in heat reservoirs are reversible. Therefore, the entropy change of the surroundings, at constant temperature $T_{surroundings}$ is

$$\Delta\mathbf{S_{surroundings}} = \frac{\mathbf{q_{surroundings}}}{T_{surroundings}} = \frac{nC_V\left(T_{system} - T_{surroundings}\right)}{T_{surroundings}} \tag{9.5}$$

By definition $nC_V\left(T_{surroundings} - T_{system}\right) = \Delta\mathbf{U_{system}}$. Therefore, we can write:

$$\Delta\mathbf{S_{surroundings}} = \frac{-\Delta\mathbf{U_{system}}}{T_{surroundings}} \tag{9.6}$$

Consequently, the entropy change of the universe can be written as

$$\Delta\mathbf{S_{universe}} = \Delta\mathbf{S_{system}} + \Delta\mathbf{S_{surroundings}} = \Delta\mathbf{S_{system}} - \frac{\Delta\mathbf{U_{system}}}{T_{surroundings}} \tag{9.7}$$

Equation (9.7) expresses the entropy change of the universe at constant temperature T in terms of thermodynamic properties of the system exclusively. In other words, we are able to write an equation for the change of entropy of the universe in terms of the parameters of the system only, i.e., the entropy change of the universe can be calculated only by knowing the system's thermodynamic parameters. Why are we able to do this? The answer is straightforward but is rarely explained: the reason is that, by definition, all energy transfer processes that occur in heat reservoirs are reversible.

There are two scenarios to consider $T_{system} > T_{surroundings}$ (Scenario I) and $T_{system} < T_{surroundings}$ Scenario II). In Scenario I both $\Delta \mathbf{U}_{system}$ and $\Delta \mathbf{S}_{system}$ are negative.[5] For a spontaneous reaction ($\Delta \mathbf{S}_{universe}$ is positive), we know $\Delta \mathbf{S}_{system} > \frac{\Delta \mathbf{U}_{system}}{T_{surroundings}}$. Using absolute numbers, we know $\left|\Delta \mathbf{S}_{system}\right| < \left|\frac{\Delta \mathbf{U}_{system}}{T_{surroundings}}\right|$. Although by now we should be used to this, I will illustrate with a numerical example.

Consider a closed system containing 1 mole of ideal gas at 100°C (373 K) in surroundings at 20°C (293 K). We know that C_V (molar heat capacity) of an ideal gas is 12.48 J.kg^{-1}.mol^{-1} (see Section 4.3.2).

We know $\Delta \mathbf{S}_{system} = (1)(12.48)\ln\frac{293}{373} = -3.0\ \text{J.K}^{-1}$, i.e., the entropy of the system decreases, as expected, since energy is flowing out by the process of heat. We know $\Delta \mathbf{S}_{surroundings} = \frac{12.48(373-293)}{293}$ $= +3.4\ \text{J.K}^{-1}$, which is positive, and larger than the entropy decrease of the system. The entropy change of the universe is $\Delta S_{universe} = \Delta S_{systen} + \Delta S_{surroundings} = (-3.0)+(+3.4) = +0.4\ \text{J.K}^{-1}$.

Let us now consider Scenario II. In Scenario II, both $\Delta \mathbf{U}_{system}$ and $\Delta \mathbf{S}_{system}$ are positive. Therefore, for a spontaneous reaction $\Delta \mathbf{S}_{system} > \frac{\Delta \mathbf{U}_{system}}{T_{surroundings}}$ (the same as Scenario I). However, as in this case, both $\Delta \mathbf{U}_{system}$ and $\Delta \mathbf{S}_{system}$ are positive; this means that in contrast to Scenario I, $\left|\Delta \mathbf{S}_{system}\right| > \left|\frac{\Delta \mathbf{U}_{system}}{T_{surroundings}}\right|$. Again, I will illustrate with a numerical example. Consider a closed system containing 1 mole of ideal gas at 20°C (293 K) in surroundings at 100°C (373 K). We know $\Delta \mathbf{S}_{system} = (1)(12.48)\ln\frac{373}{293} = +3.0\ \text{J.K}^{-1}$, i.e., the entropy of the system increases, as expected, since energy is flowing into the system by the process of heat. We know $\Delta \mathbf{S}_{surroundings} = \frac{12.48(293-373)}{373} = -2.6\ \text{J.K}^{-1}$, which is negative, and less than the entropy increase of the system. The entropy change of the universe is $\Delta \mathbf{S}_{universe} = \Delta \mathbf{S}_{system} + \Delta \mathbf{S}_{surroundings} = (+3.0)+(-2.6) = +0.4\ \text{J.K}^{-1}$.

All of the above tells us what we already know, i.e., for the entropy of the universe to increase energy must flow, by the process of heat, down a temperature gradient. In other words, spontaneous energy transfer, by the process of heat, only flows down a temperature gradient.

How can a reduction in the internal energy of a constant volume system undergoing a purely physical change[6] carry out useful work?[7] We already know one way – the heat engine.[8] We know that to achieve maximum useful work, the process must be carried out reversibly.

$$-\mathbf{w}_{max} = \mathbf{q}_H^{rev} + \mathbf{q}_C^{rev} \tag{9.8}$$

We know (using absolute values for clarity) that

$$-\left|\Delta \mathbf{U}\right| = C_V\left(T_C - T_H\right) = -\left(+\left|\mathbf{q}_H^{rev}\right| - \left|\mathbf{q}_C^{rev}\right|\right) = -\left|\mathbf{w}_{max}\right| \tag{9.9}$$

In other words, $\Delta \mathbf{U} = \mathbf{w}_{max}$. This demonstrates that the change in the internal energy of the system at constant volume is a measure of the maximum work that the system can generate, i.e., it is a measure of thermodynamic potential. For a system to be able to carry out useful work on the surroundings, it must, by definition, lose energy, i.e., $-|\Delta \mathbf{U}|$, if there is a positive flow of energy from the system to the surroundings, or using our sign convention there is a negative flow of energy to the system from the surroundings. We know the system does not work because the volume remains constant; therefore $-\Delta \mathbf{U} = -\mathbf{q}_V$, i.e., the change in internal energy is equivalent to the flow of energy from the constant volume system by the process of heat. Therefore, the maximum useful work $(-\mathbf{w})$ that can be extracted must be equal to $-\Delta \mathbf{U}$, i.e., the change in internal energy is the thermodynamic potential for constant volume systems.

Now, consider the following, if $\mathbf{w}_{rev}$ is the energy transferred to the system at constant pressure, then this energy includes a PV work component which means $\mathbf{w}^{rev}_{constant\ volume} \neq \mathbf{w}^{rev}_{constant\ pressure}$ ($\mathbf{w}^{rev}_{constant\ pressure}$ could be greater, equal, or less than $\mathbf{w}^{rev}_{constant\ volume}$ depending on whether the system's volume increases, remains the same, or decreases during the process). This subject is considered further in the next section.

9.3 ENTHALPY

9.3.1 Introduction

The energy that passes into or out of the system during a chemical reaction or physical process is the enthalpy change as a result of that reaction/process. Whether the enthalpy of the system increases (i.e., when energy is added) or decreases (because energy is given off) is a crucial factor that determines whether or not a reaction is spontaneous. Most processes of interest in chemistry (including those of life) take place at constant pressure. I have demonstrated that a system consisting of expanding gas, in a piston arrangement, pushing against the resistance of its surroundings loses energy to its surroundings by the process of work (see Section 4.3.5). When the system is expanding, we say the system carries out work on the surroundings, conversely, a contracting gas gains energy from the surroundings by process of work and we say the surroundings carry out work on the system. This PV work must be considered when considering the thermodynamics of processes at constant pressure if they involve a volume change. This can be made easier by using the concept of enthalpy which is our second thermodynamic potential. I will first introduce the concept of enthalpy and demonstrate that it is a state function.

9.3.2 Definition of Enthalpy

Consider an ideal gas system with an internal energy of $U_{initial}$. When the temperature of the system changes at constant pressure its internal energy changes to U_{final}. At constant pressure, two processes (heat and work) change the internal energy of the system and we know by the First Law of Thermodynamics:

$$\mathbf{\Delta U} = U_{final} - U_{initial} = \mathbf{q_P} + \mathbf{w} \tag{9.10}$$

where $\mathbf{q_P}$ is the flow of energy to the system by the process of heat at constant pressure, and $\mathbf{w}$ is the flow of energy to the system transferred by the process of work. Let us assume the only work done by the system is PV work, i.e., $\mathbf{w} = -P\mathbf{\Delta V}$ (remember the negative sign is a direct consequence of our sign convention[9]). Therefore:

$$\mathbf{\Delta U} = U_{final} - U_{initial} = \mathbf{q_P} + \mathbf{w} = \mathbf{q_P} + (-P\mathbf{\Delta V}) = \mathbf{q_P} - P(V_{final} - V_{initial}) \tag{9.11}$$

Equation (9.11) can be rearranged:

$$(U_{final} + PV_{final}) - (U_{initial} + PV_{initial}) = \mathbf{q_P} \tag{9.12}$$

I have already stated that $\mathbf{q_P}$ is the flow of energy to the system by the process of heat at constant pressure. Consequently $-\mathbf{q_P}$ is the flow of energy to the surroundings by the process of heat.

Now we define a new function (H) called enthalpy:

$$H = U + PV \tag{9.13}$$

Then from Equations (9.12) and (9.13) we have:

$$\mathbf{\Delta H} = H_{final} - H_{initial} = \mathbf{q_P} \tag{9.14}$$

In words: when a process occurs at constant pressure, the energy transferred into the system by the process of heat, is equal to the change in enthalpy $(\mathbf{\Delta H})$ when the only work done is mechanical PV work. Note (Equation 9.13) that when no work is done the energy transferred into the system by the process of heat, is equal to the change in internal energy $(\mathbf{\Delta U})$. Enthalpy is a state function because it is defined in terms of state functions (U, P, and V are all state functions).

We know from its definition (Equation 9.6) that the change in enthalpy during a constant pressure process is:

$$\mathbf{\Delta H} = \mathbf{\Delta U} + P\mathbf{\Delta V} \tag{9.15}$$

Equation (9.15) says that in a closed system enthalpy can increase for two reasons: either because the internal energy (U) increases, or because the system expands ($\mathbf{\Delta V}$ is positive), and work is done on the atmosphere (the surroundings) to make room for it. It is worth noting that the definition

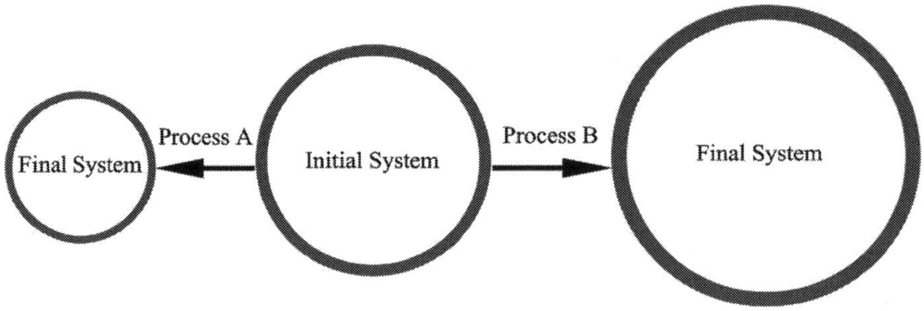

Figure 9.1 To move from the initial system to the final system we need to calculate not only $U_{final} - U_{initial}$ but also some additional energy equal to $P(V_{final} - V_{initial})$ to push the atmosphere out of the way to make room. The total energy required is the enthalpy, $\Delta H = \Delta U + P\Delta V$.

of enthalpy is not affected by the different sign conventions of chemistry and physics. This can immediately be seen if we rewrite Equations (9.10) and (9.11) using the physics sign convention, and we end up with Equation (9.18) (which is identical to Equation 9.12):

$$\mathbf{\Delta U} = U_{final} - U_{initial} = \mathbf{q_P} - w \quad (9.16)$$

$$\mathbf{\Delta U} = U_{final} - U_{initial} = \mathbf{q_P} - \mathbf{w} = \mathbf{q_P} - (\mathbf{P\Delta V}) = \mathbf{q_P} - P(V_{final} - V_{initial}) \quad (9.17)$$

$$(U_{final} + PV_{final}) - (U_{initial} + PV_{initial}) = \mathbf{q_P} \quad (9.18)$$

The following abstract may clarify matters[10] (and also see Figure 9.1):

> Instead of always talking about the energy content of a system, we can agree to always add in the work needed to make room for it (under a constant pressure, usually 1 atm). This work is PV, the pressure of the environment times the total volume of the system (that is, the total space you would need to clear out to make room for it). Adding PV onto the energy gives a quantity called the enthalpy, denoted H:
>
> $$H \equiv U + PV$$
>
> This is the total energy you would have to come up with, to create the system out of nothing and put it into this environment .... Or, put another way, if you could somehow annihilate the system, the energy could extract is not just U, but also the work (PV) done by the atmosphere as it collapses to fill the vacuum left behind.

The internal energy of an ideal gas system is the energy that must be transferred to the system, by the process of heat, to bring the system to its present state (TK) from absolute zero (0 K) when the volume of the system remains constant. In other words, the internal energy is the heat required to create the system at constant volume. Now let us consider the creation of an ideal gas system at temperature TK at constant atmospheric pressure. If the volume of the system remains constant, then the energy to create the system is its internal energy at TK (by definition). Now let us consider a process where the volume of the system increases, i.e., $+|\mathbf{\Delta V}|$. In this case, additional energy (on top of the internal energy of the final state of the system) must be supplied, by the process, of work, in pushing against the ambient (atmospheric) pressure, i.e., there is a positive flow of energy, by the process of work into the system from the surroundings. This $P\mathbf{\Delta V}$ work can be interpreted as the work that would be required to "make room" for the system if the pressure of the environment remained constant. In other words, when the volume of the system increases the energy required to create it (its enthalpy) is greater than its internal energy. Conversely, if the volume decreases during the process, then there is a negative flow of energy, by the process of work into the system from the surroundings, or, to put it another way, there is a positive flow of energy, by the process of work into the surrounding from the system. In other words, when the volume of the system decreases, the energy required to create it is less than its internal energy.

You can forget all about compression expansion work if you deal with enthalpy instead of energy. If no other types of work are being done, the change in enthalpy tells you directly the flow of energy into the system by the process of heat.

The enthalpy of an ideal gas is independent of its pressure or volume and depends only on its temperature. Real gases at common temperatures and pressures often closely approximate this behaviour, which simplifies practical thermodynamic design and analysis. As a state function, enthalpy depends only on the final configuration of internal energy, pressure, and volume, not on the path taken to achieve it.

The total enthalpy of a system cannot be measured directly because the internal energy contains components that are unknown, not easily accessible, or are not of interest in thermodynamics. In practice, a change in enthalpy is the preferred expression for measurements at constant pressure because it simplifies the description of energy transfer. Changes in internal energy and enthalpy are measured using a calorimeter.

9.3.3 Calorimetry

Calorimetry is the process of measuring the amount of energy released or absorbed, by the process of heat, during a chemical reaction. In order to measure this energy change, the process must be isolated from the surroundings so that no energy is lost. This is achieved by the use of a calorimeter, which insulates the reaction to better contain heat. To measure the change in internal energy (ΔU) a constant volume calorimeter, usually known as a bomb calorimeter is used. Constant pressure calorimeters used to measure ΔH are much simpler devices; the simplest of which is known as a coffee cup calorimeter (Figure 9.2).

Let us use the calorimeter to find the latent enthalpy of fusion of ice $\left(\Delta \mathbf{H}_{\mathbf{fusion}}^{\mathbf{ice}}\right)$. Take m_{ice} grams of ice and place it into the calorimeter containing M_{water}g of water at $T_{initial}$K. The calorimeter is stirred and after a few minutes the ice has melted, and the temperature of the water has fallen to T_{final}K. After the ice has melted, we know that the quantity of energy gained by the ice is equivalent to the energy lost by water, i.e., $|\mathbf{q}_{\mathbf{ice}}| = |\mathbf{q}_{\mathbf{water}}|$ or $\mathbf{q}_{\mathbf{ice}} + \mathbf{q}_{\mathbf{water}} = 0$. In other words, the ice cube is gaining energy (positive) while the water inside the calorimeter is losing energy (indicated by the negative sign).

We know $\mathbf{q}_{\mathbf{water}} = M_{water} \times \mathbf{c} \times \Delta \mathbf{T}$ where **c** is the specific heat of water and $\Delta \mathbf{T} = T_{final} - T_{initial}$. Knowing that $T_{final} < T_{initial}$ tells us that $\mathbf{q}_{\mathbf{water}}$ is a negative real number. The negative value of $\mathbf{q}_{\mathbf{water}}$ supports what we already know – ice melting results in a loss of energy in the water (surroundings). We assume that all the energy lost by water is gained in equal magnitude but in the opposite sign to ice, i.e., $\mathbf{q}_{\mathbf{ice}}$ is a positive real number and this indicates the ice has gained energy. By definition, $\mathbf{q}_{\mathbf{ice}} = \Delta \mathbf{H}_{\mathbf{fusion}}^{\mathbf{ice}} \times m_{ice}$. One question you might ask is why there is not a change in temperature in the previous equation to solve for $\Delta \mathbf{H}_{\mathbf{fusion}}^{\mathbf{ice}}$? The reason is straightforward, but worth mentioning – as the ice begins to melt to liquid water, the temperature remains constant even though energy is being added to the system (the ice). The temperature will remain constant until all the ice has melted into liquid water.

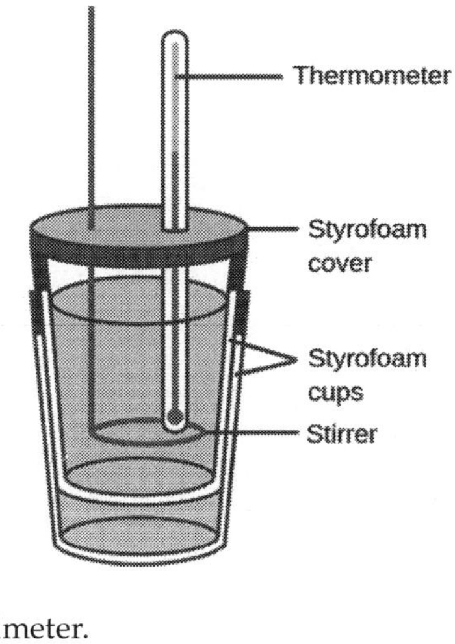

Figure 9.2 A coffee cup calorimeter.

Table 9.1: Experimental for Coffee Cup Calorimeter Determination of the Latent Enthalpy of Fusion of Ice

m_{ice}	30.0 g
M_{water}	120.0 g
$T_{initial}$	309.4 K
T_{final}	292.3 K
c	$4.18 \text{ J.g}^{-1}.\text{K}^{-1}$

$$\mathbf{q_{water}} = M_{water} \times c \times \Delta\mathbf{T} = 120.0 \times 4.18 \times (292.3 - 309.4) = -8,577 \text{ J}$$

$$\mathbf{q_{ice}} = +8,577 \text{ J}$$

$$\mathbf{q_{ice}} = m_{ice} \times \Delta\mathbf{H}^{\mathbf{ice}}_{\mathbf{fusion}} = +8,577 \text{ J} = 30.0 \times \Delta\mathbf{H}^{\mathbf{ice}}_{\mathbf{fusion}}$$

$$\Delta\mathbf{H}^{\mathbf{ice}}_{\mathbf{fusion}} = \frac{q_{ice}}{m_{ice}} = \frac{8,577}{30.0} = 286 \text{ J.g}^{-1}$$

To clarify the above calculations, I will use some real data taken from the experiment (see Table 9.1)[11]:

The accepted value[12] for the enthalpy of fusion of ice is 334 J.g^{-1} which suggests either poor experimental technique or that the coffee cup calorimeter is not an accurate piece of kit (probably a combination of the two).

I hope that this sub-section has given you some insight into calorimetric processes.

9.3.4 Enthalpy of Chemical Bonds[13]

In chemistry, experiments are often conducted at constant atmospheric pressure, and the pressure–volume work represents a small, well-defined energy exchange with the atmosphere, so that ΔH is the appropriate expression for the heat of reaction (and biological reactions in cells). The total enthalpy of a system cannot be measured directly; the *enthalpy change* of a system is measured instead. Enthalpy change is defined by the following equation: $\Delta H = H_f - H_i$ where:

- ΔH is the "enthalpy change",
- H_f is the final enthalpy of the system (in a chemical reaction, the enthalpy of the products),
- H_i is the initial enthalpy of the system (in a chemical reaction, the enthalpy of the reactants).

Apart from mentioning atoms and molecules (Section 3.6.2), there is essentially no mention of chemistry in this book. A chemical bond is an attraction between atoms, ions, or molecules that enables the formation of chemical compounds. The bond may result from the electrostatic force between oppositely charged ions as in ionic bonds or through the sharing of electrons as in covalent bonds. In essence, chemistry is concerned with processes where chemical bonds are broken and formed. A chemical bond holds two atoms together, the source of the force of attraction is a balance in the electrostatic attraction between opposite charges (i.e., electrons and the protons in the nucleus) and electrostatic repulsion of the like charges (see Figure 9.3).

Consider the following chemical reaction (the simplest chemical reaction there is):

$$H + H \rightarrow H_2 \tag{9.19}$$

If this reaction takes place at constant pressure the amount of energy transferred (**q**) to the system from the surroundings, by the process of heat, is $-436 \text{ k.J.mol}^{-1}$.

The strength of a chemical bond is the magnitude of the force required to separate its two component atoms to an infinite distance apart (Equation 9.20).

$$H_2 \rightarrow H + H \tag{9.20}$$

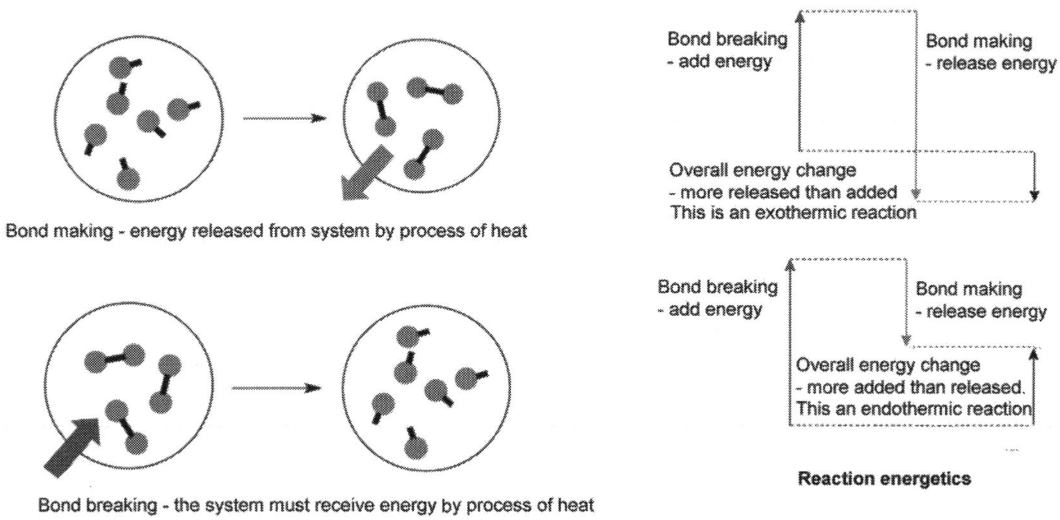

Figure 9.3 Schematic diagrams of bond formation and breaking, and reaction energetics.

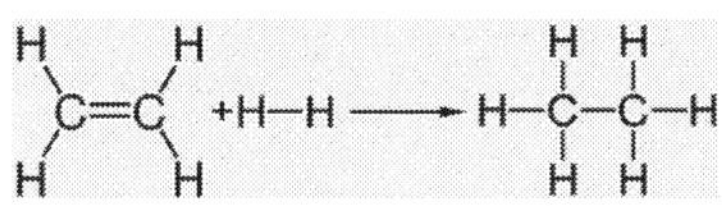

Figure 9.4 Hydrogenation of ethene.

Therefore, to break the H – H bond requires +436 k.J.mol^{-1} of energy to be transferred to the system from the surroundings, by the process of heat. Since chemistry (and biology) takes place at constant pressure and temperature, when a chemical reaction occurs the energy flow to the system, by the process of heat, is $\mathbf{q_P} = \mathbf{\Delta H}$, i.e., 436 k.J.mol^{-1} is the bond enthalpy (energy required to break the bond) and –436 k.J.mol^{-1} is the enthalpy of the reaction $H + H \rightarrow H_2$. When you measure the energy transferred to the system, by the process of heat, in a chemical reaction at constant pressure, you are measuring the change in the enthalpy of the system.

There are different ways to determine the enthalpy of a reaction, to different degrees of approximation (apart from, of course, measuring the energy absorbed by the system $(\mathbf{q_P})$ in a calorimeter). Consider for example the hydrogenation of ethene (C_2H_4) into ethane (C_2H_6) (Equation 9.21 and Figure 9.4):

$$C_2H_4 + H_2 \rightarrow C_2H_6 \tag{9.21}$$

You can just look up the standard enthalpy of the reaction (written $\mathbf{\Delta H_r^o}$ where o denotes standard conditions and r denotes reaction), where standard means under some reference conditions, like 298 K and 1 bar pressure.[14] For this reaction, we find $\mathbf{\Delta H_r^o} = -136.3 \pm 0.3$ k.J.mol^{-1}.[15] The enthalpy decreases, this means energy flows from the system to the surroundings by the process of heat.[16]

We can also calculate the enthalpy change by taking the differences of the enthalpies of the formation of the products and the reactants. The standard enthalpy of formation (written $\mathbf{\Delta H_f^o}$) is a measure of the energy released or consumed when one mole of a substance is formed by elements in their standard states (1 bar pressure and – usually –298 K). The enthalpy of formation of an element in a standard state is by definition zero and are depicted with 's', 'l' or 'g' representing solid, liquid, or gas, e.g., helium is He(g), mercury is Hg(l), and sodium is Na(s). The standard state of carbon is the allotrope graphite (not diamond) – C(s) – graphite. Confusion can arise with the gaseous elements such as hydrogen, oxygen, and nitrogen. Their standard states are the diatomic molecules H_2(g), O_2(g), and N_2(g). The source of confusion is that $\mathbf{\Delta H_f^o}$ of these molecular elements is defined as zero but we know that the $\mathbf{\Delta H_r^o}$ for chemical reactions such as $H + H \rightarrow H_2$ is not equal to zero because bond formation is an exothermic reaction. It can easily be appreciated that

Table 9.2: Average Enthalpies of Common Covalent Bonds

Average Bond Enthalpies (kJ/mol)

Single Bonds							
C—H	413	N—H	391	O—H	463	F—F	155
C—C	348	N—N	163	O—O	146		
C—N	293	N—O	201	O—F	190	Cl—F	253
C—O	358	N—F	272	O—Cl	203	Cl—Cl	242
C—F	485	N—Cl	200	O—I	234		
C—Cl	328	N—Br	243			Br—F	237
C—Br	276			S—H	339	Br—Cl	218
C—I	240	H—H	436	S—F	327	Br—Br	193
C—S	259	H—F	567	S—Cl	253		
		H—Cl	431	S—Br	218	I—Cl	208
Si—H	323	H—Br	366	S—S	266	I—Br	175
Si—Si	226	H—I	299			I—I	151
Si—C	301						
Si—O	368						
Multiple Bonds							
C=C	614	N=N	418	O_2	495		
C≡C	839	N≡N	941				
C=N	615			S=O	523		
C≡N	891			S=S	418		
C=O	799						
C≡O	1072						

'monatomic' elements in their most stable form are already formed (in the cores of stars) and a formation reaction is not necessary. What is not so clear are the standard states of the 'molecular' elements such as oxygen. The reason is straightforward – convenience.

Consider the enthalpy of the following reaction:

$$C(s)+O_2(g)\rightarrow CO_2(g) \qquad \Delta H_r^o = -393.5\ k.J.mol^{-1} \tag{9.22}$$

If we define the enthalpy of formation of molecular oxygen as zero, then the enthalpy of reaction (Equation 9.22) is the same as the enthalpy of formation of CO_2 (Equation 9.23).[17]

$$C(s)+O_2(g)\rightarrow CO_2(g) \qquad \Delta H_f^o = -393.5\ k.J.mol^{-1} \tag{9.23}$$

We could, of course, define that all elements in their monatomic state have zero enthalpies of formation. In other words, the enthalpy of formation of $CO_2(g)$ is the enthalpy change for the following reaction:

$$C(s)+2O(g)\rightarrow CO_2(g) \tag{9.24}$$

The enthalpy change for Equation (9.24) can easily be calculated from knowledge of the enthalpy change of Equation (9.23), the bond enthalpy of $O_2(g)$ (see Table 9.2) and the fact that enthalpies are state functions, i.e., they are independent of the path. The reaction $C(s)+2O(g)\rightarrow CO_2(g)$ is equivalent to $C(s)+O_2(g)\rightarrow CO_2(g)$ and $2O \rightarrow O_2$. Because enthalpies (and enthalpy changes) are state functions then we can calculate the enthalpy change for Equation (9.24) by adding the enthalpy changes of Equation (9.23) $(-393.5\ k.J.mol^{-1})$ and the enthalpy change for the formation of molecular oxygen $(-495\ k.J.mol^{-1})$:

$$C(s)+2O(g)\rightarrow CO_2(g) \qquad \Delta H_f^o = (-393.5)+(-495) = -888.5\ k.J.mol^{-1} \tag{9.25}$$

This approach is known as Hess's law which can be written as – "the enthalpy of any reaction for a specific reaction is equal to the sum of the enthalpies of reaction for any set of reactions which in sum are equivalent to the overall reaction".

The convention that all elements in their monatomic state have zero enthalpies of formation would work just as well as defining molecular elements as having zero enthalpy but the advantage of $\Delta\mathbf{H}_r^o = \Delta\mathbf{H}_f^o$ is more convenient. Unfortunately, there is a lot of misinformation out there. The following two unreferenced abstracts give a flavour:

1. "The enthalpy of formation for an element in its elemental state[18] will always be 0 because it takes no energy to form a naturally occurring compound." [This sentence – word for word – appears on many websites, it makes you want to weep.]
2. "But in the case of natural elements in their most stable state, no change in enthalpy takes place because the reactants and the product are the same. The element is already formed, so a formation reaction is not necessary. An element can't "react" to form itself."

$$O_2(g) \rightarrow O_2(g) \qquad \Delta H^o_{reaction} = 0$$

The enthalpy of formation for C_2H_4 is $\Delta\mathbf{H}_f^o = +52.4 \pm 0.5$ k.J.mol^{-1} and for C_2H_6 is $\Delta\mathbf{H}_f^o = -84.4$ ± 0.4 k.J.mol^{-1}. Combining these, the enthalpy of the reaction is $\Delta\mathbf{H}_r^o = (-84.4 \text{ k.J.mol}^{-1}) + (-52.4 \text{ k.J.mol}^{-1}) + 0 = -136.4 \text{ k.J.mol}^{-1}$ in agreement with the directly measured enthalpy of reaction. The fact that we can add and subtract formation enthalpies to get the reaction enthalpy is known as Hess's law.

Finally, we could compute the enthalpy change using bond enthalpies. There are only a finite number of relevant bonds for organic chemistry: Table 9.2 lists the average enthalpies of common covalent bolds, i.e., the energy flow into the system at constant pressure $(\mathbf{q}_P)$ to break the bond. The reaction requires:

1. The breaking of one C=C bond, four C-H bonds, and one H-H bond. This requires a flow $1 \times 614 + 4 \times 413 + 1 \times 436 = 2{,}702$ k.J of energy <u>into</u> the system.
2. The formation of one C-C bond and six C-H bonds. This requires a flow of $1 \times 348 + 6 \times 413 = 2{,}826$ k.J <u>out of</u> the system or, using our sign convention, a flow of –2826 kJ <u>into</u> the system.

Therefore, the total flow of energy into the system $(\mathbf{q}_P)$ equals $2702 - 2826 = -124$ k.J. This is the enthalpy of the reaction. This is within 10% of the measured enthalpy change of reaction value of $\Delta_{reaction}\mathbf{H}^o = -136.3$ k.J.mol^{-1} and suggests that covalent bonds between the same pair of atoms in different molecules have (slightly) different bond enthalpies.

Also keep in mind that bond enthalpies are tabulated for gases where intermolecular forces are very much less than those found in liquids and solids when the addition of bond enthalpies will be even more in error. The bottom line is:

- Using enthalpies of reaction is best (when available).
- Using enthalpies of formation and Hess's law is the most convenient compromise, giving nearly identical reaction enthalpies and requiring less tabulated data.
- Using bond enthalpies is a last resort when formation enthalpies are not available.

A natural question to ask is whether the difference between enthalpy and internal energy is significant. Let us first consider chemical reactions in solids. Some mollusc shells or corals have a layer of calcite and a layer of aragonite. These minerals are both naturally occurring forms of calcium carbonate $CaCO_3$. They both have the same chemical composition, but different crystal structures. Aragonite forms from calcite at high pressure, but at typical pressures on the surface of the earth, aragonite is unstable and turns into calcite over timescales of millions of years.

The standard enthalpy of formation of calcite is –1207.37 k.J.mol^{-1} and of aragonite is –1207.74 k.J.mol^{-1}.[19] Therefore, the transformation from aragonite to calcite involves an enthalpy increase of +0.37 k.J.mol^{-1}, i.e., it is an endothermic reaction[20]:

$$\text{Aragonite} \rightarrow \text{Calcite} \qquad \Delta H = +0.37 \text{ k.J.mol}^{-1} \tag{9.26}$$

The two minerals have different densities: aragonite is denser, at 2.95 $g.mL^{-1}$[21] than calcite at 2.71 $g.mL^{-1}$,[22] i.e., when aragonite converts into calcite, its volume expands, doing PV work. The volume change is:

$$\Delta V = \left(\frac{1}{2.71} - \frac{1}{2.95}\right) = 0.030 \text{ g.mL}^{-1} \tag{9.27}$$

One mole of $CaCO_3$ weights 100 g, and using $P = 1 \text{ bar} = 10^5 \text{ Pa} = 10^5 \text{ J.m}^{-3} = 10^{-4} \text{ k.J.mL}^{-1}$ we get

$$P\Delta V = 10^{-4} \times 0.030 \times 100 = 3.0 \times 10^{-4} \text{ k.J.mol}^{-1} \tag{9.28}$$

Therefore, in this case, $P\Delta V \ll \Delta H$ and therefore the enthalpy and internal energy changes are nearly identical. The PV work done is only a small fraction, 0.1%, of the enthalpy change.

Let us now consider a gaseous chemical reaction when the total number of molecules is not the same on both sides of a reaction. For example, consider the enthalpy change in the formation of ammonia gas, NH_3, through the reaction of hydrogen and nitrogen gases:

$$3H_2 + N_2 \rightarrow 2NH_3 \tag{9.29}$$

This reaction converts four molecules into two, so $\Delta n = -2$, and the volume decreases as the reaction proceeds. Using the ideal gas law at room temperature

$$\Delta(PV) = (\Delta n)RT = -2 \times 8.3 \times 298 = 4946.8 \text{ J} \approx -4.9 \text{ k.J} \tag{9.30}$$

This is around 5% of the measured enthalpy reaction enthalpy change of $\Delta_{reaction}H^o = -92 \text{ k.J.mol}^{-1}$, i.e., $\Delta_{reaction}U^o \approx -97 \text{ k.J.mol}^{-1}$.[23] This is small, but not so small that it can be neglected.

In summary, as a rule of thumb, enthalpies and internal energies are pretty similar for solids and liquids, differing at less than a percent level, but for gases, the difference can be relatively large. The difference between the energy and enthalpy change is essentially given by $\Delta H - \Delta U = (\Delta n)RT$ with Δn the change in the number of moles of gas. This is equal to $2.48 \text{ k.J.mol}^{-1} \times \Delta n$ at room temperature and 1 bar pressure. These enthalpy changes are included in tabulated enthalpies of formation and reaction (which are different for liquids and gases) and are also included in bond enthalpies (defined for gases and computed by averaging the formation enthalpies of various molecules with that bond type).

9.3.5 Summary

Enthalpy is the preferred expression of system energy changes in many chemical, biological, and physical measurements because a change in enthalpy takes account of energy changes to the system and surroundings because of volume changes. Just as C_V should really be called 'energy capacity' (instead of heat capacity at constant volume), then C_P should really be called 'enthalpy capacity' (instead of heat capacity at constant pressure). Enthalpy plays the role at constant pressure that internal energy does at constant volume.

Observations of natural processes led a surprising number of chemists of the late 19th century (including Berthelot and Thomsen) to conclude that all spontaneous reactions must be exothermic since:

- Objects roll downhill spontaneously (i.e., energy is "lost" from the system);
- Objects do not roll uphill spontaneously (i.e., energy does not suddenly appear from nowhere).

If this were true, all we would need to predict whether a reaction is spontaneous is the change of enthalpy. If ΔH were negative, the process should be able to occur by itself. If ΔH were positive, the reaction could not occur by itself. In other words, any process where energy is transferred from the system to the surroundings by the process of heat is spontaneous. Indeed, almost all exothermic reactions are spontaneous at standard thermodynamic conditions (1 atmosphere pressure) and 278 K. However, a number of common processes which are both endothermic and spontaneous are known. The most obvious are simple phase changes, like ice melting at room temperature. Also, many solids dissolve in water and simultaneously absorb energy from the surroundings by the process of heat. By now you should know the reason – the effect of entropy change. In the case of the melting of ice, the ordered structure of the solid ice has given way to a much less organised flowing liquid. In other words, the melting of ice is accompanied by an increase in entropy of the system. To predict the spontaneity, or otherwise of a process, we need to consider the entropy change of both the system and the surroundings, i.e., the entropy change of the universe. This is made easier by the concept of free energy. Changes in free energy are used to judge whether changes in state or chemical processes can occur spontaneously. I will reiterate and reinforce this concept in Section 9.3.1.

9.3.6 Entropy of Chemical Substances and Their Physical States

The internal energy of monatomic gases (the noble gases) is a result of the translational movement of the atoms (compare. the ideal gas). For polyatomic molecules, two additional kinds of motions are possible. One of these is rotation; in a linear molecule such as CO_2 in which the atoms are all laid out along the x-axis, the molecule can rotate along the y- and z-axes (two rotational modes), while non-linear molecules such as H_2O can rotate about all three axes (three rotational modes). In addition, polyatomic molecules can undergo internal vibrations. For freely moving molecules in a gas, the number of vibrational modes depends on both the number of atoms and the shape of the molecule, and it increases rapidly as the molecule becomes more complicated (see Figure 9.4).

At the atomic and molecular level, energy is quantised; each microparticle possesses discrete states of kinetic energy and is able to accept energy only in packets whose values correspond to the energies of one or more of these states. Polyatomic molecules can store energy in rotational and vibrational motions, and all molecules will possess translational kinetic energy at all temperatures above absolute zero. The energy difference between adjacent translational states is so minute that translational kinetic energy can be regarded as continuous (non-quantised) for most practical purposes. In translational motion, the entire molecule moves in some direction in three dimensions. In rotational motion, the entire molecule rotates around any axis, (even though it may not actually change its position translationally). In vibrational motion the atoms within a molecule have certain freedom of movement relative to each other; this displacement can be periodic motion like the vibration of a tuning fork.

The relative populations of the quantised translational, rotational, and vibrational energy states of a typical diatomic molecule are depicted by the thickness of the lines in this schematic diagram shown in Figure 9.5. The numbers at the top indicate the order-of-magnitude spacings between adjacent levels. Note the diagram is not to scale. The buff shading indicates how the internal energy at a given temperature is divided among the different energy modes. It is apparent that virtually all the internal energy resides in the translational energy mode. Notice the greatly different spacing of the three kinds of energy levels. This is important because it determines the number of energy quanta that a molecule can accept, and, as the following illustration shows, the number of different ways this energy can be distributed amongst the molecules (Figure 9.6).

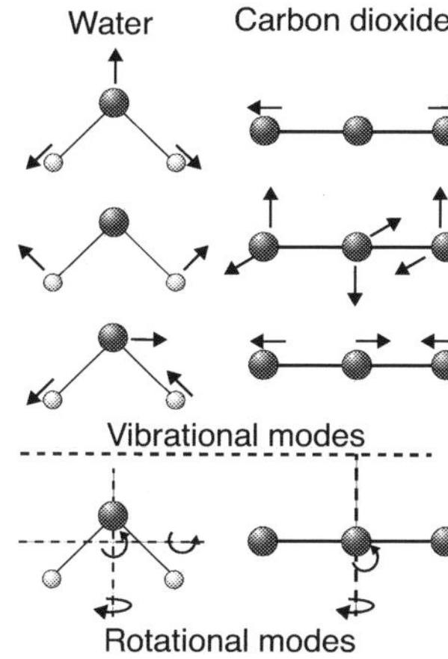

Figure 9.5 Rotational and vibrational modes of H_2O and CO_2.

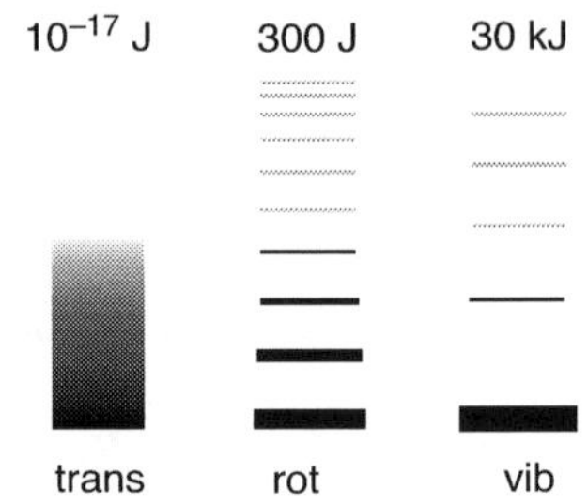

Figure 9.6 Schematic diagram representing relative populations of the three energy modes of a diatomic molecule.

Table 9.3: Standard Entropies of Some Gases at 298 K

He	126	H_2	131	CH_4	186
Ne	146	F_2	203	$H_2O(g)$	187
Ar	155	N_2	192	CO_2	213
Kr	164	O_2	205	C_2H_6	229
Xe	170	Cl_2	223	n-C_3H_8	270
				n-C_4H_{10}	310

Table 9.4: Standard Entropies of Solid Elements at 298 K

C (diamond)	2.5	Pb	51.0	Si	18.9
C (graphite)	5.7	Na	64.9	W	33.5
Fe	27.1	S(rhombic)	32.0		

The standard entropy of a substance is its entropy at 1 atm pressure. The values found in tables are normally those for 298 K and are expressed in units of $J.K^{-1}.mol^{-1}$. The values given in Table 9.3 illustrate the following points[24]:

- Although the standard internal energies and enthalpies of these substances would be zero, the entropies are not. This is because there is no absolute scale of energy, so we conventionally set the "energies of formation" of elements in their standard states to zero. Entropy, however, measures not energy itself, but its dispersal among various quantum states available to accept it, and these exist even in pure elements.
- It is apparent that entropies generally increase with molecular weight. For the noble gases, this is of course a direct reflection of the principle that translational quantum states are more closely packed in heavier molecules, allowing of them to be occupied.
- The entropies of the diatomic and polyatomic molecules show the additional effects of rotational quantum levels.

The entropies of the solid elements are strongly influenced by the manner in which the atoms are bound to one another. The contrast between diamond and graphite is particularly striking; the standard entropy of graphite, which is built up of loosely bound stacks of hexagonal sheets, is more than twice that of diamond, in which the carbon atoms are tightly locked into a three-dimensional lattice, thus affording them less opportunity to vibrate around their equilibrium positions. Looking at all the examples in Table 9.4, you will note a general inverse correlation between the hardness of a solid and its entropy: sodium, which can be cut with a knife, has almost

Table 9.5: Standard Entropies of Solids

C (diamond)	2.5
Calcium	41.4
Calcium fluoride	68.9
Calcium carbonate	92.9

Table 9.6: Comparing Standard Entropies of the Different States of Water and Benzene

Liquid benzene $C_6H_6(l)$	173
Benzene vapour $C_6H_6(g)$	269
Ice $H_2O(s)$	48
Water $H_2O(l)$	70
Steam $H_2O(g)$	189

twice the entropy of iron; the much greater entropy of lead reflects both its high atomic weight and the relative softness of this metal.

Consider Table 9.5, they all have low entropies because they are highly ordered solids, notice that the entropy increases as the solid gets more complicated.

Table 9.6 indicates how entropy changes with change of state. Notice that the benzene values are bigger than those of water-steam. This is because benzene is a more complicated molecule. There are more ways of distributing the energy in bigger molecules than smaller ones.

Finally, I want to consider a number of examples of how entropy will change during reactions or other changes.

Example 9.1

$$\mathrm{NaCl(s)+(aq)\rightarrow NaCl(aq)}$$

The above equation represents the dissolution of a crystalline solid. The solid is highly ordered; the solution is disordered. There is an increase in entropy.

Example 9.2

$$\mathrm{H_2O(g)\rightarrow H_2O(l)}$$

The water is changing from highly disordered gas into a more ordered liquid. The entropy will fall.

Example 9.3

$$\mathrm{CH_4(g)+2O_2(g)\rightarrow CO_2(g)+H_2O(l)}$$

There are three moles of gas on the left-hand side of the equation, but only one on the right. There are fewer gas molecules than you had to start with, this means that there are fewer ways of arranging the energy of the system over those molecules, and so entropy decreases. Notice that if the water had been formed as steam, you could not easily predict whether there was an increase or a decrease in entropy because there would be three molecules of gas on each side.

Example 9.4

$$CuSO_4(s) + 5H_2O(l) \rightarrow CuSO_4.5H_2O(g)$$

The presence of the five molecules of liquid water on the left-hand side means that there will be far more ways of arranging the energy of the system over those molecules than there is in the products. Entropy will decrease.

9.4 FREE ENERGY

9.4.1 Introduction

The concept of free energy, originally called available energy, was developed in the late 1800s by the American mathematical physicist Willard Gibbs and the German physicist Herman von Helmholtz. In 1873, in a footnote, Gibbs defined what he called the "available energy" of a body as such[25]:

> the greatest amount of mechanical work which can be obtained from a given quantity of a certain substance in a given initial state, without increasing its total volume or allowing heat to pass to or from external bodies, except such as at the close of the processes are left in their initial condition.

This available energy is the maximum work to do such things as breaking chemical bonds, transporting electrons and moving physical objects against a force.

It was originally thought that the maximum amount of work that could be extracted from a process was $-|\mathbf{\Delta H}|$. However, examples were found that showed that this was not the case. At and around room temperature these are nearly all physical processes. We have already come across one example – the melting of ice at temperatures above its freezing point which is an endothermic reaction. In fact, all melting and evaporative processes are endothermic and under suitable conditions are spontaneous. The "instant cold pack" is another example of an endothermic reaction. There are many possible ingredients in an instant cold pack, but they often contain solid ammonium nitrate and water. The ammonium nitrate (NH_4NO_3) is stored in a sealed plastic bag that is surrounded by water. When you pop the bag, the ammonium nitrate comes into contact with water and dissolves. When you pop the bag, the ammonium nitrate comes into contact with water and dissolves:

$$NH_4NO_3(s) \rightarrow NH_4^+(aq) + NO_3^-(aq)$$

Like other types of reactions, energy may flow in or out of the system when material dissolves in water. The energy change, at constant pressure, is known as the called the enthalpy of solution (ΔH_{soln}). From standard enthalpy of formation (ΔH_f^o) tables, we can find the enthalpies of formation of the reactant $[NH_4NO_3(s)]$ and the products [$NH_4^+(aq)$ and $NO_3^-(aq)$] – see Table 9.7. This allows us to calculate the enthalpy of solution (ΔH_{soln}^o) as +25 k.J.mol^{-1}. Because there is a flow of energy from the surroundings to the system is the reason that the ammonium nitrate water system acts as a cold pack.

If no energy flows to or from the surroundings, by the process of heat, then the external work done by the system must be equal in magnitude to the energy lost by the system by the process of heat: we know this is equal in magnitude (but opposite in sign) to the enthalpy change However, this is not the case when energy is given to or taken from the surroundings, by the process of heat. In an isothermal reaction working under reversible conditions, the energy flow from the surroundings is $T\Delta S$, if this is positive then energy is taken from the surroundings and the work done will be even greater than the magnitude of the enthalpy change (the enthalpy of reaction).

In other words, some thermodynamic processes are spontaneous because there is a flow of energy from the system to the surroundings by the process of heat $(\Delta \mathbf{H} < 0)$. Others are

Table 9.7: Enthalpy of Formation Table

ΔH_f^o	$kJ.mol^{-1}$
$NH_4NO_3(s)$	–365
$NH_4^+(aq)$	–133
$NO_3^-(aq)$	–207

spontaneous because they lead to an increase in the entropy of the system ($\Delta \mathbf{S} > 0$). What happens when one of the potential driving forces behind a chemical reaction is favourable and the other is not? We can answer this question by defining a new quantity known as free energy of the system, which reflects the balance between these forces. The beauty of the free energy concept is its ability to determine the relative importance of the enthalpy and entropy terms as driving forces behind a particular process. The change in the free energy of the system that occurs during a process measures the balance between the two driving forces that determine whether a process is spontaneous (or not). Free energy plays the same role in thermodynamic systems as potential energy does in mechanical systems.

The following extract (entitled *Meaning of* 'free' may help)[26]:

> The basic definition of 'energy' is a measure of a body's (in thermodynamics, the system's) ability to cause change. For example, when a person pushes a heavy box a few meters forward, that person exerts mechanical energy, also known as work, on the box over a distance of a few meters forward. The mathematical definition of this form of energy is the product of the force exerted on the object and the distance by which the box moved (Work=Force × Distance). Because the person changed the stationary position of the box, that person exerted energy on that box. The work exerted can also be called 'useful energy', because energy was converted from one form into the intended purpose, i.e. mechanical utilisation. For the case of the person pushing the box, the energy in the form of internal (or potential) energy obtained through metabolism was converted into work in order to push the box. This energy conversion, however, was not straightforward: while some internal energy went into pushing the box, some was diverted away (lost) in the form of heat .... For a reversible process, heat is the product of the absolute temperature T and the change in entropy S of a body .... The difference between the change in internal energy, which is ΔU, and the energy lost in the form of heat is what is called the 'useful energy' of the body, or the work of the body performed on an object. In thermodynamics, this is what is known as 'free energy'". In other words, free energy is a measure of work (useful energy) a system can perform at constant temperature.

The term free energy is frowned upon by the International Union of Pure and Applied Chemistry (IUPAC). The principal reason for this is, that despite having the units of energy, free energy is not conserved, i.e., it is not energy (more on this later). However, there are advantages in using the term and I will not be following IUPAC recommendations in this book.[27] The change in the free energy is the maximum amount of work that a thermodynamic system can perform in a process at constant temperature, and its sign indicates whether a process is spontaneous or non-spontaneous. I will be considering two types of free energy Helmholtz free energy (F) and Gibbs free energy (G). Free energy is a thermodynamic state function, like internal energy, enthalpy, and entropy. As with internal energy and enthalpy, the difference between the two free energy functions is that one (Helmholtz free energy) does not consider PV work (concerned with constant volume processes) and the other (Gibbs free energy) does (concerned with constant pressure processes).

To put it simply, free energy is a measure of the useful or process-initiating work obtainable from a thermodynamic system. When a system changes from a well-defined initial state to a well-defined final state, the free energy equals the work exchanged by the system with its surroundings, less, in the case of constant pressure processes, the $P\Delta V$ work exchanged with its surroundings to create the system. As far as we are concerned, there are two types of free energy:

1. Helmholtz free energy is the maximum amount of work obtainable from an isothermal, isochoric[28] thermodynamic system.

2. Gibbs free energy is the maximum amount of work obtainable from an isothermal, isobaric[29] thermodynamic system.

9.4.2 Helmholtz Free Energy and Gibbs Free Energy

Helmholtz free energy was originally developed by Hermann von Helmholtz[30] and is generally denoted by the letter A or F: I will use F in this book. We know that the sole criterion for spontaneous change for a real process is $\Delta S_{universe} > 0$ and we know (see Equation 9.7):

$$\Delta \mathbf{S}_{\mathbf{universe}} = \Delta \mathbf{S}_{\mathbf{system}} - \frac{\Delta \mathbf{U}_{\mathbf{system}}}{T_{surroundings}} \tag{9.31}$$

Multiplying Equation (9.31) by $-T_{surroundings}$ gives:

$$-T_{surroundings}\Delta \mathbf{S}_{\mathbf{universe}} = \Delta \mathbf{U}_{\mathbf{system}} - T_{surroundings}\Delta \mathbf{S}_{\mathbf{system}} \tag{9.32}$$

Helmholtz defined a function (F) as $F = -T_{surroundings}S_{universe}$. Rewriting Equation (9.32) in terms of F:

$$\Delta \mathbf{F} = F_{products} - F_{reactants} = \Delta \mathbf{U}_{\mathbf{system}} - T_{surroundings}\Delta \mathbf{S}_{\mathbf{system}} \tag{9.33}$$

The analogous Gibbs[31] free energy is defined as $G = H - TS = U + PV - TS$ and is the constant pressure version of Helmholtz free energy.

$$\Delta G = G_{products} - G_{reactants} = \Delta \mathbf{H}_{\mathbf{system}} - T_{surroundings}\Delta \mathbf{S}_{\mathbf{system}} \tag{9.34}$$

The Helmholtz free energy and Gibbs free energy are powerful functions that tell us whether a process is spontaneous or not, i.e., whether the process results in an increase or decrease of the entropy content of the universe. Because the Gibbs free energy is virtually ubiquitous in chemistry, from now on I will only be concerned with the Gibbs free energy which is concerned with constant pressure processes. As we shall see in the next section if the change in free energy is negative, then the entropy of the universe increases, and the process is spontaneous.

9.4.3 Free Energy Explained

In the interest of brevity, let us simplify Equation (9.34) to (where all the state functions refer to the system):

$$\Delta G = \Delta H - T\Delta S \tag{9.35}$$

Although a change in state or chemical composition nearly always results in a finite change in entropy, let us, at this stage, assume $\Delta S = 0$. Now $\Delta \mathbf{G} = \Delta \mathbf{H}$, and if $\Delta \mathbf{H}$ is a negative number (the system is losing energy by the process of heat) then, we know the energy content of the surroundings is increasing. A positive flow of energy from the system to the surroundings, by the process of heat is always accompanied by a positive of flow of entropy to the surroundings, i.e., in any exothermic reaction the entropy of the surroundings always increases. Therefore, the change of the entropy content of the surroundings $(\Delta \mathbf{S}_{\mathbf{surroundings}})$ is positive $\left(\frac{+|\Delta \mathbf{H}|}{T}\right)$.[32] We have assumed $\Delta S_{system} = 0$, therefore, $\Delta \mathbf{S}_{\mathbf{universe}} = \Delta \mathbf{S}_{\mathbf{surroundings}}$ which is positive, i.e., the entropy of the universe increases, and we know the process is spontaneous and $\Delta \mathbf{G}$ is negative $(-|\Delta \mathbf{G}|)$. We can now see that if there is no change in the entropy of the system then the energy of the system is minimised by the system in a spontaneous process, i.e., the free energy is minimised. Remember maximum useful work is equal, in magnitude, to the flow of energy from the system to the surroundings. Therefore, in this scenario $(\Delta S = 0)$, the negative sign of $\Delta \mathbf{H}$ (i.e., $-|\Delta \mathbf{H}|$) not only tells us the reaction is spontaneous, but the magnitude of $|\Delta \mathbf{G}| = |\Delta \mathbf{H}|$ indicates the maximum energy that is available to do useful work.

Let us now consider an endothermic process $(+|\Delta \mathbf{H}|)$, where $\Delta S = 0$. Now, the change in the energy content of the surroundings is $-(+|\Delta \mathbf{H}| = -|\Delta \mathbf{H}|)$ (a negative real number). Therefore, the change of the entropy content of the surroundings $(\Delta \mathbf{S}_{\mathbf{surroundings}})$ is negative $\left(\frac{-|\Delta \mathbf{H}|}{T}\right)$. By definition, $\Delta \mathbf{S}_{\mathbf{system}} = 0$, therefore, $\Delta \mathbf{S}_{\mathbf{universe}} = \Delta \mathbf{S}_{\mathbf{surroundings}}$ which is negative, i.e., the entropy of the universe decreases, and we know the process is nonspontaneous, $\Delta \mathbf{G}$ is positive $(+|\Delta \mathbf{G}|)$ and there is no useful work available from the process. This means that work must be done on the system (through some outside intervention) to force the nonspontaneous reaction to occur.

Let us now consider processes where $\Delta S \neq 0$. We now have four scenarios (I have added the subscript "system" to emphasise the fact that the state functions apply to the system (and will drop in the subsequent analysis):

1. $-|\Delta \mathbf{H}_{\text{system}}|$ and $+|\Delta \mathbf{S}_{\text{system}}|$
2. $-|\Delta \mathbf{H}_{\text{system}}|$ and $-|\Delta \mathbf{S}_{\text{system}}|$
3. $+|\Delta \mathbf{H}_{\text{system}}|$ and $-|\Delta \mathbf{S}_{\text{system}}|$
4. $+|\Delta \mathbf{H}_{\text{system}}|$ and $+|\Delta \mathbf{S}_{\text{system}}|$

Let us first consider Scenario 1. From the above analysis, we know when ΔH is negative (exothermic) process and $\Delta S = 0$ the process is spontaneous because the entropy of the universe increases. Writing Equation (9.35) using absolute values as:

$$-|\Delta \mathbf{G}| = -|\Delta \mathbf{H}| - T(+|\Delta \mathbf{S}|) \tag{9.36}$$

In this scenario, whatever the temperature (always a positive real number), $\Delta \mathbf{G}$ is always negative $(-|\Delta \mathbf{G}|)$, the change in entropy of the universe is always positive, the process is always spontaneous and the magnitude of $|\Delta \mathbf{G}|$ indicates the maximum energy that is available to do useful work.

Figure 9.7 should aid clarification. Note the position of the $T\Delta \mathbf{S}$ line (whose slope is $\Delta \mathbf{S}$ – in this case positive) which is always above the $\Delta \mathbf{H}$ line. This means that ΔG is always negative irrespective of the temperature indicating the process is spontaneous at any temperature. This is shown in Figure 9.3 by the downward pointing arrows. Please note that, for real processes, owing to the slight temperature dependence of $\Delta \mathbf{S}$ and greater temperature dependence of $\Delta \mathbf{H}$, the $T\Delta \mathbf{S}$ and $\Delta \mathbf{H}$ plots are not the straight lines as shown here. The straight-line plots shown in this section only apply for the hypothetical ideal gas.

Now let us consider Scenario 2, where the change in the system's enthalpy and entropy are both negative (see Figure 9.8). You should be able to see (Equations 9.37 and 9.38) that the spontaneity or otherwise of the process depends on T.

$$-|\Delta \mathbf{H}| - T_{\text{low}}(-|\Delta \mathbf{S}|) = -|\Delta \mathbf{H}| + T_{\text{low}}(+|\Delta \mathbf{S}|) = -|\Delta \mathbf{G}_{\text{low temperature}}| \tag{9.37}$$

$$-|\Delta \mathbf{H}| - T_{\text{high}}(-|\Delta \mathbf{S}|) = -|\Delta \mathbf{H}| + T_{\text{high}}(+|\Delta \mathbf{S}|) = +|\Delta \mathbf{G}_{\text{high temperature}}| \tag{9.38}$$

As the temperature increases the Scenario 2 process changes from spontaneous to nonspontaneous. To clarify I will use a (artificial) numerical example. Let $\Delta H = -1{,}000$ J.mol^{-1} and $\Delta S = -2$ J.K^{-1}.mol^{-1}. At 273 K $\Delta G = -454$ J˙mol^{-1}, i.e., the process is spontaneous. However, when the temperature is 1,000 K $\Delta G = 1{,}000$ J.mol^{-1}, i.e., the process is nonspontaneous. ΔG is zero when

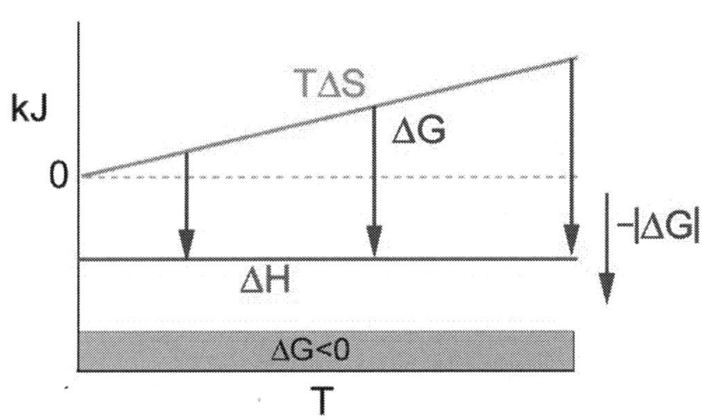

Figure 9.7 In Scenario 1, $\Delta \mathbf{H}$ is negative and $\Delta \mathbf{S}$ is positive, so $\Delta \mathbf{G}(= \Delta \mathbf{H} - T\Delta \mathbf{S})$ is negative regardless of the temperature. In other words, an exothermic process whose entropy increases is spontaneous at all temperatures.

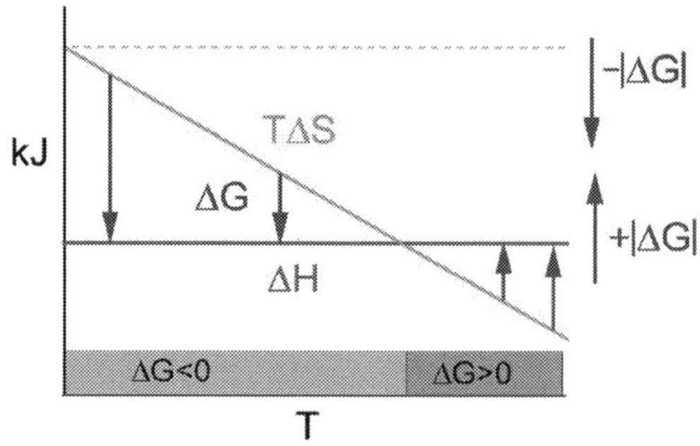

Figure 9.8 In Scenario 2 ΔG is negative only at temperatures below which $|T\Delta S| < |\Delta H|$. This means that there is a temperature $T = \frac{\Delta H}{\Delta S}$ at which the reaction is at equilibrium and the process only proceeds spontaneously below this temperature. The freezing of a liquid or the condensation of a gas are the most common examples of this scenario.

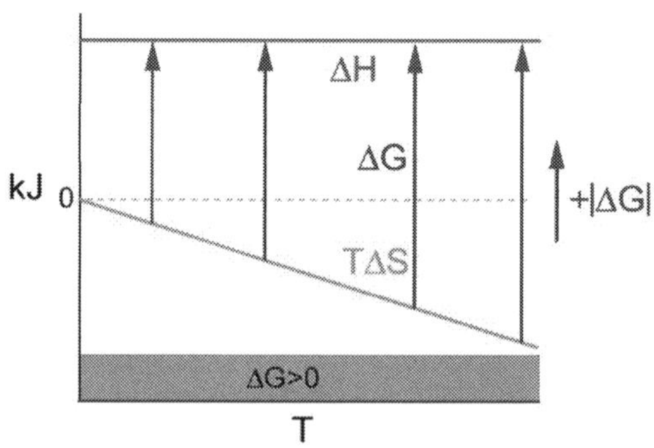

Figure 9.9 Scenario 3 is nonspontaneous at any temperature as both ΔH and ΔS are working against it.

T = 500 K; at this temperature, the process is at equilibrium, e.g., the freezing point if we are considering the freezing of a liquid.

In Scenario 3 the change in the system's enthalpy change is positive and in its entropy change is negative. Examination of Equation (9.39) and Figure 9.9 indicates the Scenario 3 process is always non-spontaneous at any temperature:

$$+|\Delta H| - T(-|\Delta S|) = +|\Delta H| + T(+|\Delta S|) = +|\Delta G| \tag{9.39}$$

Finally, let us consider Scenario 4 where the change in the systems enthalpy and entropy are both positive:

$$+|\Delta H| - T_{low}(+|\Delta S|) = +|\Delta G_{low\ temperature}| \tag{9.40}$$

$$+|\Delta H| - T_{high}(+|\Delta S|) = -|\Delta G_{high\ temperature}| \tag{9.41}$$

In contrast to Scenario 2 as the temperature increases the Scenario 4 process changes from nonspontaneous to spontaneous see Equations (9.40) and (9.41), and Figure 9.10. As before I will use a (artificial) numerical example. Let $\Delta H = +1,000\ J.mol^{-1}$ and $\Delta S = +2\ J.K^{-1}.mol^{-1}$. At 273 K $\Delta G = +454\ J.mol^{-1}$, i.e., the process is nonspontaneous. However, when the temperature is 1,000 K $\Delta G = 1,000\ J.mol^{-1}$, i.e., the process is spontaneous.

These results are summarised in Table 9.8.

We can now see that free energy is a quantity that becomes more negative during the course of any spontaneous process. Therefore, as a chemical reaction takes place, free energy will never

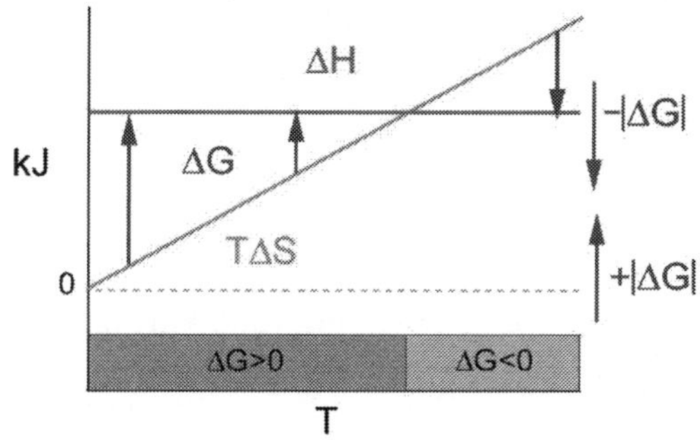

Figure 9.10 Scenario 4 is the reverse of Scenario 2. For a spontaneous process, the entropy increase must overcome the handicap of an endothermic process so that $T\Delta S| > |\Delta H|$. Since the effect of temperature is to magnify the influence of a positive ΔS, the process will be spontaneous at temperatures above $T = \frac{\Delta H}{\Delta S}$. The melting of a solid or the vaporisation of a liquid are examples (as well as many chemical processes).

Table 9.8: Enthalpy, Entropy, and Free Energy Changes

ΔH	ΔS	ΔG
Negative	Positive	Always negative
Negative	Negative	Positive at higher temperatures, negative at lower temperatures
Positive	Negative	Always positive
Positive	Positive	Negative at higher temperatures, positive at lower temperatures

become more positive. Eventually, a point is reached where any further transformation of reactants into products would cause an increase in free energy which is a "forbidden" decrease in the entropy of the universe. At this point, free energy is at minimum, and no further net change can take place; the reaction is then at equilibrium. In other words when $\Delta G = 0$ the system is at equilibrium.

Remember that ΔG is meaningful only for changes in which the temperature and pressure remain constant. These are the conditions under which most reactions are carried out in the laboratory; the system is usually open to the atmosphere (constant pressure) and we begin and end the process at room temperature (after which any energy added or which is liberated by the reaction has dissipated.

We now know that the value of **ΔG** allows us to predict the spontaneity of a physical or a chemical change. In addition, the magnitude of **ΔG** is the maximum useful non PV work that a closed system can do on its surroundings when the process occurring within the system is reversible at constant temperature and pressure. For a reversible process, we know that

$$\Delta G = \Delta H - T\Delta S = \left(w_{max}^{rev} + q_{min}^{rev}\right) - T\frac{q_{min}^{rev}}{T} = w_{max}^{rev} \tag{9.42}$$

For useful work, we know that both ΔG and w_{max}^{rev} are negative real numbers, i.e., $-|\Delta G| = -\left|w_{max}^{rev}\right|$

For an irreversible process we know that $\left|w_{max}^{rev}\right| > \left|w^{irrev}\right|$ or $-\left|w_{max}^{rev}\right| < -\left|w^{irrev}\right|$. Therefore, we know that $\Delta G < w^{irrev}$. A numerical example should clarify. If, for example, reaction $\Delta G = -10$ J then $w_{rev} = -10$ J, and $w_{rev} < w_{irrev}$, for example $w_{irrev} = -9$ J meets this inequality and $w_{irrev} = -11$ J does not. Therefore, when work is done at constant temperature, the free energy of the system is depleted to do the work. This is why free energy is called free: it is the energy available to do work.

In an insulated system (no exchange of energy by the process of heat), the internal energy of the system is used for work, but in an isothermal system, it is the free energy that is used for work. To summarise - when $\Delta\mathbf{G}$ is a negative real number, the process is spontaneous and the useful $(\mathbf{w}_{\mathbf{max}})$ that could theoretically be extracted from the system is $\Delta\mathbf{G}$. Remember for work to be "useful" $\mathbf{w}$ must be a negative real number.

This is as far as I want to go with thermodynamic potentials. The objective here was only to give an introduction.

9.4.4 Life, Free Energy and Entropy

In this book, I have not (until now) considered open systems such as living organisms. How is it possible for life to come into being in a universe governed by the Second Law of Thermodynamics? It turns out in fact, that the Second Law of Thermodynamics is not an impediment to understanding life but rather is a necessary condition for providing a complete elucidation of living processes. In his 1944 book *What is Life?* (based on a series of lectures given in 1943),[33] Austrian physicist Erwin Schrödinger argued that organisms stay alive precisely by staving off equilibrium. "How does the living organism avoid decay?" he asks. "The obvious answer is: By eating, drinking, breathing and (in the case of plants) assimilating. The technical term is metabolism" In Chapter 6 of this interesting book Schrödinger writes:

> What an organism feeds upon is negative entropy. Or, to put it less paradoxically, the essential thing in metabolism is that the organism succeeds in freeing itself from all the entropy it cannot help producing while alive.

The ultimate source of "negative entropy" on Earth, wrote Schrödinger, is the Sun. In the book, there is a section entitled "Note to Chapter 6":

> The remarks on negative entropy have met with doubt and Opposition [sic] from physicist colleagues. Let me say first, that if I had been …[34] catering for them alone I should have let the discussion turn on free energy instead. It is the more familiar notion in this context. But this highly technical term seemed linguistically too near to energy for making the average reader alive to the contrast between the two things. He is likely to take free as more or less an epitheton ornans[35] without much relevance, while actually the concept is a rather intricate one, whose relation to Boltzmann's order-disorder principle is less easy to trace than for entropy and 'entropy taken with a negative sign', which by the way is not my invention. It happens to be precisely the thing on which Boltzmann's original argument turned. But F. Simon has very pertinently pointed out to me that my simple thermodynamical considerations cannot account for our having to feed on matter 'in the extremely well ordered state of more or less complicated organic compounds' rather than on charcoal or diamond pulp. He is right. But to the lay reader I must explain that a piece of un-burnt coal or diamond, together with the amount of oxygen needed for its combustion, is also in an extremely well ordered state, as the physicist understands it. Witness to this: if you allow the reaction, the burning of the coal, to take place, a great amount of heat is produced. By giving it off to the surroundings, the system disposes of the very considerable entropy increase entailed by the reaction, and reaches a state in which it has, in point of fact, roughly the same entropy as before. Yet we could not feed on the carbon dioxide that results from the reaction. … Energy is needed to replace not only the mechanical energy of our bodily exertions, but also the heat we continually give off to the environment. And that we give off heat is not accidental, but essential. For this is precisely the manner in which we dispose of the surplus entropy we continually produce in our physical life process. This seems to suggest that the higher temperature of the warm-blooded animal includes the advantage of enabling it to get rid of its entropy at a quicker rate, so that it can afford a more intense life process.

In a Chapter entitled "The laws of thermodynamics: entropy, free energy, information and complexity" Collins et al. state[36]:

> The Nobel prize-winning thermodynamicist Ilya Prigogine described [Schrödinger's book] as 'a beautiful book', and Schrödinger's exposition not only inspired him but also the mathematician Roger Penrose. Prigogine uses the expression 'a flow of free energy carried by matter... which "feeds" the living system', while Penrose goes right back to the 'Sun's role as a source of low entropy (or 'negative entropy', in Schrödinger's terminology)'. This thermodynamic necessity for the existence of living organisms carries forward into current mainstream biology.

In the standard US text Life, W. Purves et al. explain: 'Free energy is what cells require for ... cell growth, cell division and the maintenance of cell health'.

Lehninger in his textbook "Principles of Biochemistry"[37] states that the order produced within cells as they grow and divide is more than compensated for by the disorder they create in their surroundings in the course of growth and division. He writes "living organisms preserve their internal order by taking from their surroundings free energy, in the form of nutrients or sunlight, and returning to their surroundings an equal amount of energy as heat and entropy".

9.4.5 Discussion and Summary

The change in the free energy is the maximum amount of work that a thermodynamic system can perform in a process at constant temperature, and its sign indicates whether a process is thermodynamically favourable or forbidden. Although all of the above variables have units of energy, they do not have a direct physical interpretation as energy. In other words, they are just defined variables that simplify thermodynamic calculations, i.e., they offer practical advantages to the scientist/engineer.

IUPAC argues the phrase "free energy" is doubly confusing because it is not energy that we can get for nothing (nothing is for free in this world). Therefore, to avoid misunderstanding, it is better to exclude the phrase "free energy" from the name of corresponding functions and to use the terms: Helmholtz function and Gibbs function. However, because there are so many inconsistencies in chemical nomenclature (especially in thermodynamics) I decided to continue to use the term "free energy" and trust in the reader to appreciate the subtleties.

To summarise:

- Free energy is to a constant T thermodynamic system what potential energy is to a mechanical system.
- Free energy refers to the free energy of the system only.
- Free energy can only be used in processes occurring at constant temperature.
- The Second Law of Thermodynamics is equivalent to the minimisation of free energy, and this is why free energy is powerful: because it allows us to consider the system alone.
- Free energy corresponds to the maximum possible quantity of work which can be obtained from the spontaneous process.
- The tendency to minimise free energy is entirely because of a tendency to maximise entropy. This is clearest if we write $\Delta \mathbf{G} = -T\left[\Delta \mathbf{S} + \left(-\frac{\Delta H}{T}\right)\right]$, where $\Delta \mathbf{S}$ is $\Delta \mathbf{S}_{\mathbf{system}}$ and $-\frac{\Delta H}{T}$ is $\Delta \mathbf{S}_{\mathbf{surroundings}}$.

NOTES

1 https://en.wikipedia.org/wiki/Thermodynamic_potential.

2 https://scholar.harvard.edu/files/schwartz/files/8-freeenergy.pdf.

3 For one of the more accessible introductory papers on this subject see "Thermodynamic Potentials and Natural Variables" by M. Amaku, F.A.B. Coutinho and L.N. Olivera at https://www.scielo.br/scielo.php?script=sci_arttext&pid=S1806-11172020000100420.

4 Why C_V in Equation (9.31)? The short answer is the definition of the internal energy of n moles of ideal gas at temperature T is $U = nC_VT$. This from Wikipedia (my underlining): "The internal energy of a thermodynamic system is the energy contained within it. It is the energy necessary to create or prepare the system in any given internal state. It does not include the kinetic energy of motion of the system as a whole, nor the potential energy of the system as a whole due to external force fields, including <u>the energy of displacement of the surroundings of the system</u>." See next section https://en.wikipedia.org/wiki/Internal_energy.

5 Remember energy is only transferred by the process of heat from a higher temperature to a lower temperature (unless some outside agency is involved).

6 Remember - in this context a physical change is one in which the composition of the physical object does not change, e.g., the initial and final states both consist of an ideal or real gas.

7 See Section 8.1.

8 The principal other way of converting energy transferred by the process of heat into useful work involves thermoelectricity, e.g., Seebeck effect.

9 See Section 4.3.5.

10 D.V. Schroeder, *An Introduction to Thermal Physics*, Addison-Wesley, p. 33, 2000.

11 https://www.penjiapp.com/post/understanding-coffee-cup-calorimetry.

12 CRC Handbook of Chemistry and Physics, 62nd edition.

13 The content of this sub-section is based on the content of https://scholar.harvard.edu/files/schwartz/files/8-freeenergy.pdf.

14 The bar is a metric unit of pressure, but not part of the International System of Units (SI). It is defined as exactly equal to 100,000 Pa (100 kPa), or slightly less than the current average atmospheric pressure on Earth at sea level (approximately 1.013 bar).

15 The superscript ° signifies that it is a enthalpy change that occurs under a standard state (usually 298.15 K and at 1 bar pressure).

16 This is known as an exothermic reaction; an endothermic reaction is when the surroundings transfer energy to the reaction system.

17 Remember - the standard enthalpy of formation is a measure of the energy released or consumed when one mole of a substance is created under standard conditions from its pure elements.

18 "Elemental state" does not appear to be rigorously defined. It appears to be equivalent to "standard state" under one bar pressure and a defined temperature (usually 273 K or 298 K), e.g., the elemental state of oxygen is O_2 and that of carbon is graphite is Na. This is exactly the same as I used in Section 8.4 where I wrote "most stable reference states under standard conditions".

19 https://www.uni-muenster.de/imperia/md/content/mineralogie/lecture_1_basic_thermodynamics.pdf.

20 But calcite is more stable than aragonite at 25°C and 1 atmosphere pressure, showing that a reduction in enthalpy is not a criterion for an increase in stability – see Chapter Nine Section 9.4.

21 https://en.wikipedia.org/wiki/Aragonite.

22 http://webmineral.com/data/Calcite.shtml#.YQvOuD-SmUk.

23 https://www.toppr.com/ask/question/the-reaction-of-nitrogen-with-hydrogen-to-make-ammonia-has-delta-h92-kj-n2g3h2grightarrow2nh3g-what/.

24 The data in Tables 9.3 and 9.4 are from: https://chem.libretexts.org/Bookshelves/General_Chemistry/Book%3A_Chem1_(Lower)/15%3A_Thermodynamics_of_Chemical_Equilibria/15.02%3A_Entropy_Rules.

25 https://en.wikipedia.org/wiki/Gibbs_free_energy.

26 https://en.wikipedia.org/wiki/Thermodynamic_free_energy.

27 IUPAC prefer the terms Helmholtz function and Gibbs function.

28 Isochoric – constant volume.

29 Isobaric – constant pressure.

30 "Hermann von Helmholtz (1821–1894) participated in two of the most significant developments in physics and in the philosophy of science in the 19th century: the proof that Euclidean geometry does not describe the only possible visualizable and physical space, and the shift from physics based on actions between particles at a distance to the field theory. Helmholtz achieved a staggering number of scientific results, including the formulation of energy conservation, the vortex equations for fluid dynamics, the notion of free energy in thermodynamics, and the invention of the ophthalmoscope."

31 Josiah Willard Gibbs (1839-1903) was an American mathematical physicist whose work in statistical mechanics laid the basis for the development of physical chemistry as a science. The importance of Gibbs' findings was not immediately recognised, especially in his home country (nothing new there then). His publications were considered too mathematically complex for most chemists and too scientific for many mathematicians. Albert Einstein called him "the greatest mind in American history." Gibbs's studies of thermodynamics and discoveries in statistical mechanics paved the way for many of Einstein's later discoveries. https://plato.stanford.edu/entries/hermann-helmholtz/.

32 Remember $\Delta H = q_P$ see Equation (9.14).

33 What is Life? Erwin Schrödinger. https://en.wikipedia.org/wiki/What_Is_Life%3F.

34 I have omitted the word "law" otherwise the sentence does not make sense.

35 Epithets are sometimes attached to a person's name or appear in place of his or her name, as what might be described as a glorified nickname or sobriquet. ... Other epithets can easily be omitted without serious risk of confusion, and are known as *epitheton ornans*. From https://en.wikipedia.org/wiki/Epithet#:~:text=Epithets%20are%20sometimes%20attached%20to,a%20glorified%20nickname%20or%20sobriquet.&text=Other%20epithets%20can%20easily%20be,therefore%20known%20as%20epitheton%20ornans.

36 M.W. Collins, J.A. Stasiek, J. Mikielewicz, "The laws of thermodynamics: Entropy, free energy, information and complexity", *WIT Trans. State Art Sci. Eng.* 27, © 2006. WIT Press, ISSN 1755-8336; doi:10.2495/978-1-85312-853-0/05.

37 A. Lehninger, *Principles of Biochemistry*, 2nd Edition, Worth Publishers, 1993. ISBN 0-87901-711-2.

Appendix I: Summary

This section summarises what I have told you in this book (see Prologue - "tell them what you are going to say, tell them and tell them what you have told them")

Chapters 1, 2, and 3 are, in essence, introductory chapters. Chapter 1 considered classical mechanics. The following extract gives a good overview[1]:

> Classical mechanics is the study of the motion of bodies (including the special case in which bodies remain at rest) in accordance with the general principles first enunciated by Sir Isaac Newton in his Philosophiae Naturalis Principia Mathematica (1687), commonly known as the Principia. Classical mechanics was the first branch of Physics to be discovered, and is the foundation upon which all other branches of Physics are built. Moreover, classical mechanics has many important applications in other areas of science, such as Astronomy (e.g., celestial mechanics), Chemistry (e.g., the dynamics of molecular collisions), Geology (e.g., the propagation of seismic waves, generated by earthquakes, through the Earth's crust), and Engineering (e.g., the equilibrium and stability of structures). Classical mechanics is also of great significance outside the realm of science. After all, the sequence of events leading to the discovery of classical mechanics – starting with the ground-breaking work of Copernicus, continuing with the researches of Galileo, Kepler, and Descartes, and culminating in the monumental achievements of Newton – involved the complete overthrow of the Aristotelian picture of the Universe, which had previously prevailed for more than a millennium, and its replacement by a recognizably modern picture in which humankind no longer played a privileged role.

Chapter 2 introduced the concept of energy, and Chapter 3 set down the simplest model of a physical object (which is defined and considered in Chapter 1) namely the ideal gas. It is in Chapter 3 that temperature and the concept of internal energy are first mentioned and demonstrated that the internal energy of ideal gas is directly proportional to the temperature $\left(U = \frac{3}{2}nRT\right)$. Also, in Chapter 3, I stated that when internal energy and temperature are considered, we have entered the realm of thermodynamics. In Chapter 4, I introduced the First Law of Thermodynamics which states that the change of internal energy of a system is equal to the sum of the energy flows to that system by the process of work and heat, i.e., $\Delta\mathbf{U} = \mathbf{q} + \mathbf{w}$. You should by now appreciate that for energy accountancy a sign convention is required and the one I have used is that all energy flows into a system are positive and all energy flows out of the system are negative. I also showed that U is a state function and **q** and **w** are path or process functions, i.e., there are infinite combinations of heat and work energy transfers that enable a thermodynamic system to move from one internal energy state to another state. A system and its surroundings constitute the universe, and in this book, the surroundings are divided into heat and work reservoirs. The work reservoir is just a useful device where the energy flow by the process of work either originates or finishes and most texts do not use it. However, heat reservoirs are another matter; they are considered to contain an infinite quantity of energy and to remain at a constant temperature irrespective of the quantity of energy that flows into or out of them by the process of heat. This means that all energy transfer processes that occur in the heat reservoirs are "reversible". Reversibility is a concept introduced in Section 5.5: the following is a good summary[2]:

> In thermodynamics, a reversible process is a process whose direction can be reversed to return the system to its original state by inducing infinitesimal changes to some property of the system's surroundings. Throughout the entire reversible process, the system is in thermodynamic equilibrium with its surroundings. Having been reversed, it leaves no change in either the system or the surroundings. Since it would take an infinite amount of time for the reversible process to finish, perfectly reversible processes are impossible. However, if the system undergoing the changes responds much faster than the applied change, the deviation from reversibility may be negligible.

The reversibility of heat processes in the heat reservoirs is of fundamental importance in understanding the subsequent chapters.

In Chapter 5, I set down the definition of a heat engine: "it is a machine that operates in a cycle and converts a continuous input of energy, transferred to the engine by the process of heat, into a continuous output of mechanical energy". This definition, combined with the concept of reversibility and that of a cyclical process (a cyclical process is one in which a series of events happens again and again in the same order and after each cycle the system is returned to its original condition), sets the scene for Carnot's brilliant thought experiment where a heat engine operates in a reversible cyclical process. To allow an understanding of Carnot's thought experiment, I considered: (i) how friction, turbulence, and transfer of energy down a finite temperature gradient are irreversible processes, and reversible isothermal and adiabatic processes. I then considered Carnot's thought experiment and showed how Carnot proved that there was no heat engine more efficient than Carnot's hypothetical reversible heat engine. This proof was based on two premises (i) perpetual motion machines are an impossibility and (ii) energy always flows, by the process of heat, from a hotter to a colder region unless energy is expended by the process of work (as in a heat pump or refrigerator). I had shown earlier in this chapter that had Carnot demonstrated for his reversible heat engine that its efficiency was $1+\frac{q_C}{q_H}$, the efficiency is independent of any particular working substance I then considered how this conclusion led Kelvin to develop an absolute temperature scale independent of any thermometric substance. I demonstrated that ideal gas and Kelvin's thermodynamic temperature scales are the same and that the efficiency of the reversible heat engine could also be written as $1-\frac{T_C}{T_H}$. Finally, I showed that the Carnot cycle can be extended to any reversible cycle.

In Chapter 6, I considered how, in everyday speech, heat is considered to be a substance rather than a process. In contrast, work in everyday speech and in science is considered to be a process. Analogies for heat were considered:

> To understand and quantify a physical phenomenon we must consider both intensive and extensive physical quantities. Since a difference in pressure or of electrical potential is necessary for water or electrical charge to flow, we may call the difference the driving force for the process. In the flow of water on Earth a difference of altitude, level or height is usually the cause of water flow. This suggests that a driving force may be imagined to be a difference in levels of an intensive quantity such as height, voltage or (as we shall see) temperature ...[3]

A section on the history of heat led naturally to caloric theory and how the "energeticists" rubbished it to the detriment of science.

> ... since caloric possesses the important property, essential to the natural measure of heat quantity, of remaining constant in reversible exchanges (which the energy measure of heat does not) all our equations and methods of reasoning with regard to questions of equilibrium are greatly simplified and rendered more direct if we adopt caloric as the true measure of heat quantity, and regard it as possessing energy in virtue of its temperature.[4]

The rest of Chapter 6 was spent considering caloric generation in irreversible processes. In the discussion, I set down the following quotes from Mach and Callendar:

> If we look at the matter in an unprejudiced light, we must ask if there is any scientific sense or purpose in still considering as energy a quantity of heat that can no longer be transformed into mechanical work, (for example, the heat of [an isolated] equably warmed material system).[5]
>
> Caloric is not conserved in the old sense that its total quantity remains constant in any system of exchanges under all conditions. It remains constant only in the restricted sense, provided that all exchanges are reversible or take place under conditions of equilibrium. But since in any irreversible exchanges the total quantity of caloric is always increased, the principle of the conservation of caloric can still be stated in the modified form that 'the total quantity of caloric in a system cannot be diminished by any internal reactions or by the performance of work by or on the system, provided that no heat is allowed to escape.'[6]

In other words, I am in good company when advocating the use of caloric theory in the teaching of thermodynamics.

Chapter 7 introduces the concept of entropy as defined by Clausius. I start by indicating that the caloric of Black, Lavoisier, Carnot, etc. and the entropy of Clausius are equivalent.

> The entropy introduced into physics by Clausius was, contrary to general belief, not a new physical quantity but the reconstruction of the "quantity of heat" conceived about one hundred years earlier by the Scottish chemist Black.[7]

It took Clausius over ten years "invent" entropy and because of the obscure/arcane mathematics used nobody at the time was aware that it was a resurrection of Black's heat/Carnot's entropy. In

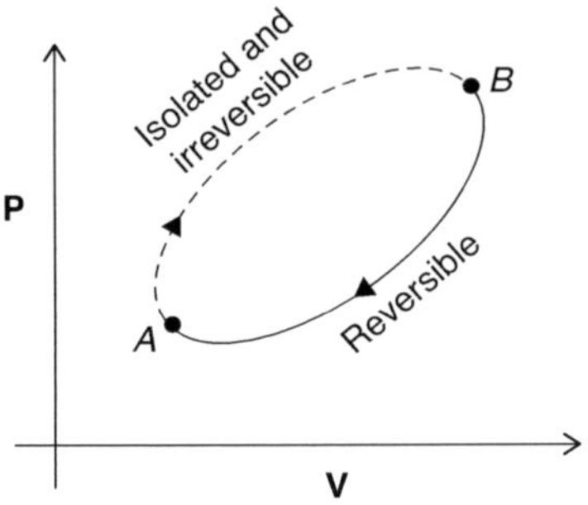

Figure A.1 Cyclic PV process in an isolated system.

Chapter 7, I introduced the concept of entropy using Carnot's analysis of the reversible heat engine cycle. This involves the concept of the cyclic integral ($\oint$) and the fact that cyclic integrals of state functions are zero I proved the Clausius Inequality which is generally written as $\oint \frac{\mathbf{dq}}{T} \leq 0$ where the equality applies to a reversible cycle and the inequality to irreversible cycles. Clausius defined a new state function which he called entropy (S) as $\int \frac{\mathbf{dq}_{\mathbf{rev}}}{T_{reservoir}}$.

I then considered a general irreversible pressure/volume process in an isolated system, leading from state A to state B represented by a dashed line in Figure A.1. The change in entropy in going from equilibrium state A to equilibrium state B is always the same irrespective of the path between A and B; this is because entropy is a state function. It makes no difference whether the path is reversible or irreversible. However, only if the path is reversible, the entropy change is given by $\int_A^B \frac{\mathbf{dq}_{\mathbf{rev}}}{T}$.

In Chapter 7, I showed, by considering thermal conduction, that entropy generation occurs whenever energy flows down a temperature gradient by the process of heat, and it is the presence of these temperature gradients that are the cause of irreversibility in any thermodynamic process. I then introduced the Second Law of Thermodynamics and considered the myriad ways the law can be presented including the Clausius statement – "The entropy of the world tends always toward a maximum" After considering entropy changes in ideal gas processes and the fact that the Second Law is equivalent to stating the impossibility of constructing a perpetual motion machine I set down some thoughts on entropy and why it is "considered one of the most difficult physical quantities to understand"? One reason for this is the near universal, but an erroneous statement that entropy is disorder (how can a quantity which has the units of energy per degree kelvin be related to disorder?)

I then considered other consequences of the Second Law including

- The eventual heat death of the universe,
- The arrow of time,
- Cause and effect.

NOTES

1 http://farside.ph.utexas.edu/teaching/301/lectures/node3.html.

2 https://en.wikipedia.org/wiki/Reversible_process_(thermodynamics).

3 In Section 3.3.4.

4 H.L. Callendar, "The caloric theory of heat and carnot's principle", Presidential Address February 10, 1911, *Proc Phys*, vol. XXIII, p. 185.

5 E. Mach, *Popular Scientific Lectures*, Cambridge University Press, p. 109, 1898.

6 H.L. Callendar, "The caloric theory of heat and carnot's principle", Presidential Address February 10, 1911, *Proc Phys*, vol. XXIII, pp. 153–189.

7 G. Falk, "Entropy, a resurrection of caloric – a look at the history of thermodynamics", *Eur. J. Phys*. 6, 108–115, 1985.

Appendix II: An Overview of Classical Thermodynamics

Thermodynamics is concerned with the energy flows and energy transformations from one system to another. These energy flow laws are captured by the fundamental principles known as the First and Second Laws of Thermodynamics. The First Law of Thermodynamics gives a precise formulation of the equivalence between heat and work and states that among all system transformations, the net system energy is conserved. Hence, energy cannot be created out of nothing and cannot be destroyed; it can merely be transferred from one form to another. The law of conservation of energy is not a mathematical truth, but rather the consequence of an inestimable number of observations made in the past 400 years or so. The first law does not tell us whether any particular process can actually occur, that is, it does not restrict the ability to convert work into heat or heat into work, except that energy must be conserved in the process. The Second Law of Thermodynamics asserts that while the system energy is always conserved, it will be degraded to a point where it cannot produce any useful work. More specifically, for any isolated cyclic process, it is impossible to extract work from heat without at the same time discarding some heat, giving rise to an increasing quantity known as entropy. While energy describes the state of a thermodynamic system, entropy is associated with the amount of lost work when thermodynamic systems transform from one state to another. Since the system entropy increases, the entropy of a thermodynamic system tends to a maximum, and thus time, as determined by system entropy increase, flows on in one direction only. Even though entropy is a physical property of matter that is not directly observable, it permeates the whole of nature, regulating the arrow of time, and is responsible for the heat death of the universe. Many mathematicians and scientists have expressed concerns about the completeness and clarity of different accounts of thermodynamics over its long and tortuous history.

The theory of classical thermodynamics was predominantly developed by Carnot, Clausius, Kelvin, Planck, and Gibbs. The pioneering work of Carnot was the first to establish the impossibility of perpetual motion of the second kind by constructing a cyclical process (now known as the Carnot cycle) involving four thermodynamically reversible processes operating between two heat reservoirs at different temperatures and showing that it is impossible to extract work from heat without at the same time discarding some heat. Carnot's main assumption (now known as Carnot's principle) was that it is impossible to perform an arbitrarily often repeatable cycle whose only effect is to produce an unlimited amount of positive work. In particular, Carnot showed that the efficiency of a reversible cycle, i.e., the ratio of the total work produced during the cycle, and the amount of heat transferred from the hot heat reservoir to the cold heat reservoir is only a function of the temperatures of the hot and the cold reservoirs, and not on the nature of the working substance. Both heat reservoirs are assumed to have an infinite source of heat so that their state is unchanged by their heat exchange with the engine (i.e., the device that performs the cycle), and hence, the engine is capable of repeating the cycle. Carnot's result (now known as Carnot's theorem) was arrived at using the caloric theory of heat.

Building on the work of Carnot (using cyclic system transformations), Clausius was the first to introduce the notion of entropy as a physical property of matter and establish the two main laws of thermodynamics involving the conservation of energy and non-conservation of entropy. Clausius postulated that it is impossible to perform a cyclic system transformation whose only effect is to transfer heat from a body at a given temperature to a body at a higher temperature. From this postulate, Clausius established the Second Law of Thermodynamics as a statement about entropy increase for adiabatically isolated systems. From this statement, Clausius goes on to state that the entropy of the universe is tending to a maximum, and the total state of the universe will inevitably approach a limiting state. Clausius' second law decrees that the usable energy (i.e., energy that can be used to do work) in the universe continually reducing.

In parallel research, Kelvin developed similar, and in some cases identical, results as Clausius, with the main difference being the absence of the concept of entropy. Kelvin's main view of thermodynamics was that of a universal irreversibility of physical phenomena occurring in nature. Building on the work of Clausius and Kelvin, Planck refined the formulation of classical thermodynamics. Planck's main theme of thermodynamics is that entropy increase is a necessary and sufficient condition for irreversibility. Without any proof (mathematical or otherwise), he goes on to conclude that every dynamical system in nature evolves in such a way that the total entropy of all of its parts increases. In the case of reversible processes, he concludes that the total entropy remains constant. Unlike the work of Clausius, Kelvin, and Planck involving cyclical system transformations, the work of Gibbs involved system equilibrium states.

Boltzmann was the first to give a probabilistic interpretation of entropy. Specifically, Boltzmann reinterpreted thermodynamics in terms of microparticles by relating the mechanical behaviour of individual particles with their thermodynamic behaviour, and he used the science of statistical mechanics to bridge between the microscopic details and the macroscopic behaviour to try and find a mechanical underpinning of the second law. The essence of Boltzmann thermodynamics is that thermodynamic systems with a constant energy level will evolve from a less probable state to a more probable state with the equilibrium system state corresponding to a state of maximum entropy (i.e., highest probability). In statistical thermodynamics, the Boltzmann entropy formula relates the entropy S of an ideal gas to the number of distinct microstates W corresponding to a given macrostate as $S = k_a \text{In} W$, where k_a is the Boltzmann constant. Even though Boltzmann was the first to link the thermodynamic entropy of a macrostate for some probability distribution of all possible microstates generated by different positions and momenta of various gas molecules, it was Planck who first stated (without proof) this entropy formula in his work on black body radiation. In addition, Planck was also the first to introduce the precise value of the Boltzmann constant to the formula; Boltzmann merely introduced the proportional logarithmic connection between the entropy S of an observed macroscopic state, to the thermodynamic probability of its occurrence W, never introducing the constant k_a to the formula. Despite its unquestionable success in computing the thermodynamic entropy of isolated systems, the theoretical justification of $S = k_a \text{In} W$ remains ambiguous and vague in most statistical thermodynamics textbooks.

In the first half of the twentieth century, many scientists and mathematical physicists expressed concerns about the completeness and clarity of the mathematical foundation of thermodynamics. In fact, many fundamental conclusions arrived at by classical thermodynamics can be viewed as paradoxical. For example, in classical thermodynamics, the notion of entropy (and temperature) is only defined for equilibrium states. However, the theory concludes that non-equilibrium states transition toward equilibrium states as a consequence of the law of entropy increase. Furthermore, classical thermodynamics is mainly restricted to systems in equilibrium. The second law infers that for any transformation occurring in an isolated system, the entropy of the final state can never be less than the entropy of the initial state. In this context, the initial and final states of the system are equilibrium states. However, by definition, an equilibrium state is a system state that has the property that whenever the state of the system starts at the equilibrium state, it will remain at the equilibrium state for all future time unless an external input acts on the system. In other words, the entropy of the system can only increase if the system is not isolated. Many aspects of classical thermodynamics are riddled with such inconsistencies, and most textbook expositions, poorly amalgamate physics with rigorous mathematics, e.g., Truesdell[1] describes the present state of the theory of thermodynamics as a "dismal swamp of obscurity"; Brush[2] remarks that "anyone who has taken a course in thermodynamics is well aware, the mathematics used in proving Clausius' theorem … [has] only the most tenuous relation to that known to mathematicians". Born[3] admits that "I tried hard to understand the classical foundations of the two theorems, as given by Clausius and Kelvin; … but I could not find the logical and mathematical root of these marvellous results"; Arnold[4] writes that "every mathematician knows it is impossible to understand an elementary course in thermodynamics". To exacerbate matters, there are also many vaguely defined terms and functions that are central to thermodynamics such as entropy, enthalpy, free energy, quasistatic, nearly in equilibrium, extensive variables, intensive variables, reversible, irreversible, etc.

NOTES

1 C. Truesdell, *The Tragicomical History of Thermodynamics*, Springer, p. 6, 1980.

2 S.G. Brush, *The Kind of Motion We Call Heat: A History of the Kinetic Theory in the Nineteenth Century*, North Holland, p. 581, 1976.

3 M. Born, *My Life: Recollections of a Nobel Laureate*, Taylor and Francis, p. 119, 1978.

4 V.I. Arnold, Contact geometry: The geometrical method of gibbs' thermodynamics. In: *Proceedings of the Gibbs Symposium*, New Haven, CT, USA, 15–17 May 1989, American Mathematical Society: Providence, RI, 1990; Page 163.

Appendix III: Taking Entropy a Bit Further

The entropy of a substance depends (i) on its state (temperature, pressure, volume, and amount); and (ii) how the substance is structured at the atomic/molecular level.

C.1 STATE OF SUBSTANCE

Temperature: There should be only one way of arranging the energy in a perfect crystal at 0 K. If $W = 1$, then $S = k_B \ln W = 0$, so that the entropy should be zero at the absolute zero of temperature. This is known as the Third Law of Thermodynamics, whch is obeyed by all solids unless some randomness of arrangement is accidentally "frozen" into the crystal. As energy is fed into the crystal with increasing temperature, we find that an increasing number of alternative ways of dividing the energy between the atoms become possible. W increases, and so does S. Without exception, the entropy of any pure substance always increases with temperature.

Volume and Pressure: The entropy of a gas always increases with increasing volume and decreases with increasing pressure. In the case of solids and liquids, the volume changes very little with the pressure, and so the entropy also changes very little.

Amount of Substance: The entropy of a substance is proportional to the amount of the substance, i.e., the entropy of 2 moles of a given substance is twice as large as the entropy of 1 mole.

Because the entropy of a substance depends on the amount of substance, the pressure, and the temperature, it is convenient to describe the entropy of a substance in terms of S^{o}_{m}, its standard molar entropy, i.e., as the entropy of 1 mole of substance at the standard pressure of 1 atm (101.3 kPa) or 1 bar (100 kPa) and a given temperature (usually 273 K or 298 K). Values of the standard molar entropies of various substances at 298 K (25°C) are given in Table C.1. Standard molar entropies can be used in much the same way as standard enthalpies of formation in order to find the entropy change ΔS^{o}_{m} for a reaction occurring at standard pressure and at 298 K. In this appendix, the tabulated values are used as examples illustrating how the atomic/molecular level structure of a particular substance affects entropy. Since standard molar entropies all refer to the same temperature, pressure, and amount of substance; any differences in the entropy values must be due to differences in the molecular structure of the various substances listed. There are two aspects of the molecular structure of a substance which affect the value of its entropy:

1. The degree to which the movement of the atoms and molecules in the structure is restricted – the less restricted this movement, the greater the entropy.
2. The mass of the atoms and molecules that are moving – the greater the mass, the larger the entropy.

I will consider each of these in turn.

C.2 RESTRICTIONS ON THE MOVEMENT OF ATOMS AND MOLECULES

We have already encountered one example of how the removal of a restriction on the motion of molecules allows an increase in the number of alternative ways in which the molecules can arrange themselves in space. This is the expansion of a gas into a vacuum. When the molecules have more freedom of movement, many more alternative arrangements are possible, and hence W is larger. As we have already calculated, doubling the volume of a gas increases its molar entropy by 5.76 $J.K^{-1}.mol^{-1}$.

A second example of the effect of relaxing a restriction on molecular motion is the case of the ring and chain forms of 1,4-butanediol. When the two ends of this molecule hydrogen bond to form a ring, the result is a fairly rigid structure (Figure C.1). However, if the restricting influence of the hydrogen bond is removed, the more flexible chain form of the molecule is capable of many more alternative configurations. As a result, we find that W for a mole of molecules in the chain form is very much larger than W for a mole of molecules in the ring form. The chain form turns out to have a molar entropy which is 41 $J.K^{-1}.mol^{-1}$ higher than that for the ring form.

C.3 MASS OF THE ATOMS AND MOLECULES

In the early 20th century, the idea that an electron traces out a definite orbit or trajectory was abandoned, and it was considered it was more useful to think of an electron in terms of a wave which

Table C.1: The Standard Molar Entropies of Selected Substances at 298.15K (25°C)

Compound	S^{o}_{m} $J^{-1}.mol^{-1}.K^{-1}$	Compound	S^{o}_{m} $J^{-1}.mol^{-1}.K^{-1}$
Monatomic Gases		**Polyatomic Gases**	
He	126.0	CH_4	186.3
Ne	146.2	C_2H_6	229.6
Ar	154.8	C_3H_8	269.9
Kr	164.0	C_4H_{10}	310.2
Xe	169.6	C_5H_{12}	348.9
Diatomic Gases		C_2H_4	219.6
H_2	130.7	N_2O_4	304.3
D_2	145.0	B_2H_6	232.0
HCl	186.9	**Solids**	
HBr	198.7	C (diamond)	2.377
HI	206.6	C (graphite)	5.74
N_2	191.6	Si	18.8
O_2	205.1	Ge	31.1
F_2	202.8	Sn (grey)	44.1
Cl_2	223.1	Pb	64.8
Br_2	245.5	Li	29.1
I_2	260.7	Na	51.2
CO	197.7	K	64.2
Triatomic Gases		Rb	69.5
H_2O	188.8	Cs	85.2
NO_2	240.1	NaF	51.5
H_2S	205.8	MgO	26.9
CO_2	213.7	AlN	20.2
SO_2	248.2	NaCl	72.1
N_2O	219.9	KCl	82.6
O_3	238.9	Mg	32.7
Liquids		Ag	42.6
Hg	76.0	I_2	116.1
Br_2	152.2	MgH_2	31.1
H_2O	69.9	AgN_3	99.2
H_2O_2	109.6		
CH_3OH	126.8		
C_2H_5OH	160.7		
C_6H_6	172.8		
BCl_3	206.3		

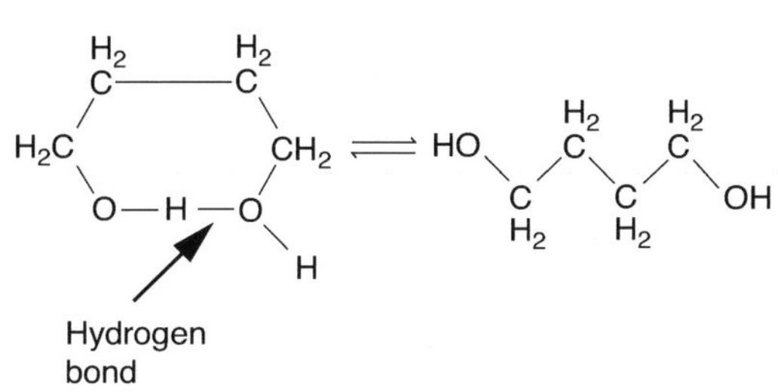

Figure C.1 The equilibrium between the ring and chain forms of 1,4-butanediol.

could fold itself around the nucleus only in certain specific three-dimensional patterns (known as orbitals instead of orbits). This approach is known as wave mechanics or quantum mechanics. The simplest way to understand the concept is an artificial example usually referred to as a "particle in a one-dimensional box". Consider the particle (e.g., an electron) to have a mass m and to be restricted in its movement in a one-dimensional box of length d into which it can just fit. This container, or box, is closed at both ends and the particle can move in only one dimension within its length.

A macroscopic object such as a billiard ball could move back and forth in this container, bouncing off either end. In a frictionless environment, the ball would oscillate indefinitely, maintaining a constant speed v and a constant kinetic energy $KE = \frac{1}{2}mv^2$. The actual magnitude of KE depends on the magnitude of the "push" the ball was given initially, to start it moving.

In 1924 by Louis de Broglie proposed that a wave of wavelength λ is associated with every particle. The larger the mass of the particle and the faster it is moving, the smaller this wavelength becomes. The relationship between wavelength, mass, and velocity is given by:

$$\lambda = \frac{h}{p} = \frac{h}{mv} \tag{1}$$

where p is the momentum of the electron $p = mv$ where m is the mass of the electron and v is its velocity), and the constant of proportionality h is known as Planck's constant $\left(h = 6.626 \times 10^{-34}\,\text{J.s}\right)$.

The wave-mechanical view no longer pictures the particle as the oscillating billiard ball. Instead, we must think of the electron as having a behaviour similar to that of the guitar string anchored at both ends in our one-dimensional box. When the string is attached in this manner, only those waves or vibrations in which the ends of the string do not move are possible. The length d of the box can thus correspond to a single half wavelength, to two half wavelengths, to three half wavelengths, etc., but *not* to any intermediate situation. In other words, the length of the tube must correspond to an integral number of half wavelengths, or

$$d = n\frac{\lambda}{2} \tag{2}$$

where n is an integer (i.e., $n = 1, 2, 3, 4, \ldots$. If $n = 1$, then $d = \frac{\lambda}{2}$; if $n = 2$, $d = \lambda$; and so on. Rearranging Equation [2].

$$\lambda = \frac{2d}{n} \tag{c}$$

From Equation [a] and [c], we can write:

$$v = \frac{nh}{2md} \tag{d}$$

The kinetic energy of the wave-particle $(E_k)$is given by:

$$E_k = \frac{1}{2}mv^2 = \frac{1}{2}m\left(\frac{nh}{2md}\right)^2 = n^2\left(\frac{h^2}{8md^2}\right) \tag{e}$$

Since the value of n is restricted to positive whole numbers, we can see that the kinetic energy of the electron can have only certain values and not others. If the microparticle is an electron ($m = 9.1 \times 10^{-31}$ kg) and the one-dimensional box is about the size of an atom ($d = 1 \times 10^{-10}$ m), the allowed values of the energy are given by

$$E_k = n^2\left(\frac{h^2}{8md^2}\right) = n^2 \times \frac{\left(6.624 \times 10^{-34}\,\text{J.s}\right)^2}{\left(8 \times 9.1 \times 10^{-31}\,\text{kg}\right) \times \left(1 \times 10^{-20}\,\text{m}^2\right)} = n^2 \times 6.0 \times 10^{-18}\,\text{J} \tag{f}$$

Therefore:

If n = 1, $E_k = 6.0 \times 10^{-18}\,\text{J}$

If n = 2, $E_k = 12.0 \times 10^{-18}\,\text{J}$

If n = 3, $E_k = 54 \times 10^{-18}\,\text{J}$

and so on.

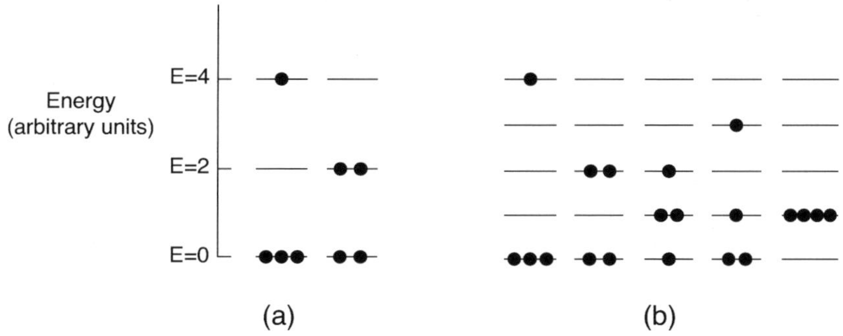

Figure C.2 Effect of microparticle mass on entropy. (a) The energy levels available to lower mass microparticles are widely spaced. If the spacing is two energy units, and if four units of energy are supplied to four microparticles, only two arrangements are possible: either one microparticle has all the energy, or two microparticles have two units each. (b) For heavier microparticles, the energy levels are closer together. If the spacing is one unit, and four units of energy are supplied, then five arrangements are possible

This shows that by treating the electron as a wave, its energy is automatically restricted to certain specific values and not those in between, i.e., the energy of the electron is quantised. One more property of the electron arises out of the wave analogy. If the electron in a box behaves like a guitar string, we can no longer state that the electron is located at a specific position within the box or is moving in one direction or the other. In fact, the electron seems to be everywhere in the box at once, and all that can be said is that the wave has a certain intensity at any point along the box. This intensity is larger in some places than in others and is always zero at both ends of the box.

From Equation [f] we can see that the larger the mass m of a particle, the smaller the value of its energy E_k for a given value of the quantum number n. Furthermore, the larger the value of m, the smaller the difference between E_{k2} (for n = 2) and E_{k1} (for n = 1). This is true for rotational, vibrational, or translational energy. The greater the mass of a particle, the closer together its energy levels.

The effect of the closeness of energy levels on the entropy is shown in Figure C.2. The more closely spaced the levels, the more different ways the same quantity of energy can be distributed among them. This applies in general for any number of particles and any quantity of energy. Therefore, the heavier the molecules of a substance, the larger its molar entropy. This effect is quite obvious among the noble gases—their molar entropies increase steadily with molar (and hence molecular) mass.

C.3.1 Solids, Liquids, and Gases

There is a general increase in the molar entropy as we move from solids to liquids to gases. In a solid, the molecules are only capable of vibrations around fixed points, but when a solid melts the molecules have more freedom of movement (rotational and to a limited extent translational). Therefore, when a solid melts, the molar entropy of the substance increases. When a liquid vaporises, the restrictions on the molecules' ability to move around are relaxed almost completely and a larger increase in the entropy occurs. The usual examples to demonstrate this is the entropy of fusion of ice (i.e., the entropy change when liquid water freezes at 273 K (0°C) and the entropy of vaporisation of water (i.e., the entropy change when liquid water evaporates to form water vapour) at 373 K (100°C). As both of these processes occur at constant temperature the processes are reversible the entropy changes can be calculated from the respective enthalpies of fusion and vaporisation[1]:

$$\Delta S_{fusion} = \frac{\Delta H_{fusion}}{273} = \frac{+6010}{273} = +22\ J.K^{-1} \quad [aa]$$

$$\Delta S_{vaporisation} = \frac{\Delta H_{vaporisation}}{373} = \frac{+40670}{373} = +109\ J.K^{-1} \quad [bb]$$

It is important to realise that fusion and vaporisation are endothermic processes and that the reverse processes (solidification and condensation) are exothermic processes, i.e., enthalpies of fusion and vaporisation are positive real numbers, and the enthalpies of solidification and condensation are negative real numbers. This means entropy increases during melting and vaporisation (Equations [aa] and [bb]) and decreases during solidification and condensation (Equations [cc] and [dd]).

$$\Delta S_{solidification} = \frac{\Delta H_{solidification}}{273} = \frac{-6010}{273} = -22\ J.K^{-1} \quad [cc]$$

$$\Delta S_{condensation} = \frac{\Delta H_{vaporisation}}{373} = \frac{-40670}{373} = -109\ J.K^{-1} \quad [dd]$$

The cooling effect accompanying vaporisation is apparent when we get out of a swimming pool: as the water evaporates from our skin, energy is removed from our bodies by the process of heat. The evaporation of water through perspiring keeps our bodies cool on hot days and during strenuous exercise. Conversely, the reverse effect is the reason that steam can cause severe burns, because when it comes in contact with skin, it condenses, releasing considerable energy that is transferred to and through the skin by the process of heat.

C.3.2 Molecular Complexity

Another trend is the higher molar entropy of substances with more complex molecules. To some extent, this is due to the mass since on the whole more complex molecules are heavier than simpler ones. However, there is an increase of entropy with complexity when we compare atoms molecules of very similar masses:

Substance	Argon (Ar)	Fluorine (F_2)	Carbon Dioxide (CO_2)	Propane (C_3H_8)
S^o_m $J.K^{-1}.mol^{-1}$	155	202.7	213.6	269.9
Atomic weight	40	38	44	44
Number of atoms	1	2	3	11

The more atoms there are in a molecule, the more ways the molecule can change its shape by vibrating. In consequence, there are more ways in which energy can be distributed among a more complex molecule.

C.3.3 Strength of Bonding

Another trend in entropy (which is most noticeable in the case of solids) is the decrease in the entropy as the forces between the atoms, molecules, or ions increase.

Substance	Sodium Fluoride (NaF)	Magnesium Oxide (MgO)	Aluminium Nitride (AlN)
S^o_m $J.K^{-1}.mol^{-1}$	155	202.7	213.6
Atomic weight	42.0	40.3	41.0

Sodium fluoride, magnesium oxide, and aluminium nitride are isoelectronic.[2] Since there are only small differences in atomic weights, the entropy decrease can only be ascribed to the increase in the bond strength as we move from the singly charged ions Na^+ and F^- through the doubly charged ions Mg^{2+} and O^{2-}, to the triply charged ions $A1^{3+}$ and N^{3-}.

NOTES

1 Enthalpy data from https://chem.libretexts.org/Bookshelves/General_Chemistry/Map%3A_A_Molecular_Approach_(Tro)/18%3A_Gibbs_Energy_and_Thermodynamics/18.04%3A_Entropy_Changes_Associated_with_State_Changes.

2 Isoelectronic refers to atoms, ions, or molecules that have the same electronic structure and the same number of valence electrons. Isoelectronic chemical compounds typically display similar chemical properties.

Index

Note: **Bold** page numbers refer to tables; *italic* page numbers refer to figures and page numbers followed by "n" denote endnotes.

MIX
Papier | Fördert gute Waldnutzung
FSC® C083411

Kolibri 360 GmbH
Ferdinand-Jühlke-Straße 7
99095 Erfurt, Deutschland
produktsicherheit@kolibri360.de